PROGRESS IN
Nucleic Acid Research and Molecular Biology
Volume 72

PROGRESS IN
Nucleic Acid Research and Molecular Biology

edited by

KIVIE MOLDAVE
Department of Molecular Biology and Biochemistry
University of California, Irvine
Irvine, California

Volume 72

An imprint of Elsevier Science

Amsterdam Boston London New York Oxford Paris
San Diego San Francisco Singapore Sydney Tokyo

This book is printed on acid-free paper.

Copyright © 2002, Elsevier Science (USA).

All Rights Reserved.
No part of this publication may be reproduced or transmitted in any form or by any means, electronic or mechanical, including photocopy, recording, or any information storage and retrieval system, without permission in writing from the Publisher.

The appearance of the code at the bottom of the first page of a chapter in this book indicates the Publisher's consent that copies of the chapter may be made for personal or internal use of specific clients. This consent is given on the condition, however, that the copier pay the stated per copy fee through the Copyright Clearance Center, Inc. (222 Rosewood Drive, Danvers, Massachusetts 01923), for copying beyond that permitted by Sections 107 or 108 of the U.S. Copyright Law. This consent does not extend to other kinds of copying, such as copying for general distribution, for advertising or promotional purposes, for creating new collective works, or for resale. Copy fees for pre-2002 chapters are as shown on the title pages. If no fee code appears on the title page, the copy fee is the same as for current chapters.
0079-6603/2002 $35.00

Explicit permission from Academic Press is not required to reproduce a maximum of two figures or tables from an Academic Press chapter in another scientific or research publication provided that the material has not been credited to another source and that full credit to the Academic Press chapter is given.

Academic Press
An imprint of Elsevier Science
525 B Street, Suite 1900, San Diego, California 92101-4495, USA
http://www.academicpress.com

Academic Press
84 Theobalds Road, London WC1X 8RR, UK
http://www.academicpress.com

International Standard Book Number: 0-12-540072-1

PRINTED IN THE UNITED STATES OF AMERICA
02 03 04 05 06 07 MM 9 8 7 6 5 4 3 2 1

Contents

SOME ARTICLES PLANNED FOR FUTURE VOLUMES ix

Viral Strategies of Translation Initiation: Ribosomal Shunt and Reinitiation 1

Lyubov A. Ryabova, Mikhail Pooggin, and Thomas Hohn

I. Introduction 2
II. Main Initiation Strategies in Eukaryotes 4
III. Translation and Translation-Dependent Strategies of Pararetroviruses and Retroviruses 7
IV. Shunting Mechanisms 12
V. Polycistronic Translation Strategies 21
VI. Outlook 28
References 29

Initiation of Eukaryotic DNA Replication: Regulation and Mechanisms 41

Heinz-Peter Nasheuer, Richard Smith, Christina Bauerschmidt, Frank Grosse, and Klaus Weisshart   41

I. The Eukaryotic Cell Cycle 42
II. Factors Required for the Initiation of DNA Replication 53
III. The Organization of Replication-Initiation Factors on Chromatin 69
IV. Cell Cycle Control by Checkpoints 73
V. Outlook 80
References 81

Deoxyribonucleotide Synthesis in Anaerobic Microorganisms: The Class III Ribonucleotide Reductase .. 95

Marc Fontecave, Etienne Mulliez, and Derek T. Logan

I. Introduction 96
II. The Anaerobic Ribonucleotide Reductase: A Multicomponent System ... 99
III. The nrdD Protein: The Reductase Component 99

IV. The nrdG Protein: The Activase 114
V. Gene Organization and Regulation 120
VI. The Anaerobic RNR: The Link between the RNA World and the DNA World 120
References 125

Regulation of Pathways of mRNA Destabilization and Stabilization 129

Robin E. Dodson and David J. Shapiro

I. Life and Half-Life of mRNA 130
II. Degradation of mRNA through the General Pathway; Deadenylation-Dependent mRNA Decay 131
III. Special Pathways for Regulating the Stability of mRNAs 133
IV. Regulation of Vitellogenin and Albumin mRNA Stability 142
References 159

Jasmonates and Octadecanoids: Signals in Plant Stress Responses and Development 165

Claus Wasternack and Bettina Hause

I. Introduction 166
II. Occurrence of Jasmonates and Octadecanoids 168
III. Biosynthesis of Jasmonates and Octadecanoids 171
IV. Jasmonate-Induced Gene Expression 180
V. Jasmonates in Stress Response and Signal Transduction Pathways 185
VI. Jasmonates and Octadecanoids in Plant Development 196
VII. Concluding Remarks 205
References 206

The Ubiquitous Nature of RNA Chaperone Proteins 223

Gaël Cristofari and Jean-Luc Darlix

I. Introduction 224
II. RNA Structure and the Folding Problem 226
III. RNA-Binding Proteins 229
IV. Investigating the Biochemical Properties of Nucleic Acid Chaperone Proteins 234
V. How Do RNA Chaperone Proteins Work? 246

VI. Conclusions and Future Prospects	258
References	263

Mechanisms of Basal and Kinase-Inducible Transcription Activation by CREB 269

Patrick G. Quinn

I. Introduction	270
II. Distinct CREB Activation Domains Mediate Constitutive and Kinase-Inducible Transcription	272
III. A Concerted Mechanism of Transcription Activation Involving Stimulation of Sequential Steps in Transcription Initiation by the CAD and P-KID	284
IV. Perspectives	298
References	299

eIF4A: The Godfather of the DEAD Box Helicases 307

George W. Rogers, Jr., Anton A. Komar, and William C. Merrick

I. eIF4A: The Protein	308
II. The Biology of eIF4A	312
III. eIF4A: Biochemical Properties	313
IV. Influence of Other Proteins on eIF4A Function	319
V. Does Any of This Biochemistry Make Sense or Is There a Contradiction Somewhere?	321
VI. The Role of the DEAD Box Sequences	323
VII. The Function of the Helicase Activity of eIF4A	324
VIII. eIF4A and the 80S Initiation Pathway	326
IX. Lessons Learned from eIF4A	328
References	329

CTD Phosphatase: Role in RNA Polymerase II Cycling and the Regulation of Transcript Elongation 333

Patrick S. Lin, Nicholas F. Marshall, and Michael E. Dahmus

I. Historical Overview	334
II. General Properties of CTD Phosphatase	337
III. RNAP II Recycling Mediated by CTD Phosphatase	346
IV. Involvement of CTD Phosphatase in the Regulation of Transcript Elongation	349
V. Perspectives and Future Directions	357
References	359

Translational Control of Gene Expression: Role of IRESs and Consequences for Cell Transformation and Angiogenesis . . 367
Anne-Catherine Prats and Hervé Prats

I. Introduction	369
II. Background: Translation Initiation in Eukaryotes	371
III. The Murine Leukemia Virus: The Retroviral Genomic mRNA Codes for Two Gag–Pol Polyproteins from Two CUG and AUG Initiation Codons	378
IV. The FGF-2 mRNA: How Cap-Dependent and IRES-Dependent Translation of a Single mRNA Leads to Expression of Five Isoforms with Distinct Localizations and Functions	383
V. VEGF mRNA: Two Distinct IRESs Control Alternative Initiation of the Translation of Two Isoforms	393
VI. C-*myc* mRNA: A Natural Multicistronic Messenger	396
VII. IRES-Dependent Translational Control: IRES Activity Enhancement in Transformed Cells	398
VIII. IRES Tissue Specificity *in Vivo*	399
IX. FGF-2 Translational Silencing by Tumor Suppressor p53	400
X. Concluding Remarks	403
References	405

Structure and Function of $\phi29$ Hexameric RNA That Drives the Viral DNA-Packaging Motor: Review 415
Peixuan Guo

I. Introduction	416
II. Approaches and Strategies for the Study of pRNA	419
III. Studies of pRNA Structure	430
IV. pRNA–Protein Interactions	445
V. Effect of Mono- and Divalent Cations on pRNA Dimer Formation, Procapsid Binding, and Viral Assembly	448
VI. Functions of pRNA	449
VII. Significance and Application of the Study of pRNA	457
VIII. Concluding Remarks	461
References	462

INDEX 473

Some Articles Planned for Future Volumes

Tandem CCCH Zinc Finger Proteins in the Regulation of mRNA Turnover
  PERRY BLACKSHEAR

Initiation and Recombination: Early and Late Events in the Replication of Herpes Simplex Virus
  PAUL E. BOEHMER

Dynamic O-Glycosylation of Nuclear and Cytosolic Proteins: a New Paradigm for Metabolic Control of Signal Transduction and Transcription
  GERALD W. HART AND KAZUO KAMEMURA

DNA-Protein Interactions Involved in the Initiation and Termination of Plasmid Rolling Circle Replication
  SALEEM A. KAHN, T.-L. CHANG, M. G. KRAMER, AND M. ESPINOSA

FGF3: A Gene With a Finely Tuned Spatiotemporal Pattern of Expression during Development
  CHRISTIAN LAVIALLE

Specificity and Diversity in DNA Recognition by *E. coli* Cyclic AMP Receptor Protein
  JAMES C. LEE

Steroid Signaling in Procaryotes
  EDMOND MASER

The Occurrence and Functions of the Double-Stranded RNA Binding Motif: a Genome-Wide Survey
  MICHAEL B. MATHEWS

Transcriptional Control of Multidrug Resistance in the Yeast Saccharomyces
  SCOTT MOYE-ROWLEY

Oxygen Sensing and Oxygen-Regulated Gene Expression in Yeast
  ROBERT O. POYTON

Protein Kinase CK2-Linked Gene Expression Control
  WALTER PYERIN AND KARIN ACKERMANN

Ribonucleases in Cancer Chemotherapy
  ROBERT T. RAINES, P. A. LELAND, M. C. HERBERT, AND K. E. STANISZEWSKI

T7 RNA Polymerase: Structure and Mechanism
  RUI J. SOUSA

Some Articles Planned for Future Volumes

Broad Specificity of Serine/Arginine (SR)-Rich Proteins Involved in the Regulation of Alternative Splicing of Premessenger RNA
  JAMES STEVENIN, CYRIL BOURGEOIS, AND FABRICE LEJUNE

DNA Double-Strand Break Repair in Eukaryotic Cells
  PATRICK SUNG, LUMIR KREJCI, AND ALAN TOMKINSON

Fidelity Mechanism of DNA Polymerase β
  JOANN B. SWEASY

Molecular Mimicry in the Decoding of Translational Stop Signals
  WARREN P. TATE, D. J. SCARLETT, M. A. AMIRI, L. L. MAJOR K. K. MCCOUGHAN, AND E. S. POOLE

Protein Tyrosyl Phosphatases in T-Cell Activation: Implications in HIV Transcriptional Activity
  MICHAEL J. TREMBLAY, MICHEL QUALLET, AND BENIOT BARBEAU

FOXO Forkhead Transcription Factors in Insulin and Growth Factor Action
  TERRY UNTERMAN, SHAODONG GUO, AND XIAOHUI ZHANG

HIV Transcriptional Regulation in the Context of Chromatin
  ERIC VERDIN

Viral Strategies of Translation Initiation: Ribosomal Shunt and Reinitiation

LYUBOV A. RYABOVA,
MIKHAIL M. POOGGIN,
AND THOMAS HOHN

Friedrich Miescher-Institute
CH-4002, Basel, Switzerland

I. Introduction	2
II. Main Initiation Strategies in Eukaryotes	4
A. Cap-Dependent Scanning	4
B. Leaky Scanning	5
C. Internal Initiation	6
D. Reinitiation	6
E. Shunting	6
III. Translation and Translation-Dependent Strategies of Pararetroviruses and Retroviruses	7
A. The Interplay of Translation and Packaging of Viral RNA	7
B. Retroelement Gag–Pol Translation	10
IV. Shunting Mechanisms	12
A. sORF-Dependent Shunt in Caulimoviridae	12
B. Shunt in Animal Viruses	17
C. Candidates for Shunting in Cellular mRNAs	21
V. Polycistronic Translation Strategies	21
A. Translation of Polycistronic mRNA via Leaky Scanning in Bacilliform Caulimoviridae	22
B. Activated Reinitiation in Icosahedral Caulimoviridae	23
C. Model of TAV-Activated Reinitiation	27
VI. Outlook	28
References	29

    Due to the compactness of their genomes, viruses are well suited to the study of basic expression mechanisms, including details of transcription, RNA processing, transport, and translation. In fact, most basic principles of these processes were first described in viral systems. Furthermore, viruses seem not to respect basic rules, and cases of "abnormal" expression strategies are quiet common, although such strategies are usually also finally observed in rare cases of cellular gene expression. Concerning translation, viruses most often violate Kozak's original rule that eukaryotic translation starts from a capped monocistronic mRNA and involves linear scanning to find the first suitable start codon. Thus, many viral cases have been described where translation is initiated from noncapped RNA, using an internal ribosome entry site. This review centers on other viral translation strategies, namely shunting and virus-controlled reinitiation as first described in plant pararetroviruses (Caulimoviridae). In shunting, major parts of a complex leader are bypassed and not melted by scanning ribosomes. In the Caulimoviridae, this process is coupled to reinitiation after translation of a small open reading frame; in other cases, it is possibly initiated upon pausing of the scanning ribosome. Most of the Caulimoviridae produce polycistronic mRNAs. Two basic mechanisms are used for their translation. Alternative translation of the downstream open reading frames in the bacilliform Caulimoviridae occurs by a leaky scanning mechanism, and reinitiation of polycistronic translation in many of the icosahedral Caulimoviridae is enabled by the action of a viral transactivator. Both of these processes are discussed here in detail and compared to related processes in other viruses and cells. © 2002, Elsevier Science (USA).

I. Introduction

Eukaryotic chromatin seems to be very uneconomically used. Large regions are devoid of genes, large distances usually separate individual genes, and large and multiple introns separate the true coding regions of a gene. The situation is fundamentally different with virus genomes. These have evolved to optimize the available genomic space to allow a high density of coding information and *cis*-acting control elements such as promoters, polyadenylators, and replicators (*1*). Noncoding regions are minimized and DNA sequences are often multiply used, that is, several overlapping reading phases of viral DNA are used to encode proteins, and coding regions also contain *cis*-elements. Often, polyproteins are produced that are later cleaved to individual functional proteins, which reduces the numbers of promoters and polyadenylators on the DNA as well as on 3'- and 5'-untranslated regions on the RNA. Apparently, this justifies the cost of the extra proteinase required for polyprotein processing. Furthermore, virus RNAs are often multiply used, either through alternative splicing or by various mechanisms of polycistronic translation. In addition some viral proteins are multifunctional.

The class of reverse-transcribing elements (*2*), that is, retroviruses, pararetroviruses, and retrotransposons, although related in type and number of genes, have evolved a variety of different strategies of gene economy (Fig. 1). Hepadnaviruses have the highest degree of gene and signal overlap: The envelope-coding region is completely included in a different reading phase of the *pol* gene. Hepadnaviruses also make use of multiple transcription initiation from a single promoter. This leads to two types of capsid protein and three types of envelope protein, distinguished by different N-termini. Complex retroviruses, such as human immunodeficiency virus-1 (HIV-1), are specialists in alternative splicing. This leads to six different mRNAs originating from a single original

VIRAL STRATEGIES OF TRANSLATION INITIATION

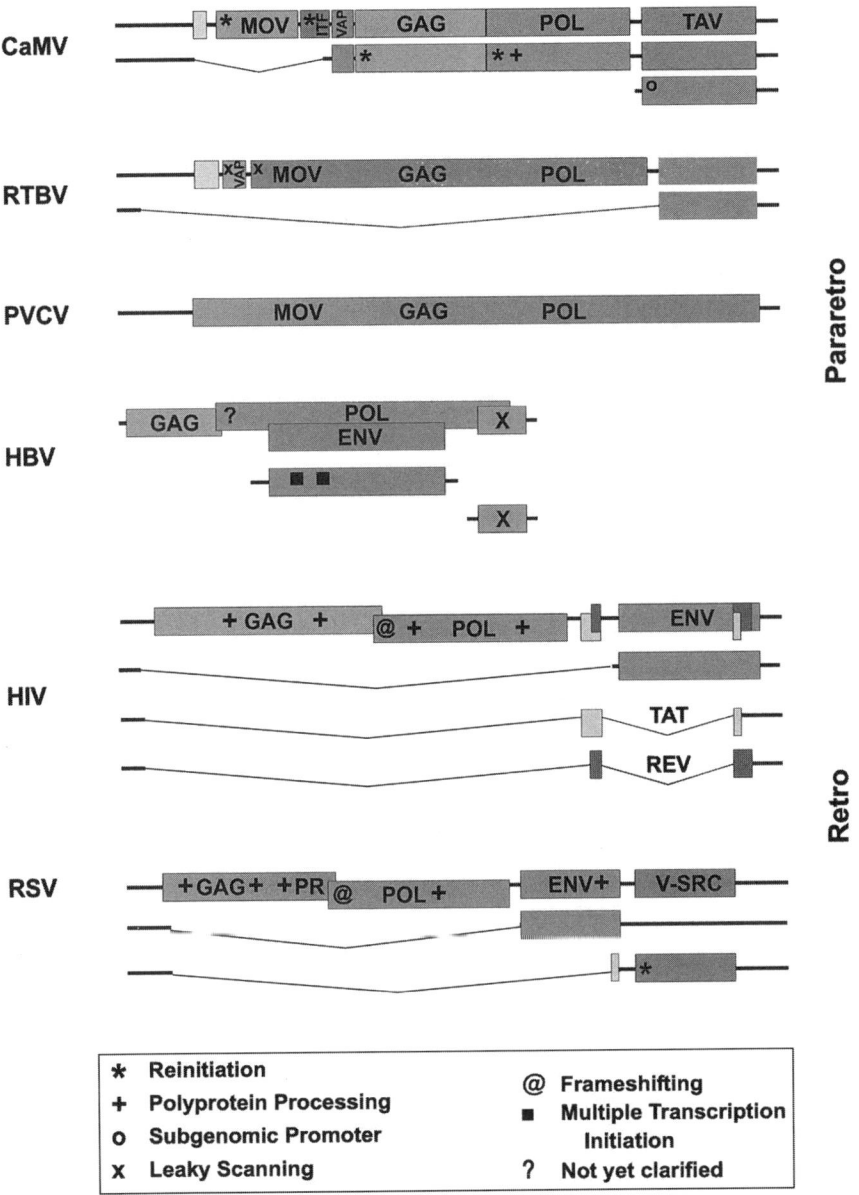

FIG. 1. Schematic representation of pararetro- and retroviral genomes and their expression strategies. The most representative RNAs and features are shown. For more complete information see Refs. 3 and 192. The processing sites of the PVCV polyprotein are not known. HBV Pol expression does not occur by frameshifting; conflicting results exist regarding whether it is translated via leaky scanning/reinitiation or internal ribosome entry as well.

transcript. This alternative splicing is controlled by a specialized virus protein, Rev, that determines the timing of transport of viral RNA from the nucleus to the cytoplasm (3). Retroviruses also use a frameshift or stop-codon readthrough strategy to produce Gag and Gag–Pol polyproteins from a single RNA, and a polyprotein strategy whereby the Gag protein is processed to yield at least three different structural proteins: the matrix, capsid, and nucleocapsid proteins. In addition, the Gag–Pol polyprotein is processed to yield protease, reverse transcriptase/RNAse H, and integrase. Finally, some of the Caulimoviridae have developed highly sophisticated polycistronic translation mechanisms to produce the maximum number of proteins from a single RNA.

Even among the Caulimoviridae, expression strategies can differ greatly. Petunia vein clearing virus produces a single polyprotein (4), from which the individual proteins are thought to be derived. Badnaviruses and rice tungro bacilliform virus (RTBV) produce a large polyprotein and two small individual proteins by multiple leaky scanning translation (5); the additional open reading frame (ORF) in RTBV is expressed from a spliced RNA. Finally, cauliflower mosaic virus (CaMV) and its closest relatives use a subgenomic promoter for the transactivator/viroplasmin (TAV) protein, splicing and polycistronic translation for expression of six of its ORFs, and the polyprotein strategy to produce protease and reverse transcriptase/RNAse H from a single precursor. Furthermore, at least one of its proteins, TAV, is multifunctional: It both acts as a translation reinitiation activator and directs assembly of virus particles. The "virion-associated protein" (VAP) in Caulimoviridae might also be multifunctional. In addition to being required for insect transmission, it is essential for insect-independent plant infection. Another interesting feature of the Caulimoviridae is that the leader of the pregenomic RNA is long and highly structured and includes packaging and replication signals. The secondary structure of this leader is not melted during the pretranslational scanning process, since most of the leader is bypassed by scanning ribosomes, a process that has been termed shunting.

II. Main Initiation Strategies in Eukaryotes

Mechanisms of translation initiation have been intensively studied and grouped into several categories that differ in the pathway by which ribosomal subunits reach the initiator AUG codon (reviewed by Jackson in Ref. 6).

A. Cap-Dependent Scanning

The majority of eukaryotic mRNAs are monocistronic. Their 5'-leader sequences are capped, short, and unstructured, and they direct initiation from the AUG codon nearest to the capped 5'-end via a cap-dependent ribosome-scanning mechanism (7, 8). The 40S ribosomal subunit, together with associated initiation factors, engages the mRNA at the capped 5'-end and migrates linearly until it encounters the first AUG codon. This occurs in several steps: the m^7 GpppX cap at the 5'-end of most mRNAs is recognized by subunit 4E of eukaryotic initiation factor (eIF) 4F, which contains in addition eIF4G, eIF4A (an ATP-dependent helicase), and ATP. Ribosomes separate into 40S and 60S subunits before binding mRNA. The 40S subunit binds eIF3 and the eIF2–GTP–initiation methionyl transfer RNA (Met-tRNAi) ternary complex (for review, see Refs. 9, 10). eIF5 has been implicated in bridging eIF3 and eIF2 (11). The complex requires also eIF1 and eIF1A to begin scanning to search for the correct AUG start codon (12). At the AUG codon, which is recognized by base pairing with the anticodon in Met-tRNAi, the 40S subunit stops, a 60S subunit joins to form an 80S ribosome in the presence of factor eIF5B (13), and elongation begins.

The mechanism of cap-dependent scanning has been elaborated in mammalian and yeast cell systems, but most of its features are also valid for plants (14, 15). Scanning can be strongly affected by cis-acting elements and trans-acting factors: (1) strong secondary structures (below −50 kcal/mole) completely block scanning, causing stalling of 40S ribosomes at the 5'-side of the hairpin (16–19); (2) multiple AUG start sites upstream of the main ORF reduce efficiency of downstream initiation (20, 21); and (3) RNA-binding proteins can specifically block scanning or 40S loading on the capped 5'-end (22, 23).

About 10% of eukaryotic mRNAs (particularly those encoding growth-related factors, tumor suppressors, transcription factors, and protooncogenes) and many viral RNAs contain long, often structured leaders with one or multiple small ORFs upstream of the main coding region, which would interfere with the normal scanning process. To account for this, alternative modes of translation initiation have been proposed that allow the presence of inhibitory elements in the leader to be overcome: (1) "leaky scanning" (8), (2) internal initiation (24–26), (3) reinitiation after small ORF translation (see reviews by Hinnebusch, Ref. 27, and Morris and Geballe, Ref. 28), and (4) nonlinear ribosomal scanning or ribosomal shunting (29–34).

B. Leaky Scanning

Leaky scanning allows the bypass of the upstream AUG codon in at least two cases. A start codon located close (<10 nucleotides, nt) to the cap site (35–37) or an AUG in a suboptimal nucleotide context for initiation (20) (or a non-AUG start codon in an optimal context) is poorly recognized by ribosomes. The latter effect can be suppressed by a structural element located 14 nt downstream of the start codon (38). Thus, in some cases, upstream ORFs might be bypassed by leaky scanning, obviating the need for reinitiation at the downstream AUG start site.

C. Internal Initiation

Internal initiation involves direct binding of the 40S ribosome to a region (internal ribosome entry site, IRES) at, or upstream of, the authentic AUG codon, and is mainly used by viruses which have developed a strategy to shut down cap-dependent initiation in a host cell (for review see Refs. *39, 40, 10*). This mechanism does not require a cap structure on the mRNA and thus does not depend on eIF4E and the N-terminal part of eIF4G containing the eIF4E-binding domain (*41*). Therefore, in contrast to cap-dependent initiation, internal initiation is not impaired by cleavage of eIF4G by viral proteases. The best-characterized IRESs are from picornaviruses and hepatitis C virus (HCV), classical swine fever virus (CSFV), and bovine viral diarrhea virus (BVDV) (*42, 43*).

D. Reinitiation

After translating an upstream ORF, terminating ribosomes might remain associated with the mRNA, continue scanning, and initiate at a downstream start codon with an efficiency depending both on mRNA *cis*-elements and *trans*-acting factors (*28*). Reinitiation requires the recruitment of initiation factors and, in general, can occur in eukaryotic RNAs when the upstream ORF is short (2–30 codons), with reinitiation frequency increasing with the distance between the short ORF (sORF) and the "main" ORF (*44, 27, 45, 46*). The requirement for scanning 40S ribosomes to recruit initiation factors, including the ternary complex eIF2–GTP–Met-tRNAi, is the limiting step for reinitiation of translation (*27*). The decrease of reinitiation frequency with the increase of the length of the sORF suggests that, during the short translation event, some initiation factors remain associated with the translational machinery after termination of translation and can be used for reinitiation (*44, 19*).

Scanning posttermination ribosomes might be affected by the mRNA structure surrounding the stop codon of the sORF (*47*), the structure of the sORF-encoded peptide (*48, 49*), and, in some special cases, by degradation of the mRNA (*50, 51*).

After translating an upstream ORF longer than about 30 codons, the terminating ribosome is believed to dissociate from the mRNA. However, there are a few reports suggesting reinitiation at AUG codons located a short distance upstream of the long ORF termination codon in artificial RNA constructs in mammalian cells (*52, 53*).

E. Shunting

Ribosome shunting, first described in CaMV, is a nonlinear scanning mechanism in which initially scanning ribosomes are transferred directly from a 5′-donor site to a 3′-acceptor site without linear scanning of an intervening region (*30*). Shunting will be described later in detail.

III. Translation and Translation-Dependent Strategies of Pararetroviruses and Retroviruses

The plant viral family Caulimoviridae and its animal counterpart Hepadnaviridae, including the well-known human hepatitis B virus (HBV), have been proposed to be classified into a suborder Pararetrovirinae of the order Retrovirales of reverse-transcribing elements (2). The members of Pararetrovirinae—also known as pararetroviruses—contain DNA in mature virions, in contrast to the RNA-containing retroviral virions. It is worth mentioning that virions of spumaviruses—a separate genus of Retrovirinae (or retroviruses)—can also contain DNA (54). This, together with other striking similarities to pararetroviruses (reviewed in Refs. 55, 56), places them between the pararetro- and orthoretrovirus suborders of the Retrovirales. The DNA of pararetroviruses is targeted to the nucleus of infected cells, where host RNA polymerase II transcribes a terminally redundant pregenomic RNA, which is eventually reverse-transcribed by the viral reverse transcriptase complex to produce the DNA genome. Retroviruses and pararetroviruses use their (pre)genomic RNA and its spliced versions as polycistronic mRNAs that can be translated into more than one protein. Strategies of polycistronic translation include: (1) frameshifting (in most retroviruses; reviewed in Refs. 3, 57), (2) leaky scanning [e.g., RTBV (5); HIV-1 (58)], and (3) reinitiation activated by a viral transactivator [discovered first in CaMV (59); see below]. In the controversial case of the HBV Pol ORF (the second cistron following the overlapping Core ORF), translation is believed to occur by leaky scanning combined with reinitiation after translation of a short ORF within the Core ORF (60–62) (see Fig. 1). However, alternative initiation mechanisms such as the internal ribosome entry suggested in earlier studies (63) and ribosome shunting have not been convincingly ruled out. Moreover, in duck hepatitis B virus, which represents a distinct genus of Hepadnaviridae, a mechanism other than leaky scanning, with some, but not all features of internal, initiation, has been implicated in Pol translation (64).

A. The Interplay of Translation and Packaging of Viral RNA

To regulate the multiple usage of pregenomic RNA for replication as well as translation, pararetroviruses have adapted to versatile properties of the eukaryotic translation machinery. In HBV, passage of the translating 80S ribosome (but not the scanning 40S ribosomal subunit) through the RNA packaging signal (ε) disrupts its secondary structure, thereby preventing encapsidation/reverse transcription of the pre-Core RNA (65) (Fig. 2). On the other hand, the reduced ability of the 40S ribosome to scan through such secondary structures (18) accounts for efficient functioning of the ε signal for packaging of the shorter Core

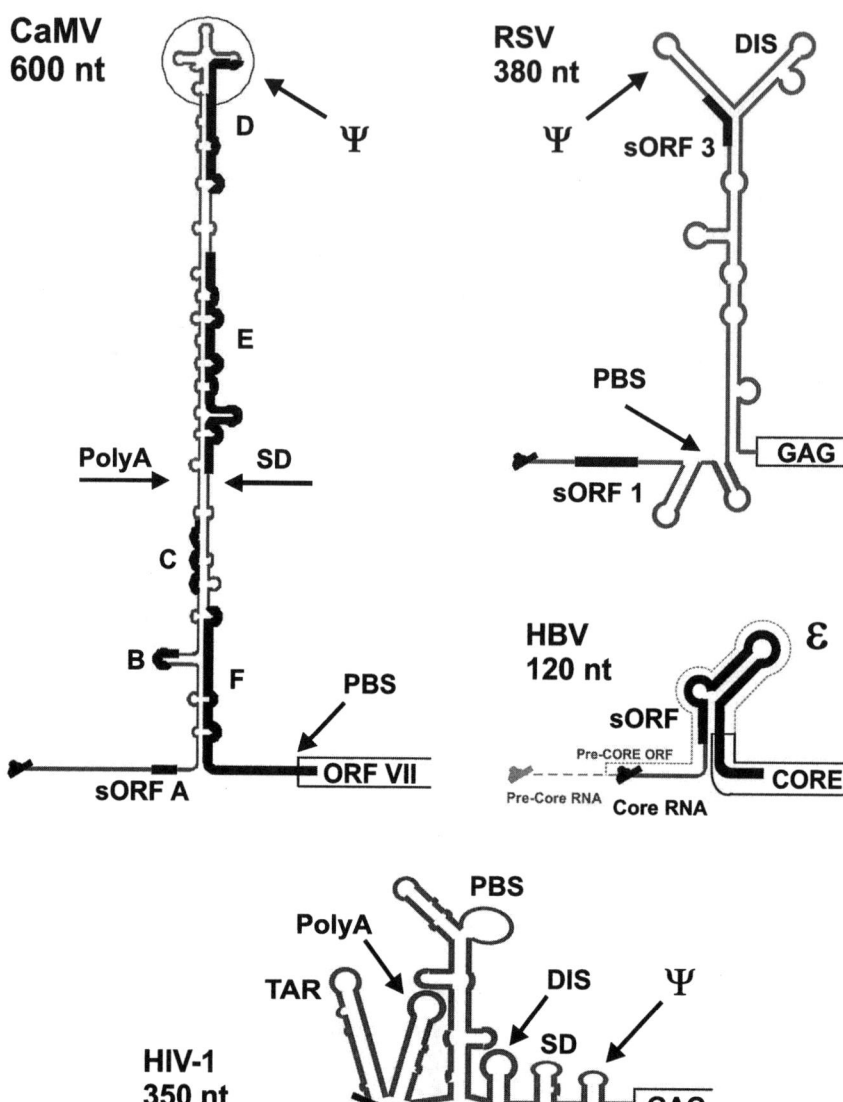

FIG. 2. RNA leaders of pararetro- and retroviruses. The most representative leaders and features are shown. DIS, Dimerization signal; PBS, primer-binding site; polyA, polyadenylation signal; SD, splice donor; TAR, Tat-responding element; Ψ or ε, packaging signal.

RNA, which lacks the 5' portion with the AUG start codon of the pre-Core ORF (65) (see Fig. 2). The ε structure-mediated retardation of the scanning ribosome heading toward the Core start codon might allow the formation of the ε–Pol complex and initiation of reverse transcription on a subpopulation of Core RNAs. Moreover, possible collision of the Pol-nascent DNA complex with the scanning ribosome may help to relocate this complex to the 3'-end of Core RNA in the poorly understood process of template switching that is required for a complete cycle of reverse transcription (66–68). Interestingly, despite its negative effect on scanning, and therefore on initiation of polycistronic translation of the Core RNA, the ε structure plays a positive role in pre-Core RNA translation, because it ensures the recognition of the upstream pre-Core AUG, which is in a suboptimal context (69), in accordance with Kozak's rule (38). This should ensure complete exclusion of pre-Core RNA from the packaging process.

Plant pararetroviruses such as CaMV and RTBV exploit the shunting property of the scanning ribosome (31, 70) (see below in greater detail) to allow the bypass of a large region of the pregenomic RNA leader containing a putative packaging signal (for CaMV, see Fig. 2). We speculate that regulation of shunting efficiency may affect conformations of the intervening structure and thereby modulate the interaction of the viral coat protein and the putative packaging signal. In the case of CaMV, this specific RNA–coat protein interaction has been demonstrated by using the yeast three-hybrid system (71). Computer-aided analysis of the pregenomic RNA leaders with regard to the consensus shunt configuration (see below) predicts that ribosome shunting operates in most plant pararetroviruses sequenced so far (Ref. 72 and our unpublished results for *Mirabilis* mosaic virus). Similar configurations of shunt elements have also been recognized in the (pre)genomic RNAs of HBV and human foamy spumavirus (our unpublished predictions).

The sorting of genomic RNA for translation and replication has been studied in retroviruses. In some studies, the existence of two separate pools of RNA, one for translation, the other for packaging, was proposed (e.g., 73). However, in many complex retroviruses, such as lentiviruses (e.g., HIV-1) and alfaretroviruses (e.g., Rous sarcoma virus, RSV), RNA for packaging seems to be sequestered away from the pool of translated RNAs (see, e.g., Ref. 74). The case of RSV illustrates the complexity of possible interplay between the host translation machinery and viral *cis*- and *trans*-acting factors involved in RNA packaging. The 379-nt-long leader of RSV RNA preceding the Gag ORF contains three phylogenetically conserved sORFs and extensive secondary structure (Fig. 2). The 5'-proximal sORF, sORF 1, is the major ribosome-binding site (75, 76). sORF 3 is also translated, and the efficiency of this translation event is thought to regulate a conformational switch in the structure of the RNA packaging signal, Ψ (77–79). The conflicting reinitiation/leaky scanning models of sORF 3 and Gag ORF translation deduced from those early studies have been recently revisited by the group of J.-L. Darlix (80), who identified a bipartite IRES driving the two initiation events. A new model proposes that the translating ribosomes stall at sORF 1, most likely due to the stable downstream structure and/or a particular feature of the encoded peptide (MAGPLIP), which resembles the mammalian S-AdoMet decarboxylase regulatory peptide (MAGDIS), which blocks proper termination (28, 81) thereby preventing reinitiation at downstream start codons. On the other hand, the Gag AUG start codon is recognized by ribosomes entering internally at the 3′ IRES. Thus, both events would preserve the intervening structure containing the Ψ signal. In addition, ribosomes can enter at a somewhat weaker IRES, roughly mapped to the intervening region upstream of the sORF 3 AUG start codon. This latter ribosome binding followed by sORF 3 translation might interfere with Gag binding to the Ψ signal. The same study (80) also proposes a mechanism for translation initiation on the RSV spliced v-Src mRNA. The splicing event creates a new sORF of 30 nucleotides, where the Gag ATG is in-frame with a stop codon located behind the splice junction (see Fig. 1). Presumably, IRES-driven recognition of the Gag AUG would lead to translation of this sORF followed by reinitiation at the v-Src AUG. In fact, the 66-nt region between the sORF and the v-Src ORF favored efficient reinitiation when placed after a large cistron on an artificial bicistronic mRNA (80), thus supporting the reinitiation model.

The current revision of RSV translation mechanisms is in line with identification of IRESs in lentiviruses simian immunodeficiency virus (SIV; 82) and HIV-1 (83) as well as in other genera of Retroviridae [murine leukemia virus (84), murine sarcoma virus (85), avian reticuloendotheliosis virus (86)]. However, in these cases the usage of viral RNA for translation and replication may be regulated by means differing from those proposed for RSV. For example, the HIV-1 IRES is located just downstream of the Gag ORF start codon (83) and the preceding leader sequence containing the Ψ signal does not seem to exhibit any IRES activity (87) (Fig. 2). Scanning ribosomes migrating through the Ψ signal may interfere with the packaging process, unless they are retarded by the leader secondary structure as in the case of HBV Core RNA described above.

Notably, the structural configuration of the RSV RNA leader with a 5′-proximal sORF terminating in front of an extended stem–loop structure resembles the shunt configuration found in plant pararetroviruses (72). The experimental evidence does not necessarily exclude the possible occurrence of shunt-dependent initiation events on RSV RNA as suggested in earlier work by Darlix and coworkers (75).

B. Retroelement Gag–Pol Translation

A tandem array of Gag (or Core, in the case of HBV) and Pol genes represents the core of reverse-transcribing elements that provides the structural and enzymatic functions essential for replication. The expression of Gag and Pol, and their ratio, are tightly regulated at different stages of replication. Translational control plays a major role in this regulation. In most retroviruses, Pol is expressed from genomic RNA as a Gag–Pol fusion protein (e.g., 88), which is thought to be coassembled into the virion together with the much more abundant Gag protein, and eventually processed to release the mature reverse transcriptase. The correct ratio of Gag protein to Gag–Pol protein is critical for viral infectivity (e.g., 20:1 for HIV-1; 89). The mechanisms driving Gag–Pol translation in retroviruses are (1) programmed ribosome "−1" frameshifting (reviewed in Refs. 90, 91) and (2) stop-codon readthrough [in type C retroviruses, e.g., UAG readthrough in murine leukemia virus (92); reviewed in Ref. 93]. In both cases, 5–10% of the translating ribosomes produce the Gag–Pol fusion protein, and the others terminate at the Gag stop codon. A structural stem–loop or pseudoknot element downstream of the "slippery" heptamer in the case of frameshifting, or the suppressed stop codon in the case of readthrough, is believed to pause 80S translating ribosomes to increase the efficiency of these events about three orders of magnitude above background. Interestingly, transient pausing of translating (or scanning) ribosomes at various structural elements in mRNA seems to be a common feature of unusual translation mechanisms such as frameshifting, readthrough, hopping (see Ref. 94 and references therein), and shunting (see below).

In hepadnaviruses and many plant pararetroviruses, the Gag and Pol genes are uncoupled and Pol is translated from its own AUG start codon (95–97). However, different translation strategies are used to initiate Pol translation. Whereas hepadnaviruses seem to rely on leaky scanning or internal entry abilities of the 40S ribosome (as discussed above), plant caulimoviruses, including CaMV and two other genera of Caulimoviridae, code for a transactivator protein (transactivator/viroplasmin, TAV) that allows polycistronic translation of viral RNAs by a reinitiation mechanism (see below). In CaMV, the Gag and Pol genes (ORFs IV and V, respectively) are located internally both on the pregenomic RNA and its spliced derivatives. Their expression in plant protoplasts requires the presence of TAV (98). ORF IV (Gag) is, most likely, translated as a second cistron from one of the three spliced RNAs (99), although its translation from pregenomic RNA is not excluded. The mechanism of Pol translation has not been investigated in sufficient detail to draw any definitive conclusion (97; also M. Schultze, unpublished PhD thesis, University of Basel, Basel, Switzerland). Theoretically, ORF V (Pol) could be translated in the presence of TAV from the ORF IV-expressing spliced RNA as the third cistron, or from pregenomic RNA as the sixth cistron. Alternatively, a monocistronic RNA species (either subgenomic or spliced RNA) for ORF V might exist. Indeed, there have been occasional, unconfirmed reports of a 22S RNA in CaMV-infected plants (e.g., 100). In this regard, an additional link has been established between plant caulimoviruses and animal spumaviruses (see above); in human foamy virus, Gag and Pol translational events are also uncoupled (101) and Pol is expressed from a separate, spliced RNA (102).

In contrast to the three genera of Caulimoviridae coding for TAV (CaMV-like, soybean chlorotic mottle virus-like, and, possibly, cassava vein mosaic virus-like viruses; see Ref. *103*), the three other genera of plant pararetroviruses (RTBV-like, badna-viruses, and PVCV-like viruses) do not possess any homology to TAV consensus sequences. Strikingly, their genome organization further differs in that they encode Gag and Pol within a single polyprotein, thus predicting a 1:1 ratio upon proteolytic processing. The paradox of the apparent surplus of Pol protein that might be expressed by these viruses remains to be understood. Interestingly, the *Schizosaccharomyces pombe* retrotransposon Tf1 also contains a single ORF for Gag and Pol without any obvious means for overexpressing Gag protein (*104*).

IV. Shunting Mechanisms
A. sORF-Dependent Shunt in Caulimoviridae

The RNA leaders of CaMV and other members of the Caulimoviridae contain all three types of elements inhibitory for scanning: (1) a low-energy elongated hairpin (*105*), (2) several sORFs (*72*), and (3) a putative packaging signal (*72*) that, in the case of CaMV, interacts with the viral coat protein (*71*). These would make translation initiation at a downstream ORF difficult (Figs. 2 and 3A). However, translation downstream of the leader of CaMV occurs with reasonable efficiency in plant protoplasts (*106, 30*), in several different *in vitro* systems [wheat germ extract (*107*), reticulocyte lysate (*81*), and yeast extract (*108*)], and in transgenic plants (*109*). This cannot be explained by internal ribosome entry, since initiation at the AUG of ORF VII downstream of the CaMV leader is strongly cap-dependent (*107, 110*).

Originally, it was found that certain parts of the leader support downstream translation, whereas others are inhibitory (*30*). This analysis suggested a mechanism termed the "ribosomal shunt" by which ribosomes bypass the inhibitory leader elements during normal cap-dependent scanning (*30, 31*). Antisense oligonucleotides strongly inhibited translation if directed against 3′- and 5′-proximal parts of the leader, but not if directed against the central region (*107*), again indicating bypass of the central part of the leader.

The ribosomal shunt hypothesis was confirmed in a series of transient expression experiments in plant protoplasts. A strong stem structure inhibited downstream translation if positioned close to the cap site, whereas it had a little effect if placed within the central part of the leader. This result shows that the capped 5′-end is involved in initiation, excludes internal ribosome entry, and suggests that the central stem structure is bypassed. A dicistronic mRNA with

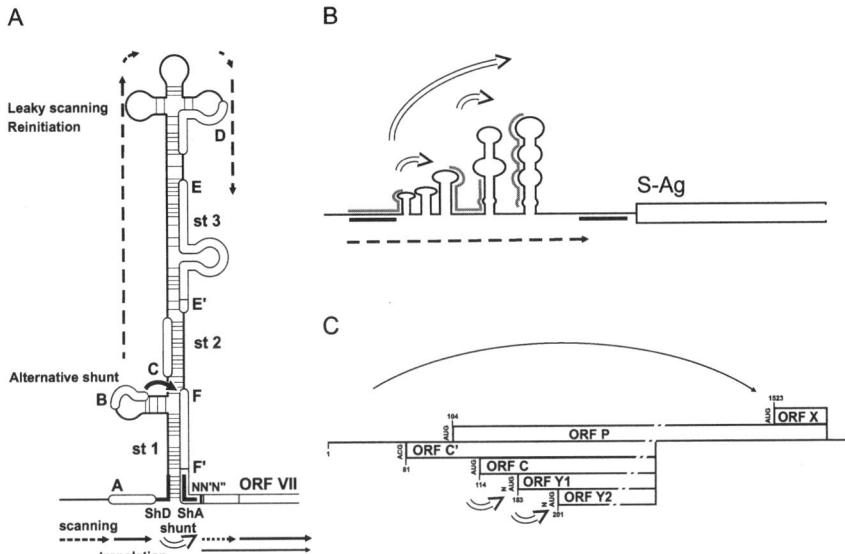

FIG. 3. Schematic representation of shunt-mediating leaders of (A) CaMV, (B) adenovirus, and (C) paramixovirus. (A) The secondary structure of the CaMV leader followed by ORF VII. Black solid lines indicate the positions of the shunt donor (ShD) and shunt acceptor (ShA) sites. Arrows show migration of ribosomes by scanning (dashed), translation (black), main shunting (open), and alternative pathways (solid black). CaMV initiation strategies: Main mechanism, shunt via stem section 1 (st 1; open arrow); alternative shunting via stem section 2 (st 2; solid black arrow) and linear ribosome migration along the leader via leaky scanning and reinitiation (dashed black arrows). Reinitiation can occur at non-AUG codons (N, N', N''), but the majority of shunting ribosomes reinitiate at the AUG of ORF VII. (B) The secondary structure of the tripartite leader (33). Arrows show migration of ribosomes by linear scanning (dashed) and possible shunting pathways (open). The locations of the complementarity regions are schematically shown (thin lines) on the leader structure, as are potential shunt take-off and landing sites (solid black lines). (C) Linear schematic presentation of the Sendai virus P/C mRNA. ORFs are shown as boxes. The AUG or introduced non-AUG (N) start sites are indicated. Arrows show AUGs recognized by shunting ribosomes initiating at Y1 and Y2 ORFs (open) and ORF X (solid black).

a β-glucuronidase (GUS) ORF inserted into the central portion and a chloramphenicol acetyltransferase (CAT) ORF downstream of the leader resulted in translation of both ORFs. Thus, one population of ribosomes might scan into the central hairpin to reach GUS (see later), whereas a second subpopulation might shunt the hairpin to translate CAT (31).

The ribosomal shunt was reconstructed by providing the leader in two RNA molecules transcribed from separate plasmids (31). The first, spanning residues 1–300, contains the shunt "take-off site" and the second contains the "shunt landing site" (see below) and a reporter ORF. If these two RNA molecules anneal, the elongated hairpin could form, reconstituting the shunt structure. That the shunt can indeed occur from these two separate molecules was demonstrated in plant protoplasts, albeit with lowered efficiency (31).

1. "TAKE-OFF" AND "LANDING" SITES

Extensive analysis of the effects of insertions of a strong stem interfering with scanning ribosomes, as well as start codon insertions, revealed shunt take-off and landing sites flanking the bypassed region of the leader. It is assumed that formation of the leader hairpin structure promotes shunting by bringing the shunt landing site upstream of the first main ORF into close spatial proximity with a shunt take-off site downstream of sORF A, the first 5'-proximal sORF (110). Indeed, the CaMV shunt landing site has been mapped on the 3'-proximal part of the elongated hairpin to between positions 548 and 562 (31). In RTBV, the shunt landing site has been precisely mapped to just downstream of the leader hairpin (70). CaMV sORF A turns out to be a key element promoting shunting in CaMV and other pararetroviruses. The first indication of the importance of sORF A for shunting came from experiments where replacement of the sORF A AUG start codon with an UUG codon nearly abolished translation downstream of the CaMV leader (31). Extensive mutational analysis of sORFs within the 35S RNA leader, separately and in different combinations, revealed that only mutations in sORF A frequently reverted on passage of the respective CaMV mutants *in planta* (see Table 1 in Ref. 111). In addition, mutations leading to the disruption of the elongated hairpin also reverted. Moreover, a striking correlation has been found between the efficiency of ribosomal shunt and viral infectivity; point mutations in sORF A which reduced the level of shunt-dependent expression also reduced infectivity of the virus in turnip plants (112). Mutational analysis of sORF A confirmed its important role in ribosomal shunt in different *in vitro* systems (81, 108) and in plant protoplasts (113).

Comparison of sORF A and several artificial sORFs of different length revealed that the optimal length of the 5'-proximal sORF for efficient shunting was between 2 and 10 codons; longer sORFs significantly reduced shunt-mediated translation (108, 110, 113). The peptide sequence of sORF A in general was not important for shunting or infectivity of the virus (112, 114). A notable exception is that a one-codon (start–stop) ORF does not promote shunting at all (113). Note, however, that in special cases the sORF of optimal length promotes shunting with only a low efficiency (81).

Another critical parameter was found to be the spacing between the stop codon of sORF A and the base of the leader hairpin. For optimal shunting this distance should be about 5–10 nucleotides (110, 113). Moving the position of sORF A closer to the hairpin base would allow a corresponding number of additional stem base pairs to melt during translation of sORF A, and, as a consequence, the shunt landing site is shifted to a position further upstream (*81*). Shifting of sORF A further away from the hairpin base precludes melting of the hairpin; the landing site is moved to position further downstream, and the shunting efficiency is strongly diminished. Thus, the position of the shunt landing site depends on the take-off site, which is itself determined by the end of the 5′-proximal sORF.

2. Role of the Secondary Structure Between the Take-off and Landing Sites

The stem structure downstream of the 5′-proximal sORF supports near-optimal shunting; a reduction in the number of base pairs reduced its efficiency (*110, 115*). In vitro analysis of the strength of the CaMV elongated hairpin, in particular its most stable lower part (so-called stem section 1), suggested the importance of base pairings rather than primary sequence of stem section 1. Physical disruption of stem section 1 by mutations in the left arm of the hairpin structure resulted in a significant reduction of shunt expression, but this could be restored by compensatory second-site mutations that restored secondary structure (*110, 113, 115*). Moreover, an artificial strong stem positioned downstream of sORF A was well able to replace the CaMV hairpin, and an artificial shunt structure could be assembled from the Kozak stem mentioned above with an artificial sORF upstream (*108, 116*). It was therefore proposed that the combination of sORF followed by a stem could be a universal signal promoting the bypass of internal leader region. In line with this, the sORF/stem combination, but little of the primary sequence, is conserved in most of the Caulimoviridae (*72*).

3. Fidelity of Initiation by Shunting Ribosomes

Shunting ribosomes seem to initiate with low fidelity, since they can initiate at non-AUG start codons if such codons are positioned within or near the shunt landing site. After some scanning in the 3′-direction, fidelity improves significantly and non-AUG codons are no longer well recognized (*81*). These results indicate that, upon landing, shunting ribosomes might be deficient in factors responsible for correct AUG start codon recognition. The situation is similar to that in the case of IRES-mediated initiation, where enhanced recognition of non-AUG codons placed at the entry site has been reported (*117*).

In the case of RTBV, ORF I lacks a proper AUG initiation codon and shunt-dependent translation is initiated at an AUU codon (*70*). Shunting ribosomes most likely land directly on this non-AUG codon, apparently without scanning. An AUG start codon introduced at the position of the AUU codon gives a six- to sevenfold increase in the level of ORF I expression, whereas an AUG placed 9 nt upstream of the authentic start site is not recognized. The efficiency of ORF I start codon recognition is low, and effective leaky scanning through the AUU codon as well as through further downstream start codons in suboptimal context allows polycistronic translation of three of the four large ORFs on the RTBV RNA (5) (see Fig. 1).

4. Shunting Is a Type of Reinitiation

Translation downstream of the CaMV leader could occur either via reinitiation by shunting ribosomes or via an IRES induced after 5′-proximal sORF translation. To distinguish between these two possibilities, sORF A was replaced by an artificial, six-amino-acid-long sORF, MAGDIS, which is known to be inhibitory for downstream start codon recognition (49, 118). This sORF caused stalling of the translating ribosome near the sORF stop codon (81), thus preventing scanning toward the 3′-end and downstream reinitiation in a bicistronic construct. The stalling event strongly represses expression of a downstream reporter gene in wheat germ extract, reticulocyte lysate, and plant protoplasts, in both linear and shunt-mediating constructs, suggesting that shunting involves a reinitiation process (81). The inhibitory effect of the MAGDIS sORF demonstrates that the 5′-proximal sORF must be translated before shunting occurs. Mutant sORFs MAGDI or MAGRIS, which did not cause stalling, allowed efficient progression toward the 3′-end. Thus, CaMV-type shunting can be considered as modified reinitiation.

Further support for this hypothesis comes from the fact that the CaMV-encoded reinitiation factor TAV significantly enhances shunt-dependent expression in plant protoplasts (31, 113).

The host protein requirement for shunting is not clear. However, the efficiency of shunting varies significantly between protoplasts originating from host and nonhost plants (106). On the other hand, shunting functions well in reticulocyte lysate and yeast, indicating that plant-specific factors are not required.

For the CaMV-type shunt mechanism, the 5′-proximal sORF translation event leads to the bypass of the structural element downstream. This could indicate that the short translation event might selectively remove some of the canonical initiation factors associated with the 48S initiation complex that may be inhibitory for ribosomal shunt, such as the initiation factor eIF4F-associated eIF4A helicase, and eIF4B, which could melt the secondary structure of the stem. However, other factors might still be associated with the ribosome to promote shunting, for example, eIF3.

The current shunt model (Fig. 3A) can be summarized as follows: (1) Ribosomes start cap-dependent linear scanning from the cap until they reach a 5′-proximal sORF start codon located just upstream of a strong structural element. (2) The sORF is translated and terminated, rendering 40S ribosomes shunt- and reinitiation-competent. (3) The ribosomes bypass the structured region and are able to reinitiate at an AUG or a non-AUG start codon immediately downstream or resume scanning and reinitiate at an AUG start codon located further downstream (81, 113).

5. Other Initiation Events Operating on the CaMV Leader

After translation of sORF A, a fraction of ribosomes shunt, but others can continue to migrate linearly toward the center of the CaMV leader (*113, 119*). Here, they can recognize some of the central sORFs, reinitiate according to a distance-dependent mechanism, then dissociate from the mRNA after termination of translation (Fig. 3A; *119*). In reticulocyte lysate, the fraction of shunting ribosomes greatly exceeds the fraction of linearly migrating ribosomes (*119*).

The arrangement of an sORF followed by a stem structure occurs twice in the CaMV leader; if the main shunt is impaired or ribosomes manage to scan through stem section 1, the combination of sORF B and stem section 2 can support shunting *in vivo* and *in vitro* (*119*) (Fig. 3A). The function of the second shunt might just be to contribute to a tighter protection of the central leader structure, which is thought to be required for packaging. This alternative shunt pathway, although not essential, may increase viral fitness *in planta*.

B. Shunt in Animal Viruses

sORF-dependent shunting has not been reported for viruses other than Caulimoviridae. The molecular details of shunting or of related phenomena proposed for other viruses remain obscure. However, three intensively studied cases of shunting might shed some light on general mechanism of ribosomal shunting.

1. Adenovirus

The adenovirus tripartite leader is a 200-nt-long noncoding region, which facilitates translation of viral mRNAs late in infection (Fig. 3B). The adenovirus late mRNAs transcribed from the major late promoter all contain the tripartite leader. Structural analysis of the leader predicted that the 5′-end is unstructured, whereas the 3′-part includes several moderately stable stem–loops (*120*) (Fig. 3B). As was shown for Caulimoviridae, translation initiation starts at the 5′-end of the tripartite leader, which does not function as an IRES (*121*). In uninfected cells, initiation downstream of the tripartite leader occurs by both scanning and shunting mechanisms with about similar efficiencies (*33*), again resembling the situation in CaMV, where a significant proportion of scanning ribosomes migrates toward the central part of the leader (*113, 119*).

The shunting phenomenon in adenovirus has been demonstrated mainly by insertion of strong hairpin structures (−80 kcal/mole) or AUG codons within the leader, either close to its 5′-end or in the middle part, where both types of insertion would affect the level of translation at the downstream ORF if this region is scanned (*33*). These insertions are inhibitory in the unstructured leader regions comprising the first 80 nt and about last 35 nt and, accordingly, these regions are believed to be scanned linearly. In contrast, insertions into the structured middle part of the leader are not inhibitory and therefore are considered to be potentially shunted. However, it was found recently that this structured part contains elements essential for efficient shunting. These regions are complementary to the 3′-end of 18S ribosomal RNA (*34*) (see Fig. 3B). The first of these is located within the first 80 nucleotides of the tripartite leader, while the other two are located in the central structured domain. A high level of ribosomal shunting was found to be dependent on the presence of any two of these three regions. Some kind of redundancy exists, since the removal of each of these regions separately had little influence on shunting and scanning efficiency. However, the combination of the second and the third complementarities are functionally dominant, that is, their combined removal significantly impaired shunting. Thus, the shunt take-off and landing sites are apparently degenerate and might be affected by the particular complementarity region present.

The functional role of the complementarity regions in ribosomal shunt remains to be understood. Despite lack of direct evidence for prokaryotic Shine–Dalgarno-type interactions in eukaryotes, a growing number of examples are accumulating where interactions between 5′-noncoding mRNA and 18S RNA regions either promote internal initiation (*122*) or facilitate the binding of 40S ribosomes to RNA leaders (S. Zhanibekova, R. Akbergenov, and B. Iskakov, personal communication). Similar sequences are found at uORF4 of the yeast GCN4 leader (*47*) and close to the shunt landing site in CaMV (*98*). These complementarity regions might induce direct binding of ribosomes or rebinding of dissociated scanning ribosomes, or they might cause stalling of scanning ribosomes migrating within these regions. In this regard, an additional link might be established between adenovirus and CaMV shunt: In both cases shunting might be induced by stalling, mediated by the complementarity region in the case of adenovirus and the stem structure in the case of CaMV.

Consistent with this idea, replacement of sORF A by one of the complementarity regions from the adenovirus leader promotes shunting in the context of the CaMV leader with an efficiency similar to that of the wild-type leader (L. Ryabova and T. Hohn, unpublished). The use of 18S RNA complementarities as shunt elements might be universal, since similar complementarities in the case of another adenovirus RNA, the IVa2 mRNA, which is synthesized during the intermediate phase of viral infection, promote bypass of internal leader regions during late stages of viral infection (*34*).

It is interesting that the basal shunting activity identified in the absence of all three complementarity elements is found to be approximately 5% of the shunting efficiency of the original tripartite leader and considered to represent a basic shunt, the mechanism of which is unknown (*34*). The tripartite leader confers the ability to eliminate or gradually reduce the normal requirement for eIF4E or eIF4F (*121, 123*). Late adenovirus mRNAs have a low requirement for eIF4F and are preferentially translated late in infection or under heat shock conditions when eIF4F is inactivated (*123*). Significantly, ribosomal shunt mechanisms in adenovirus also function exclusively in conditions when eIF4F is inactivated, during late adenovirus infection or heat shock conditions (*33*).

Like the CaMV shunt, which is activated by TAV (*112*), adenovirus shunt can also be influenced by a viral protein, L4-100K, which is involved in the preferential translation of adenovirus late mRNAs (*124*). Its mechanism of action in the shunt process and whether this is specific for adenovirus late RNA translation remain to be clarified.

Two molecular models have been suggested to explain the phenomenon of ribosome shunt in adenovirus (*125*):

1. *Dissociation model.* The loss of eIF4F RNA unwinding activity would accelerate a high dissociation off-rate of the 40S ribosomes loaded at the 5′-unstructured end, followed by direct rebinding of these ribosomes nearby, or at the start codon, via sequential leader–18S RNA interactions.
2. *Nondissociating or translocation model.* The lack of unwinding activity would block the 40S complex from further scanning into the structural region and promote instead translocation of the tethered 40S complex to the start codon via RNA–RNA interactions or initiation factors recruited to the shunting elements.

2. SENDAI VIRUS

The Sendai virus, a paramyxovirus, contains a nonsegmented minus-strand RNA genome of 15.3 kb, from which six mRNAs are transcribed (*126*). One of these, the polycistronic P/C mRNA, contains two overlapping ORFs and is known to initiate protein synthesis at multiple start codons (*127*). The order of these initiation start sites from the capped 5′-end is ACG^{81} of the C′ ORF, AUG^{104} of the P, V, and W ORFs (moderate initiation context), AUG^{114} of the C ORF, $AUGs^{183/201}$ of the Y1/Y2 ORFs (these two AUGs are in the weakest initiation context), and AUG^{1523} of the X ORF (Fig. 3C). Three proteins (P, V, and W) contain the same N-terminal 317 amino acids (aa), the X protein represents approximately the C-terminal 95 aa of the 568-aa-long P protein, and C, C′, Y1, and Y2 (215, 204, 183, and 175 aa, respectively) represent a nested set of four C proteins with a common C-terminus (*32*, *128*, *129*). Translation of the P/C mRNA is cap-dependent. Accordingly, its function is inhibited by cap analogs, insertion of a 5′-proximal stem–loop, and poliovirus infection (*130*). ORFs C′, P, and C are translated by linear leaky scanning, while the Y1/Y2, and presumably X, start sites are accessed by shunting.

No specific shunt "take-off" site could be identified by deletion mutagenesis. Moreover, artificial unstructured leaders fused to the start site of C′ ORF mediated efficient shunting, leading to normal levels of Y1/Y2 expression (*130*).

Both Y1 and Y2 expression yields are not affected by mutation of these AUGs to ACGs, suggesting that the fidelity of start codon recognition is strongly reduced. This is reminiscent of the situation found in CaMV and RTBV, although in the latter cases AUGs in the landing site are much more efficiently recognized than non-AUGs. AUGs inserted in two positions in front of the Y1 and Y2 start sites were not recognized as start sites and thus had little effect on the yield of Y1/Y2 proteins, suggesting that shunting ribosomes are directed precisely to the non-AUGs, much as in the case of CaMV and RTBV. Thus, the shunt landing site might be located between the two start sites of the Y1 and Y2 ORFs.

It can be speculated that shunting on Sendai virus RNA might also be induced by stalling of the scanning ribosome, which may be caused by the possible combination of unknown leader *cis*-elements and downstream initiation events at the C′, P, and C initiation codons. Pausing of the ribosome during an initiation event (*131*), further induced, for example, by secondary structure, might create an obstacle for the following ribosome; such pausing then would induce shunting of this ribosome past the stalled first ribosome, allowing the former to initiate further downstream.

Another candidate ORF, the AUG of which is reached by discontinuous ribosome scanning, appeared to be ORF X located at the 3′-end of the P/C mRNA (*29*) (see Fig. 3C).

3. PAPILLOMAVIRUSES

Papillomaviruses are small, double-stranded DNA (dsDNA) viruses that replicate extrachromosomally in the nuclei of infected cells (reviewed in Ref. *132*). Primary transcripts of early and late virus RNAs overlap and are processed by complex splicing and polyadenylation events. Human papillomavirus RNAs, like those of the Caulimoviridae, appear to be polycistronic, that is, encode more than one functional protein (*133–135*). However, posttranscriptional gene regulation in this case is poorly understood and requires further investigation.

Two pathways of initiation of translation have been suggested for expression of the polycistronic papillomavirus RNA containing three long ORFs, E6, E7, and E1, reading from the 5′-end: "extreme" leaky scanning for expression of E7 and ribosomal shunt for E1 (*136, 137*). Despite the very tight arrangement of ORFs E6, E7, and E1 (only 6-bp intergenic space) and 5′-dependent expression of all these ORFs, translational coupling via readthrough or reinitiation is apparently not used. Using the established criterion of examining the effect of insertion of stable stem–loops and upstream strong start codons, both linear scanning to reach E7 [without initiating at any of the 13 AUGs preceding the E7 ORF (*137*)] and nonlinear scanning (ribosomal shunt) for E1 initiation (*136*) have been demonstrated. In the latter case, the shunt "take-off" site is presumably located within the first half of ORF E6, and shunt "landing" would occur somewhere just upstream of the ORF E1 AUG. This indicates that shunting is not a randomly organized process in papillomavirus, although the specific set of conditions, such as secondary structure or protein requirements, that can trigger the bypass mechanism are not known.

4. Possible Combinations of Shunting and Internal Initiation

Hepatitis C virus mRNAs under control of the HCV IRES were inefficiently translated compared with capped and polyadenylated mRNAs. Addition of a cap and a polyA tail on the HCV mRNAs revealed that these structures interacted with the HCV IRES in a synergistic manner to load ribosomes onto the HCV mRNAs, thereby strongly enhancing translation. The positive effect of the cap and the polyA tail on initiation of translation at the initiator AUG embedded in the HCV IRES might have been the result of a discontinuous scanning, or shunting, mechanism where ribosomes are translocated from the cap site to the IRES (*138*).

Internal initiation in poliovirus might also include a shunting step to bypass the intervening region between the upstream AUG of the IRES used for direct ribosome binding and the downstream AUG used as the translation start site (*139*).

A combination of internal ribosome entry, leaky scanning, and/or shunting was also proposed for the expression of ORF 3b from polycistronic porcine transmissible gastroenteritis coronavirus (TGEV) mRNA 3 (*140*).

C. Candidates for Shunting in Cellular mRNAs

The type of shunting mechanism employed by adenovirus might also be used by the cellular mRNAs encoding hsp70 and c-*fos* (*34*); the leader sequence of both these RNAs contains a single element of complementarity to 18S rRNA and promotes initiation downstream via shunting and scanning mechanisms (*34*).

One unconfirmed report suggested an "internal ribosome repositioning" mechanism to explain differential expression of Myc1 and Myc2 isoforms in the human c-*myc* 5'-UTR (*141*). However, alternative mechanisms such as internal initiation have not been ruled out (*142, 143*).

V. Polycistronic Translation Strategies

After translating an upstream ORF of more than about 30 codons, the terminating ribosome is believed to dissociate from the eukaryotic mRNA in a scanning- and reinitiation-incompetent state. Although we have some knowledge of the first steps of the termination process in eukaryotes and the factors associated with it, our understanding of termination is far from complete. The ribosomal release factor (RRM), which triggers 50S or 70S ribosome release from the mRNA in prokaryotes (*144*), is not found in eukaryotes. Thus, the fate of the eukaryotic ribosome after translation termination remains unclear. However, following sORF translation, some ribosomes can be assumed to remain associated with the mRNA, continue scanning, and reinitiate. Very little is known about factor requirements for reinitiation of translation. The GCN4 case clarified the role of eIF2 in recognition of the second start site for the reinitiating 40S subunit (27). These experiments also suggest that eIF2 is not essential for the 40S scanning process in yeast. eIF3 would apparently also be required to promote binding of the ternary complex to the 40S ribosome. The role of eIF3 and other canonical initiation factors in reinitiation remains to be clarified.

Several groups have reported the existence of potentially bicistronic gene structures in eukaryotes, which could require reinitiation steps for their expression. Phylogenetic analysis of the SNRPN (small nuclear ribonucleoprotein N, SmN) mRNA reveals a highly conserved coding sequence SNURF (SNRPN upstream reading frame, 71 amino acids in length) upstream of the SNRPN ORF; both SNURF and SNRPN are produced from a bicistronic transcript in normal human and mouse tissues (145). In mammals, active LINE 1 elements of retroviral origin produce two proteins from a single transcript (146–148). Several bicistronic messages have been reported in *Drosophila melanogaster*, at the *stoned* locus (149), at the *Adh* locus (150), and from the *mei-218* gene (151). Interestingly, the overlapping stop–start codon UGAUG, found in embryonic exons as a result of alternative splicing during development, converts the monocistronic adult-type message encoding glutamic acid decarboxylase (GAD) into a bicistronic one coding for a 25-kDa leader peptide and a 44-kDa enzymatically active truncated GAD (152). It remains to be seen if any of these mRNAs use internal initiation, reinitiation, or leaky scanning. Also, for each case of a potentially polycistronic RNA, careful analysis of additional promoter regions, splice sites, and degradation mechanisms is required (153).

The best-studied examples of polycistronic mRNAs are found in the plant viral family Caulimoviridae, which contain up to seven long ORFs within a single RNA. However, the strategy employed to effect polycistronic translation differs in RTBV-like viruses, which use a leaky scanning mechanism, and CaMV-like viruses, which use a reinitiation mechanism activated by a viral protein.

A. Translation of Polycistronic mRNA via Leaky Scanning in Baciliform Caulimoviridae

The pregenomic RNA of RTBV and other baciliform Caulimoviridae contains four or three main ORFs, respectively, and serves for translation of at least three proteins, the products of ORFs I, II, and III (5). In RTBV, the splicing event between the first sORF and ORF IV provides a mRNA for the production of the ORF IV protein (154) (see Fig. 1). The first ORF recognized after shunting in RTBV is ORF I, which begins with the non-AUG start codon AUU (70). According to the Kozak scanning model, this non-AUG codon could be easily bypassed by linear migrating ribosomes or shunting ribosomes landing upstream of ORF I. A remarkable property of ORFs I and II is the complete lack of AUGs in any reading phase to impede scanning ribosomes and preclude downstream initiation. The start codon of ORF II is in an unfavorable sequence context and about half of the scanning ribosomes ignore it and initiate at the start codon of ORF III, which is in a strong initiation context (5). Thus, ribosomes loaded at the capped 5'-end of the RTBV RNA first shunt the leader sequence and then scan over a distance of about 900 nt upstream of ORF III, confirming the reported high processivity of scanning ribosomes *in vitro* (155). Mutation of the ORF I AUU codon to create a strong initiation codon leads to a drastic decrease in expression of ORFs II and III *in vivo* and *in vitro* (5).

B. Activated Reinitiation in Icosahedral Caulimoviridae

The pregenomic RNA of CaMV and its internally spliced derivatives (99) serve as polycistronic mRNAs for a number of viral proteins. There are several lines of evidence that these polycistronic RNAs are indeed used for viral protein production in CaMV-infected plants. First, only one subgenomic RNA, the 19S RNA encoding the transactivator/viroplasmin protein (TAV), has been identified. The other ORFs are translated from the 35S RNA or its spliced versions. Second, ORFs I–V are tightly arranged and often overlap by a single base, suggesting that their translation might be linked. Indeed, some mutations within ORFs VII and II are polar, typical for polycistronic prokaryotic mRNAs, meaning that they affect translation of the following ORF (156, 157).

1. POLYCISTRONIC TRANSLATION IN ICOSAHEDRAL CAULIMOVIRIDAE IS UNDER THE CONTROL OF TAV

Translation of the 35S RNA is 5'-end-dependent and the first translatable ORF is sORF A within the leader. Thus, translation of major ORFs on the CaMV RNA requires reinitiation. Reinitiation on the polycistronic 35S RNA and its spliced versions is activated in the presence of TAV (59, 98, 99), which is expressed early in infection. Transient expression of derivatives of the CaMV genome, each containing the upstream sequence of a specific ORF fused to the CAT or GUS reporter genes, was shown in plant protoplasts (98). A reporter gene fusion to the start codon of ORFs VII and II was expressed at 15% of the level of the monocistronic control, but at 100% if a second plasmid expressing the TAV ORF was provided; fusions to the start codons of ORFs I, III, IV, and V were expressed only in the presence of TAV (100% for ORF I, 33% for ORF III, 50% for ORF IV, and 3% for ORF V). TAV-mediated transactivation of expression of all the major ORFs on the polycistronic pregenomic RNA has been also reported for figwort mosaic virus (FMV; 158–160) and peanut chlorotic streak virus (161), and TAV of CaMV and FMV can reciprocally activate polycistronic translation in these viruses (158).

The use of partially or completely artificial constructs has confirmed the role of TAV in general activation of polycistronic translation (19, 45). When bicistronic constructs, for example, consisting of ORFs VII and CAT, or GUS and CAT, were analyzed in plant protoplasts, the level of expression of the second ORF was always under TAV control, whereas expression of the first ORF was not affected by TAV. Specific cis sequence signals are not required for transactivation of second ORF translation, since TAV can activate reinitiation after translation of any first ORF in an artificial bicistronic RNA (19, 45). Notably, the presence of a sORF (optimally around 30 codons) upstream of both long ORFs strongly enhances the process, activating expression of the first ORF 2-fold and the second 8- to 10-fold. A long overlap of the major ORFs (130 nt) inhibits transactivation, whereas a short overlap (17 nt) is permissible. In caulimoviruses a short overlap between two ORFs, via an AUGA, is very common, although expression of the downstream ORF is TAV-dependent as well (19). A stem structure at the cap site inhibits expression of both reporters, whereas a stem between the two ORFs inhibits only expression of the second (19).

Similar results were obtained in yeast when the 35S RNA promoter was used to direct bicistronic RNA transcription (162) and in transgenic plants (163). Transgenic plants expressing TAV exhibit chlorosis and other virus-like symptoms (164–168). TAV expression in these plants also correlates with changes in plant morphology and development, suggesting that TAV might transactivate translation from as-yet-unknown complex mRNAs encoding regulatory proteins involved in development.

2. TAV AND ITS INTERACTIONS WITH THE HOST
TRANSLATIONAL MACHINERY

TAV is mainly expressed from the 19S subgenomic RNA, but it is also able to transactivate its own expression from the 35S RNA (169, 170). TAV is very abundant in the cytoplasm of infected cells and forms a dense matrix, the so-called inclusion bodies or viroplasm. In early phases of infection, these are surrounded by polyribosomes (171–173). These "tethered" ribosomes could direct CaMV translation products into the inclusion bodies, where all proteins expressed from CaMV RNAs are found (174, 175), including heterologous nonviral proteins encoded by transgenic CaMV RNA (176).

TAV has many functions in the life cycle of the virus. In addition to polycistronic translation, it controls virus assembly and replication and determines the host range (see Ref. 177 for review). Virion assembly might be guided by the interaction of TAV with coat protein (178). Moreover, it appears that TAV is involved in the stabilization of other viral gene products (179).

A basal level of transactivation activity has been demonstrated to be associated with the central portion of the TAV protein (miniTAV or MAV; 180) (Fig. 4) in protoplasts transfected with a 100-fold excess of DNA encoding this

VIRAL STRATEGIES OF TRANSLATION INITIATION

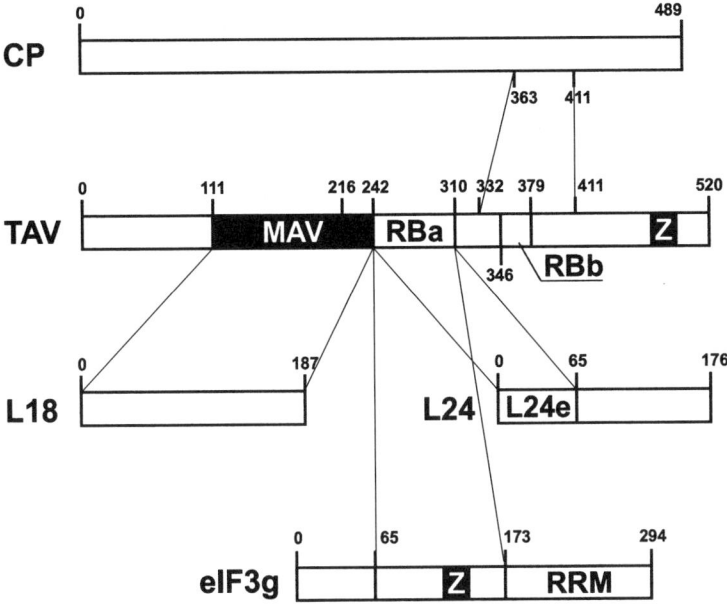

FIG. 4. Protein–protein interactions between TAV and its viral and host partners. CaMV coat protein (CP), ribosomal proteins L18 and L24, and eIF3g are shown as boxes. The interacting regions of the two proteins are connected by thin lines. The MAV, RBa, RBb, and Zn finger of TAV are shown. The L24 region of homology with archaebacterial L24e is indicated. The RRM and Zn finger of eIF3g are also indicated.

polypeptide. In addition to the defect in transactivation, MAV has so far only been found to be active in *Nicotiana plumbaginifolia* protoplasts, in contrast to full-length TAV, which is active in protoplasts of a number of dicot and monocot plants (59, 158). The C-terminal part of TAV (CTAV; Fig. 4) can efficiently inhibit the transactivation activity of the entire protein, suggesting the CTAV is able to sequester host factors that are essential for transactivation activity (180). Notably, the N-terminal region located upstream of the MAV domain was found also to affect TAV-mediated transactivation activity (181).

TAV transactivates polycistronic translation by association with host translational machinery, namely polysomes and associated proteins. Indeed, TAV was found in polysomes isolated from CaMV-infected turnip plants, and it cosediments with polysomes isolated from healthy turnip plants (182). This interaction is apparently mediated by interaction with key components of the host translational machinery, namely the 60S large ribosomal subunit via at least two ribosomal proteins, L24 (182) and L18 (183), and with initiation factor eIF3 via its subunit g (182). The L18 ribosomal protein interacts with the MAV domain of

NTAV, while both the central segment of eIF3g, including the Zn-finger motif, and the N-terminal half of ribosomal protein L24, compete for the same binding site within one of the RNA-binding domains (RBa) of CTAV (Fig. 4).

It is remarkable that both ribosomal subunits and eIF3 can be found in a complex with TAV *in vitro*, but TAV association with 40S is indirect, that is, mediated by eIF3. In contrast, eIF3 association with 60S is mediated by TAV. Thus, the existence of two ternary complexes, TAV/eIF3/40S and eIF3/TAV/60S, might play an essential role in the transactivation process. Significant accumulation of eIF3 in polysomes isolated from CaMV-infected plants correlates well with TAV accumulation, whereas only traces of eIF3 were found in polysomes isolated from healthy turnip plants (*182*).

Transient overexpression of eIF3g has a strong negative effect on TAV-mediated polycistronic translation in plant protoplasts (*182*), which correlates well with the reported *in vitro* interaction between TAV and complete eIF3, suggesting that a TAV/eIF3 complex is active in transactivation.

Archaebacterial homologs of L24 and L18 are located on the interface and the external surface of the 60S ribosome, respectively (*184, 185*). In eukaryotes, the location of L24 in the main factor-binding site on the interface of the 60S ribosome has also been reported (*186*; R. Beckmann, personal communication). This suggests that, in the eukaryotic 60S ribosome, these proteins are located too far apart to interact with the same molecule of TAV. Thus, the 60S ribosome is presumably capable of binding two TAV molecules simultaneously on its external and internal surfaces.

These observations can explain why transient overexpression of L24 has a strong positive effect on TAV-mediated polycistronic translation in plant protoplasts (*182*). Interaction of TAV with the main factor-binding site on the 60S ribosome via L24 might indeed lead to inhibition of protein synthesis during the late phase of viral infection, promoting a switch to viral assembly. On the other hand, L24 might have an extraribosomal function and a separate protein might be involved in 60S subunit turnover, as shown previously for yeast L24 (*187*). The ribosomal protein L18, due to its location on the external surface of the 60S ribosome, might interact with other partners such as TAV without affecting 80S ribosome formation.

The accumulation of TAV as the main matrix protein in viral inclusion bodies together with the coaccumulation of proteins derived from the polycistronic CaMV RNA suggests that TAV-enhanced translation occurs on the inclusion body surface. The TAV molecules that form the inclusion bodies might present a perfect surface for ribosome and eIF3 recruitment; the concomitant increase in local concentration of these factors might contribute to the increased reinitiation rate observed in the presence of TAV. RNA-binding activities of TAV, mediated by its ssRNA- and DNA-binding domains within its C-terminal part (*180*), and the dsRNA-binding domain within the MAV region (*188*) might further enhance local TAV concentration in polysomes (Fig. 4).

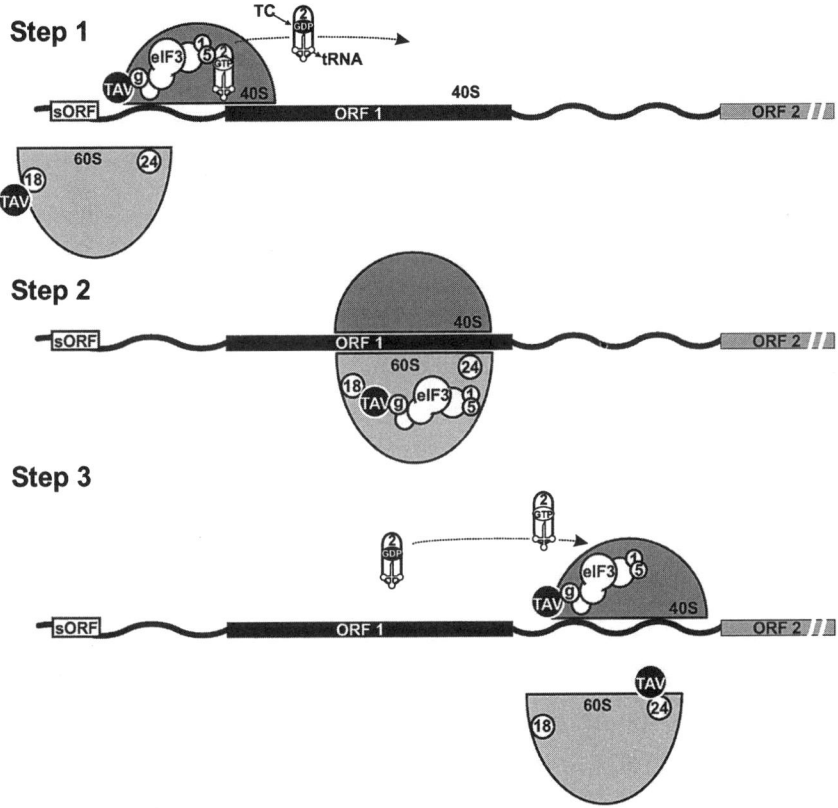

FIG. 5. Model of TAV function during translation of the polycistronic RNA. Steps 1–3 are described in the text. eIF3, its subunit g, eIF1 (1), eIF2 (2), eIF5 (5), tRNA, L24 (24), L18 (18), and TAV are indicated. Recycling of eIF2 is shown by a dotted arrow.

C. Model of TAV-Activated Reinitiation

Despite the abundance of data on TAV molecular interactions with the host translational machinery, we do not yet understand the exact mechanism of TAV-mediated reinitiation of translation. However, the accumulated data allow us to offer the following three-step working model (Fig. 5).

Step 1. Primary TAV acquisition by eIF3. Several observations suggest that TAV is preferentially sequestered by the translational machinery during sORF translation. (1) Ribosomes often translate an sORF and then reinitiate efficiently downstream of it (*44, 28*). This suggests that some of the eIFs are still associated with the translational machinery after sORF translation, but not after translation of larger ORFs. (2) Polycistronic translation under TAV control is enhanced by an sORF (or CaMV ORF VII) positioned in the leader (*19, 45*). (3) TAV interacts with eIF3 as well as with eIF3 bound to the 40S ribosome (*182*). (4) eIF3 might be associated with the 40S ribosome without interfering with the 80S ribosome formation (*189*). We assume that after sORF translation, the 40S ribosome including certain eIFs remains loosely attached to the mRNA and can resume scanning. We further speculate that eIF3 is one of the remaining factors, whereas eIF2/GDP, and perhaps also eIF4F and eIF4B, are preferentially released. eIF2/GTP/Met-tRNAi can then be reacquired after resumption of scanning via the bridging function of eIF5 interacting with eIF3c (*190*) and TAV could be acquired through its association with eIF3 via its subunit g, perhaps even docking at the same position as other eIF3-interacting proteins. Yeast eIF3g interacts with eIF4B (*191*), and we have confirmed that this interaction also occurs in plants (H.-S. Park, T. Hohn, and L. A. Ryabova, unpublished); TAV could thus be sequestered by eIF3g upon release of eIF4B.

Step 2. Preservation of reinitiation-competent ribosomes during translation of the first large ORF. During translation of a longer ORF in the absence of TAV, eIF3 will be released during the translation elongation step, leaving ribosomes incompetent for reinitiation. In the presence of TAV, however, the interaction of eIF3 with the translating ribosome might be stabilized. We do not know in this case whether eIF3 holds its original position or is moved to a new location within the translation machinery. An exciting possibility is that eIF3 is delivered to the 60S ribosome via the L18 ribosomal protein, which is located at the 60S external surface. We were able to show the existence of both TAV/eIF3/40S and eIF3/TAV/60S complexes *in vitro*. This raises the possibility that eIF3 is transferred to the 60S ribosome apparently via the L18/TAV interaction. We do not think that the alternative L24/TAV interaction is involved at this step, since L24 and eIF3g compete for the same binding site on TAV and L24 is located at the 60S/40S interface.

Step 3. Reinitiation at the second large ORF. We postulate that during the translation termination process, the binding site on the 40S ribosome for eIF3 becomes available again and eIF3 could be transferred back to this site, leading to reacquisition of eIF2/GTP/Met-tRNAi and scanning to find the second ORF start codon.

VI. Outlook

The translation process is the most important step in transforming genetic into functional information. The interplay of initiation factors and regulatory regions on the leaders and trailers of RNA is the most important parameter controlling translation, both quantitatively and qualitatively. Although the basic translation initiation factors and some details of the translation process are known, control of translation initiation varies from case to case and is controlled by environmental factors, much as is transcription. Scientists try to find rules that make the behavior of biological entities and processes predictable. However, also in biology, rules are there to be broken and often the exceptions lead to deeper understanding of the rule. Viruses are among the best model organisms for studying these parameters due to the rapid evolution of their translation control mechanisms and the high density of regulatory sequences located on their genome. Members of the Caulimoviridae are excellent examples of such viral rule-breakers, circumventing linear scanning by shunting and allowing reinitiation by interactions of a viral protein with the translational machinery. The future will tell whether similar mechanisms occur also during translation of cellular messenger RNAs.

ACKNOWLEDGMENTS

We thank H. Rothnie, K. Kobayashi, and J. Fütterer for critical reading of the manuscript. This work was supported by the Novartis Research Foundation, an INTAS grant to L.R. and T.H., and a FEBS fellowship to M.M.P.

REFERENCES

1. G. Drugeon, S. Urcuqui-Inchima, M. Milner, G. Kadare, R. P. Valle, A. Voyatzakis, and A. L. Haenni, The Strategies of plant virus gene expression: models of economy. *Plant Sci.*, in press.
2. R. Hull, Classification of reverse transcribing elements: a discussion document. *Arch. Virol.* **144**, 209–213 (1999).
3. J. M. Coffin, S. H. Hughes, and H. Varmus, "Retroviruses." Cold Spring Harbor Laboratory Press, Cold Spring Harbor, New York, 1997.
4. K. R. Richert-Pöggeler and T. Hohn, Petunia vein clearing virus (PVCV): a potential retrovirus in petunia. *Eur. J. Cell Biol.* **79**, 391 (2000).
5. J. Fütterer, H. M. Rothnie, T. Hohn, and I. Potrykus, Rice tungro bacilliform virus open reading frames II and III are translated from polycistronic pregenomic RNA by leaky scanning. *J. Virol.* **71**, 7984–7989 (1997).
6. R. J. Jackson, Comparative view of initiation site selection mechanisms. In "Translational Control of Gene Expression" (N. Sonenberg, J. W. B. Hershey, and M. B. Mathews, eds.), pp. 127–183. Cold Spring Harbor Laboratory Press, Cold Spring Harbor, New York, 2000.
7. M. Kozak, The scanning model for translation an update. *J. Cell Biol.* **108**, 229–241 (1989).
8. M. Kozak, Initiation of translation in prokaryotes and eukaryotes. *Gene* **234**, 187–208 (1999).
9. J. W. B. Hershey and W. C. Merrick, Pathway and mechanisms of initiation of protein synthesis. In "Translational Control of Gene Expression" (N. Sonenberg, J. W. B. Hershey, and M. B. Mathews, eds.), pp. 33–88. Cold Spring Harbor Laboratory Press, Cold Spring Harbor, New York, 2000.
10. T. V. Pestova, V. G. Kolupaeva, I. B. Lomakin, E. V. Pilipenko, I. N. Shatsky, V. I. Agol, and C. U. Hellen, Molecular mechanisms of translation initiation in eukaryotes. *Proc. Natl. Acad. Sci. USA* **98**, 7029–7036 (2001).

11. A. Bandyopadhyay and U. Maitra, Cloning and characterization of the p42 subunit of mammalian translation initiation factor 3 (eIF3): demonstration that eIF3 interacts with eIF5 in mammalian cells. *Nucleic Acids Res.* **27**, 1331–1337 (1999).
12. T. V. Pestova, S. I. Borukhov, and C. U. Hellen, Eukaryotic ribosomes require initiation factors 1 and 1A to locate initiation codons. *Nature* **394**, 854–859 (1998).
13. T. V. Pestova, I. B. Lomakin, J. H. Lee, S. K. Choi, T. E. Dever, and C. U. Hellen, The joining of ribosomal subunits in eukaryotes requires eIF5B. *Nature* **403**, 332–335 (2000).
14. K. S. Browning, The plant translational apparatus. *Plant Mol. Biol.* **32**, 107–144 (1996).
15. J. Fütterer and T. Hohn, Translation in plants rules and exceptions. *Plant Mol. Biol.* **32**, 159–189 (1997).
16. J. Pelletier and N. Sonenberg, Internal binding of eukaryotic ribosomes on poliovirus RNA: Translation in HeLa cell extracts. *J. Virol.* **63**, 441–443 (1989).
17. M. Kozak, Influence of mRNA secondary structure on initiation by eukaryotic ribosomes. *Proc. Natl. Acad. Sci. USA* **83**, 2850–2854 (1986).
18. M. Kozak, Circumstances and mechanisms of inhibition of translation by secondary structure in eukaryotic mRNAs. *Mol. Cell. Biol.* **9**, 5134–5142 (1989).
19. J. Fütterer and T. Hohn, Translation of a polycistronic mRNA in presence of the cauliflower mosaic virus transactivator protein. *EMBO J.* **10**, 3887–3896 (1991).
20. M. Kozak, Context effects and inefficient initiation at non-AUG codons in eukaryotic cell-free translation systems. *Mol. Cell. Biol.* **9**, 5073–5080 (1989).
21. M. Kozak, Regulation of translation in eukaryotic systems. *Annu. Rev. Cell Biol.* **8**, 197–225 (1992).
22. R. Stripecke, C. C. Oliveira, J. E. McCarthy, and M. W. Hentze, Proteins binding to 5′ untranslated region sites: a general mechanism for translational regulation of mRNAs in human and yeast cells. *Mol. Cell. Biol.* **14**, 5898–5909 (1994).
23. E. Paraskeva, N. K. Gray, B. Schlager, K. Wehr, and M. W. Hentze, Ribosomal pausing and scanning arrest as mechanisms of translational regulation from cap distal iron responsive elements. *Mol. Cell. Biol.* **19**, 807–816 (1999).
24. J. Pelletier and N. Sonenberg, Internal initiation of translation of eukaryotic mRNA directed by a sequence derived from poliovirus RNA. *Nature* **334**, 320–325 (1988).
25. A. Kaminski, M. T. Howell, and R. J. Jackson, Initiation of encephalomyocarditis virus RNA translation: the authentic initiation site is not selected by a scanning mechanism. *EMBO J.* **9**, 3753–3759 (1990).
26. G. J. Belsham, Dual initiation sites of protein synthesis on foot and mouth disease virus RNA are selected following internal entry and scanning of ribosomes in vivo. *EMBO J.* **11**, 1105–1110 (1992).
27. A. G. Hinnebusch, Translational regulation of yeast GCN4. A window on factors that control initiator-trna binding to the ribosome. *J. Biol. Chem.* **272**, 21661–21664 (1997).
28. D. R. Morris and A. P. Geballe, Upstream open reading frames as regulators of mRNA translation. *Mol. Cell. Biol.* **20**, 8635–8642 (2000).
29. J. Curran and D. Kolakofsky, Scanning independent ribosomal initiation of the Sendai virus X protein. *EMBO J.* **7**, 2869–2874 (1988).
30. J. Fütterer, K. Gordon, H. Sanfacon, J. M. Bonneville, and T. Hohn, Positive and negative control of translation by the leader of cauliflower mosaic virus pregenomic 35S RNA. *EMBO J.* **9**, 1697–1707 (1990).
31. J. Fütterer, Z. Kiss-László, and T. Hohn, Nonlinear ribosome migration on cauliflower mosaic virus 35S RNA. *Cell* **73**, 789–802 (1993).
32. J. Curran and D. Kolakofsky, Scanning independent ribosomal initiation of the Sendai virus Y proteins in vitro and in vivo. *EMBO J.* **8**, 521–526 (1989).
33. A. Yueh and R. J. Schneider, Selective translation initiation by ribosome jumping in adenovirus infected and heat-shocked cells. *Genes Dev.* **10**, 1557–1567 (1996).

34. A. Yueh and R. J. Schneider, Translation by ribosome shunting on adenovirus and hsp70 mRNAs facilitated by complementarity to 18S rRNA. *Genes Dev.* **14,** 414–421 (2000).
35. S. A. Sedman, G. W. Gelembiuk, and J. E. Mertz, Translation initiation at a downstream AUG occurs with increased efficiency when the upstream AUG is located very close to the 5′ cap. *J. Virol.* **64,** 453–457 (1990).
36. M. Kozak, A short leader sequence impairs the fidelity of initiation by eukaryotic ribosomes. *Gene Expression* **1,** 111–115 (1991).
37. H. Ruan, J. R. Hill, S. Fatemie-Nainie, and D. R. Morris, Cell-specific translational regulation of S adenosylmethionine decarboxylase mRNA. Influence of the structure of the 5′ transcript leader on regulation by the upstream open reading frame. *J. Biol. Chem.* **269,** 17905–17910 (1994).
38. M. Kozak, Downstream secondary structure facilitates recognition of initiator codons by eukaryotic ribosomes. *Proc. Natl. Acad. Sci. USA* **87,** 8301–8305 (1990).
39. R. J. Jackson and A. Kaminski, Internal initiation of translation in eukaryotes: the picornavirus paradigm and beyond. *RNA* **1,** 985–1000 (1995).
40. M. Holcik, N. Sonenberg, and R. G. Korneluk, Internal ribosome initiation of translation and the control of cell death. *Trends Genet.* **16,** 469–473 (2000).
41. T. V. Pestova, I. N. Shatsky, and C. U. Hellen, Functional dissection of eukaryotic initiation factor 4F: the 4A subunit and the central domain of the 4G subunit are sufficient to mediate internal entry of 43S preinitiation complexes. *Mol. Cell. Biol.* **16,** 6870–6878 (1996).
42. G. J. Belsham and N. Sonenberg, Picornavirus RNA translation: roles for cellular proteins. *Trends Microbiol.* **8,** 330–335 (2000).
43. K. Tsukiyama-Kohara, N. Iizuka, M. Kohara, and A. Nomoto, Internal ribosome entry site within hepatitis C virus RNA. *J. Virol.* **66,** 1476–1483 (1992).
44. M. Kozak, Effects of intereistronic length on the efficiency of reinitiation by eukaryotic ribosomes. *Mol. Cell. Biol.* **7,** 3438–3445 (1987).
45. J. Fütterer and T. Hohn, Role of an upstream open reading frame in the translation of polycistronic mRNA in plant cells. *Nucleic Acids Res.* **20,** 3851–3857 (1992).
46. B. G. Luukkonen, W. Tan, and S. Schwartz, Efficiency of reinitiation of translation on human immunodeficiency virus type 1 mRNAs is determined by the length of the upstream open reading frame and by intercistronic distance. *J. Virol.* **69,** 4086–4094 (1995).
47. P. F. Miller and A. G. Hinnebusch, Sequences that surround the stop codons of upstream open reading frames in GCN4 mRNA determine their distinct functions in translation control. *Genes Dev.* **3,** 1217–1225 (1989).
48. P. S. Lovett and E. J. Rogers, Ribosome regulation by the nascent peptide. *Microbiol. Rev.* **60,** 366–385 (1996).
49. J. R. Hill and D. R. Morris, Cell specific translational regulation of S-adenosylmethionine decarboxylase mRNA. Dependence on translation and coding capacity of the cis-acting upstream open reading frame. *J. Biol. Chem.* **268,** 726–731 (1993).
50. M. J. Ruiz-Echevarria, K. Czaplinski, and S. W. Peltz, Making sense of nonsense in yeast. *Trends Biochem. Sci.* **21,** 433–438 (1996).
51. C. Vilela, C. V. Ramirez, B. Linz, C. Rodrigues-Pousada, and J. E. McCarthy, Post termination ribosome interactions with the 5′ UTR modulate yeast mRNA stability. *EMBO J.* **18,** 3139–3152 (1999).
52. D. S. Peabody, S. Subramani, and P. Berg, Effect of upstream reading frames on translation efficiency in simian virus 40 recombinants. *Mol. Cell. Biol.* **6,** 2704–2711 (1986).
53. D. S. Peabody and P. Berg, Termination reinitiation occurs in the translation of mammalian cell mRNAs. *Mol. Cell. Biol.* **6,** 2695–2703 (1986).
54. S. F. Yu, M. D. Sullivan, and M. L. Linial, Evidence that the human foamy virus genome is DNA. *J. Virol.* **73,** 1565–1572 (1999).

55. M. L. Linial, Foamy viruses are unconventional retroviruses. *J. Virol.* **73**, 1747–1755 (1999).
56. C. H. Lecellier and A. Saib, Foamy viruses: between retroviruses and pararetroviruses. *Virology* **271**, 1–8 (2000).
57. P. J. Farabaugh, Translational frameshifting: implicatinos for the mechanism of translational frame maintenance. *Prog. Nucleic Acid Res. Mol. Biol.* **64**, 131–170 (2000).
58. S. Schwartz, B. K. Felber, and G. N. Pavlakis, Mechanism of translation of monocistronic and multicistronic human immunodeficiency virus type 1 mRNAs. *Mol. Cell. Biol.* **12**, 207–219 (1992).
59. J. M. Bonneville, H. Sanfacon, J. Fütterer, and T. Hohn, Posttranscriptional transactivation in cauliflower mosaic virus. *Cell* **59**, 1135–1143 (1989).
60. C.-G. Lin and S. J. Lo, Evidence for the involvement of a ribosomal leaky scanning mechanism in the translation of the hepatitis B virus pol-gene from the viral pregenome RNA. *Virology* **188**, 342–352 (1992).
61. N. Fouillot, S. Tlouzeau, J. M. Rossignol, and O. Jean-Jean, Translation of the hepatitis B virus P gene by ribosomal scanning as an alternative to internal initiation. *J. Virol.* **67**, 4886–4895 (1993).
62. W. L. Hwang and T. S. Su, Translational regulation of hepatitis B virus polymerase gene by termination-reinitiation of an upstream minicistron in a length dependent manner. *J. Gen. Virol.* **79**(Pt 9), 2181–2189 (1998).
63. O. Jean-Jean, T. Weimer, A. M. deRecondo, H. Will, and J.-M. Rossignol, Internal entry of ribosomes and ribosomal scanning involved in HBV P gene expression. *J. Virol.* **63**, 5451–5454 (1989).
64. L.-J. Chang, D. V. Ganem, and H. E. Varmus, Mechanism of translation of the hepadnaviral polymerase (P) gene. *Proc. Natl. Acad. Sci. USA* **87**, 5158–5162 (1990).
65. M. Nassal, M. Junker-Niepmann, and H. Schaller, Translational inactivation of RNA function: Diserimination against a subset of genomic transcripts during HBV nucleocapsid assembly. *Cell* **63**, 1357–1363 (1990).
66. J. E. Tavis and D. V. Ganem, Expression of functional hepatitis B virus polymerase in yeast reveals it to be the sole viral protein required for correct initiation of reverse transcription. *Proc. Natl. Acad. Sci. USA* **90**, 4107–4111 (1993).
67. J. Beck and M. Nassal, Formation of a functional hepatitis B virus replication initiation complex involves a major structural alteration in the RNA template. *Mol. Cell. Biol.* **18**, 6265–6272 (1998).
68. T. C. Ho, K. S. Jeng, C. P. Hu, and C. Chang, Effects of genomic length on translocation of hepatitis B virus polymerase-linked oligomer. *J. Virol.* **74**, 9010–9018 (2000).
69. W. L. Hwang and T. S. Su, The encapsidation signal of hepatitis B virus facilitates preC AUG recognition resulting in inefficient translation of the downstream genes. *J. Gen. Virol.* **80**, 1769–1776 (1999).
70. J. Fütterer, I. Potrykus, Y. Bao, L. Li, T. M. Burns, R. Hull, and T. Hohn, Position-dependent ATT initiation during plant pararetrovirus Rice Tungro Bacilliform Virus translation. *J. Virol.* **70**, 2999–3010 (1996).
71. O. Guerra-Peraza, M. de Tapia, T. Hohn, and M. Hemmings-Mieszczak, Interaction of the cauliflower mosaic virus coat protein with the pregenomic RNA leader. *J. Virol.* **74**, 2067–2072 (2000).
72. M. M. Pooggin, J. Fütterer, K. G. Skryabin, and T. Hohn, A short open reading frame terminating in front of a stable hairpin is the conserved feature in pregenomic RNA leaders of plant pararetroviruses. *J. Gen. Virol.* **80**, 2217–2228 (1999).
73. L. I. Messer, J. G. Levin, and S. K. Chattopadhyay, Metabolism of viral RNA in murine leukemia virus infected cells; evidence for differential stability of viral message and virion precursor RNA. *J. Virol.* **40**, 683–690 (1981).

74. N. Dorman and A. Lever, Comparison of viral genomic RNA sorting mechanisms in human immunodeficiency virus type 1 (HIV-1), HIV-2, and Moloney murine leukemia virus. *J. Virol.* **74,** 11413–11417 (2000).
75. J. L. Darlix, M. Zuker, and P. F. Spahr, Structure function relationship of Rous sarcoma virus leader RNA. *Nucleic Acids Res.* **10,** 5183–5196 (1982).
76. O. Donzé and P.-F. Spahr, Role of the open reading frames of Rous sarcoma virus leader RNA in translation and genome packaging. *EMBO J.* **11,** 3747–3757 (1992).
77. A. Moustakas, T. S. Sonstegard, and P. B. Hackett, Effects of the open reading frames in the Rous sarcoma virus leader RNA on translation. *J. Virol.* **67,** 4350–4357 (1993).
78. O. Donzé, P. Damay, and P. F. Spahr, The first and third uORFs in RSV leader RNA are efficiently translated: implications for translational regulation and viral RNA packaging. *Nucleic Acids Res.* **23,** 861–868 (1995).
79. T. S. Sonstegard and P. B. Hackett, Autogenous regulation or RNA translation and packaging by Rous sarcoma virus Pr76gag. *J. Virol.* **70,** 6642–6652 (1996).
80. C. Deffaud and J. L. Darlix, Rous sarcoma virus translation revisited: characterization of an internal ribosome entry segment in the 5' leader of the genomic RNA. *J. Virol.* **74,** 11581–11588 (2000).
81. L. A. Ryabova and T. Hohn, Ribosome shunting in the cauliflower mosaic virus 35S RNA leader is a special case of reinitiation of translation functioning in plant and animal systems. *Genes Dev.* **14,** 817–829 (2000).
82. T. Ohlmann, M. Lopez-Lastra, and J. L. Darlix, An internal ribosome entry segment promotes translation of the simian immunodeficiency virus genomic RNA. *J. Biol. Chem.* **275,** 11899–11906 (2000).
83. C. B. Buck, X. Shen, M. A. Egan, T. C. Pierson, C. M. Walker, and R. F. Siliciano, The human immunodeficiency virus type 1 gag gene encodes an internal ribosome entry site. *J. Virol.* **75,** 181–191 (2001).
84. C. Berlioz and J. L. Darlix, An internal ribosomal entry mechanism promotes translation of murine leukemia virus gag polyprotein precursors. *J. Virol.* **69,** 2214–2222 (1995).
85. C. Berlioz, C. Torrent, and J. L. Darlix, An internal ribosomal entry signal in the rat VL30 region of the Harvey murine sarcoma virus leader and its use in dicistronic retroviral vectors. *J. Virol.* **69,** 6400–6407 (1995).
86. M. Lopez-Lastra, C. Gabus, and J. L. Darlix, Characterization of an internal ribosomal entry segment within the 5' leader of avian reticuloendotheliosis virus type A RNA and development of novel MLV-REV-based retroviral vectors. *Hum. Gene Ther.* **8,** 1855–1865 (1997).
87. G. Miele, A. Mouland, G. P. Harrison, E. Cohen, and A. M. Lever, The human immunodeficiency virus type 1–5' packaging signal structure affects translation but does not function as an internal ribosome entry site structure. *J. Virol.* **70,** 944–951 (1996).
88. T. Jacks, M. D. Power, F. R. Masiarz, P. A. Luciw, P. J. Barr, and H. E. Varmus, Characterization of ribosomal frameshifting in HIV-1 gag-pol expression. *Nature* **331,** 280–283 (1988).
89. M. Shehu-Xhilaga, S. M. Crowe, and J. Mak, Maintenance of the Gag/Gag-Pol ratio is important for human immunodeficiency virus type 1 RNA dimerization and viral infectivity. *J. Virol.* **75,** 1834–1841 (2001).
90. P. J. Farabaugh, Programmed translational frameshifting. *Microbiol. Rev.* **60,** 103–134 (1996).
91. R. F. Gesteland and J. F. Atkins, Recoding: dynamic reprogramming of translation. *Annu. Rev. Biochem.* **65,** 741–768 (1996).
92. L. Philipson, P. Andersson, U. Olshevsky, R. Weinberg, D. Baltimore, and R. Gesteland, Translation of MuLV and MSV RNAs in nuclease treated reticulocyte extracts: enhancement of the gag-pol polypeptide with yeast suppressor tRNA. *Cell* **13,** 189–199 (1978).
93. A. Rein and J. G. Levin, Readthrough suppression in the mammalian type C retroviruses and what it has taught us. *New Biol.* **4,** 283–289 (1992).

94. A. J. Herr, N. M. Wills, C. C. Nelson, R. F. Gesteland, and J. F. Atkins, Drop off during ribosome hopping. *J. Mol. Biol.* **311**, 445–452 (2001).
95. L.-J. Chang, P. Pryciak, D. V. Ganem, and H. E. Varmus, Biosynthesis of the reverse transcriptase of hepatitis B viruses involves de novo translational initiation not ribosomal frameshifting. *Nature* **337**, 364–367 (1989).
96. H.-J. Schlicht, J. Salfeld, and H. Schaller, Synthesis and encapsidation of duck hepatitis B virus reverse transcriptase do not require formation of core polymerase fusion proteins. *Cell* **56**, 85–92 (1989).
97. M. Schultze, T. Hohn, and J. Jiricny, The reverse transcriptase gene of CaMV is translated separately from the capsid gene. *EMBO J.* **9**, 1177–1185 (1990).
98. J. Fütterer, J. M. Bonneville, K. Gordon, M. DeTapia, S. Karlsson, and T. Hohn, Expression from polycistronic cauliflower mosaic virus pregenomic RNA. *In* "Posttranscriptional Control of Gene Expression" (J. E. G. McCarthy and M. F. Tuite, eds.), pp. 349–357. Springer, Berlin, 1990.
99. Z. Kiss-László, S. Blanc, and T. Hohn, Splicing of Cauliflower Mosaic Virus 35S RNA is essential for viral Infectivity. *EMBO J.* **14**, 3552–3562 (1995).
100. A. L. Plant, S. N. Covey, and D. Grierson, Detection of a subgenomic mRNA for gene V, the putative reverse transcriptase gene of cauliflower mosaic virus. *Nucleic Acids Res.* **13**, 8305–8321 (1985).
101. J. Enssle, I. Jordan, B. Mauer, and A. Rethwilm, Foamy virus reverse transcriptase is expressed independently from the Gag protein. *Proc. Natl. Acad. Sci. USA* **93**, 4137–4141 (1996).
102. S. F. Yu, D. N. Baldwin, S. R. Gwynn, S. Yendapalli, and M. L. Linial, Human foamy virus replication: a pathway distinct from that of retroviruses and hepadnaviruses. *Science* **271**, 1579–1582 (1996).
103. A. de Kochko, B. Verdaguer, N. Taylor, R. Carcamo, R. N. Beachy, and C. Fauquet, Cassava Vein mosaic virus (CsVMV), type species for a new genus of plant double stranded DNA viruses? *Arch. Virol.* **143**, 945–962 (1998).
104. H. L. Levin, D. C. Weaver, and J. D. Boeke, Novel gene expression mechanism in a fission yeast retroelement: Tf1 proteins are derived from a single primary translation-product. *EMBO J.* **12**, 4885–4895 (1993).
105. M. Hemmings-Mieszczak, G. Steger, and T. Hohn, Alternative structures of the cauliflower mosaic virus 35 S RNA leader: implications for viral expression and replication. *J. Mol. Biol.* **267**, 1075–1088 (1997).
106. J. Fütterer, K. Gordon, P. Pfeiffer, H. Sanfacon, B. Pisan, J. M. Bonneville, and T. Hohn, Differential inhibition of downstream gene expression by the CaMV 35S RNA leader. *Virus Genes* **3**, 45–55 (1989).
107. W. Schmidt-Puchta, D. Dominguez, D. Lewetag, and T. Hohn, Plant ribosome shunting in vitro. *Nucleic Acids Res.* **25**, 2854–2860 (1997).
108. M. Hemmings-Mieszczak, T. Hohn, and T. Preiss, Termination and peptide release at the upstream open reading frame are required for downstream translation on synthetic shunt-competent mRNA leaders. *Mol. Cell. Biol.* **20**, 6212–6223 (2000).
109. N. Schärer-Hernàndez and T. Hohn, Nonlinear ribosome migration on cauliflower mosaic virus 35S RNA in transgenic tobacco plants. *Virology* **242**, 403–413 (1998).
110. D. I. Dominguez, L. A. Ryabova, M. M. Pooggin, W. Schmidt-Puchta, J. Fütterer, and T. Hohn, Ribosome shunting in cauliflower mosaic virus. Identification of an essential and sufficient structural element. *J. Biol. Chem.* **273**, 3669–3678 (1998).
111. M. M. Pooggin, T. Hohn, and J. Futterer, Forced evolution reveals the importance of short open reading frame A and secondary structure in the cauliflower mosaic virus 35S RNA leader. *J. Virol.* **72**, 4157–4169 (1998).

112. M. M. Pooggin, J. Futterer, K. G. Skryabin, and T. Hohn, Ribosome shunt is essential for infectivity of cauliflower mosaic virus. *Proc. Natl. Acad. Sci. USA* **98,** 886–891 (2001).
113. M. M. Pooggin, T. Hohn, and J. Futterer, Role of a short open reading frame in ribosome shunt on the cauliflower mosaic virus RNA leader. *J. Biol. Chem.* **275,** 17288–17296 (2000).
114. T. Hohn, S. Corsten, D. Dominguez, J. Futterer, D. Kirk, M. Hemmings-Mieszczak, M. Pooggin, N. Scharer-Hernandez, and L. Ryabova, Shunting is a translation strategy used by plant pararetroviruses (Caulimoviridae). *Micron* **32,** 51–57 (2001).
115. M. Hemmings-Mieszczak, G. Steger, and T. Hohn, Regulation of CaMV 35 S RNA translation is mediated by a stable hairpin in the leader. *RNA* **4,** 101–111 (1998).
116. M. Hemmings-Mieszczak and T. Hohn, A stable hairpin preceded by a short open reading frame promotes nonlinear ribosome migration on a synthetic mRNA leader. *RNA* **5,** 1149–1157 (1999).
117. J. E. Reynolds, A. Kaminski, H. J. Kettinen, K. Grace, B. E. Clarke, A. R. Carroll, D. J. Rowlands, and R. J. Jackson, Unique features of internal initiation of hepatitis C virus RNA translation. *EMBO J.* **14,** 6010–6020 (1995).
118. J. R. Hill and D. R. Morris, Cell-specific translation of S-adenosylmethionine decarboxylase mRNA. Regulation by the 5′ transcript leader. *J. Biol. Chem.* **267,** 21886–21893 (1992).
119. L. A. Ryabova, M. M. Pooggin, D. I. Dominguez, and T. Hohn, Continuous and discontinuous ribosome scanning on the cauliflower mosaic virus 35 S RNA leader is controlled by short open reading frames. *J. Biol. Chem.* **275,** 37278–37284 (2000).
120. Y. Zhang, P. J. Dolph, and R. J. Schneider, Secondary structure analysis of adenovirus tripartite leader. *J. Biol. Chem.* **264,** 10679–10684 (1989).
121. P. J. Dolph, J. T. Huang, and R. J. Schneider, Translation by the adenovirus tripartite leader: elements which determine independence from cap-binding protein complex. *J. Virol.* **64,** 2669–2677 (1990).
122. G. C. Owens, S. A. Chappell, V. P. Mauro, and G. M. Edelman, Identification of two short internal ribosome entry sites selected from libraries of random oligonucleotides. *Proc. Natl. Acad. Sci. USA* **98,** 1471–1476 (2001).
123. Y. Zhang, D. Feigenblum, and R. J. Schneider, A late adenovirus factor induces eIF 4E dephosphorylation and inhibition of cell protein synthesis. *J. Virol.* **68,** 7040–7050 (1994).
124. R. Cuesta, Q. Xi, and R. J. Schneider, Adenovirus-specific translation by displacement of kinase Mnk1 from cap initiation complex eIF4F. *EMBO J.* **19,** 3465–3474 (2000).
125. R. J. Schneider, Adenovirus inhibition of cellular protein synthesis and preferential translation of viral mRNAs. *In* "Translational Control of Gene Expression" (N. Sonenberg, M. B. Hershey, and M. B. Mathews, eds.), pp. 901–914. Cold Spring Harbor Press, Cold Spring Harbor, New York, 2000.
126. J. Curran and D. Kolakofsky, Replication of paramyxoviruses. *Adv. Virus Res.* **54,** 403–422 (1999).
127. C. Giorgi, B. M. Blumberg, and D. Kolakofsky, Sendai virus contains overlapping genes expressed from a single mRNA. *Cell* **35,** 829–836 (1983).
128. K. C. Gupta and S. Patwardhan, ACG, the initiator codon for a Sendai virus protein. *J. Biol. Chem.* **263,** 8553–8556 (1998).
129. S. Patwardhan and K. C. Gupta, Translation initiation potential of the 5′ proximal AUGs of the polycistronic P/C mRNA of Sendai virus. A multipurpose vector for site specific mutagenesis. *J. Biol. Chem.* **263,** 4907–4913 (1988).
130. P. Latorre, D. Kolakofsky, and J. Curran, Sendai virus Y proteins are initiated by a ribosomal shunt. *Mol. Cell. Biol.* **18,** 5021–5031 (1998).
131. S. L. Wolin and P. Walter, Ribosome pausing and stacking during translation of a eukaryotic mRNA. *EMBO J.* **7,** 3559–3569 (1988).

132. P. L. Stern and M. A. Stanley, "Human Papillomaviruses and Cervical Cancer." Oxford University Press, Oxford, 1994.
133. M. S. Barbosa and F. O. Wettstein, E2 of cottontail rabbit papillomavirus is a nuclear phosphoprotein translated from an mRNA encoding multiple open reading frames. *J. Virol.* **62,** 3242–3249 (1988).
134. B. Roggenbuck, P. M. Larsen, S. J. Fey, D. Bartsch, L. Gissmann, and E. Schwarz, Human papillomavirus type 18 E6*, E6, and E7 protein synthesis in cell free translation systems and comparison of E6 and E7 in vitro translation products to proteins immunoprecipitated from human epithelial cells. *J. Virol.* **65,** 5068–5072 (1991).
135. S. N. Stacey, D. Jordan, P. J. Snijders, M. Mackett, J. M. Walboomers, and J. R. Arrand, Translation of the human papillomavirus type 16 E7 oncoprotein from bicistronic mRNA is independent of splicing events within the E6 open reading frame. *J. Virol.* **69,** 7023–7031 (1995).
136. M. Remm, A. Remm, and M. Ustav, Human papillomavirus type 18 E1 protein is translated from polycistronic mRNA by a discontinuous scanning mechanism. *J. Virol.* **73,** 3062–3070 (1999).
137. S. N. Stacey, D. Jordan, A. J. Williamson, M. Brown, J. H. Coote, and J. R. Arrand, Leaky scanning is the predominant mechanism for translation of human papillomavirus type 16 E7 oncoprotein from E6/E7 bicistronic mRNA. *J. Virol.* **74,** 7284–7297 (2000).
138. L. Wiklund, K. Spangberg, L. Goobar-Larsson, and S. Schwartz, Cap and polyA tail enhance translation initiation at the hepatitis C virus internal ribosome entry site by a discontinuous scanning, or shunting, mechanism. *J. Hum. Virol.* **4,** 74–84 (2001).
139. T. A. Poyry, M. W. Hentze, and R. J. Jackson, Construction of regulatable picornavirus IRESes as a test of current models of the mechanism of internal translation initiation. *RNA* **7,** 647–660 (2001).
140. J. B. O'Connor and D. A. Brian, Downstream ribosomal entry for translation of coronavirus TGEV gene 3b. *Virology* **269,** 172–182 (2000).
141. P. S. Carter, M. Jarquin-Pardo, and A. De Benedetti, Differential expression of Myc1 and Myc2 isoforms in cells transformed by eIF4E: evidence for internal ribosome entry site. *Oncogene* **18,** 4326–4335 (1999).
142. C. Nanbru, I. Lafon, S. Audigier, M. C. Gensac, S. Vagner, G. Huez, and A. C. Prats, Alternative translation of the proto oncogene c-myc by an internal ribosome entry site. *J. Biol. Chem.* **272,** 32061–32066 (1997).
143. C. Nanbru, A. C. Prats, L. Droogmans, P. Defrance, G. Huez, and V. Kruys, Translation of the human c-myc P0 tricistronic mRNA involves two independent internal ribosome entry sites. *Oncogene* **20,** 4270–4280 (2001).
144. L. Janosi, S. Mottagui-Tabar, L. A. Isaksson, Y. Sekine, E. Ohtsubo, S. Zhang, S. Goon, S. Nelken, M. Shuda, and A. Kaji, Evidence for in vivo ribosome recycling, the fourth step in protein biosynthesis. *EMBO J.* **17,** 1141–1151 (1998).
145. T. A. Gray, S. Saitoh, and R. D. Nicholls, An imprinted, mammalian bicistronic transcript encodes two independent proteins. *Proc. Natl. Acad. Sci. USA* **96,** 5616–5621 (1999).
146. H. Ilves, O. Kahre, and M. Speek, Translation of the rat LINE bicistronic RNAs in vitro involves ribosomal reinitiation instead of frameshifting. *Mol. Cell. Biol.* **12,** 4242–4248 (1992).
147. J. P. McMillan and M. F. Singer, Translation of the human LINE-1 element, L1Hs. *Proc. Natl. Acad. Sci. USA* **90,** 11533–11537 (1993).
148. K. Bouhidel, C. Terzian, and H. Pinon, The full length transcript of the I factor, a LINE element of Drosophila melanogaster, is a potential bicistronic RNA messenger. *Nucleic Acids Res.* **22,** 2370–2374 (1994).

149. J. Andrews, M. Smith, J. Merakovsky, M. Coulson, F. Hannan, and L. E. Kelly, The stoned locus of Drosophila melanogaster produces a dicistronic transcript and encodes two distinct polypeptides. *Genetics* **143**, 1699–1711 (1996).
150. S. Brogna and M. Ashburner, The Adh-related gene of Drosophila melanogaster is expressed as a functional dicistronic messenger RNA: multigenic transcription in higher organisms. *EMBO J.* **16**, 2023–2031 (1997).
151. H. Liu, J. K. Jang, J. Graham, K. Nycz, and K. S. McKim, Two genes required for meiotic recombination in Drosophila are expressed from a dicistronic message. *Genetics* **154**, 1735–1746 (2000).
152. G. Szabo, Z. Katarova, and R. Greenspan, Distinct protein forms are produced from alternatively spliced bicistronic glutamic acid decarboxylase mRNAs during development. *Mol. Cell. Biol.* **14**, 7535–7545 (1994).
153. M. Kozak, New ways of initiating translation in eukaryotes? *Mol. Cell. Biol.* **21**, 1899–1907 (2001).
154. J. Fütterer, I. Potrykus, M. P. Valles-Brau, I. Dasgupta, R. Hull, and T. Hohn, Splicing in a plant pararetrovirus. *Virology* **198**, 663–670 (1994).
155. M. Kozak, Primer extension analysis of eukaryotic ribosome mRNA complexes. *Nucleic Acids Res.* **26**, 4853–4859 (1998).
156. L. K. Dixon and T. Hohn, Initiation of translation of the cauliflower mosaic virus genome from a polycistronic mRNA: evidence from deletion mutagenesis. *EMBO J.* **3**, 2731–2736 (1984).
157. B. Gronenborn, The molecular buiology of cauliflower mosaic virus and its application as plant vector. *In* "Plant DNA Infections Agents" (T. Hohn and J. Schell, eds.), pp. 1–29. Spring-Verlag, Vienna, 1987.
158. S. Gowda, F. C. Wu, H. B. Scholthof, and R. J. Shepherd, Gene VI of figwort mosaic virus (caulimo virus group) functions in posttranscriptional expression of genes on the full length RNA transcript. *Proc. Natl. Acad. Sci. USA* **86**, 9203–9207 (1989).
159. S. Gowda, F. C. Wu, H. B. Scholthof, and R. J. Shepherd, Gene VI of figwort mosaic virus activates expression of internal cistrons of the full-length polycistronic transcript. *In* "Viral Genes and Plant Pathogenesis" (T. P. Pirone and J. G. Shaw, eds.), pp. 79–88. New York, Springer-Verlag, 1990.
160. H. B. Scholthof, S. Gowda, F. C. Wu, and R. J. Shepherd, The full-length transcript of a caulimovirus is a polycistronic mRNA whose genes are transactivated by the product of gene VI. *J. Virol.* **66**, 3131–3139 (1992).
161. I. B. Maiti, R. D. Richins, and R. J. Shepherd, Gene expression regulated by gene VI of caulimovirus: transactivation of downstream genes of transcripts by gene VI of peanut chlorotic streak virus in transgenic tobacco. *Virus Res.* **57**, 113–124 (1998).
162. Y. S. Sha, E. P. Broglio, J. F. Cannon, and J. E. Schoelz, Expression of a plant viral polycistronic mRNA in yeast, Saccharomyces cerivisiae, mediated by a plant virus translation transactivator. *Proc. Natl. Acad. Sci. USA* **92**, 8911–8915 (1995).
163. C. Zijlstra and T. Hohn, Cauliflower mosaic virus gene VI controls translation from dicistronic expression units in transgenic arabidopsis plants. *Plant Cell* **4**, 1471–1484 (1992).
164. E. Balázs, Diseases symptoms in transgenic tobacco induced by integrated gene VI of cauliflower mosaic virus. *Virus Genes* **3**, 205–211 (1990).
165. G. A. Baughman, J. D. Jacobs, and S. H. Howell, Cauliflower mosaic virus gene VI produces a symptomatic phenotype in transgenic tobacco plants. *Proc. Natl. Acad. Sci. USA* **85**, 733–737 (1988).
166. K. B. Goldberg, J. M. Kiernan, and R. J. Shepherd, A disease syndrome associated with expression of gene VI of caulimovirus may be a non host reaction. *Mol. Plant Microbe Interact.* **4**, 182–189 (1991).

167. H. Takahashi, K. Shimamoto, and Y. Ehara, Cauliflower mosaic virus gene VI causes growth suppression, development of necrotic spots and expression of defence-related genes in transgenic tobacco plants. *Mol. Gen. Genet.* **216**, 188–194. (1989).
168. C. Zijlstra, N. Schärer-Hernandez, S. Gal, and T. Hohn, Arabidopsis thaliana expressing the Cauliflower Mosaic Virus ORF VI transgene has a late flowering phenotype. *Virus Genes* **13**, 5–17 (1996).
169. H. B. Scholthof, F. C. Wu, S. Gowda, and R. J. Shepherd, Regulation of caulimovirus gene expression and the involvement of cis-acting elements on both viral transcripts. *Virology* **190**, 403–412 (1992).
170. M. Driesen, R.-M. Benito-Moreno, T. Hohn, and J. Fütterer, Transcription from the CaMV 19S promoter and autocatalysis of translation from CaMV RNA. *Virology* **195**, 203–210 (1993).
171. I. Furusawa, N. Yamaoka, T. Okuno, M. Yamamoto, M. Kohno, and H. Kunoh, Infection of turnip brassica-rapa oultivar perviribis protoplasts with cauliflower mosaic virus. *J. Gen. Virol.* **48**, 431–435. (1980).
172. E. W. Kitajima, J. A. Lauritis, and H. Swift, Fine structure of zinnial leaf tissues infected with dahlia mosaic virus. *Virology* **39**, 240–249 (1969).
173. R. H. Lawson and S. S. Hearon, Ultrastructure of carnation etched ring virus-infected Saponaria vaccaria and Dianthus caryophyllus. *J. Ultrastruct. Res.* **48**, 201–215 (1974).
174. L. Givord, C. Xiong, M. Giband, I. Koenig, T. Hohn, G. Lebeurier, and L. Hirth, A second cauliflower mosaic virus gene product influences the structure of the viral inclusion body. *EMBO J.* **3**, 1423–1427 (1984).
175. J. Martinez-Izquierdo, J. Fütterer, and T. Hohn, Protein encoded by ORFI of cauliflower mosaic virus is part of the viral inclusion body. *Virology* **160**, 527–530 (1987).
176. G. A. De Zoeten, J. R. Penswick, M. A. Horisberger, P. Ahl, M. Schultze, and T. Hohn, The expression, localization, and effect of a human interferon in plants. *Virology* **172**, 213–222 (1989).
177. H. M. Rothnie, Y. Chapdelaine, and T. Hohn, Rararetroviruses and retroviruses: a comparative review of viral structure and gene expression strategies. *Adv. Virus Res.* **44**, 1–67 (1994).
178. A. Himmelbach, Y. Chapdelaine, and T. Hohn, Interaction between Cauliflower mosaic virus inclusion body protein and capsid protein implications for viral assembly. *Virology* **217**, 147–157 (1996).
179. K. Kobayashi, S. Tsuge, H. Nakayashiki, K. Mise, and I. Furusawa, Requirement of cauliflower mosaic virus open reading frame VI product for viral gene expression and multiplication in turnip protoplasts. *Microbiol. Immunol.* **42**, 377–386 (1998).
180. M. de Tapia, A. Himmelbach, and T. Hohn, Molecular dissection of the cauliflower mosaic virus translational transactivator. *EMBO J.* **12**, 3305–3314 (1993).
181. E. P. Broglio, Mutational analysis of cauliflower mosaic virus gene VI: changes in host range, symptoms, and discovery of transactivation-positive, noninfectious mutants. *Mol. Plant Microbe Interact.* **8**, 755–760 (1995).
182. H.-S. Park, A. Himmelbach, K. Browning, T. Hohn, and L. A. Ryabova, A plant viral "reinitiation" factor interacts with the host translational machinery. *Cell* **106**, 723–733 (2001).
183. V. Leh, P. Yot, and M. Keller, The cauliflower mosaic virus translational transactivator interacts with the 60S ribosomal subunit protein L 18 of Arabidopsis thaliana. *Virology* **266**, 1–7 (2000).
184. T. Hatakeyama, F. Kaufmann, B. Schroeter, and T. Hatakeyama, Primary structures of five ribosomal proteins from the archaebacterium. Halobacterium marismortui and their structural relationship to eubacterial and eukaryotic ribosomal proteins. *Eur. J. Biochem.* **185**, 685–693 (1989).
185. N. Ban, P. Nissen, J. Hansen, P. B. Moore, and T. A. Steitz, The complete atomic structure of the large ribosomal subunit at 2.4 A resolution. *Science* **289**, 905–920 (2000).

186. M. J. Marion and C. Marion, Localization of ribosomal proteins on the surface of mammalian 60S ribosomal subunits by means of immobilized enzymes. Correlation with chemical cross-linking data. *Biochem. Biophys. Res. Commun.* **149,** 1077–1083 (1987).
187. D. M. Baronas-Lowell and J. R. Warner, Ribosomal protein L30 is dispensable in the yeast Saccharomyces cerevisiae. *Mol. Cell. Biol.* **10,** 5235–5243 (1990).
188. S. M. Cerritelli, O. Y. Fedoroff, B. R. Reid, and R. J. Crouch, A common 40 amino acid motif in eukaryotic RNases H-1 and caulimovirus ORF VI proteins binds to duplex RNAs. *Nucleic Acids Res.* **26,** 1834–1840 (1998).
189. S. Srivastava, A. Verschoor, and J. Frank, Eukaryotic initiation factor 3 does not prevent association through physical blockage of the ribosomal subunit-subunit interface. *J. Mol. Biol.* **226,** 301–304 (1992).
190. K. Asano, J. Clayton, A. Shalev, and A. G. Hinnebusch, A multifactor complex of eukaryotic initiation factors, eIF1, eIF2, eIF3, eIF5, and initiator tRNA(Met) is an important translation initiation intermediate in vivo. *Genes Dev.* **14,** 2534–2546 (1992).
191. H. P. Vornlocher, P. Hanachi, S. Ribeiro, and J. W. Hershey, A 110-kilodalton subunit of translation initiation factor eIF3 and an associated 135-kilodalton protein are encoded by the Saccharomyces cerevisiae TIF32 and TIF31 genes. *J. Biol. Chem.* **274,** 16802–16812 (1999).
192. M. Nassal and H. Schaller, Hepatitis B virus replication. *Trends Microbiol.* **1,** 221–228 (1993).

Initiation of Eukaryotic DNA Replication: Regulation and Mechanisms

HEINZ-PETER NASHEUER,
RICHARD SMITH,
CHRISTINA BAUERSCHMIDT,
FRANK GROSSE, AND
KLAUS WEISSHART[†]

Institut für Molekulare Biotechnologie
Abteilung Biochemie
D-07745 Jena, Germany

I. The Eukaryotic Cell Cycle	42
A. Regulation of Cell Cycle Progression by Protein Kinases and Phosphatases	43
B. Transcriptional Regulation of Cell Cycle Progression	46
C. Proteolysis as a Control Mechanism of the Eukaryotic Cell Cycle	49
II. Factors Required for the Initiation of DNA Replication	53
A. Origins of Eukaryotic DNA Replication............................	53
B. Proteins Involved in the Initiation of DNA Replication	54
III. The Organization of Replication-Initiation Factors on Chromatin..........	69
A. The Assembly of the Prereplicative Complex	70
B. The Activation of Origins: From the Prereplicative Complex to the Initiation Complex	71
C. The Initiation of DNA Replication on the Leading and Lagging Strands: DNA Polymerase α–Primase Holds the Key.................	72
D. The Elongation Reaction of Eukaryotic DNA Replication	72
IV. Cell Cycle Control by Checkpoints	73
A. The Sensors: DNA Replication Machinery-like Complexes and the BRCT Family	74
B. The Transducing Phosphoinositol-3-Kinase-like Complexes and the Coiled-Coil Proteins	77
C. The Checkpoint Effector Kinases Chk1 and Chk2	78
D. A Network of Proteins Is Involved in Checkpoint Control	78
V. Outlook	80
References	81

The accurate and timely duplication of the genome is a major task for eukaryotic cells. This process requires the cooperation of multiple factors to ensure the stability of the genetic information of each cell. Mutations, rearrangements,

---

[†]Present address: Carl Zeiss Jena GmbH, D-07745 Jena, Germany.

or loss of chromosomes can be detrimental to a single cell as well as to the whole organism, causing failures, disease, or death. Because of the size of eukaryotic genomes, chromosomal duplication is accomplished in a multiparallel process. In human somatic cells between 10,000 and 100,000 parallel synthesis sites are present. This raises fundamental problems for eukaryotic cells to coordinate the start of DNA replication at each origin and to prevent replication of already duplicated DNA regions. Since these general phenomena were recognized in the middle of the 20th century the regulation and mechanisms of the initiation of eukaryotic DNA replication have been intensively investigated. These studies were carried out to find the essential factors involved in the process and to determine their functions during DNA replication. These studies gave rise to a model of the organization and the coordination of DNA replication within the eukaryotic cell. The elegant experiments carried out by Rao and Johnson (1970) (1), who fused cells in different phases of the cell cycle, showed that G1 cells are competent for replication of their chromosomes, but lack a specific diffusible factor required to activate their replicaton machinery and showed that G2 cells are incompetent for DNA replication. These findings suggested that eukaryotic cells exist in two states. In G1 phase, cells are competent to initiate DNA replication, which is subsequently triggered in S phase. After completion of S phase, cells in G2 are no longer able to initiate DNA replication and they require a transition through mitosis to reenable initiation of DNA replication to take place in the next S phase. The *Xenopus* cell-free replication system has proved a good model system in which to study DNA replication *in vitro* as well as the mechanism preventing rereplication within a single cell cycle (2). Studies using this system resulted in the development of a model postulating the existence of a replication licensing factor, which binds to chromatin before the G1–S transition and which is displaced during replication (2, 3). These results were supported by genetic and biochemical experiments in *Saccharomyces cerevisiae* (budding yeast) and *Schizosaccharomyces pombe* (fission yeast) (4, 5). The investigation of cell division cycle mutants and the budding yeast origin of replication resulted in the concept of a prereplicative and a postreplicative complex of initiation proteins (6–9). These three individual concepts have recently started to merge and it has become obvious that initiation in eukaryotes is generally governed by the same ubiquitous mechanisms. © 2002, Elsevier Science (USA).

I. The Eukaryotic Cell Cycle

The mitotic cell cycle represents a model for the temporal sequence of processes such as duplication of the genomic information and cell division in growing eukaryotic cells. The cell cycle is divided into four parts, the G1, S, G2, and M phases (10): The first phase, G1 (G for gap), begins with the end of cell division and is characterized by cell growth. In addition, during G1, sequential growth factor-dependent regulatory steps take place. After the G1 phase is completed cells accurately duplicate their chromosomal DNA in the S (synthesis) phase. After another growth phase (G2 phase) cells divide their content, including the genetic information, into two daughter cells during M phase (mitosis).

In order to ensure a synchronized temporal sequence cells have established regulatory mechanisms which monitor progression of the cell cycle. Depending on the environment, the dividing cells can also reversibly leave the cell cycle and remain in a resting status. This "decision" takes place at a specific time window and, depending on two general concepts, is either called "start" in yeast or "restriction point" in mammalian cells (4, 11). After a change of the growth conditions these cells are able to restart cell growth and enter a new cell cycle in G1 phase. With reference to the cell cycle phases this resting status is also called G0 phase.

The doubling of chromosomal DNA is an essential prerequisite for cell division (2, 10, 12). In particular it must be guaranteed that the genome is replicated exactly once per cell cycle and this is largely regulated at the initiation step of eukaryotic DNA replication (2, 13). The concepts of the initiation of eukaryotic DNA replication have been extended and improved in recent years by the discovery of further details and new proteins as essential components of the initiation reaction.

A. Regulation of Cell Cycle Progression by Protein Kinases and Phosphatases

1. CYCLIN-DEPENDENT KINASES RULE THE EUKARYOTIC CELL CYCLE

The cyclin–cyclin-dependent kinase (Cdk) complexes are necessary for controlling the eukaryotic cell cycle. Cdks are serine and threonine kinases (14). Their activity is controlled in a temporal order and they are activated and deactivated according to their function in the cell cycle. Furthermore, they are essential for controlling each individual cell cycle phase and the transition from one phase to the next (14, 15). Cyclins are the positive regulatory subunits and the Cdks are the catalytic polypeptides. The first of the cyclin genes to be induced is cyclin E in late G1, followed by cyclin A in early and cyclin B in late S phase, with a peak expression of the latter in G2. At their N-termini, mitotic cyclins possess a so-called destruction box, which is essential for phase-specific ubiquitin-dependent degradation (16).

In budding and fission yeast (*Saccharomyces cerevisiae* and *Schizosaccharomyces pombe*, respectively) only a single Cdk gene exists, and the Cdk protein associates with various cyclins (14, 15). The Cdks are functionally and structurally conserved. However, in higher eukaryotes several Cdk genes and their products have been found. In G1 phase the cyclin D–Cdk4/6 and cyclin E–Cdk2 complexes are active, whereas in S phase the cyclin A–Cdk2 and cyclin A–Cdk1 complexes become activated and cyclin E–Cdk2 is inactivated. During the cell cycle the concentration of cyclin B–Cdk1 increases and reaches its maximum during mitosis. These Cdks seem to interact with replication proteins in an ordered manner for the activation of initiation. Cyclin E–Cdk2 binds to DNA polymerase α in G1, whereas cyclin A–Cdk2 interacts with the enzyme complex in S phase (*17*). The cyclin E kinase only regulates cell cycle progression in a positive way, whereas the cyclin A kinases excute positive as well as negative functions during S phase (*17–20*). Similarly, in budding yeast, the G1 cyclins (ScCln1p and ScCln2p; for an explanation of nomenclature see legend to Fig. 2) positively modulate cell cycle progression in G1, whereas the B-type cyclins positively and negatively regulate cell cycle progression in S phase. Budding yeast has six related cyclins which can be fitted in three groups with functionally redundant members. The cyclins ScClb5p and ScClb6p are first active in early S phase, ScClb3p and ScClb4p function later in the cell cycle, whereas the cyclins ScClb1p and ScClb2p control the entry in mitosis (*21*). In fission yeast the expression of the mitotic cyclin cdc13 is sufficient to control the whole cell cycle, since both G1 cyclin genes cig1 and cig2 can be deleted and the mutant cell is still viable (*22*).

Cdk activity can be controlled by various mechanisms. First, Cdks are only active when they are associated with a cyclin. Second, specific conserved amino acids, tyrosine and threonine, in the N-termini of Cdks must be dephosphorylated and a threonine in their C-termini must be phosphorylated for the enzymes to retain their activity. Third, Cdks are inactivated through interaction with specific Cdk inhibitors (CKIs; reviewed in Ref. 23). Cyclins as well as CKIs are themselves subject to regulation by phosphorylation in a number of ways. These mechanisms include the changing of protein conformation, the modification of binding sites, or, as in the case of cyclins, the marking of the proteins for ubiquitin-dependent proteolytic degradation (*19, 23–25*). However, a recent report suggests another regulatory switch mechanism. Cdk-dependent phosphorylation of CKI XlXic1p is not required for its ubiquitination. The recruitment of cyclin E–Cdk2 and XlXic1p to replication origins via XlCdc6p is essential for both inactivation and degradation of XlXic1p just before the onset of DNA replication (*24, 26*).

The regulation of Cdk activity by phosphorylation is a fairly well understood process. The activation requires both kinases and phosphatases. The dephosphorylation of the N-terminal tyrosine and threonine residues is essential for kinase activity and this is accomplished by the Cdc25p phosphatase. The phosphorylation control is ultimately linked to cell cycle progress and to certain checkpoint controls (for details, see below). In fission yeast SpCdc25p is negatively controlled by both the checkpoint effector kinases SpChk1p and SpChk2p, which are activated upon DNA damage, resulting in a cell cycle block (*27*). In human cells three HsCdc25p phosphatases (A, B, and C) exist (*28*). The HsChk2p kinase is possibly involved in the inhibition of all three phosphatases, whereas HsChk1p seems to act specifically on HsCdc25Cp. In S phase the consequence of HsCdc25Ap phosphorylation is that cyclin A–Cdk2 is downregulated, whereas in G2 phosphorylation of HsCdc25Cp keeps cyclin B–Cdk1 in its inactive form. In the latter case, the process has begun to be understood. The phosphorylated form of phosphatase HsCdc25p binds to various 14-3-3 proteins that are known to interact with phosphoserines of other proteins (29). This results in a sequestration of Cdc25p to the cytosol, where it is degraded (30). As such the phosphatase is prevented from dephosphorylating and activating Cdk1 in the nucleus (31). On a second control level Cdc25-dependent activation is counteracted by the Wee1p and Mik1p kinases, which phosphorylate Cdks on residues T14 and Y15, holding them in an inactive state (32, 33). In budding yeast the Cdk1 kinase is not subject to regulation by phosphorylation, but other control mechanisms have evolved (see Section IV).

2. Dbf4p–Cdc7p KINASE ACTIVATES ORIGINS
   OF DNA REPLICATION

The Dbf4p–Cdc7p kinase, also called DDK (Dbf4-dependent kinase), is a serine–threonine kinase (34, 35). It consists of the catalytic subunit Cdc7p and the regulatory subunit Dbf4p, which are conserved from yeast to mammals (34–38). Throughout the cell cycle its concentration is constant, but the kinase activity peaks at the transition from G1 to S phase (39). The ScDBF4 gene was initially identified as a suppressor of a temperature-sensitive *cdc7* mutant (34, 35). Later it was shown that for enzyme activity Cdc7p must be associated with its unstable regulatory subunit Dbf4p. Phosphorylation of a threonine in the C-terminus of Cdc7p, for example, T367 of human Cdc7p, is required for kinase activity and this site is most likely recognized by cyclin E–Cdk2 or cyclin A–Cdk2 (38). Mutations of the ScDbf4p–ScCdc7p kinase cause an arrest of the cells in late G1/early S phase, just prior to the initiation of DNA replication. DDK is probably not a general activator of S phase, but it seems to act at individual origins of replication (40). There is a controversy concerning the time sequence and the dependence of DDK actions. Whereas the kinase is loaded onto chromatin via the origin recognition complex (ORC) and is independent of Mcm (minichromosome maintenance) proteins in budding yeast, the converse occurs in *Xenopus* extracts (39, 41). In addition the sequential activation of the kinase is still under discussion; DDK acts after Cdks in budding yeast, whereas XlDDK executes its function before Cdk action (39, 41). The Mcm complex is most likely one of the physiological substrates of DDK (9, 42). There are differences in the preference of DDK for the Mcm proteins *in vitro*. Mcm2p, for example, is an excellent substrate, whereas Mcm4p and Mcm6p are phosphorylated to lesser extents (38).

3. PROTEIN PHOSPHATASE 2A (PP2A)

Protein phosphatase 2A (PP2A) belongs to the family of dual-specific (serine–threonine) phosphatases, which are sensitive to okadaic acid and do not require divalent cations for their activity. Mammalian PP2A dephosphorylates a broad spectrum of proteins and has many fundamental functions in eukaryotic cells, such as the regulation of cell division, intracellular signal transduction, and gene expression (reviewed in Ref. 43). The PP2A holoenzyme consists of three subunits, the 36-kDa catalytic C subunit, the 65-kDa structural A subunit, and a variable regulatory B subunit. The C and A subunits together form the so-called core complex, which has enzyme activity. In combination with the various B subunits they form a large number of PP2A subtypes, which exhibit different subcellular localization and substrate specificities (reviewed in Ref. 44). In mammalian cells the core enzyme and the holoenzyme are present in equal amounts (45). PP2A is important for the entry of S. cerevisiae cells into mitosis (46). In extracts from Xenopus laevis oocytes (Xenopus extracts) PP2A is essential for DNA replication, and in particular it plays a role during initiation (47). In human cells the p48/B subunit undergoes specific interactions with HsCdc6p and these are probably responsible for dephosphorylation of the latter. In line with a role of PP2A in negatively regulating cell cycle progression is the observation that overexpression of p48 arrests cells at the G1–S phase transition (48). Besides these functions PP2A is probably involved in the coordination of the nuclear and centrosome cycles (49).

B. Transcriptional Regulation of Cell Cycle Progression

Some of the proteins that drive or function in processes linked to the cell cycle are periodically expressed to ensure their presence at exactly the time when they are needed (50). Paradigms are the cyclins, which are activators of Cdk kinases. Their expression profile reflects the phase-specific regulatory role which they play during cell cycle progression (23; see above). Other important cell-cycle-regulated genes code for replication factors that are induced maximally in S phase and transcription factors that themselves control defined sets of genes (50). Connections between cell cycle progression and transcriptional regulation are now well established in mammalian and yeast cells (51). Control is exerted mostly by phosphorylation of key proteins involved in transcription that can be classified as specific and general transcription factors or as coactivators and corepressors. Modification of these proteins affects their binding specificities, localization within the cell, or stability.

1. THE E2F FAMILY OF TRANSCRIPTION FACTORS COOPERATES WITH THE pRB FAMILY OF COACTIVATORS AND COREPRESSORS

The transcription factor E2F regulates several proteins that are required for proliferation (Fig. 1). It exists in two different states, one that activates and the other that represses transcription. In its transcriptionally active form E2F is a heterodimer composed of an E2F and a DP family member. At least five members of the E2F family (E2F-1 to E2F-5) and three DP proteins (DP-1 to

INITIATION OF EUKARYOTIC DNA REPLICATION

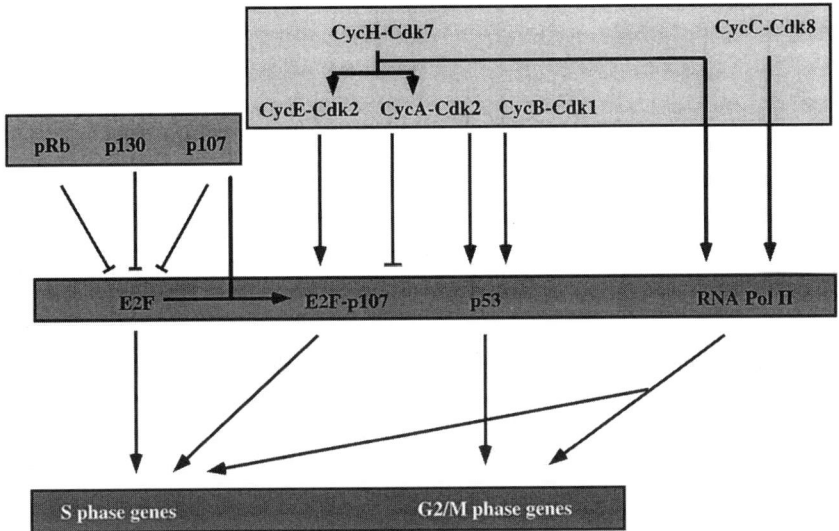

FIG. 1. Transcriptional control. The E2F transcription factor is kept silenced during G1 by association with RB-family members (pRb, p107, and p130). In early S phase, the E2F–p107 complex is activated by cyclin E–Cdk2 and induces many genes the products of which are involved in DNA metabolism and replication. Once cyclin A–Cdk2 becomes active at late S phase, the transcription complex is inactivated. Both cyclin A and cyclin E kinase complexes are activated by the cyclin H–Cdk7 complex. In addition, cyclin A–cdk2 positively regulates the p53 transcription factor, which will induce G2-specific gene products. In mitosis, cyclin B–Cdk1 is the active kinase which stimulates the p53 transcription factor to activate M phase-specific genes. The process of transcription is also controlled by the cyclin C–Cdk8 complex, which stimulates the RNA polymerase II, and hence transcription of S and G2/M phase-specific genes in general.

DP-3) have been identified (52). The dimers can be part of even larger complexes, containing the retinoblastoma protein pRb or its cousins p107 and p130, and a cyclin–Cdk complex. Each family member shows binding preferences. E2F-1 to E2F-3 bind to the pRb, whereas E2F-4 and E2F-5 form complexes with p107 and p130, or only p130, respectively. These complexes exists at specific cell cycle stages: E2F–p130 is found in G0 and early G1 cells, E2F–p107 in mid G1, and E2F–pRb in late G1 and they are transcriptionally inactive or repress transcription. Subsequently the disassembly of these complexes allows the association of E2F and DP family members, which can take place with the dimeric transcription factor bound to DNA. This complex formation leads to active gene transcription in late G1 (53, 54).

The E2F–p107 complex associates with cyclin E–Cdk2 at the end of G1. Since p107 is directly bound by cyclin–Cdk complexes its function might consist in recruiting these kinases to the E2F transcription factors, thereby mediating their phosphorylation. Phosphorylation of E2F by cyclin E–Cdk 2 contributes to the relief of E2F-mediated gene repression at late G1 (55). During S phase cyclin E is exchanged by cyclin A concomitant with a downregulation of E2F transcriptional activity (Fig. 1) (18, 56). Other phosphorylation events cooperate in the activation of the E2F transcription factor. When p107 is hyperphosphorylated by cyclin D–Cdk4 its interaction with the E2F subunit is prevented and the ternary complex disrupted, thus liberating transcritionally active E2F (57). Free E2F accumulates during G1 and S phase additionally by *de novo* synthesis (58). Phosphorylation of pRb and related proteins by both cyclin D–Cdk4 and cyclin E–Cdk2 at the end of G1 will prevent this free E2F from becoming complexed and inactivated (59). This results in a burst of E2F activity at late G1. The important part E2F plays during the transition from G1 to S relates to its activation of key players in the replication process such as dihydrofolate reductase (DHFR), thymidine kinase (TK), and DNA polymerase α (60–63). Besides silencing S-phase genes by blocking the E2F protein, pRb also recruits proteins which modify chromatin structure, such as histone deacetylases, or alter the mobility of nucleosomes (64). Among those proteins that are targeted to promotor regions are the protein SUV39H1, which methylates lysine residues in histone H3 protein and heterochromatin protein 1 (HP-1), which binds the methylated H3 and induces a heterochromatin-like structure, thereby preventing transcription (65).

2. THE SBF AND MBF TRANSCRIPTION FACTOR COMPLEXES IN YEAST

In budding yeast the regulation of the transcription factor ScSwi4p closely resembles that of E2F (66). In its active form it associates with a regulatory subunit called ScSwi6p to form the SBF (SCB-binding factor) complex, which binds to the SCB (Swi cell cycle box) regulatory element in the promotor region of several genes including the G1 cyclins (ScCln1p and ScCln2p) (67). SBF is activated in a positive feedback loop through phosphorylation by Cdk1 complexed to both the G1 and S phase-specific cyclins, which ultimately leads to G1 and S phase-specific gene expression, respectively. On the other hand, Cdk1 complexed to G2 cyclins (ScClb1p–ScClb4p) inactivates the SBF transcription factor in perhaps the same way that the activity of E2F is repressed in humans (68). A second set of S phase-specific genes is controlled by the MBF transcription factor, a dimer of ScMbp1p and the same regulatory subunit ScSwi6p that is part of the SBF complex (69, 70). This factor binds the MCP (MluI cell cycle box) regulatory sequence, which is present in the promoters of genes coding for the S-phase-specific cyclins ScClb5p and ScClb6p, and enzymes required for DNA synthesis including all three replicative DNA polymerases, both DNA primase subunits, and ScCdc6p, all of which regulate entry into S phase (71, 72). SBF- and MBF-like activities have also been detected in fission yeast, suggesting that linking transcriptional control to the cell cycle is a highly conserved feature. In fission yeast SpCdc10p, SpRes1p, and SpSct1p are most similar to ScSwi6p, ScSwi4p, and ScMbp1p, respectively (69, 70).

C. Proteolysis as a Control Mechanism of the Eukaryotic Cell Cycle

Cell cycle transitions require degradative processes which remove proteins involved in specific cell cycle activities when they are no longer required (73). For this reason particular cell cycle proteins are tagged on lysine residues by polyubiquitin chains and are eliminated by the 26S proteasome. The attachment of ubiquitin is a sequential process (74). First, ubiquitin forms a thioester bond with a ubiquitin-activating enzyme (E1) and then is transesterified to a ubiquitin-conjugating enzyme (E2, Ubc). By the action of a ubiquitin ligase (E3) the ubiquitin is then transferred to a specific substrate. Whereas there seems to be only one E1 protein, a variety of E2 and E3 proteins participate to ubiquinate different substrates. E3 activities are sometimes provided by complex protein assemblies and they represent the major determinants of substrate specifity. Not only are cyclins and their inhibitors regulated by proteolysis, so are also various other proteins of the DNA replication and segregation machinery (Fig. 2). The cell cycle-relevant ubiquitin substrates can be classified according to the role that they play (75). The first class comprises proteins whose destruction is absolutely required to drive cell cycle progression such as ScSic1p, ScPds1p, and B-type cyclins in S. cerevisiae (the latter correspond to cyclin A and B in higher eukaryotes). The other class consists of proteins whose destruction is not necessary for cell cycle progression but is necessary for keeping up cell cycle homeostasis; these include Cdc6p and G1 cyclins. Cell cycle proteins are in nearly all cases first marked by phosphorylation, which then allows them to be recognized by the ubiquination system (76). Another level of regulation is imposed by the temporally controlled activation of the ubiquitination system itself, in most cases also by phosphorylation. Since phosphorylation events are induced in a cell cycle-specific manner and the phosphorylation state of the substrate might determine its subcellular localization, turnover is regulated in time and space (77).

1. Regulated Proteolysis at the Transition of G0 to G1

Controlled proteolysis is also involved when quiescent cells resume proliferation. In resting cells the cyclin-dependent kinase inhibitor (CKI) p27 prevents cells from entering the cell cycle by binding to and inhibiting cyclin D–Cdk2 and

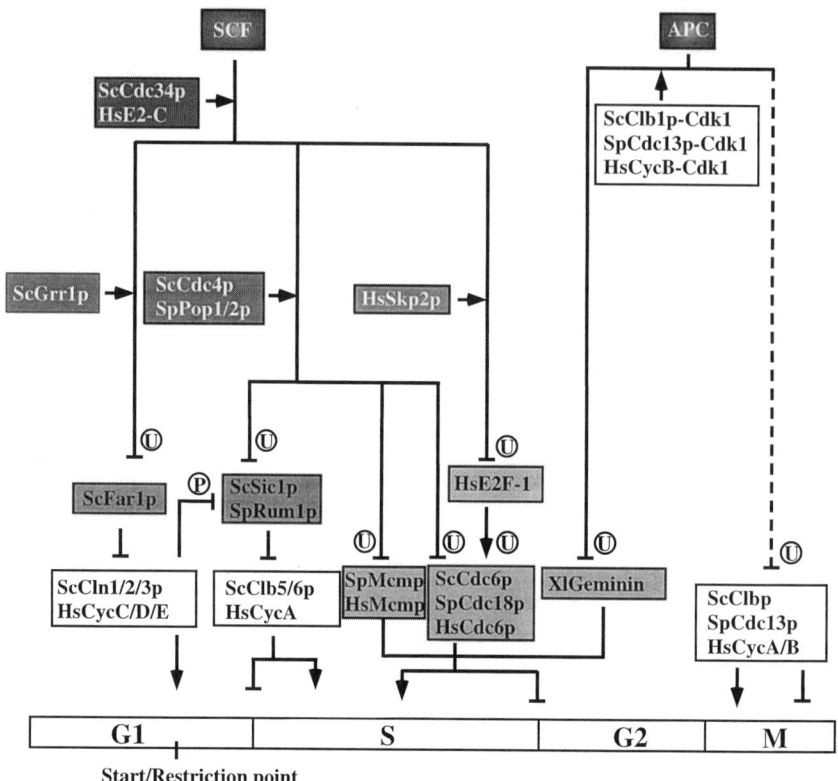

FIG. 2. Proteolytic control. Proteolytic control during G1, S, and G2 phases of the cell cycle is mainly exerted by the SCF complex, which acts in concert with a special Ubc (ubiquitin-conjugating enzyme, E2) such as ScCdc34p, in budding yeast and E2-C in humans. Various F-proteins, for example ScGrr1p, ScCdc4p, and HsSkp2p, confer specifity to the complex. The first two of these F-box proteins target CKIs, for example, ScFar1p and ScSic1p, which in turn inhibit different cyclin–Cdk complexes, which leads to a block in G1 or S phase. In G1 phase, a major control event is established in nearly all eukaryotic cells called "start" in budding yeast and "restriction point" in mammalian cells. SCF acts not only on the cyclin–Cdk, but also directly or indirectly via regulation of the transcription factor E2F on replication proteins such as Cdc6 and Mcm proteins. Progression through M phase is mainly controlled by the APC (anaphase promoting complex). The APC also influence progression from S to G2 by targeting the replication factor Geminin. *Nomenclature:* When discussing *S. cerevisiae*, we use the standard nomenclature where a wild-type gene is written in capital letters, for example, CDC6, and a mutant gene is written in lower case italic, for example, *cdc6*. For *S. pombe*, we follow the usual convention where the name of a gene is given in lower case roman, for example, cdc18. To emphasize that we are discussing a protein, the letter p is added to the name of the gene, where the first letter is capitalized and the rest are lower case, for example, Cdc6p. Furthermore, when comparing proteins from different organisms, we prefix the name of the gene/protein with a two-letter abbreviation, for example, ScCdc6p/SpCdc18p. We use the abbreviations Sc, *S. cerevisiae;* Sp, *S. pombe;* Nc, *Neurospora crassa;* Dm, *Drosophila melanogaster;* Xl, *Xenopus laevis;* Mm, *mouse musculus;* Hs, *Homo sapiens.* Ⓤ and Ⓟ in the figure indicate that cellular functions are regulated by ubiquitination and phosphorylation, respectively.

cyclin D–Cdk4 (Fig. 2). However, after a trigger the Ubc2 and Ubc3 ubiquitin-conjugating proteins mark p27 for degradation via the proteasome pathway (78–80). Another level of control is achieved by a protein called Jab1 (81). Jab1 binds to p27 and directs its movement from the nucleus to the cytoplasm (82). This export occurs in a proteasome-dependent manner, suggesting that the proteasome controls the turnover of an important export factor in the nucleus. Since Jab1 is part of the so-called COP9 complex, which contains kinases, it is likely that phosphorylation and export are coordinated processes (83). The relocalization of p27 is absolutely necessary for its destruction. A similar dependence on export-mediated proteolysis was observed for cyclin D and p53 (84, 85). Since the stability is controlled by phosphorylation-dependent ubiquination, it is likely that their turnover is regulated by SCF complexes that act as E3 ligases (see below; 86).

2. REGULATED PROTEOLYSIS AT THE G1–S TRANSITION

The budding yeast Cdk complexes, which promote S phase and which are synthesized during G1, are inactivated through association with the CKI ScSic1p until they are required (Fig. 2) (87). ScSic1p is phosphorylated by G1 Cdk complexes, which become active in late G1 and are resistant to ScSic1p inhibition, thereby marking the protein for subsequent ubiquination (88, 89). Ubiquination is brought about by ScCdc34p, an E2 ubiquitin-conjugating enzyme, and the F-box protein ScCdc4p, which form a complex with ScCdc53p (or Cullin) and ScSkip1p providing the E3 ubiquitin ligase activity. This complex is termed SCF for ScSkp1p/Cullin/F-box protein. ScSkp1p and the F-box protein interact through the F-box motif and ScCdc53p binds to the ScCdc34p, an E2 ubiquitin-conjugating enzyme (87, 89, 90). Another component is the ScRbx1/ScRoc1 protein, which contains a domain called the R-box and plays a role in recruiting ScCdc34p (91–93). As such the SCF structurally resembles the von Hippel–Lindau (VHL) tumor suppressor complex, which, besides ScRbx1p/ScRoc1p, contains elongin C, Cul2, and the VHL protein, with the latter acting analogously to the F-box proteins (91, 94).

The substrate specifity of the different SCF complexes is determined by the different F-box proteins, which bind to target proteins containing the so-called PEST sequences (rich in proline, glutamic acid, serine, and threonine). In all cases determined to date, the SCF complex recognizes phosphorylated proteins. SCF with the F-box protein ScCdc4p, referred to as the SCF^{Cdc4} complex, for example, binds only to phosphorylated CKI ScSic1p, but not to the unphosphorylated form (95). In budding yeast ScSic1p, which binds and inhibits mitotic Cdks, is degraded at the G1–S boundary together with ScCdc6p, which by then has completed its part in the assembly of replication complexes. This results in a trigger for the initiation of DNA synthesis (87). In human cells cyclin E is degraded in an HsCdc4p-dependent manner and HsCdc4p is mutated in breast cancer cell lines (96). A similar mechanism operates in *Xenopus*, where the CKI XlXic1p, an inhibitor of cyclin E–Cdk2, is targeted for destruction by an ScCdc34p homolog (97). In fission yeast two Cdc4 homologs, SpPop1p and SpPop2p, regulate DNA replication through phosphorylation-dependent degradation of either SpRum1p or SpCdc18p (98).

SCF complexes also play a role in response to extracellular signals. The CKI ScFar1p inhibits G1 Cdks if induced by mating pheromones in budding yeast. Phosphorylation of ScFar1p by these Cdks targets it for destruction (99). Proteolysis of budding yeast G1 cyclins (ScCln1p and ScCln2p) requires yet another SCF, which contains the F-box protein ScGrr1p instead of ScCdc4p (89). SCF^{Grr1} is specific for the phosphorylated forms of both ScCln1p and ScCln2p. Although active throughout the cell cycle, SCF complexes work mostly during the G1 and S phases of the cell cycle to inactivate G1 cyclins, replication proteins such as ScCdc6p, and transcription factors such as the mammalian E2F family (100, 101). Occasionally, however, SCF complexes work during the M phase, and the Wee1p kinase, which inactivates Cdk1 activity by tyrosine phosphorylation, is controlled by an SFC complex (32, 102).

3. A Glimpse at the Regulated Proteolysis in Mitosis

The E3 ligase in proteolytic processes during mitosis is the anaphase-promoting complex (APC), also called the cyclosome, which targets proteins containing a so-called destruction box (D-box) which is essential for proteolytic degradation (73). APC function is required for both entry into anaphase and transition to G1. Substrate specifity is thereby transferred to the APC by ScCdc20p and ScHct1p/ScCdh1p, respectively (103, 104). Both ScHdc1/ScCdh1 and ScCdc20/ScSlp1 can be regarded as substrate-specific activators of the APC. They belong to a family of proteins containing so-called WD40 [a unit containing about 40 residues with tryptophan (W) and aspartic acid (D) at defined positions] and leucine-rich repeat (LRR) motifs, which are involved in binding of phosphorylated protein regions. In addition to being a regulator Cdc20p is also a substrate of the APC (105, 106). Cdc20p is degraded in a D-box-dependent manner late in mitosis and in G1.

The APC has a complex structure and contains at least 8–12 proteins (107–109). Two members of the Ubc family, Ubc4 and Ubc9, probably act as ubiquitin-conjugating enzymes in association with the APC (73, 110). The human APC colocalizes to the centrosome and to the mitotic spindle (108). The APC is activated by either cyclin B–Cdk1 or Polo-like kinases (106). The APC also has important functions in controlling DNA replication. In *Xenopus laevis* it targets geminin, an important factor which prevents replication until exit from mitosis, for degradation (111–115).

II. Factors Required for the Initiation of DNA Replication

A. Origins of Eukaryotic DNA Replication

In bacteria, bacteriophages, and animal and human viruses the initiation of DNA replication requires *cis*-acting elements on the chromosomal DNA called origins of DNA replication (*116, 117*). In eukaryotes the composition of such origins has been under debate for a long time. The basic elements of such an origin of DNA replication are well known in *S. cerevisiae*, but only recently have some specific origins been defined more precisely in higher eukaryotes (*117–122*). Since the origins of DNA replication of *S. cerevisiae* are best understood, its functional elements will be predominately discussed here. The budding yeast origin is called ARS (autonomously replicating sequence) and consists of different sequence elements. The A element, also called ACS (ARS consensus sequence), is an A/T-rich sequence, which was first described as an element required for replication of plasmid DNA in budding yeast (*120, 123*). Additional elements adjacent to A were found later by a genetic screen with ARS1 and were named B elements (B1–B3). The sequences B1 and B2 are also essential and together with an A element form the core origin (*124*). This core origin is bound by essential replication factors such as the origin recognition complex (ORC) and is essential for the unwinding of the origin (*125, 126*). In the neighborhood of the core origin the B3 element is bound by proteins such as transcription factors which stimulate the DNA replication activity of the core origin (*120*). This structure of the budding yeast origin is reminiscent of the organization found in bacteria and eukaryotic viruses (*116, 117, 120, 127*).

The composition of other eukaryotic origins seem to be more complicated. In very rapidly replicating cells such as embryonic cells from *Xenopus laevis* or *Drosophila melanogaster* there are no specific sequence requirements at all (*128, 129*). In contrast to these findings several sequences which are required for the initiation of DNA replication were recently characterized in higher eukaryotes, although the existence of large regions used to start DNA replication, called initiation zones, is still under discussion (for review see Refs. *119, 127, 130*). In some cases, such as the DHFR origin, the β-globin locus, the c-Myc promoter, the origin within mouse ribosomal genes, and the 3′-end of the lamin B2 gene, the functional components have been intensively studied (*131–135*). The lamin B2 origin, for example, is well defined and a short region has been identified where the orientation of Okazaki fragments switches and which defines the probable startpoint of the initiation of DNA replication (*131*). It seems that specific factors bind to these DNA elements in a cell cycle-dependent fashion

(136). The sequence requirements of higher organisms, even those of fission yeast, are more relaxed than those in budding yeast, where a small discrete region of just 100–200 bp specifically governs all the activities required for the initiation of DNA replication (117, 124), but several elements of origins in higher eukaryotes seem to have specific functions (131).

B. Proteins Involved in the Initiation of DNA Replication

1. THE ORIGIN RECOGNITION COMPLEX

In addition to these *cis*-acting sequences the initiation of DNA replication requires several *trans*-acting proteins which cooperate in a temporal fashion to start the replication of the leading strand. In budding yeast such proteins were first identified as factors that specifically bind the origin sequences in an ATP-dependent manner and were collectively called the origin recognition complex (ORC) (125, 126). The complex consists of six subunits (Orc1p to Orc6p), which are all essential for cell growth in yeast (123, 126, 137, 138). ORC and Cdc6p belong to the AAA+ ATPase family (see below; 139). Proteins related to budding yeast ORC (ScORC) subunits have been discovered in all eukaryotes studied so far. These polypeptides were identified either by protein purification or sequence homology, which was often followed by cloning of the cDNAs and expression of the proteins. It is more than likely that the function of this protein complex is conserved in the whole eukaryotic kingdom (123).

ScORC as well as its binding sites at origins are the best characterized of these special protein–DNA complexes and we will initially focus on this system. ScORC interacts with the core origin and specifically requires the A and the adjacent B1 elements. Mutations of the DNA-binding elements which reduce the binding of ScORC *in vitro* also cause a decrease in DNA replication activity *in vivo* (125). During the entire cell cycle ScORC binds to the origin DNA as determined by genomic footprinting and chromatin immunoprecipitation (CHIP) assays (6, 140, 141). Until recently it was not clear which subunits directly contact the DNA strands, but the subunits ScOrc1p, ScOrc2p, and ScOrc4p are most likely to be involved, since they were crosslinked to the A element (142). The ScORC–DNA interaction requires ATP and might occur through the two subunits ScOrc1p and ScOrc5p; it has a K_d in the range of 20–100 nmol/l. The ATP- and DNA-binding activities of ScOrc1p are interdependent, since formation of an ATP–Orc1p complex requires the presence of a wild-type ScORC recognition sequence (143). ScOrc1p coordinates ATP hydrolysis by the ORC, and a mutation in Orc1p which results in the ability to bind, but not to hydrolyze ATP is lethal in budding yeast (144). The three subunits ScOrc1p, ScOrc2p, and ScOrc6p, but not the other subunits, are phosphorylated *in vitro* by the S phase-specific ScClb5p/ScCdc28p cyclin–Cdk complex (145). The phosphorylation is most likely involved in the control of the initiation of DNA replication (146).

The binding of ScORC has been intensively studied by footprinting and crosslinking methods (6, 140, 141). These investigations showed that the binding pattern of ORC varied throughout the cell cycle (6). In G1-arrested cells the ScORC recognizes and protects an extended region, whereas in S and M phase the ScORC-protected sequences are more concentrated at the central elements and an additional characteristic hypersensitive site is detectable. The latter is quite similar to the footprinting pattern of purified ScORC *in vitro*. These findings suggested that ScORC binding changes before and after the initiation of DNA replication and the terms "prereplicative complex," present in G1, and "postreplicative complex," present in S, G2, and M phases (pre-RC and post-RC, respectively), were introduced (6). It is thought that the ScORC–DNA complex serves as a landing platform for additional initiation factors, which in return modulate the DNA-binding activity of ScORC (144, 145, 147). This hypothesis stems from experiments in which proteins and origin DNA were chemically crosslinked *in vivo* and then precipitated with specific antibodies (140, 141, 148). Recently the biochemical properties of these proteins have started to emerge. In genetic and biochemical studies ScCdc6p interacts with ScORC (145, 149). The protein modulates the binding of ScORC to DNA by increasing sequence-specific DNA binding of the latter (145). ScCdc6p induces conformational changes in ScOrc1p, ScOrc2p, and ScOrc6p and so probably reduces nonspecific DNA binding of ScORC.

The studies of ORC in other organisms such as *Drosophila*, *Xenopus*, mouse, and humans revealed that the functions and the amino acid sequences of these proteins are highly conserved (147, 150–169). However, these studies also provided new clues and presented variations among organisms. HsOrc1 and HsOrc6 do not stably associate with the other subunits in higher eukaryotes and the HsOrc2, HsOrc3, HsOrc4, and HsOrc5 subunits form a stable subcomplex (153, 170). Recent studies suggest that in higher eukaryotes ORC does not bind as a heterohexamer complex, but the subunit HsOrc1 can dissociate from the protein–DNA complex in a cell cycle-dependent manner which leads ultimately to its proteolytic degradation (158, 160).

2. Cdc6/Cdc18 Protein and the Competence to Replicate Chromosomal DNA

ScCdc6p (cell division cycle protein 6) of budding yeast and the homologous protein in fission yeast, SpCdc18p, are unstable proteins, and their synthesis as well as degradation are strictly controlled (4, 5, 171–173). ScCdc6p/SpCdc18p is a member of the AAA(+) superfamily of ATPases, which includes replication factors such as subunits of the origin recognition complex (ORC), Orc1p, Orc4p and Orc5, minichromosome maintenance (Mcm) proteins, and the prokaryotic as well as eukaryotic clamp loading proteins (139, 174, 175). All these proteins contain a number of conserved regions and share a common three-dimensional architecture. The protein loading function of Cdc6p is consistent with the idea that it loads the Mcm proteins onto chromatin (176). The conserved Walker A and B motifs of Cdc6p are involved in nucleotide metabolism and are essential for these Cdc6p activities *in vivo* (177, 178).

In budding yeast the CDC6 gene is only expressed in mitosis in parallel with the SIC1 gene. ScCdc6p and ScSic1p are predominantly present in yeast cells during the G1 phase (179–181). Both proteins bind to mitotic Cdk complexes and cooperate to inactivate them *in vivo* and *in vitro* (179, 182, 183). This inhibitory activity is located in the N-terminus of ScCdc6p, which is not required for its replication functions (100, 182). However, in contrast to ScSic1p and SpCdc18p, overexpression of ScCdc6p in G2 or expression of a stabilized version of ScCdc6p does not cause rereplication in yeast (146, 179, 184). In late G1 phase ScCdc6p as well as ScSic1p are phosphorylated by G1 cyclin–Cdk1 and the modification of both yeast proteins results in a ubiquitin-dependent proteolytic degradation (100, 146). In contrast to budding yeast, Cdc6p of higher eukaryotes such as XlCdc6p and HsCdc6p are phosphorylated and then exported from the cell nucleus (185, 186). This phosphorylation probably occurs via S phase-specific cyclin–Cdk complexes. ScCdc6p induces a concomitant change in the conformation of ORC and increases the DNA-binding specificity of the protein complex by inhibiting nonspecific DNA binding (145). Mutations in the Cdc6p Walker A and Walker B motifs or nonhydrolyzable ATP homologs inhibit these activities of Cdc6p (177, 178). These data suggest that Cdc6p modifies ORC function at DNA replication origins and may be an essential determinant of origin specificity. HsCdc6p interacts with PP2A via the B subunit p48. Overexpression of p48 arrests cells at the G1 to S phase transition, which suggests that phosphorylation of HsCdc6p is most likely required for entry of human cells into S phase (48). In addition, studies in *Xenopus* extracts suggested that XlCdc6p acts as a nuclear receptor for cyclin E–Cdk2 attachment to the chromatin and that this interaction is essential for the activation of pre-RC and the initiation of DNA replication (24, 26). Phosphorylation mutants of HsCdc6 support this view, since phosphorylation of HsCdc6p is absolutely required in a late step in the transition of pre-RC to the initiation complex or during the initiation reaction itself (187).

3. THE PROTEINS Cdt1p, GEMININ, AND Mcm10p AND THEIR BINDING TO CHROMATIN

The inititation of DNA replication requires a large number of factors. In fission yeast expression of several of these proteins, such as SpCdt1p, is under control of the transcription factor SpCdc10p (188). The protein SpCdt1p (Cdc10 target 1) causes overreplication of cellular chromosomes in the presence of slight overexpression of SpCdc18 (113, 189). Cdc6p/Cdc18p and Cdt1p cooperate to load a functional Mcm2p–7p complex onto the chromatin (111, 114, 115, 190).

However, both proteins can only bind chromatin if ORC is present (*111, 145*). Cdt1p together with the Mcm2p–7p complex is a component of the so-called licensing system and the activity of XlCdt1p greatly overlaps with the activity of licensing factor B (*114, 191*). After the formation of the initiation complex in early S phase Cdt1p forms a stable complex with the protein geminin, which prevents a renewed association of Cdt1p with the chromatin until mitosis. Geminin is then proteolytically degraded in mitosis and Cdt1 can again support pre-RC formation in the following G1 phase (*112, 114, 115, 190*). In contrast to fission yeast and higher eukaryotes Cdt1- and geminin-like proteins have yet not been described in budding yeast, but factors with related functions are probably present (*7*). A functionally conserved analog of Cdt1p was recently discovered in budding yeast (S. Tanaka and J. Diffley, personal communication).

Mutation of replication factor Mcm10p (minichromosome maintenance protein 10) significantly reduces the frequency of initiation, which suggests that the protein is involved in the initiation step (*192*). Although ScMcm10p was found in a similar genetic screen as were the ScMcm2–7 proteins, it has no structural similarity to them. Expression of a mutant form of *mcm10* rescues *cdc46* (*mcm5*) and *cdc47* (*mcm7*), which are also suppressed by the expression of mutant *cdc45* (*9, 193*). These results suggest that Cdc45p and Mcm10p act in similar pathways during the initiation of DNA replication. Mcm10p binds to chromatin throughout the cell cycle in an ORC-independent manner. It most likely influences chromatin structure and, like ORC, its presence on chromatin is a prerequisite for the loading of the Mcm2p–7p complexes. Furthermore, the protein is probably involved in the release of origin bound factors (see Fig. 6) (*193*). After the ScMcm2p–7p complex has been loaded onto chromatin, an interaction between ScMcm10p and the Mcm2p–7p complex can be found. During the transition of the prereplicative complex (pre-RC) to the initiation complex (IC) this interaction is probably abolished by ScCdc45p binding to chromatin (*9, 193*). The connection of the concepts IC and post-RC is still under discussion, since IC formation describes the activation of replication factors during initiation, whereas the appearance of post-RC depends on a change of the ORC footprint.

4. The Mcm2p–7p Complex, a Key Component for the Initiation of DNA Replication

The minichromosome maintenance (Mcm) proteins were first described as factors which, when they are mutated, cause a reduced stability of plasmids in budding yeast (for review see Refs. *123, 194, 195*). Furthermore, some of them were also identified as proteins which are involved in cell cycle progression (*196*). The six polypeptides Mcm2p to Mcm7p form a family of related proteins which are highly conserved from yeast to human and related proteins are also found in archaea (Fig. 3) (*195, 197, 198*). The Mcm proteins play a central

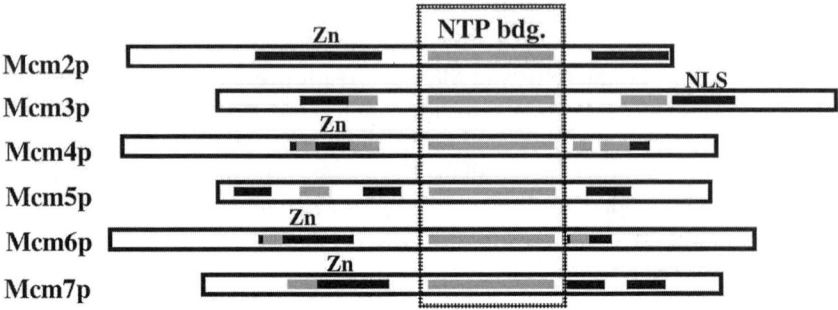

FIG. 3. Sequence comparison of the six Mcm proteins Mcm2p to Mcm7p. The six Mcm (minichromosome maintenance) proteins assemble into a stable heterohexameric complex. The amino acid sequences of each of these proteins are highly conserved in all eukaryotes. They even show regions of conservation with archaeal proteins (marked in grey). The conserved amino acids of each Mcm protein are presented in black (sequence comparison adapted from Ref. 195 with permission from the *Annual Review of Biochemistry*, vol. 68, 1999 by Annual Reviews www.AnnualReviews.org). The nucleotide triphosphate binding (NTP bdg.) region of each Mcm protein is boxed. Abbreviations: Zn, putative zink finger; NLS, nuclear localization sequence.

role in the initiation and elongation step of DNA replication. Moreover, they fulfill a regulatory function during cellular DNA replication (*3, 6, 7, 9, 123, 194, 195*). The six proteins Mcm2p to Mcm7p are part of the so-called licensing system, which coordinates chromosomal replication and prevents replication of newly duplicated DNA within the same cell cycle (*191, 199, 200*). The functions of these six Mcm proteins are not redundant, since deletion or mutation of each of them is detrimental to the cell (*195, 196*). All of them are members of the AAA(+) superfamily of ATPases (*139, 174, 175*). The ATPase activity of the Mcm2p to Mcm7p complex is subject to complex control, which reflects the importance of its function to the cell (A. Schwacha and S. Bell, personal communication).

At the end of mitosis the Mcm2p–7p complex associates with the chromatin until the beginning of the next S phase. The loading onto chromatin depends on ORC, Mcm10p, Cdc6p, and Cdt1p (*114, 115, 190, 201*). During S phase the Mcm proteins dissociate from the chromatin and the replicated DNA loses its ability to reduplicate (*202*). Although this process follows similar rules in all eukaryotes investigated, there is a major difference between budding yeast and the other eukaryotes. In *S. cerevisiae* the Mcm complex is exported to the cytosol, whereas it is translocated from the chromatin to the nucleoplasm in *S. pombe* and in higher eukaryotes (*123, 194, 195, 201*). This implies that in mitosis the ScMcm2p–7p complex must be transported into the nucleus, whereas the Mcm complex of other eukaryotes is merely redistributed within the nucleus (*123, 194, 195, 201*). A mutant ScMcm2p–7p complex which is not exported from the nucleus of budding yeast does not cause reinitiation of a replicated chromosomal region (146). As is the case for Cdc6p, Cdt1p, and geminin the operation of switches from the establishment of replication competence to the initiation of DNA replication followed by a rereplication block is accomplished by slightly different mechanisms that depend not only on the nature of the organism, but also on its developmental stage (7).

In all eukaryotes the six Mcm proteins assemble *in vivo* into a stable heterohexameric complex with a molecular mass of about 560 kDa which can be purified from a variety of organisms. This suggests approximately stoichiometric amounts of each subunit. However, other complexes could be purified from HeLa cells and *Xenopus* extracts, such as an Mcm4p, Mcm6p, Mcm7p complex, which can be loosely associated with Mcm2p, as well as a complex consisting of Mcm3p and Mcm5p (203, 204). Recently, it was reported that the Mcm4p, Mcm6p, Mcm7p complex isolated from HeLa cells as well as the recombinant mouse and fission yeast proteins possess a low, nonprocessive DNA helicase activity (203, 205). It was also shown that the fission yeast heterohexameric Mcm4p, Mcm6p, Mcm7p helicase requires fork structures for processive activity (206). However, such a subcomplex can no longer support licensing of DNA replication in the *Xenopus* cell-free system (204). These results suggest that at least some of the Mcm proteins have enzymatic activities, such as ATPase and helicase, in addition to their role in the licensing reaction, which requires the heterohexameric Mcm2p to Mcm7p complex. The helicase activity of the Mcm proteins is controlled by several factors. The addition of Mcm2p, which can bind histones, to the Mcm4p, Mcm6p, Mcm7p helicase changes the conformation of the protein complex as determined by ultracentrifugation and gel filtration and inhibits the enzyme activity (203, 205, 207). An inhibition of the enzyme is also attained through its phosphorylation by cyclin A–Cdk2 (208). Multiple phosphorylation of Mcm4p is required for the inactivation of the helicase activity, and mutation of potential phosphorylation sites abrogated protein phosphorylation and inhibition of the enzyme complex by cyclin A–Cdk2 (209). On the other hand, phosphorylation of the Mcm2p–7p complex, possibly by Dbf4p–Cdc7p, leads to structural modifications and to the formation of the enzymatically active Mcm4p, Mcm6p, Mcm7p subcomplex (9, 38). This transformation might be a prerequisite for melting the double-stranded (ds) origin DNA and loading of Cdc45p (210). Recent findings in the cell-free *Xenopus* replication system support the view that the helicase activity of the XlMcm2p–7p complex is highly coordinated and tightly controlled at multiple levels. In addition, they suggest that XlCdc45p is probably involved as an essential factor in this regulatory network to control the helicase function of the Mcm complex (S. Mimura and H. Takisawa, personal communication).

Despite the discovery that an enzymatic function is associated with the Mcm proteins, it is still under discussion whether they represent the long-sought replicative DNA helicase, the counterparts of which are the *Escherichia coli*

DnaB or the SV40 (simian virus 40) large T antigen (*116, 211*). In budding yeast all six proteins are required for proper initiation and elongation during the DNA replication process (*8, 9*). The DNA unwinding activity, however, can only be detected with a subset of the Mcm proteins (*203, 205, 207*). In addition, the *Xenopus* cell-free system also requires all six proteins for licensing and Mcm4p, Mcm6p, and Mcm7p are not sufficient for this function (*204*). Genetic and biochemical findings point out that either additional factors are required for helicase activity of the Mcm2p–7p complex or that they have additional, yet unknown cellular functions. For example, it may be that the Mcm proteins are involved in intranuclear transport events to deliver replication proteins or DNA to replication centers. This hypothesis is supported by reports of prokaryotes where proteins with helicase sequence motifs and even helicase activity are essential for transport processes rather than unwinding of dsDNA (*212*). Likewise in eukaryotes and prokaryotes there are a large number of "helicases" and some of them might not be employed in the unwinding reaction which occurs during DNA replication, DNA repair, and DNA recombination.

The Mcm proteins contain conserved nucleotide-binding domains and these motifs suggest yet another possible role for Mcm proteins, which is their involvement in remodeling of the chromatin (*195*). Indeed, HsMcm2p binds to histones (*203*). Remodeling of the chromatin might be a prerequisite for the formation of a DNA bubble structure at an origin prior to the initiation step and subsequently this activity may be required for replication fork progression. In an iteration of these thoughts, the Mcm2p-7p complex might also play an essential role in the ordered loading of specific limiting factors onto the pre-RC or the initiation complex, thereby allowing the licensing of DNA replication. Interestingly, Mcm proteins, like ORC subunits, have sequence similarities with clamp loaders and they interact with replication proteins such as DNA polymerase α-primase (*17, 139, 174, 175*). However, the opposite action might also apply. In analogy to eukaryotic RNA helicases, which remove proteins from RNA (*213*), the Mcm2p–7p complex might disrupt protein–DNA interactions within the chromatin before DNA is replicated and might work as such as a reverse "clamp loader." Regardless of which cellular functions the Mcm proteins will finally turn out to perform, it is noteworthy that the amount of Mcm2p–7p loaded onto the chromatin is about 10 times higher than that of ORC (*3*). This clearly implies additional roles for these proteins. Mcm proteins are targets of checkpoint control systems and might relay crucial signals to prevent replication upon DNA damage. Alternatively, the additional copies of the Mcm complex attached to chromatin might be required for reinitiation at specific unreplicated regions after the replication fork has been stalled at DNA lesions. However, as long as no checkpoint function is activated during S phase (see below for more details) these extra copies are inactivated and removed by the passing replication fork.

5. THE REPLICATION FACTOR Cdc45 PROTEIN AND THE TRANSITION FROM G1 TO S PHASE

The CDC45 gene was discovered and its product was initially characterized in *S. cerevisiae* (ScCDC45) (*196, 214, 215*). Thereafter, homologous proteins were identified in other eukaryotes ranging from *S. pombe* (sna41p/SpCdc45p) to human (HsCdc45p) (*216–221*). Cdc45p in *S. pombe*, also refered to as sna41, was originally isolated as a suppressor of *nda4* mutant [equivalent to MCM5 of *S. cerevisiae* (*219*)]. The human homolog of ScCdc45p, named Cdc45L, also called HsCdc45p, is localized on chromosome 22q11.2 close to the DiGeorge critical region (DGCR) (*217, 220*). In budding yeast the protein level of ScCdc45p is constant throughout the cell cycle and the protein exclusively localizes in the nucleus, whereas in human cells its distribution varies in a cell cycle-dependent manner (*215, 220, 222*). Immunofluorescence and fractionated cell extraction methods showed that during G1 and S phase HsCdc45p is predominantly in the nucleus, whereas during G2 phase the protein is equally distributed between nucleus and cytoplasm without a significant change in the protein level (*220*; C. Bauerschmidt, F. Grosse, and H.-P. Nasheuer, unpublished results). The picture is different in *Drosophila*. Immunofluorescence experiments showed that, during the interphase, DmCdc45p is associated with chromatin, and that during prophase it also partitions into the nucleoplasm. In metaphase the quantity of the protein dramatically decreases, resulting in undetectable levels in anaphase and telophase (*216*).

To perform its function Cdc45p is part of a multiprotein network of replication factors. In fission yeast the subunits of the ScMcm complex and ScCdc45p cooperate. The phenotype of a *cdc45* mutant could be suppressed by the expression of mutant forms of *cdc46* (*mcm5*) and *cdc47* (*mcm7*), which are also rescued by an *mcm10* mutant (*196*). The functional cooperation of Cdc45p and the Mcm complex is highly conserved and recent findings show that Cdc45p and Mcm10p probably act in the same pathways. Some interactions of ScCdc45p with components of the Mcm2p–7p complex, such as with Mcm2p, only exist in the late G1 phase and in the S phase and depend on the Dbf4–Cdc7 kinase (*210, 215, 223–225*). The protein also interacts with the Mcm7 protein. In budding yeast the essential protein ScSld3p interacts with ScCdc45p during the entire cell cycle and binds to origins in S phase. Mutational analysis suggested that ScSld3p is required for the interaction of ScCdc45p and ScMcm2p (*226*). By comparing the interactions of Cdc45p with other replication factors it seems that they are not fully conserved in eukaryotes. For instance, ScCdc45p forms complexes with RPA. Furthermore, recruitment of DNA polymerases to the chromatin requires active ScCdc45p (*141*). Subsequently, an interaction with the catalytic subunit of DNA polymerase ε was described in budding yeast (*210*). Although ScCdc45p does not interact with DNA polymerase α–primase (Pol–Prim), a mutational analysis showed that only active ScCdc45p supports chromatin binding of Pol–Prim, indicating that it is indirectly needed for this process (*210*).

SpCdc45p (Sna41p) is involved in DNA replication and interacts with Pol–Prim during the entire cell cycle (*227*). Although it is not required for the loading of Pol–Prim onto the chromatin of fission yeast, the presence of SpCdc45p/Sna41p stabilizes the chromatin association of Pol–Prim (*227*). Furthermore, physical contacts of SpCdc45p with SpMis5p, the fission yeast equivalent of ScMcm6p, were determined during S phase (*227*). In contrast, coimmunoprecipitation experiments failed to detect interactions of XlCd45p with XlMcm7p and PCNA, but a weak binding of XlCdc45p to XlMcm5p was found (*218*). In addition, the same authors determined that Pol–Prim binds to XlCdc45 and requires XlCdc45p for loading onto chromatin. Walter and Newport reported an ordered sequence of chromatin binding by replication proteins (*228*): First, XlCdc45p binds, which is followed by the attachment of XlRPA, and then XlPol–Prim. In accordance with what was found in the SV40 system, an unwinding assay revealed that *E. coli* SSB (single-strand binding protein) stabilized ssDNA and allowed the unwinding of ori-DNA, but that XlRPA was required for the recruitment of Pol–Prim to chromatin (*228, 229*). Binding of XlCdc45p to chromatin depends on phosphorylation of XlMcm2p–7p by XlDDK (*39*). In *Drosophila* DmCdc45p forms a complex with the proteins DmMcm2p, DmMcm3p, DmMcm4p, and DmMcm5p. However, no interactions have been found between DmCdc45p and DmPol–Prim, or with DmMcm6p, DmOrc2p, and DmOrc5p. *In vitro* experiments with HsCdc45p revealed its interaction with Mcm7p and the second largest subunit of Pol–Prim (*230*).

Cdc45p is most likely involved in the initiation of DNA replication (*141, 215, 224*), but is not a component of the pre-RC itself. In budding yeast temperature-sensitive *cdc45* mutant strains arrest at the G1–S phase transition immediately prior to the initiation of DNA replication. In *Xenopus* oocyte extracts XlCdc45p is needed for the initiation of DNA replication after formation of the nuclear structure. The finding that the same *mcm* mutants suppress mutant *cdc45* and *mcm10* suggests that both proteins target the same pathway and interact with the same regions of these MCM proteins. Since Mcm10p stabilizes the pre-RC in a premature state, Cdc45p most likely activates the pre-RC complex. Cdc45p also is critical during the conversion of the pre-RC to the initiation complex (*9, 193, 231*). In addition, Cdc45p is not only essential for the G1–S transition, but it also has an essential role in the elongation phase of DNA replication (*232*). These functions of Cdc45p in the replication of chromosomal DNA raise a question about their coordination. Recently it was shown that the proteins

INITIATION OF EUKARYOTIC DNA REPLICATION

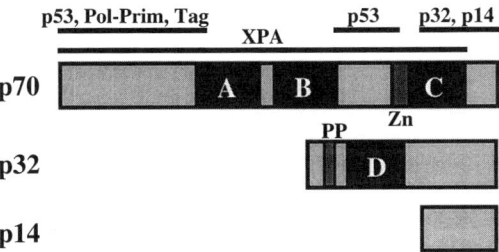

FIG. 4. Subunit structure of replication protein A, the eukaryotic single-strand DNA-binding protein. The three subunits of replication protein A (RPA) contain four ssDNA-binding domains A–D. The largest subunit of RPA, p70, interacts with the tumor suppressor protein p53, DNA polymerase α–primase (Pol–Prim), SV40 large T antigen (Tag), xeroderma pigmentosum protein A (XPA; DNA repair protein), and the smaller subunits p32 and p14. Additional abbreviations: PP, region of extended phosphorylation; Zn, zinc-finger domain.

ScCdc45p and ScDbf4p-ScCdc7p kinase (DDK) probably act at the same time point in the early stages of DNA replication (222). *In vitro* experiments showed that DDK phosphorylates ScCdc45p (233). Furthermore, the S phase-specific Cdks are also important for the interaction of ScCdc45p with chromatin (233). These data point to a so-called "double trigger" or double activation model in which the DDK and Cdk seem to act sequentially to allow the initiation of DNA replication at the correct sequence and at the right time (233).

6. REPLICATION PROTEIN A, THE EUKARYOTIC SINGLE-STRANDED DNA-BINDING PROTEIN

The replication protein A (RPA; also called RF-A and, if purified from human cells, HSSB) is a protein complex that consists of the three polypeptides p70, p32, and p14 (Fig. 4). The protein complex binds to ssDNA and has essential functions in all types of DNA metabolism (for review see Refs. *25, 234, 235*). Its function, subunit structure, and posttranslational modifications are highly conserved in all eukaryotes so far examined. The RPA complex was first described as a factor required for SV40 DNA replication and budding yeast DNA recombination (*236–239*; for review see Refs. *138, 240–242*). Subsequently it was determined that RPA also stimulates DNA polymerase α–primase (Pol–Prim), several DNA helicases, and DNA polymerases δ and ε (*243–251*). Furthermore, RPA increases the fidelity of DNA replication and might function as a "fidelity clamp" for DNA polymerase α (*252–254*). Specific protein–protein interactions occur between RPA and Pol–Prim during primosome assembly (*228, 229, 244, 255*). RPA also plays a role in the stability of telomere sequences (*256*).

In addition to DNA replication, RPA is also required for nucleotide excision repair (NER; *257–264* for review see Refs. *265–267*), base excision repair

(BER; 268–270), and DNA recombination (237, 271–277; for review see Refs. 234, 235, 266). Recently a novel activity of RPA was discovered. On template primer systems the RPA complex and especially its middle-sized subunit interacts with the 3′- and 5′-ends of a primer that is hybridized to a template (234, 278–281). During DNA repair this activity of RPA is most likely involved in directing the XPF and XPG endonucleases to excise the strand containing a lesion (278). Whether the interaction of RPA with the 5′- and 3′-ends of primers has any function during viral or cellular DNA replication is still unclear. However, in the replication of SV40 DNA the RPA subunits specifically interact with the 3′-ends of newly synthesized, short RNA–DNA replication products. Therefore, it was suggested that these interactions might be involved in monitoring the size of the products of Pol–Prim during Okazaki fragment synthesis (281).

The largest subunit of RPA, p70, is the main subunit required for ssDNA binding (Fig. 4) (234, 235, 239, 247, 282). p70 interacts with several proteins: Pol–Prim, BPV-E1, SV40 TAg, Cdc45p, transcription factors such as p53 and VP16, as well as other proteins (210, 229, 234, 255, 283–290). The polypeptide is composed of multiple domains (291, 292): The C-terminal domain is involved in complex formation with the two smaller subunits, whereas the N-terminus interacts with other cellular and viral proteins. A central region that can be divided into three subdomains is involved in protein–DNA interactions and its binding to DNA was structurally resolved (293–296). In addition the C-terminus of p70 most likely also interacts specifically with ssDNA and the Zn finger has a major role during particular repair processes as determined by biochemical and structural investigations (296–298).

In budding yeast the middle subunit, p32, is encoded by an essential gene (Fig. 4) (299, 300). In recent years some indications as to its function have started to emerge. p32 interacts with ssDNA and the 3′-end of a primer as well as with various proteins, which include Pol–Prim (234, 235, 296, 298, 301). During DNA replication a high degree of phosphorylation of RPA was reported by several groups (302–312). In human cells the p32 subunit is also phosphorylated in response to DNA damage and replication stress and might be a direct target for the ATM and ATR kinases, respectively (306, 313). RPA also represents a substrate for DNA-PK, a kinase that is activated by double-strand breaks (234). Correspondingly, the ScMec1p kinase is responsible for RPA phosphorylation after radiation treatment of budding yeast cells (302). Extracts of human cells treated with UV irradiation are unable to support SV40 DNA replication *in vitro* and this could be a consequence of a reduced ability of RPA to form complexes with Pol–Prim upon phosphorylation (234). Since RPA is required for replication and NER, phosphorylation might alter the preference of the protein for partners with which it will interact (234, 235). However, no direct function of p32 phosphorylation either during DNA replication or in connection with other

DNA metabolic events has yet been described. Only recently it was reported that under specific conditions the RPA complex becomes disrupted upon phosphorylation of the p32 subunit (290). In cells treated with hypoxia the p70 and p32 subunits differentially associated with chromatin (H. J. Riedinger and H. Probst, personal communication). The function of the smallest subunit of RPA is still not known. It seems to interact with the middle and the large subunits, but the latter interaction appears to require the binding of p14 to p32 (234, 302, 309).

7. DNA POLYMERASE α–PRIMASE, THE INITIATOR ENZYME COMPLEX

DNA polymerase α–primase (Pol–Prim) is a central component in the initiation of cellular DNA replication, since it is the only enzyme that can start eukaryotic DNA replication *de novo* (for review see Refs. 25, 138, 314–316). The function, amino acid sequence, and subunit composition are highly conserved from yeast to human (316).To carry out its cellular functions Pol–Prim needs to interact with other replication initiation proteins such as RPA, Cdc45p, Mcm2, mouse polyomavirus, and SV40 T antigen, as well as with HPV11, HPV18, and BPV E1 (17, 218, 229, 231, 255, 288, 317–325). The interactions of all four subunits with other proteins are most likely required to initiate origin-dependent DNA replication (229). In addition, it was recently reported that Pol–Prim binds to cell cycle regulator proteins and tumor suppressor proteins such as Cdks, PP2A, retinoblastoma protein (pRb), and p53 (17, 19, 288, 326). These interactions might be involved in the checkpoint functions of the enzyme complex in higher eukaryotes. Additional protein–protein interactions might be responsible for the checkpoint function of Pol–Prim in budding and fission yeast (327–330).

In addition to its enzyme activities during DNA replication, Pol–Prim appears to play a key role in coordinating DNA replication, DNA repair, and cell cycle checkpoints (328, 331, 332). Recent evidence suggests that the DNA polymerase and the primase fulfill essential functions in homologous DNA recombination (331). The primase subunits on their own might also be involved in the coordination of DNA repair and DNA replication, since mutations in the primase genes abrogate these functions in budding yeast and Pol–Prim interacts with repair proteins such as poly(ADP-ribose)polymerase (329, 333–335). Furthermore, it has specific functions in telomere replication, because mutants of DNA polymerase α–primase subunits interfered with telomere stability (336–339). Recently direct functions of Pol–Prim in telomere maintenance were studied by synthesizing telomeric sequences *in vitro* (340–342). The report of a new primase activity-carrying protein GANP (germinal center-associated nuclear protein) in human B cells raises the question of whether in specific cells or for specialized functions another initiator enzyme exists (343).

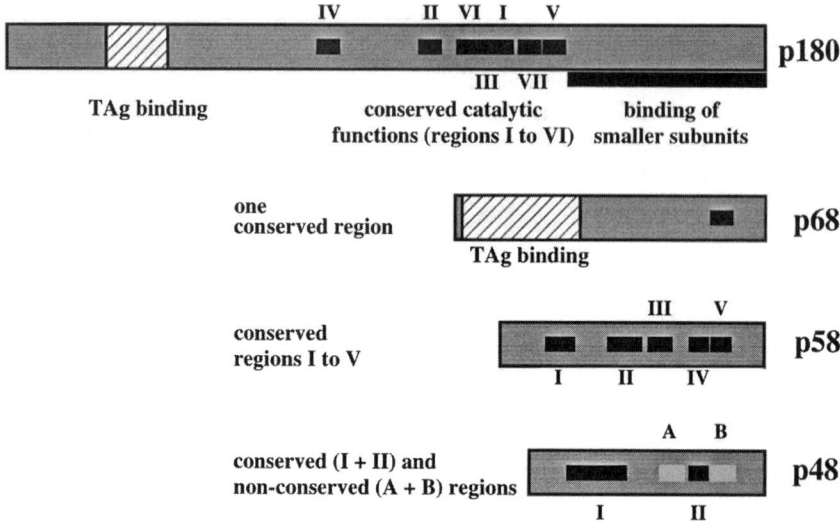

FIG. 5. Subunit structure of DNA polymerase α–primase. In addition to its enzyme activities during DNA replication, Pol–Prim appears to play a key role in coordinating DNA replication, DNA repair, and cell cycle checkpoints. The four subunits of the enzyme complex, p180, p68, p58, and p48, have catalytic and noncatalytic functions. Each subunit contains regions which are conserved from yeast to humans, marked in roman numerals. Each subunit interacts with the viral initiator protein SV40 large T antigen (TAg). These sites are mapped in p180 and p68. The C-terminus of p180 binds to the three smaller subunits. One of the two less conserved regions, A and B, in mammalian p48 is involved in the regulation of polyoma virus DNA replication.

Pol–Prim contains four subunits, p180, p68, p58, and p48 (Fig. 5). The p180 subunit carries the DNA polymerase activity, whereas the p48 subunit contains the catalytic center for the RNA polymerase (primase) activity (Fig. 5) (344–352; for review see Refs. 314–316, 353). The p58 subunit stabilizes the DNA primase activity of the p48 subunit and fully functional DNA primase activity requires both subunits (276, 315, 349, 351, 353). It was reported that mouse p58 is required for the nuclear import of p48 (354). Each subunit contains regions which are highly conserved between eukaryotic primases and the p48 subunit even has homologies to archaeal primases (Fig. 5) (344, 352, 355–357). The two conserved aspartates (D109 and D111) in region I of human p48 together with aspartate D306 in region II are involved in metal ion binding (344). In addition, region II contains an arginine at position 304, which probably contacts the synthesized primer. The neighboring, least conserved region B controls species specificity of mouse polyomavirus DNA replication *in vitro* (322, 358). Both the subunits contain stretches of homology to the class X nucleotidyl-transferase superfamily, which includes enzymes such as DNA polymerase β (pol β) and terminal transferase (Fig. 5, regions II of p48 and p58) (*317, 352, 353, 358*). Region II of p58 also shows similarity to the helix–hairpin–helix motif described by Shao and Grishin (*359*) and might be involved in DNA binding. The homology of the polymerases is reflected in their three-dimensional structures, since those of pol β and an archaeal primase p40 are similar. In contrast the sequence and structure of this archaeal primase are unrelated to those of the *E. coli* and *Bacillus stearothermophilus* primases (*315, 344, 360–362*).

The largest subunit, p180, which belongs to the class B type of DNA polymerases (*316, 363*), contains seven sequence motifs conserved between pol α homologs (Fig. 5) (*352, 364, 365*). The most highly conserved region, I, contains two aspartate residues, which are probably involved in the binding of Mg^{2+}, whereas the second most conserved region, II, probably interacts with the incoming dNTP and with the template primer (*366–368*). The p180 subunit interacts with the cellular and viral initiator proteins RPA, SV40 T antigen, HPV11, and BPV E1 as well as with the tumor suppressor protein p53 (*229, 288, 318, 324*). The polypeptide is phosphorylated by various serine and threonine kinases (*369–372*). The second largest subunit, also called B subunit, has molecular masses ranging from 66 to 86 kDa (*346, 352, 356, 373–375*). Although the polypeptide has no detectable enzyme activity, it is essential in yeast. No kinases apart from Cdks phosphorylate the B subunit (*369, 376*). Recently it was reported that mouse p68 is required for the nuclear import of p180 (*354*). Earlier reports described the interaction of human p 68 with the viral initiator proteins SV40 T antigen, HPV11, and HPV18 E1 (Fig. 5) (*229, 318, 323, 373*).

The Pol–Prim complex plays a central role in leading and lagging strand synthesis (*138, 314, 316, 328, 377*). The subunits p48 and p58 synthesize oligoribonucleotides, called RNA primers, to initiate the leading strand DNA synthesis of genomic DNA (*138, 314, 316, 328, 377*). In addition, the primase activity of Pol–Prim synthesizes primers for discontinuous DNA synthesis of the lagging strand (*229, 351, 378–381*). After Pol–Prim has synthesized the primer RNA molecules at the origin of DNA replication, a polymerase switch takes place, and these primers are elongated by the DNA polymerase activity of p180 (*138, 241, 314, 377*). A similar sequence of reactions takes place for synthesis of Okazaki fragments on the lagging strand. The primase synthesizes an RNA primer consisting of approximately 10 nucleotides (nt), which is elongated by the DNA polymerase activity of the enzyme complex to approximately 40 nt. Since Pol–Prim is not very processive, it dissociates after the synthesis of the DNA–RNA primer and the replication factor C (RFC) prevents its reassociation (*382–386*).

The role of Pol–Prim in the synthesis of initiation molecules for DNA replication is underlined by its regulation during the cell cycle. Pol–Prim is a target of posttranslational modification (Fig. 5). The p180 and p68 subunits in human cells at G2/M, the fission yeast p180 subunit in late S, and the p68 homolog in budding yeast at G1/S are cell cycle-dependently phosphorylated (*17, 19, 328, 369, 376, 387, 388*). Phosphorylation of the p58 and p48 subunits was also observed, but it seems not to be cell cycle-dependent (*369, 376*). Phosphopeptide maps of human p180 and p68 indicate that a Cdk is most likely responsible for the modification (*328, 369, 389, 390*). Recent results suggested that p180 is also a substrate of DDK *in vitro* (*371*). In human cells two immunologically distinct subpopulations exist which are differentially phosphorylated (*17*). The hypophosphorylated Pol–Prim is present in the nucleus from G1 to early S phase and colocalizes with Mcm2p, but not with the incorporated nucleotide analog BrdU, whereas the localization of the hyperphosphorylated form overlaps with newly synthesized DNA (*17*). Only the hypophosphorylated Pol–Prim population binds to the initiator proteins Mcm2p and SV40 T antigen. The cell cycle regulator proteins PP2A and cyclin E–Cdk2 bind to the hypophosphorylated Pol–Prim in G1 phase, whereas the hyperphosphorylated form interacts with cyclin A-dependent kinases in S and G2 phase (*17*). The phosphorylation of Pol–Prim by cyclin A–Cdk2, but not by cyclin E-Cdk2 prevents its interactions with the viral initiator SV40 T antigen and abolishes origin-dependent initiation (*17, 19, 389, 390*). The inhibitory effect of Cdks is counteracted by PP2A or by mutating potential Cdk phosphorylation sites of p180 and p68, respectively (*17, 19, 390*).

8. DNA Topoisomerase I

Eukaryotic topoisomerases belong to a class of enzymes that change the topology of DNA and that can reduce topological stress introduced during DNA replication (for review see Refs. *391–393*). Depending on their catalytic mechanism, they are classified as type I or type II enzymes: Type I (topo I) is independent of ATP, but its activity is stimulated by Mg^{2+}. Since topo I transiently cleaves one strand of duplex DNA and ligates the DNA during the reaction (nicking–closing reaction), the topological number is reduced by one in each nicking–closing cycle. Type I enzymes are active as monomers. Resolution of the crystal structure of eukaryotic topo I homologs, including the human and viral enzymes, led to the elucidation of the catalytic mechanism at a molecular level (*394–399*). Since topo I is a major anticancer drug target, this opened the possibility of rational drug design. The anticancer activity of these drugs results from the trapping of topo I cleavage complexes such that the enzyme is reversibly inactivated as it cleaves DNA (*397, 400*).

The 91-kDa human topo I consists of four major domains. The highly charged N-terminal region, which interacts with other proteins, has a molecular mass of about 24 kDa. The sizes of the conserved core and the C-terminal domain are about 56 and 6 kDa, respectively, and these are connected by a 7-kDa, positively charged linker domain (*394–397*). The conserved C-terminus of topo I contains the active site (Tyr723 of the human enzyme). The catalytic intermediate of eukaryotic topo I, which belongs to the type IB subfamily, is a cleavage complex in which the C-terminal tyrosine attacks a DNA phosphodiester bond and forms a covalent bond with the 3′-phosphorus of the cleavage site while a free 5′-hydroxyl is generated (*391–393*). The covalent bond between Tyr723 and the 3′-phosphorus requires a 5′-hydroxyl group belonging either to the same DNA (relaxation of DNA supercoiling and relief of torsional strain) or to another DNA molecule (recombination). The conserved His632 is required for enzyme activity and probably stabilizes the transition state by carrying hydrogen bonds to a nonbridging oxygen (*399*).

Topo I is an abundant enzyme which is involved in several nuclear processes, such as transcription, DNA recombination, DNA repair, and DNA replication (*401–406*). It can relax DNA supercoiling and relieve torsional strain during DNA processing. It can also perform intermolecular religation leading to DNA recombination (*391–393*). The role of topo I in initiation of DNA replication is underlined by the observation that the viral replication protein SV40 T antigen physically interacts with topo I (*407, 408*). In human cells one topo I gene and two pseudogenes have been found as well as two topo III genes, which code for members of the IA type (*391, 393*; data not shown). In budding yeast there is one gene coding for topo I belonging to the IB subfamily and another one coding for topo III of the IA subfamily (*391*). In budding yeast none of the topo I genes is essential, since yeast cells with a deleted topo IA and IB gene are viable, whereas in fission yeast deletion of the type IB gene is lethal due to defects in cell division (*391, 393*). In contrast to the situation in budding yeast, topo I is also an essential enzyme in higher eukaryotes (*393, 401, 409*).

III. The Organization of Replication-Initiation Factors on Chromatin

Initiation of chromosomal DNA replication in both eukaryotic and prokaryotic cells is achieved by the stepwise assembly of protein complexes at origins of DNA replication (*12, 116, 410–413*). In bacteria regulation and mechanism of the assembly are well known, but these events are only rudimentarily understood in eukaryotic cells. This situation has been in part due to the long lack of a biochemically defined origin-dependent system of eukaryotic chromosomal DNA replication. Fortunately, viral systems have helped substantially to define and purify various replication factors and made us aware of the importance that protein–protein interactions have for the duplication of DNA. Since viruses usually depend on multifunctional initiator proteins, it does not come as a surprise that only a few cellular components of the initiation complex have been identified in these systems. Only recently has the combination of genetically tractable organisms such as yeast and *Drosophila* with cell-free biochemical systems as well as cell-biological methods led to the identification of the major players in the initiation process. Integrating all these data has enabled scientists to develop a general model for the early steps of DNA replication. In this review we will mainly focus on the conserved features. However, one should bear in mind that there might be considerable differences when looking at different developmental stages of an organism or when comparing model systems based on different evolutionary stages.

The establishment of functional replication forks requires multiple protein–DNA and protein–protein interactions, which are converted from an inactive prereplicative to an active initiation state in a carefully regulated manner. The activity of initiator proteins is tightly controlled at various levels and by different mechanisms. Some of these proteins are transcriptionally induced, others are tagged with ubiquitin for their timely destruction, yet others are modified by phosphorylation or actively transported in and out of the nucleus. Specialized surveillance systems are also in operation. Checkpoint controls permanently monitor DNA for damage or aberrant structures and delay or stop cell cycle progression. To deal with the enormous size of their genomes eukaryotic cells initiate chromosome replication at multiple origins of replication. Such a battery of start sites allows a multiparallel and rapid duplication of the chromosomal DNA. The activation of an origin is determined by its location and the chromatin structure it is embedded in. Origins in transcriptionally active euchromatin are, for example, activated ealier in S phase than origins surrounded by hetreochromatic regions. To achieve stability of the genomic information each chromosomal region must not replicate more than once per cell cycle. The term "licensing of replication origins" was coined to describe the observation made in *Xenopus* egg extracts that chromosomes become competent for one round of DNA replication in G1 phase, but subsequently lose this competence for initiating or "licensing" DNA replication once again within the same cell cycle.

A. The Assembly of the Prereplicative Complex

Origins can only be activated if they are bound by a protein complex called ORC (Fig. 6, see color insert). ORC remains largely associated with origins throughout the cell cycle, so it certainly is not the direct activator for origin "firing." It rather serves as a landing platform for additional proteins that themselves represent docking sites for still others. The next players entering the scene are Cdt1p, which largely resembles an activity referred to as RLF-B (replication licensing factor B), and Cdc6p. Both proteins interact with each other and specifically with the ORC–DNA complex. Once tethered to chromatin these proteins

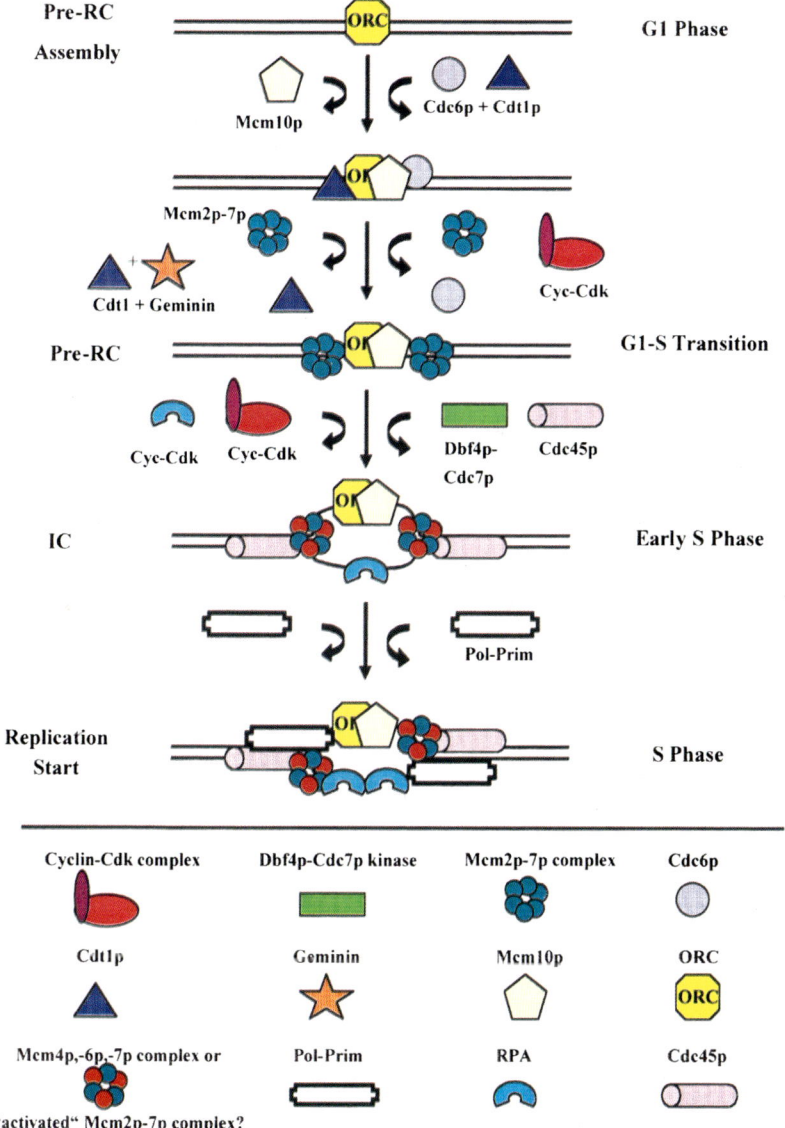

FIG. 2.6. The initiation of eukaryotic DNA replication. The formation of the prereplicative complex (pre-RC) in G1 phase, its conversion in the initiation complex (IC) at the G1—S phase transition, and the beginning of DNA replication are summarized. The following abbreviations are used: ORC, origin recognition complex; Cdt, Cdc10 target 1; Mcm, minichromosome maintenance; Cdc, cell division cycle; Dbf4, regulatory subunit of Cdc7p; Cdk, cyclin-dependent kinase; RPA, replication protein A (the eukaryotic ssDNA-binding protein); Pol—Prim, DNA polymerase α-primase.

INITIATION OF EUKARYOTIC DNA REPLICATION

recruit the heterohexameric Mcm2p–7p complex, also known as RLF-M (replication licensing factor M). The so-formed assembly is called the prereplicative complex (pre-RC). It is essential for the initiation of DNA replication and plays a central role in limiting initiation to once per cell cycle. Only when the functions of all three, ORC, Cdc6p, and Mcm2p–7p, are severely impaired does rereplication occur, as was powerfully demonstrated in a budding yeast strain engineered to express a Cdk phosphorylation-negative ScORC mutant, an N-terminal deletion mutant of ScCdc6p, and a ScMcm2p–7p mutant which permanently remains in the nucleus (146). In addition, the protein Mcm10p helps to retain the Mcm2p–7p complex in the pre-RC bound to the origin.

B. The Activation of Origins: From the Prereplicative Complex to the Initiation Complex

The sequential progress of the prereplication to the inititation state requires the establishment of sophisticated molecular switches. The cooperation of kinase activities and specific proteolysis events regulates these switches from "licensing" to "firing" of eukaryotic origins and their subsequent resetting in a cell cycle-dependent manner. These cellular mechanisms ensure that reloading of the Mcm2p–7p complex does not occur after the initiation of DNA replication. The pre-RC assembles in late mitosis to early G1 phase and is converted into the initiation complex (IC) at the G1–S transition. In all eukaryotes studied so far, two classes of kinases, the Cdk and DDK complexes, fulfill essential control functions in this activation. Cdc6p recruits Cdks to the chromatin, whereas DDK interacts with the chromatin via the Mcm2p–7p complex. Once attached to chromatin the main and possibly only function of DDK is to phosphorylate Mcm2p. This modification results in a conformational change of the Mcm2p–7p complex and the local melting of the origin. Cdk target(s) have not been well characterized, but it is clear that their activity, like that of DDK, is crucial for the loading of Cdc45p onto chromatin (Fig. 6). For DNA replication to occur the Mcm2p–7p complex must be released from origin sequences and its interactions with Mcm10p have to be dissolved. Cdc45p might assist in this process, since both ScMcm10p and ScCdc45p interact with the same sites on ScMcm5p and ScMcm7p (193, 196). As a consequence of this release a helicase activity of the Mcm complex becomes apparent, but the mechanism through which this occurs, either by activation of the Mcm2p–7p complex or by the formation of an Mcm4p, Mcm6p, Mcm7p subcomplex, is unknown. This helicase activity could be involved in a more extensive melting of dsDNA at the origin (203, 210). As a consequence of IC formation, Cdt1p and Cdc6p are removed from the chromatin. In yeast, this is accomplished by proteolytic degradation. Inactivation in higher eukaryotes operates on different levels. Cdc6p is actively exported into the cytosol, whereas Cdt1 is associated with the inhibitor geminin at the time of fork progression.

C. The Initiation of DNA Replication on the Leading and Lagging Strands: DNA Polymerase α–Primase Holds the Key

RPA will then enter the complex, possibly through attraction by the ssDNA regions which have been created. As the fork is established and progresses the action of topoisomerase I (topo I) is required to release the torsional stress that will be built ahead of the fork. The initiation complex is likely to contain Mcm2p–7p, Cdc45p, RPA, and possibly topo I (summarized in Fig. 6). The activation of the Mcm2p–7p complex, the unwindung of origin dsDNA, and the recruitment of RPA, topo I, and Cdc45p are all essential steps culminating in the activation of Pol–Prim to synthesize the first RNA–DNA at an origin on the leading strand (Fig. 6; for more details see above). The RNA primer is about 10 nt long and elongated by the DNA polymerase activity to a strech of about 40 nt. After initiating the leading strand, the first primer is also synthesized and elongated on the lagging strand. Since the lagging strand will be replicated discontinuously, such primer synthesis and elongation steps are performed repeatedly. The synthesis products are known as Okazaki fragments.

D. The Elongation Reaction of Eukaryotic DNA Replication

Once RNA primer synthesis and elongation are completed, a polymerase switch occurs at the end of the growing strand and the replication factor C (RFC) enters the picture. This protein complex is evolutionarily conserved from yeast to humans and consists of five subunits, Rfc1p to Rfc5p, with molecular masses of 140, 40, 38, 37, and 36 kDa, respectively (138). The main functions of RFC are to remove DNA polymerase α–primase from the newly synthesized RNA–DNA primers and to load proliferating cell nuclear antigen (PCNA) in an ATP-dependent reaction onto the 3′-end of a primer hybridized to an ssDNA template (384–386). It therefore acts as the clamp loader. PCNA is a 36-kDa protein that assembles into homotrimers. These trimers form rings around the DNA and interact with the DNA polymerases δ and ε (pol δ and pol ε) at the 3′-junction of a template primer as well as with various other proteins (reviewed in Ref. 377). They increase the processivity of these polymerases by preventing their premature dissociation from the DNA. Because its presence is restricted to rapidly proliferating cells, it is used in diagnostics as a tumor marker. PCNA and RFC have been referred to according to their function as the polymerase clamp and the clamp loader, respectively. Polymerase switches repeatedly occur in an identical manner after completion of each Okazaki fragment. It is probably pol δ that is involved in filling in the gap between the Okazaki fragments (138). In the next step the Okazaki fragments

INITIATION OF EUKARYOTIC DNA REPLICATION   73

have to mature, RNAses are activated to remove the RNA and the resulting gap is filled by a DNA polymerase, and finally the Okazaki fragments are ligated by a ligase activity.

Pol δ is probably a heterohexamer with subunits with molecular masses of 125, 66, 48, and 12 kDa, respectively (*314, 414, 415*). Pol δ possesses a 3'–5' exonuclease activity with a proofreading function (*314, 416*). Mutation of the exonuclease activity in budding yeast increased the cellular mutation rate by a factor of about 10 and such an exonuclease-negative mutation of pol δ results in a higher frequency of tumors in mice (*416, 417*). In cellular DNA synthesis this polymerase is involved in leading and lagging strand synthesis (*314, 377*). Pol ε probably consists of four subunits (reviewed in Ref. *377*). The large, 261-kDa subunit carries the catalytic activity and a 3'–5' exonuclease, also with a proofreading function (*314, 418*), and is involved in checkpoint control (see below). In budding yeast, mutations which abolished the enzyme activity of both exonucleases proved to be lethal, suggesting that both exonucleases cooperate to maintain the integrity of genomic information (*314, 419*). The occurrence of replication errors or the arrest of the replication fork at lesions introduces "replication stress," which has to be resolved before the cell cycle can progress.

IV. Cell Cycle Control by Checkpoints

Surveillance systems can act to interrupt cell cycle progression after damage of the genome or replication stress is sensed (*420*). The existence of such checkpoints is defined empirically as loss-of-function mutations that relieve the dependence of a certain event on the completion of a preceding event (*421*). Failure to induce a transient cell cycle arrest leads to accumulation of mutations, which ultimately induce cell death or cancer (*422*). Albeit defined previously as nonessential regulatory pathways to arrest the cell cycle, it has become clear that these checkpoints are also involved in the control of repair mechanisms, transcription, telomere length, and apoptosis (*423–427*). Therefore, several checkpoint genes are essential for cell survival, demonstrating that checkpoint pathways are also important factors for physiological processes (*428, 429*).

DNA-damaging agents fall into two main types, endogenous and exogenous. The major endogenous sources are water, oxygen, and DNA replication errors. Hydrolysis leads to deamination of DNA bases (converting one base into another), depurination, and depyrimidation. Reactive oxygen species such as superoxide radicals or covalent attachment of chemical groups also modify bases or the phosphodiester backbone (*430*). Misincorporation of base pairs will lead to mismatches that, when not repaired, will be propagated as transitions and transversions. Exogeneous agents include ultraviolet (UV) light and ionizing irradiation (γ and X rays). UV predominantly induces pyrimidine dimerization, but can also damage individual bases. Ionizing radiation produces potentially lethal double- and single-strand breaks in DNA or gives rise to active oxygen species (431).

A variety of DNA lesions can damage the cellular genetic information and induce a complex set of repair mechanisms to rectify them. The activation of such repair processes is the ultimate goal of the checkpoint systems. Once the integrity of the DNA is again established signals are sent to continue cell cycle progression. The major repair systems comprise mismatch repair (MMR), nucleotide excision repair (NER), base excision repair (BER), and double-strand break (DSB) repair. Misincorporation of nucleotides is corrected by the mismatch repair system, which discriminates between the parental and daughter strands (432). The BER system allows for the correction of modified bases and alterations in the phosphodiester backbone caused by oxidative damage (433), whereas NER predominantly acts on bulky, helix-distorting lesions such as those induced by UV light and carcinogens (434). The DSB repair pathways work to repair breaks in ssDNA and dsDNA, the latter being repaired by nonhomologous end joining (NHEJ) or by homologous recombination (430, 435). DNA is constantly monitored for the presence of lesions. Damage to DNA (genotoxic stress) is sensed by the damage checkpoints that induce the cell cycle to pause during progression from G1 to S phase (at the G1 checkpoint), progression through S phase (the intrinsic S phase checkpoint); and progression from G2 to mitosis (the G2 checkpoint). Stalled replication forks (replication stress) activate the replication checkpoint, which delays DNA replication. Although every organism has evolved some unique features, checkpoints are highly conserved among species and in general utilize the same set of homologous proteins. Checkpoint systems can be regarded as signal transduction pathways or more accurately as networks of interacting pathways (see Fig. 7) consisting of sensors, transducers, and effectors (436).

A. The Sensors: DNA Replication Machinery-like Complexes and the BRCT Family

Sensors have not been identified conclusively, but are likely to bind aberrant DNA structures. Candidates, which show similarities to the DNA replication machinery, fall into two groups, termed the checkpoint loading complex (CLC) and the checkpoint sliding complex (CSC; Table I) (33). For example, fission yeast and human Rad1p, Rad9p, and Hus1p form a heterotrimeric CSC complex which is related in structure to homotrimeric PCNA (437). Both Rad9p and Hus1p are phosphorylated after DNA damage. This phosphorylation depends on the SpRad3p kinase or its human homolog ATR. A similar trimeric complex consisting of ScRad17p, ScDdc1p, and ScMec3p is found in budding

INITIATION OF EUKARYOTIC DNA REPLICATION

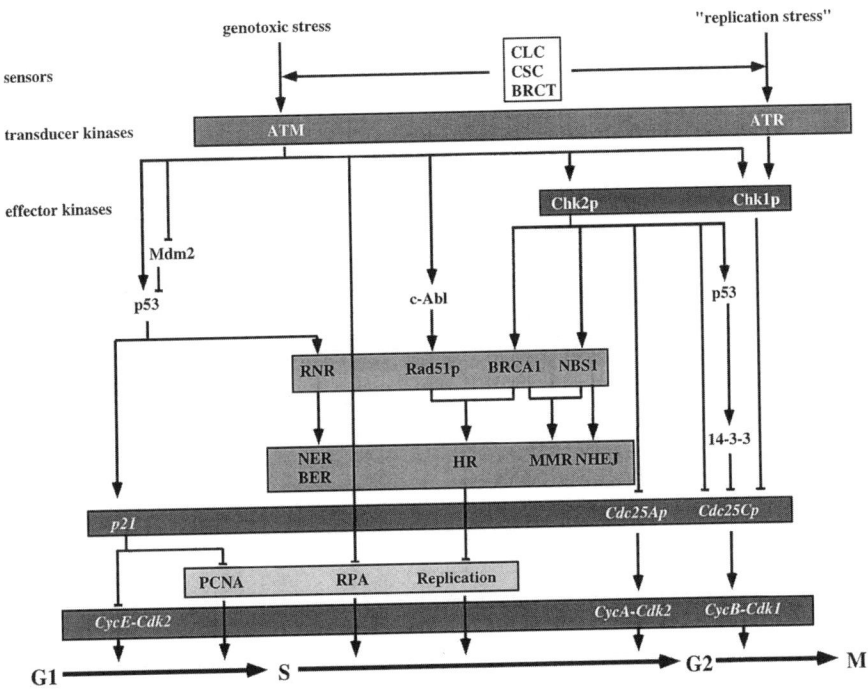

FIG. 7. Checkpoint control. Central functions of ATM and ATR in cell cycle checkpoints after DNA damage. Both kinases are activated in dependence on checkpoint loading complexes (CLC), checkpoint sliding clamps (CSC), and BRCT family members (BRCT, proteins with homologies to the C-terminal region of BRCA1). Similar pathways exist in budding and fission yeast. In G1 phase, activated ATM directly phosphorylates p53 and two proteins, Chk2 and Mdm2, which control the activity of p53. All phosphorylation events lead to an inhibition of the binding of Mdm2 to p53 and result in an increase in the level of p53 protein. The p53 protein induces the CKI p21, which inhibits cyclin E–Cdk2, and causes arrest in the G1 phase before the restriction point. In S phase, activated ATM also phosphorylates NBS1, c-Abl, BRCA1, and Chk2. NBS1 exists in a complex with HsMre11, HsRad50, and BRCA1, which is involved in nonhomologous end joining (NHEJ) during double-strand break repair (DSB). c-Abl and Chk2 activate the Rad51 and BRCA1 proteins, respectively, that are involved in homologous recombination (HR). BRCA1 together with Rad51 and NBS1 acts also in DNA mismatch repair (MMR). p53 induces the activation of both Gadd45 and ribonucleotide reductase (RNR), which both stimulate nucleotide (NER) and base (BER) excision repair. Repair processes actively slow down the replication machinery by competing for protein factors shared in both processes. A second level of control is executed by the direct inhibition of replication proteins such as PCNA by CKI p21 or RPA by ATM. Chk2 also phosphorylates and thereby inactivates the Cdc25A phosphatase, which is needed for the activation of the cyclin A–Cdk2 complex. As a consequence, progression of S phase is stalled. In G2 phase, Chk2 and Chk1 are activated by ATM and ATR, respectively, and they directly inhibit the Cdc25C phosphatase. The latter is also inhibited by binding to 14-3-3 proteins that are induced by the action of p53. Inactivation of the phosphatase prevents activation of the cyclin B–Cdk1 complex and hence to a block in G2.

TABLE I
DNA DAMAGE CHECKPOINT PROTEINS[a]

Class	Fission yeast	Budding yeast	Human	Comments
Checkpoint loading Complex (CLC), RFC-like	Rad17p Rfc3p, other subunits?	Rad24p Rfc2p–5p	Rad17p Rfc2p–5p	— —
Checkpoint sliding Clamp (CSC), PCNA-like	Rad1p Rad9p Hus1p	Rad17p Ddc1p Mec3p	Rad1p Rad9p Hus1p	$5' \to 3'$ Exo P P
BRCT proteins	Rhp9p/Crb2p Rad4p/Cut5p	Rad9p Dbp11p	BRCA1	P
PI3K-like proteins	Rad3p Tel1p	Mec1p Tel1p	ATR ATM	Kinase Kinase
Coiled-coil proteins	Rad26p	Ddc2p/Lcd1	—	P by Rad3
Effector kinases	Chk1p Cds1p	Chk1p Rad53p	Chk1p Chk2p/Cds1p	Kinase, P Kinase, P

[a]Abbreviations: RFC, replication factor C; PCNA, proliferating cell nuclear antigen; BRCT, C-terminal region of BRCA1; PI3K, phosphotidylinositol-3-kinase; Exo, exonuclease; P, damage-induced phosphorylation; Rad, radiation mutant; Chk, checkpoint kinase; Hus1p, hydroxy urea-sensitive protein 1.

yeast and is under the regulation of the ScMec1p kinase, the ATM and SpRad3p homolog. Analogously to the five-subunit replication factor C (RFC), budding yeast ScRad24p and its human homolog HsRad17p assemble a complex with the four small subunits (Rfc2p to Rfc5p) of RFC called the CLC complex (438). In fission yeast SpRad17p was also shown to interact with SpRfc3p (439). In analogy to its replication counterparts the SpRad17p-containing CLC complex might bind to the damaged site and assist loading of the Rad1p–Rad9p–Hus1p/CSC complex. Since mutations in RFCs have been shown to be defective in the DNA replication checkpoint, the CLC might help to maintain the replication checkpoint during S-phase (440, 441). The $3' \to 5'$ exonuclease activity associated with SpRad1p might then be involved in the processing of the damaged site (442).

The BRCT proteins contain a domain, termed BRCT, with homologies to the C-terminal region of BRCA1, the human breast cancer susceptibility gene (Table I). This domain promotes protein–protein interactions. Fission yeast has two checkpoint proteins belonging to this class, SpRad4p and SpRhp9p, which are also known as SpCut5p and SpCrb2p, respectively. SpRad4p is required for both DNA replication and checkpoint control (443, 444). The second protein, SpRhp9p, is phosphorylated in a SpRad3p-, CLC-, and CSC-dependent manner and is the fission yeast equivalent to the budding yeast ScRad9p (445). Phosphorylated ScRad9p multimerizes and interacts with the downstream effector

INITIATION OF EUKARYOTIC DNA REPLICATION   77

kinase ScRad53p, the SpChk2p homolog (446, 447). To overcome the cell cycle arrest induced by DNA damage after repair, SpRhp9p has to be phosphorylated by Cdk1 (448). In vertebrate cells, the BRCA1 protein, which most likely resembles SpRhp9p, is phosphorylated as a result of DNA damage and is needed for the G2 arrest (449).

B. The Transducing Phosphoinositol-3-Kinase-like Complexes and the Coiled-Coil Proteins

SpRad3p is a protein kinase and belongs to the phosphoinositol-3-kinase (PI3K) familiy (Table I). It exists in a complex with the coiled-coil protein SpRad26p, which it phosphorylates after the cell has been exposed to ionizing radiation independently of the CLC and CSC complexes (445). The SpRad3p–SpRad26p complex is likely to play a role in damage recognition, including that of double-strand breaks. In contrast, SpRad26p phosphorylation in response to replication blocks depends on the RFC-like, SpRad1p-containing and the PCNA-like, Rad17-containing complexes (445). A similar phosphorylation dependence was observed with the budding yeast homolog ScDdc2p, also called ScLdc1p (450). Regardless of the pathway for transmitting the signal after SpRad3p-mediated activation of the SpChk1p and SpChk2p (SpCds1p) kinases, the functions of both the CLC and CSC complexes are still required (451, 452).

In vertebrates the role of SpRad3p is divided between the two SpRad3p homologs ATM (ataxia telangiectasia-mutated) and ATR (ataxia telangiectasia and Rad-3-related), which possess some overlapping, but nonredundant functions (423). Ionizing radiation causes double-strand breaks (DSBs) and induces the ATM kinase, the key regulator of all three damage checkpoints (Fig. 7). Mutations in this protein cause the autosomal-recessive inherited disorder ataxia telangiectasia, which is characterized by a complex clinical phenotype (435). ATR is an essential gene and participates in replication checkpoint responses that are induced by UV irradiation and hydroxyurea treatment, the latter inhibits DNA replication by depleting the cellular nucleotide pool (453, 454). ATR most likely recognizes primed DNA replication intermediates or alternatively proteins that are enriched at single-stranded regions when the replication fork is stalled (32, 228, 455). Since SpRad3p can be directly activated, it might itself sense structural aberrations in DNA without the help of the CLC and CSC (33, 445). Like the clamp loader RFC, the CLC complex might load CSC, which in turn targets ATR to the distorted site. ATR phosphorylates and thereby activates the SpChk1p kinase, which enters the pathway by inactivating the HsCdc25Cp phosphatase. For the G2 checkpoint ATM is essential only when the cells are damaged in G2. When cells are ionized in G1 or S phase and then escape to G2, the ATM-related protein ATR is responsible for the G2 delay.

C. The Checkpoint Effector Kinases Chk1 and Chk2

Two effector kinases, SpChk1p and SpChk2p, also called SpCds1p, act downstream of the SpRad3p kinase in fission yeast (456, 457). SpChk1p is phosphorylated by SpRad3p predominantly in late S and G2 phase, and thus is dependent on the CLC and on the CSC (Table I) (451). The phosphorylated form of SpChk1p binds to 14-3-3 proteins, which are thought to facilitate protein–protein interactions (27). SpChk2p acts predominantly at the replication checkpoint. Its activation is dependent on SpRad3p, SpRad26p, CLC, and CSC. Once activated it leads to a slowing down of replication (456, 458).

In budding yeast, ScRad53p, the SpChk2p homolog, also plays a central role in the replication checkpoint. It is phosphorylated and activated by DNA alterations that are sensed by the CLC and the CSC complexes as well as by DNA polymerase ε (459–461). Activated ScRad53p kinase then modulates the activity of its target proteins. The protein inhibits the ScSwi4p/ScSwi6p transcription complex, which is required for expression of G1 cyclins, stalling the cells in G1 phase (462). ScRad53p also negatively acts on the cellular ability to initiate replication and delays progression of S phase. The large subunit of RPA and the ScRfc5 subunit, which are involved in both DNA replication and damage repair, are S phase targets of ScRad53p kinase (273, 330, 441, 463). One of the prime outcomes of ScRad53p action is to prevent stalled replication forks from collapsing (464, 465). A similar situation is found in human cells. Here Chk2 is activated predominantly by ATM upon DNA damage, whereas ATR phosphorylates Chk1 when replication stress is sensed (28, 466). In addition, ScRad53p stimulates ScPds1p, which in turn inhibits progression through G2 phase. Thus, the ScRad53p kinase is a general inhibitor of cell cycle progression in G1, S, and G2 phases by modifying cell cycle-specific target proteins.

D. A Network of Proteins Is Involved in Checkpoint Control

In *S. pombe* and higher eukaryotes checkpoints make use of the modification of Cdks to control progression of the cell cycle (see above). Cdk activity is determined by the phosphorylation state of the residues T14 and Y15, which is positively regulated by the Cdc25p phosphatase and negatively by the activation of the Wee1p and Mik1p kinases as well as by the Chk-dependent stabilization (32, 467, 468). In its active form, both sites have to be dephosphorylated. In budding yeast ScCdk1 is not regulated by phosphorylation and another mechanism of ScChk1p action has evolved. ScChk1p prevents anaphase by stabilizing the sister-chromatid-cohesion protein ScPds1p. In addition ScRad53p, the SpChk2p homolog, downregulates the ScCdc5p polo-like kinases to prevent cyclin destruction and hence mitotic exit (469).

1. P53

Additional factors participate in this protein network. ATM is responsible for the cell cycle arrest in G1 by inducing increased cellular levels of p53 in human cells (*435*). This is brought about by phosphorylation both of p53 by HsChk2p and of a protein called Mdm2p, an important regulator of p53 degradation, by ATM. Both phosphorylation events disrupt the Mdm2p–p53 complex, thus preventing Mdm2p from shuttling p53 out of the nucleus into the cytoplasm where it is degraded (see Fig. 7) (*470*). ATM can itself phosphorylate p53, a modification that modulates the transactivation of target genes by the transcription factor p53 (*471*). The stabilized and activated p53 turns on the production of a key inhibitor of cell cycle progression, the p21 protein. p21 acts directly on the cell cycle machinery by binding to and inhibiting cyclin E–Cdk2 activity and hence progression from G1 to S phase (*23, 472*). The p53-dependent G1 arrest occurs before the restriction point, the moment when the cell becomes committed to progress through the cell cycle (*473*). Recent genetic data in mouse show that the tumor suppressors p53 and BRCA1 cooperate in apoptosis, cell cycle control, and tumorigenesis (*474*).

Cells damaged in G1 and entering S phase or G2 are still under control of the p53-dependent checkpoint. p21 not only inhibits Cdks, but also complexes PCNA, which acts as a clamp for the leading strand pol δ and pol ε (*475*). Thus, in S phase the DNA replication machinery itself is directly inhibited. In addition, p53 induces the transcription of a special subunit of ribonucleotide reductase (RNR) that is involved in replenishing the nucleotide pool (*476*). Since RNR, known to be important for damage-response pathways, with this particular subunit is found exclusively in the nucleus, repair processes, especially NER and BER, are stimulated, which in turn compete with the replication machinery at sites of damage (*477*). p53 also leads to transcriptional activation of the 14-3-3σ protein (*478, 479*). The 14-3-3σ protein induces nuclear exclusion of cyclin B–Cdk1, thus leading to a block in the cell cycle at G2 (*27*). An alternative scenario with the same outcome has evolved in budding yeast. Here ScRad53p, after activation by ScMec1p, phophorylates and thereby stimulates the ScDun1p kinase. ScDun1p in turn phosphorylates and inhibits the ability of ScCrt1p to repress various RNR genes (*480*).

2. NBS1

ATM-deficient cells fail to arrest DNA synthesis after exposure to ionizing radiation, a phenomenon called radioresistant DNA synthesis (RDS) (*481*). The prime target in this S phase checkpoint by ATM is the NBS1 protein, which is mutated in a disorder known as Nijmegen breakage syndrome (NBS) (*482*). NBS1 forms a complex with HsRad50p and HsMre11p that is recruited to double-strand breaks (DSBs) and helps to repair them (*482, 483*). The NBS1/HsRad50p/HsMre11p complex can be part of an even larger complex called BASC

(BRCA1-associated genome surveillance complex), which also contains mismatch repair proteins HsMhs2p/HsMhs6p and HsMlh2p as well as Bloom's helicase (BLM) (*484, 485*). This complex is believed to sense altered DNA structures, leading to the activation of ATM. ATM also activates the HsChk2p kinase, which acts on both the NBS1 and the BRCA1 proteins (*423*). BRCA1 is also part of a complex that contains HsRad51p and HsRad52p, which are involved in homologous recombination (*482*). HsRad51p is activated by the c-Abl kinase, which itself is a target of ATM activation, and binds to Rad52p (*27, 486*). The NBS1/HsRad50p/HsMre11p complex communicates with DNA-PK, the key player in the NHEJ process (*482*). By the activation of repair processes ATM ultimately leads to a delay in the replication process (*487*).

DNA replication initiates from large numbers of replication origins throughout S phase according to a temporal program. Early in S phase the activation of late-firing replication origins is blocked by an active process which requires an intact DNA damage checkpoint control. A functional checkpoint is also required for DNA replication forks to replicate DNA damaged with alkylating agents such as methyl methanesulfonate (*465*). Since various drugs involved in cancer therapies cause DNA damage and since it is widely believed that most tumor cells have defects in DNA damage response, the understanding of the mechanisms of the checkpoints which control late-origin firing and replication fork stability might have important implications for cancer treatment.

V. Outlook

In recent years the mechanisms governing processes involved in the maintenance of the genetic integrity of the organism have been identified at a molecular level. These detailed studies have allowed the development of working models and have yielded insights into both the mechanism and regulation of DNA replication. In the near future several discrepancies between the various model systems will be resolved and put into a general outline. Further evidence is likely to emerge that various organisms have evolved similar, but not identical strategies to ensure a single round of DNA replication per cell cycle. In addition, the understanding of the regulated amplification of chromosomes in specific cells will enable us to understand similar processes that occur unregulated in certain tumor cells. The study of the activities of the Mcm2 to Mcm7 proteins will be an important field, with several open questions awaiting solution. One of the mysteries still concerns the replicative helicase.

In addition to ORC, Mcm proteins, and Cdc6p, cyclin–Cdks regulate various cellular proteins which control rereplication. Application of this knowledge in medicine will benefit humans, since these factors are potential targets for combatting the deregulated growth of cancer cells. Furthermore, studying these proteins in tumor cells will give rise to new tools for classifying and monitoring tumors and carcinomas. The understanding of the mechanisms that regulate initiation of DNA replication will allow the development of new drugs interfering with uncontrolled growth of cells. This is a current focus of biomedical science.

ACKNOWLEDGMENTS

We thank S. Bell, J. Diffley, S. Kearsey, T. Krude, S. Mimura, A. Schwacha, S. Tanaka, H. Takisawa, and J. Walter for discussions and for providing information ahead of publication, as well as the other attendants at the Cold Spring Harbor Meeting 2001 on Eukaryotic DNA Replication for inspiration with new ideas. We especially thank T. Kelly and B. Stillman for organizing this exciting meeting. This work was supported by the European Community (CT97-0125) and the Deutsche Forschungsgemeinschaft. The Institut für Molekulare Biotechnologie is a member of the Gottfried-Wilhelm-Leibniz Gemeinschaft and is funded by the Land Thüringen and the German federal government.

REFERENCES

1. P. N. Rao and R. T. Johnson, *Nature* **225**, 159–164 (1970).
2. J. J. Blow and R. A. Laskey, *Nature* **332**, 546–548 (1988).
3. J. J. Blow, *EMBO J.* **20**, 3293–3297 (2001).
4. L. H. Hartwell, *J. Mol. Biol.* **59**, 183–194 (1971).
5. K. Nasmyth and P. Nurse, *Mol. Gen. Genet.* **182**, 119–124 (1981).
6. J. F. Diffley, J. H. Cocker, S. J. Dowell, and A. Rowley, *Cell* **78**, 303–316 (1994).
7. J. F. Diffley, *Curr. Biol.* **11**, R367–R370 (2001).
8. K. Labib and J. F. Diffley, *Curr. Opin. Genet. Dev.* **11**, 64–70 (2001).
9. M. Lei and B. K. Tye, *J. Cell Sci.* **114**, 1447–1454 (2001).
10. K. Nasmyth, *Science* **274**, 1643–1645 (2001).
11. A. B. Pardee, *Proc. Natl. Acad. Sci. USA* **71**, 1286–1290 (1974).
12. B. Stillman, *Science* **274**, 1659–1664 (1996).
13. J. J. Li, *Curr. Biol.* **5**, 472–475 (1995).
14. D. O. Morgan, *Annu. Rev. Cell Dev. Biol.* **13**, 261–291 (1997).
15. J. Pines, *Nat. Cell Biol.* **1**, E73–E79 (1999).
16. M. Glotzer, *Curr. Biol.* **5**, 970–972 (1995).
17. S. Dehde, G. Rohaldy, O. Schub, H. P. Nasheuer, W. Bohn, J. Chemnitz, W. Deppert, and I. Dornreiter, *Mol. Cell. Biol.* **21**, 2581–2593 (2001).
18. W. Krek, M. E. Ewen, S. Shirodkar, Z. Arany, W. G. Kaelin, Jr., and D. M. Livingston, *Cell* **78**, 161–172 (1994).
19. O. Schub, G. Rohaly, R. W. P. Smith, A. Schneider, S. Dehde, I. Dornreiter, and H. P. Nasheuer, *J. Biol. Chem.* **276**, 38076–38083 (2001).
20. F. Sprenger, N. Yakubovich, and P. H. O'Farrell, *Curr. Biol.* **7**, 488–499 (1997).
21. K. Nasmyth, in "DNA Replication in Eukaryotic Cells" (M. L. DePamphilis, ed.), pp. 331–386. Cold Spring Harbor Laboratory Press, Cold Spring Harbor Laboratory, New York, 1996.
22. D. L. Fisher and P. Nurse, *EMBO J.* **15**, 850–860 (1996).

23. C. J. Sherr and J. M. Roberts, *Genes Dev.* **13**, 1501–1512 (1999).
24. L. Furstenthal, B. K. Kaiser, C. Swanson, and P. K. Jackson, *J. Cell Biol.* **152**, 1267–1278 (2001).
25. R. W. P. Smith and H. P. Nasheuer, in "Recent Research Developments in Virology," Vol. 2 (S. G. Pandalai, ed.), pp. 67–92. Transworld Research Network, Trivandrum, India, 2000.
26. L. Furstenthal, C. Swanson, B. K. Kaiser, A. G. Eldridge, and P. K. Jackson, *Nat. Cell Biol.* **3**, 715–722 (2001).
27. L. Chen, T. H. Liu, and N. C. Walworth, *Genes Dev.* **13**, 675–685 (1999).
28. S. Matsuoka, M. Huang, and S. J. Elledge, *Science* **282**, 1893–1897 (1998).
29. A. Aitken, D. Jones, Y. Soneji, and S. Howell, *Biochem. Soc. Trans.* **23**, 605–611 (1995).
30. N. Mailand, J. Falck, C. Lukas, R. G. Syljuasen, M. Welcker, J. Bartek, and J. Lukas, *Science* **288**, 1425–1429 (2000).
31. C. Y. Peng, P. R. Graves, R. S. Thoma, Z. Wu, A. S. Shaw, and H. Piwnica-Worms, *Science* **277**, 1501–1505 (1997).
32. W. M. Michael and J. Newport, *Science* **282**, 1886–1889 (1998).
33. M. J. O'Connell, N. C. Walworth, and A. M. Carr, *Trends Cell Biol.* **10**, 296–303 (2000).
34. L. H. Johnston, H. Masai, and A. Sugino, *Trends Cell Biol.* **9**, 249–527 (1999).
35. L. H. Johnston, H. Masai, and A. Sugino, *Prog. Cell Cycle Res.* **4**, 61–69 (2000).
36. T. Faul, C. Staib, I. Nanda, M. Schmid, and F. Grummt, *Chromosoma* **108**, 26–31 (1999).
37. M. Lepke, V. Putter, C. Staib, M. Kneissl, C. Berger, K. Hoehn, I. Nanda, M. Schmid, and F. Grummt, *Mol. Gen. Genet.* **262**, 220–229 (1999).
38. H. Masai, E. Matsui, Z. You, Y. Ishimi, K. Tamai, and K. Arai, *J. Biol. Chem.* **275**, 29042–29052 (2000).
39. P. Jares and J. J. Blow, *Genes Dev.* **14**, 1528–1540 (2000).
40. K. Bousset and J. F. Diffley, *Genes Dev.* **12**, 480–490 (1998).
41. P. Pasero, B. P. Duncker, E. Schwob, and S. M. Gasser, *Genes Dev.* **13**, 2159–2176 (1999).
42. M. Lei, Y. Kawasaki, M. R. Young, M. Kihara, A. Sugino, and B. K. Tye, *Genes Dev.* **11**, 3365–3374 (1997).
43. T. A. Millward, S. Zolnierowicz, and B. A. Hemmings, *Trends Biochem. Sci.* **24**, 186–191 (1999).
44. S. Wera and B. A. Hemmings, *Biochem. J.* **311**, 17–29 (1995).
45. E. Kremmer, K. Ohst, J. Kiefer, N. Brewis, and G. Walter, *Mol. Cell. Biol.* **17**, 1692–1701 (1997).
46. F. C. Lin and K. T. Arndt, *EMBO J.* **14**, 2745–2759 (1995).
47. X. H. Lin, J. Walter, K. Scheidtmann, K. Ohst, J. Newport, and G. Walter, *Proc. Natl. Acad. Sci. USA* **95**, 14693–14698 (1998).
48. Z. Yan, S. A. Fedorov, M. C. Mumby, and R. S. Williams, *Mol. Cell Biol.* **20**, 1021–1029 (2000).
49. H. A. Snaith, C. G. Armstrong, Y. Guo, K. Kaiser, and P. T. Cohen, *J. Cell Sci.* **109**, 3001–3012 (1996).
50. N. B. La Thangue, *Curr. Opin. Cell Biol.* **6**, 443–450 (1994).
51. B. D. Dynlacht, *Nature* **389**, 149–152 (1997).
52. R. Müller, *Trends Genet.* **11**, 173–178 (1995).
53. N. B. La Thangue, *Trends Biochem. Sci.* **19**, 108–114 (1994).
54. S. J. Weintraub, C. A. Prater, and D. C. Dean, *Nature* **358**, 259–261 (1992).
55. R. Fagan, K. J. Flint, and N. Jones, *Cell* **78**, 799–811 (1994).
56. B. D. Dynlacht, O. Flores, J. A. Lees, and E. Harlow, *Genes Dev.* **8**, 1772–1786 (1994).
57. M. Hatakeyama, J. A. Brill, G. R. Fink, and R. A. Weinberg, *Genes Dev.* **8**, 1759–17571 (1994).
58. D. G. Johnson, K. Ohtani, and J. R. Nevins, *Genes Dev.* **8**, 1514–1525 (1994).
59. J. K. Schwarz, S. H. Devoto, E. J. Smith, S. P. Chellappan, L. Jakoi, and J. R. Nevins, *EMBO J.* **12**, 1013–1020 (1993).

60. D. G. Johnson, J. K. Schwarz, W. D. Cress, and J. R. Nevins, *Nature* **365**, 349–352 (1993).
61. H. T. Liu, C. W. Gibson, R. R. Hirschhorn, S. Rittling, R. Baserga, and W. E. Mercer, *J. Biol. Chem.* **260**, 3269–3274 (1985).
62. E. Ogris, H. Rotheneder, I. Mudrak, A. Pichler, and E. Wintersberger, *J. Virol.* **67**, 1765–1771 (1993).
63. B. E. Pearson, H. P. Nasheuer, and T. S. Wang, *Mol. Cell. Biol.* **11**, 2081–2095 (1991).
64. C. Wang, M. Fu, S. Mani, S. Wadler, A. M. Senderowicz, and R. G. Pestell, *Front. Biosci.* **6**, D610–D629 (2001).
65. S. J. Nielsen, R. Schneider, U. M. Bauer, A. J. Bannister, A. Morrison, D. O'Carroll, R. Firestein, M. Cleary, T. Jenuwein, R. E. Herrera, et al., *Nature* **412**, 561–565 (2001).
66. C. Koch, A. Schleiffer, G. Ammerer, and K. Nasmyth, *Genes Dev.* **10**, 129–141 (1996).
67. L. Breeden and K. Nasmyth, *Cell* **48**, 389–397 (1987).
68. A. Amon, M. Tyers, B. Futcher, and K. Nasmyth, *Cell* **74**, 993–1007 (1993).
69. N. F. Lowndes, A. L. Johnson, L. Breeden, and L. H. Johnston, *Nature* **357**, 505–508 (1992).
70. N. F. Lowndes, C. J. McInerny, A. L. Johnson, P. A. Fantes, and L. H. Johnston, *Nature* **355**, 449–453 (1992).
71. B. J. Andrews and I. Herskowitz, *J. Biol. Chem.* **265**, 14057–14060 (1990).
72. E. Schwob and K. Nasmyth, *Genes Dev.* **7**, 1160–1175 (1993).
73. R. W. King, R. J. Deshaies, J. M. Peters, and M. W. Kirschner, *Science* **274**, 1652–1659 (1996).
74. M. Hochstrasser, *Annu. Rev. Genet.* **30**, 405–439 (1996).
75. D. M. Koepp, J. W. Harper, and S. J. Elledge, *Cell* **97**, 431–434 (1999).
76. A. Hershko and A. Ciechanover, *Annu. Rev. Biochem.* **67**, 425–479 (1998).
77. M. Scheffner, *Nature* **398**, 103–104 (1999).
78. M. Pagano, S. W. Tam, A. M. Theodoras, P. Beer-Romero, G. Del Sal, V. Chau, P. R. Yew, G. F. Draetta, and M. Rolfe, *Science* **269**, 682–685 (1995).
79. R. J. Sheaff and J. M. Roberts, *Chem. Biol.* **3**, 869–873 (1996).
80. J. Vlach, S. Hennecke, and B. Amati, *EMBO J.* **16**, 5334–5344 (1997).
81. F. X. Claret, M. Hibi, S. Dhut, T. Toda, and M. Karin, *Nature* **383**, 453–457 (1996).
82. K. Tomoda, Y. Kubota, and J. Kato, *Nature* **398**, 160–165 (1999).
83. K. Hofmann and P. Bucher, *Trends Biochem. Sci.* **23**, 204–205 (1998).
84. J. A. Diehl, M. Cheng, M. F. Roussel, and C. J. Sherr, *Genes Dev.* **12**, 3499–3511 (1998).
85. D. A. Freedman and A. J. Levine, *Mol. Cell. Biol.* **18**, 7288–7293.
86. S. J. Elledge and J. W. Harper, *Biochim. Biophys. Acta* **1377**, M61–M70 (1998).
87. E. Schwob, T. Böhm, M. D. Mendenhall, and K. Nasmyth, *Cell* **79**, 233–244 (1994).
88. B. L. Schneider, Q. H. Yang, and A. B. Futcher, *Science* **272**, 560–562 (1996).
89. D. Skowyra, K. L. Craig, M. Tyers, S. J. Elledge, and J. W. Harper, *Cell* **91**, 209–219 (1997).
90. R. M. Feldman, C. C. Correll, K. B. Kaplan, and R. J. Deshaies, *Cell* **91**, 221–230 (1997).
91. T. Kamura, D. M. Koepp, M. N. Conrad, D. Skowyra, R. J. Moreland, O. Iliopoulos, W. S. Lane, W. G. Kaelin, Jr., S. J. Elledge, R. C. Conaway, et al., *Science* **284**, 657–661 (1999).
92. D. Skowyra, D. M. Koepp, T. Kamura, M. N. Conrad, R. C. Conaway, J. W. Conaway, S. J. Elledge, and J. W. Harper, *Science* **284**, 662–665 (1999).
93. P. Tan, S. Y. Fuchs, A. Chen, K. Wu, C. Gomez, Z. Ronai, and Z. Q. Pan, *Mol. Cells* **3**, 527–533 (1999).
94. C. E. Stebbins, W. G. Kaelin, Jr., and N. P. Pavletich, *Science* **284**, 455–461 (1999).
95. C. Bai, P. Sen, K. Hofmann, L. Ma, M. Goebl, J. W. Harper, and S. J. Elledge, *Cell* **86**, 263–274 (1996).
96. H. Strohmaier, C. H. Spruck, P. Kaiser, K. A. Won, O. Sangfelt, and S. I. Reed, *Nature* **413**, 316–322 (2001).
97. P. R. Yew and M. W. Kirschner, *Science* **277**, 1672–1676 (1997).
98. K. Kominami, I. Ochotorena, and T. Toda, *Genes Cells* **3**, 721–735 (1998).

99. S. Henchoz, Y. Chi, B. Catarin, I. Herskowitz, R. J. Deshaies, and M. Peter, *Genes Dev.* **11**, 3046–3060 (1997).
100. L. S. Drury, G. Perkins, and J. F. Diffley, *EMBO J.* **16**, 5966–5976 (1997).
101. A. Marti, C. Wirbelauer, M. Scheffner, and W. Krek, *Nat. Cell Biol.* **1**, 14–19 (1999).
102. P. Kaiser, R. A. Sia, E. G. Bardes, D. J. Lew, and S. I. Reed, *Genes Dev.* **12**, 2587–2597 (1998).
103. M. Schwab, A. S. Lutum, and W. Seufert, *Cell* **90**, 683–693 (1997).
104. R. Visintin, S. Prinz, and A. Amon, *Science* **278**, 460–463 (1997).
105. G. Fang, H. Yu, and M. W. Kirschner, *Mol. Cells* **2**, 163–171 (1998).
106. M. Shirayama, W. Zachariae, R. Ciosk, and K. Nasmyth, *EMBO J.* **17**, 1336–1349 (1998).
107. F. M. Townsley and J. V. Ruderman, *Trends Cell. Biol.* **8**, 238–244 (1998).
108. S. Tugendreich, J. Tomkiel, W. Earnshaw, and P. Hieter, *Cell* **81**, 261–268 (1995).
109. W. Zachariae, M. Schwab, K. Nasmyth, and W. Seufert, *Science* **282**, 1721–1724 (1998).
110. H. Yu, J. M. Peters, R. W. King, A. M. Page, P. Hieter, and M. W. Kirschner, *Science* **279**, 1219–1222 (1998).
111. D. Maiorano, J. Moreau, and M. Mechali, *Nature* **404**, 622–625 (2000).
112. T. J. McGarry and M. W. Kirschner, *Cell* **93**, 1043–1053 (1998).
113. H. Nishitani, Z. Lygerou, T. Nishimoto, and P. Nurse, *Nature* **404**, 625–628 (2000).
114. S. Tada, A. Li, D. Maiorano, M. Mechali, and J. J. Blow, *Nat. Cell Biol.* **3**, 107–113 (2001).
115. J. A. Wohlschlegel, B. T. Dwyer, S. K. Dhar, C. Cvetic, J. C. Walter, and A. Dutta, *Science* **290**, 2309–2312 (2000).
116. A. Kornberg and T. Baker, DNA Replication, 2nd Edition, W. H. Freeman & Company, New York, 1992.
117. M. Mechali, *Nat. Rev. Genet.* **2**, 640–645 (2001).
118. J. A. Bogan, D. A. Natale, and M. L. Depamphilis, *J. Cell. Physiol.* **184**, 139–150 (2000).
119. M. L. DePamphilis, *Bioessays* **21**, 5–16 (1999).
120. C. S. Newlon, in "DNA Replication in Eukaryotic Cells" (M. L. DePamphilis, ed.), pp. 873–914. Cold Spring Harbor Laboratory Press, Cold Spring Harbor Laboratory, New York, 1996.
121. J. M. Ortega and M. L. DePamphilis, *J. Cell Sci.* **111**, 3663–3673 (1998).
122. T. Kobayashi, T. Rein, and M. L. DePamphilis, *Mol. Cell. Biol.* **18**, 3266–3277 (1998).
123. T. J. Kelly and G. W. Brown, *Annu. Rev. Biochem.* **69**, 829–880 (2000).
124. Y. Marahrens and B. Stillman, *Science* **255**, 817–823 (1992).
125. S. P. Bell and B. Stillman, *Nature* **357**, 128–134 (1992).
126. S. P. Bell, R. Kobayashi, and B. Stillman, *Science* **262**, 1844–1849 (1993).
127. M. L. DePamphilis, *Cell* **52**, 635–638 (1988).
128. J. J. Blow, P. J. Gillespie, D. Francis, and D. A. Jackson, *J. Cell Biol.* **152**, 15–25 (2001).
129. M. Mechali and S. Kearsey, *Cell* **38**, 55–64 (1984).
130. A. Falaschi, *Trends Genet.* **16**, 88–92 (2000).
131. G. Abdurashidova, M. Deganuto, R. Klima, S. Riva, G. Biamonti, M. Giacca, and A. Falaschi, *Science* **287**, 2023–2026 (2000).
132. M. I. Aladjem, L. W. Rodewald, J. L. Kolman, and G. M. Wahl, *Science* **281**, 1005–1009 (1998).
133. A. L. Altman and E. Fanning, *Mol. Cell. Biol.* **21**, 1098–1110 (2001).
134. C. Berger, A. Horlebein, E. Gogel, and F. Grummt, *Chromosoma* **106**, 479–484 (1997).
135. C. McWhinney and M. Leffak, *Nucleic Acids Res.* **18**, 1233–1242 (1990).
136. E. de Stanchina, D. Gabellini, P. Norio, M. Giacca, F. A. Peverali, S. Riva, A. Falaschi, and G. Biamonti, *J. Mol. Biol.* **299**, 667–680 (2000).
137. S. Loo, C. A. Fox, J. Rine, R. Kobayashi, B. Stillman, and S. Bell, *Mol. Biol. Cell* **6**, 741–756 (1995).
138. S. Waga and B. Stillman, *Annu. Rev. Biochem.* **67**, 721–751 (1998).
139. D. G. Lee and S. P. Bell, *Curr. Opin. Cell Biol.* **12**, 280–285 (2000).
140. O. M. Aparicio, D. M. Weinstein, and S. P. Bell, *Cell* **91**, 59–69 (1997).

141. O. M. Aparicio, A. M. Stout, and S. P. Bell, *Proc. Natl. Acad. Sci. USA* **96,** 9130–9135 (1999).
142. D. G. Lee and S. P. Bell, *Mol. Cell. Biol.* **17,** 7159–7168 (1997).
143. R. D. Klemm, R. J. Austin, and S. P. Bell, *Cell* **88,** 493–502 (1997).
144. R. D. Klemm and S. P. Bell, *Proc. Natl. Acad. Sci. USA* **98,** 8361–8367 (2001).
145. T. Mizushima, N. Takahashi, and B. Stillman, *Genes Dev.* **14,** 1631–1641 (2000).
146. V. Q. Nguyen, C. Co, and J. J. Li, *Nature* **411,** 1068–1073 (2001).
147. X. H. Hua, H. Yan, and J. Newport, *J. Cell Biol.* **137,** 183–192 (1997).
148. V. Orlando, *Trends Biochem. Sci.* **25,** 99–104 (2000).
149. C. Liang, M. Weinreich, and B. Stillman, *Cell* **81,** 667–676 (1995).
150. P. B. Carpenter, P. R. Mueller, and W. G. Dunphy, *Nature* **379,** 357–360 (1996).
151. I. Chesnokov, M. Gossen, D. Remus, and M. Botchan, *Genes Dev.* **13,** 1289–1296 (1999).
152. S. K. Dhar, K. Yoshida, Y. Machida, P. Khaira, B. Chaudhuri, J. A. Wohlschlegel, M. Leffak, J. Yates, and A. Dutta, *Cell* **106,** 287–296 (2001).
153. S. K. Dhar, L. Delmolino, and A. Dutta, *J. Biol. Chem.* **276,** 29067–29071 (2001).
154. A. E. Ehrenhofer-Murray, M. Gossen, D. T. Pak, M. R. Botchan, and J. Rine, *Science* **270,** 1671–1674 (1995).
155. P. J. Gillespie and J. J. Blow, *Nucleic Acids Res.* **28,** 472–480 (2000).
156. M. Gossen, D. T. S. Pak, S. K. Hansen, J. K. Acharya, and M. R. Botchan, *Science* **270,** 1674–1677 (1995).
157. X. H. Hua and J. Newport, *J. Cell Biol.* **140,** 271–281 (1998).
158. S. Kreitz, M. Ritzi, M. Baack, and R. Knippers, *J. Biol. Chem.* **276,** 6337–6342 (2001).
159. M. A. Madine, M. Swietlik, C. Pelizon, P. Romanowski, A. D. Mills, and R. A. Laskey, *J. Struct. Biol.* **129,** 198–210 (2000).
160. D. A. Natale, C. J. Li, W. H. Sun, and M. L. DePamphilis, *EMBO J.* **19,** 2728–2738 (2000).
161. D. G. Quintana, K. C. Thome, Z. H. Hou, A. H. Ligon, C. C. Morton, and A. Dutta, *J. Biol. Chem.* **273,** 27137–27145 (1998).
162. D. G. Quintana, Z. Hou, K. C. Thome, M. Hendricks, P. Saha, and A. Dutta, *J. Biol. Chem.* **272,** 28247–28251 (1997).
163. P. Romanowski, M. A. Madine, A. Rowles, J. J. Blow, and R. A. Laskey, *Curr. Biol.* **6,** 1416–1425 (1996).
164. P. Romanowski, J. Marr, M. A. Madine, A. Rowles, J. J. Blow, J. Gautier, and R. A. Laskey, *J. Biol. Chem.* **275,** 4239–4243 (2000).
165. K. C. Thome, S. K. Dhar, D. G. Quintana, L. Delmolino, A. Shahsafaei, and A. Dutta, *J. Biol. Chem.* **275,** 35233–35241 (2000).
166. T. Tugal, X. H. Zou-Yang, K. Gavin, D. Pappin, B. Canas, R. Kobayashi, T. Hunt, and B. Stillman, *J. Biol. Chem.* **273,** 32421–32429 (1998).
167. J. Walter and J. W. Newport, *Science* **275,** 993–995 (1997).
168. J. Walter, L. Sun, and J. Newport, *Mol. Cells* **1,** 519–529 (1998).
169. P. Zisimopoulou, C. Staib, I. Nanda, M. Schmid, and F. Grummt, *Mol. Gen. Genet.* **260,** 295–299 (1998).
170. S. Vashee, P. Simancek, M. D. Challberg, and T. J. Kelly, *J. Biol. Chem.* **276,** 26666–26673 (2001).
171. T. J. Kelly, G. S. Martin, S. L. Forsburg, R. J. Stephen, A. Russo, and P. Nurse, *Cell* **74,** 371–382 (1993).
172. J. Liszieuicz, A. Godany, D. V. Agoston, and H. Küntzel, *Nucleic Acids Res.* **16,** 11507–11520 (1988).
173. C. Zhou, S. H. Huang, and A. Y. Jong, *J. Biol. Chem.* **264,** 9022–9029 (1989).
174. J. Liu, C. L. Smith, D. DeRyckere, K. DeAngelis, G. S. Martin, and J. M. Berger, *Mol. Cells* **6,** 637–648 (2000).
175. A. Schepers and J. F. Diffley, *J. Mol. Biol.* **308,** 597–608 (2001).

176. S. Donovan, J. Harwood, L. S. Drury, and J. F. Diffley, *Proc. Natl. Acad. Sci. USA* **94**, 5611–5616 (1997).
177. U. Herbig, C. A. Marlar, and E. Fanning, *Mol. Biol. Cell* **10**, 2631–2645 (1999).
178. M. Weinreich, C. Liang, and B. Stillman, *Proc. Natl. Acad. Sci. USA* **96**, 441–446 (1999).
179. C. Dahmann, J. F. X. Diffley, and K. A. Nasmyth, *Curr. Biol.* **5**, 1257–1269 (1995).
180. L. S. Drury, G. Perkins, and J. F. Diffley, *Curr. Biol.* **10**, 231–240 (2000).
181. W. Zwerschke, H. W. Rottjakob, and H. Küntzel, *J. Biol. Chem.* **269**, 23351–23356 (1994).
182. A. Calzada, M. Sacristan, E. Sanchez, and A. Bueno, *Nature* **412**, 355–358 (2001).
183. S. Elsasser, F. Lou, B. Wang, J. L. Campbell, and A. Jong, *Mol. Biol. Cell* **7**, 1723–1735 (1996).
184. H. Nishitani and P. Nurse, *Cell* **83**, 397–405 (1995).
185. C. Pelizon, M. A. Madine, P. Romanowski, and R. A. Laskey, *Genes Dev.* **14**, 2526–2533 (2000).
186. A. Rowles, S. Tada, and J. J. Blow, *J. Cell Sci.* **112**, 2011–2018 (1999).
187. U. Herbig, J. W. Griffith, and E. Fanning, *Mol. Biol. Cell* **11**, 4117–4130 (2000).
188. J. F. Hofmann and D. Beach, *EMBO J.* **13**, 425–434 (1994).
189. Z. Lygerou and P. Nurse, *Science* **290**, 2271–2273 (2000).
190. M. Madine and R. Laskey, *Nat. Cell Biol.* **3**, E49–E50 (2001).
191. J. P. J. Chong, H. M. Mahbubani, C. Y. Khoo, and J. J. Blow, *Nature* **375**, 418–421 (1995).
192. A. M. Merchant, Y. Kawasaki, Y. Chen, M. Lei, and B. K. Tye, *Mol. Cell. Biol.* **17**, 3261–3271 (1997).
193. L. Homesley, M. Lei, Y. Kawasaki, S. Sawyer, T. Christensen, and B. K. Tye, *Genes Dev.* **14**, 913–926 (2000).
194. M. Ritzi and R. Knippers, *Gene* **245**, 13–20 (2000).
195. B. K. Tye, *Annu. Rev. Biochem.* **68**, 649–686 (1999).
196. K. M. Hennessy, A. Lee, E. Chen, and D. Botstein, *Genes Dev.* **5**, 958–969 (1991).
197. J. P. Chong, M. K. Hayashi, M. N. Simon, R. M. Xu, and B. Stillman, *Proc. Natl. Acad. Sci. USA* **97**, 1530–1535 (2000).
198. Z. Kelman, J. K. Lee, and J. Hurwitz, *Proc. Natl. Acad. Sci. USA* **96**, 14783–14788 (1999).
199. Y. Kubota, S. Mimura, S. Nishimoto, H. Takisawa, and H. Nojima, *Cell* **81**, 601–609 (1995).
200. M. A. Madine, C.-Y. Khoo, A. D. Mills, and R. A. Laskey, *Nature* **375**, 421–424 (1995).
201. S. E. Kearsey, S. Montgomery, K. Labib, and K. Lindner, *EMBO J.* **19**, 1681–1690 (2000).
202. K. Labib, J. F. Diffley, and S. E. Kearsey, *Nat. Cell Biol.* **1**, 415–422 (1999).
203. Y. Ishimi, Y. Komamura, Z. You, and H. Kimura, *J. Biol. Chem.* **273**, 8369–8375 (1998).
204. P. Thömmes, Y. Kubota, H. Takisawa, and J. J. Blow, *EMBO J.* **16**, 3312–3319 (1997).
205. J. K. Lee and J. Hurwitz, *J. Biol. Chem.* **275**, 18871–18878 (2000).
206. J. K. Lee and J. Hurwitz, *Proc. Natl. Acad. Sci. USA* **98**, 54–59 (2001).
207. Z. You, Y. Komamura, and Y. Ishimi, *Mol. Cell. Biol.* **19**, 8003–8015 (1999).
208. Y. Ishimi, Y. Komamura-Kohno, Z. You, A. Omori, and M. Kitagawa, *J. Biol. Chem.* **275**, 16235–16241 (2000).
209. Y. Ishimi and Y. Komamura-Kohno, *J. Biol. Chem.* **276**, 34428–34433 (2001).
210. L. Zou and B. Stillman, *Mol. Cell. Biol.* **20**, 3086–3096 (2000).
211. E. Fanning and R. Knippers, *Annu. Rev. Biochem.* **61**, 55–85 (1992).
212. E. H. Egelman, *Curr. Biol.* **11**, R103–R105 (2001).
213. E. Jankowsky, C. H. Gross, S. Shuman, and A. M. Pyle, *Science* **291**, 121–125 (2001).
214. C. F. Hardy, *Gene* **187**, 239–246 (1997).
215. B. Hopwood and S. Dalton, *Proc. Natl. Acad. Sci. USA* **93**, 12309–12314 (1996).
216. D. Loebel, H. Huikeshoven, and S. Cotterill, *Nucleic Acids Res.* **28**, 3897–3903 (2000).
217. J. M. McKie, R. B. Wadey, H. F. Sutherland, C. L. Taylor, and P. J. Scambler, *Genome Res.* **8**, 834–841 (1998).
218. S. Mimura and H. Takisawa, *EMBO J.* **17**, 5699–5707 (1998).

219. S. Miyake and S. Yamashita, *Genes Cells* **3**, 157–166 (1998).
220. P. Saha, K. C. Thome, R. Yamaguchi, Z. Hou, S. Weremowicz, and A. Dutta, *J. Biol. Chem.* **273**, 18205–18209 (1998).
221. K. Yoshida, F. Kuo, E. L. George, A. H. Sharpe, and A. Dutta, *Mol. Cell. Biol.* **21**, 4598–4603 (2001).
222. J. C. Owens, C. S. Detweiler, and J. J. Li, *Proc. Natl. Acad. Sci. USA* **94**, 12521–12526 (1997).
223. S. Dalton and B. Hopwood, *Mol. Cell. Biol.* **17**, 5867–5875 (1997).
224. L. Zou, J. Mitchell, and B. Stillman, *Mol. Cell. Biol.* **17**, 553–563 (1997).
225. L. Zou and B. Stillman, *Science* **280**, 593–596 (1998).
226. Y. Kamimura, Y. S. Tak, A. Sugino, and H. Araki, *EMBO J.* **20**, 2097–2107 (2001).
227. M. Uchiyama, D. Griffiths, K. Arai Ki, and H. Masai, *J. Biol. Chem.* **276**, 26189–26196 (2001).
228. J. Walter and J. Newport, *Mol. Cells.* **5**, 617–627 (2000).
229. K. Weisshart, H. Förster, E. Kremmer, B. Schlott, F. Grosse, and H. P. Nasheuer, *J. Biol. Chem.* **275**, 17328–17337 (2000).
230. I. Kukimoto, H. Igaki, and T. Kanda, *Eur. J. Biochem.* **265**, 936–943 (1999).
231. S. Mimura, T. Masuda, T. Matsui, and H. Takisawa, *Genes Cells* **5**, 439–452 (2000).
232. J. A. Tercero, K. Labib, and J. F. Diffley, *EMBO J.* **19**, 2082–2093 (2000).
233. R. Nougarede, F. Della Seta, P. Zarzov, and E. Schwob, *Mol. Cell. Biol.* **20**, 3795–3806 (2000).
234. C. Iftode, Y. Daniely, and J. A. Borowiec, *Crit. Rev. Biochem. Mol. Biol.* **24**, 141–180 (1999).
235. M. S. Wold, *Annu. Rev. Biochem.* **66**, 61–92 (1997).
236. M. P. Fairman and B. Stillman, *EMBO J.* **7**, 1211–1218 (1988).
237. W.-D. Heyer, M. R. S. Rao, L. F. Erdile, T. J. Kelly, and R. D. Kolodner, *EMBO J.* **9**, 2321–2329 (1990).
238. C. R. Wobbe, L. Weissbach, J. A. Borowiec, F. B. Dean, Y. Murakami, P. Bullock, and J. Hurwitz, *Proc. Natl. Acad. Sci. USA* **84**, 1834–1838 (1987).
239. M. S. Wold and T. J. Kelly, *Proc. Natl. Acad. Sci. USA* **85**, 2523–2527 (1988).
240. M. Challberg and T. J. Kelly, *Annu. Rev. Biochem.* **58**, 671–717 (1989).
241. J. Hurwitz, F. B. Dean, A. D. Kwong, and S.-H. Lee, *J. Biol. Chem.* **265**, 18043–18046 (1990).
242. B. Stillman, *Annu. Rev. Cell Biol.* **5**, 197–245 (1989).
243. A. Atrazhev, S. Zhang, and F. Grosse, *Eur. J. Biochem.* **210**, 855–865 (1992).
244. K. A. Braun, Y. Lao, Z. He, C. J. Ingles, and M. S. Wold, *Biochemistry* **36**, 8443–8454 (1997).
245. R. M. Brosh, Jr., J. L. Li, M. K. Kenny, J. K. Karow, M. P. Cooper, R. P. Kureekattil, I. D. Hickson, and V. A. Bohr, *J. Biol. Chem.* **275**, 23500–23508 (2000).
246. A. Constantinou, M. Tarsounas, J. K. Karow, R. M. Brosh, V. A. Bohr, I. D. Hickson, and S. C. West, *EMBO Rep.* **1**, 80–84 (2000).
247. L. F. Erdile, W.-D. Heyer, R. Kolodner, and T. J. Kelly, *J. Biol. Chem.* **266**, 12090–12098 (1991).
248. M. K. Kenny, U. Schlegel, H. Furneaux, and J. Hurwitz, *J. Biol. Chem.* **265**, 7693–7700 (1990).
249. I. Ohsugi, Y. Tokutake, N. Suzuki, T. Ide, M. Sugimoto, and Y. Furuichi, *Nucleic Acids Res.* **28**, 3642–3648 (2000).
250. Y. S. Seo, S. H. Lee, and J. Hurwitz, *J. Biol. Chem.* **266**, 13161–13170 (1991).
251. P. Thömmes, E. Ferrari, R. Jessberger, and U. Hübscher, *J. Biol. Chem.* **267**, 6063–6073 (1992).
252. M. P. Carty, Y. Ishimi, A. S. Levine, and K. Dixon, *Mutat. Res.* **232**, 141–153 (1990).
253. M. P. Carty, A. S. Levine, and K. Dixon, *Mutat. Res.* **274**, 29–43 (1992).
254. G. Maga, I. Frouin, S. Spadari, and U. Hübscher, *J. Biol. Chem.* **276**, 18235–18242 (2001).
255. I. Dornreiter, L. F. Erdile, I. U. Gilbert, D. von Winkler, T. J. Kelly, and E. Fanning, *EMBO J.* **11**, 769–776 (1992).
256. J. Smith, H. Zou, and R. Rothstein, *Biochimie* **82**, 71–78 (2000).

257. A. Aboussekhra, M. Biggerstaff, M. K. Shivji, J. A. Vilpo, V. Moncollin, V. N. Podust, M. Protic, U. Hübscher, J. M. Egly, and R. D. Wood, *Cell* **80**, 859–868 (1995).
258. D. Coverley, M. K. Kenny, D. P. Lane, and R. D. Wood, *Nucleic Acids Res.* **20**, 3873–3880 (1992).
259. D. Coverley, M. K. Kenny, M. Munn, W. D. Rupp, D. P. Lane, and R. D. Wood, *Nature* **349**, 538–541 (1991).
260. S. N. Guzder, Y. Habraken, P. Sung, L. Prakash, and S. Prakash, *J. Biol. Chem.* **270**, 12973–12976 (1995).
261. Z. He, J. M. Wong, H. S. Maniar, S. J. Brill, and C. J. Ingles, *J. Biol. Chem.* **271**, 28243–28249 (1996).
262. L. Li, X. Lu, C. A. Peterson, and R. J. Legerski, *Mol. Cell. Biol.* **15**, 5396–5402 (1995).
263. T. Matsuda, M. Saijo, I. Kuraoka, T. Kobayashi, Y. Nakatsu, A. Nagai, T. Enjoji, C. Masutani, K. Sugasawa, F. Hanaoka, et al., *J. Biol. Chem.* **270**, 4152–4157 (1995).
264. M. Wakasugi, M. Shimizu, H. Morioka, S. Linn, O. Nikaido, and T. Matsunaga, *J. Biol. Chem.* **276**, 15434–15440 (2001).
265. D. P. Batty and R. D. Wood, *Gene* **241**, 193–204 (2000).
266. R. D. Kolodner and G. T. Marsischky, *Curr. Opin. Genet. Dev.* **9**, 89–96 (1999).
267. S. Prakash and L. Prakash, *Mutat. Res.* **451**, 13–24 (2000).
268. M. S. DeMott, S. Zigman, and R. A. Bambara, *J. Biol. Chem.* **273**, 27492–27498 (1998).
269. G. L. Dianov, B. R. Jensen, M. K. Kenny, and V. A. Bohr, *Biochemistry* **38**, 11021–11025 (1999).
270. M. Stucki, B. Pascucci, E. Parlanti, P. Fortini, S. H. Wilson, U. Hübscher, and E. Dogliotti, *Oncogene* **17**, 835–843 (1998).
271. P. Baumann and S. C. West, *EMBO J.* **16**, 5198–5206 (1997).
272. P. Baumann and S. C. West, *J. Mol. Biol.* **291**, 363–374 (1999).
273. M. P. Longhese, H. Neecke, V. Paciotti, G. Lucchini, and P. Plevani, *Nucleic Acids Res.* **24**, 3533–3537 (1996).
274. M. J. McIlwraith, E. Van Dyck, J. Y. Masson, A. Z. Stasiak, A. Stasiak, and S. C. West, *J. Mol. Biol.* **304**, 151–164 (2000).
275. S. Sigurdsson, K. Trujillo, B. Song, S. Stratton, and P. Sung, *J. Biol. Chem.* **276**, 8798–8806 (2001).
276. C. Santocanale, H. Neecke, M. P. Longhese, G. Lucchini, and P. Plevani, *J. Mol. Biol.* **254**, 595–607 (1995).
277. B. Song and P. Sung, *J. Biol. Chem.* **275**, 15895–15904 (2000).
278. W. L. de Laat, E. Appeldoorn, K. Sugasawa, E. Weterings, N. G. Jaspers, and J. H. Hoeijmakers, *Genes Dev.* **12**, 2598–2609 (1998).
279. O. I. Lavrik, D. M. Kolpashchikov, H. P. Nasheuer, K. Weisshart, and A. Favre, *FEBS Lett.* **441**, 186–190 (1998).
280. O. I. Lavrik, H. P. Nasheuer, K. Weisshart, M. S. Wold, R. Prasad, W. A. Beard, S. H. Wilson, and A. Favre, *Nucleic Acids Res.* **26**, 602–607 (1998).
281. G. Mass, T. Nethanel, and G. Kaufmann, *Mol. Cell. Biol.* **18**, 6399–6407 (1998).
282. H. P. Nasheuer, D. von Winkler, C. Schneider, I. Dornreiter, I. Gilbert, and E. Fanning, *Chromosoma* **102**, S52–S59 (1992).
283. N. A. Abramova, J. Russell, M. Botchan, and R. Li, *Proc. Natl. Acad. Sci. USA* **94**, 7186–7191 (1997).
284. A. Dutta, J. M. Ruppert, J. C. Aster, and E. Winchester, *Nature* **365**, 79–82 (1993).
285. Y. Han, Y. M. Loo, K. T. Militello, and T. Melendy, *J. Virol.* **73**, 4899–4907 (1999).
286. Z. He, B. T. Brinton, J. Greenblatt, J. A. Hassell, and C. J. Ingles, *Cell* **73**, 1223–1232 (1993).
287. S. G. Huang, K. Weisshart, I. Gilbert, and E. Fanning, *Biochemistry* **37**, 15345–15352 (1998).

288. C. Kühn, F. Müller, C. Melle, H. P. Nasheuer, F. Janus, W. Deppert, and F. Grosse, *Oncogene* **18**, 769–774 (1999).
289. R. Li and M. R. Botchan, *Cell* **73**, 1207–1221 (1993).
290. K. Treuner, A. Okuyama, R. Knippers, and F. O. Fackelmayer, *Nucleic Acids Res.* **27**, 1499–1504 (1999).
291. X. V. Gomes and M. S. Wold, *J. Biol. Chem.* **270**, 4534–4543 (1995).
292. X. V. Gomes, L. A. Henricksen, and M. S. Wold, *Biochemistry* **35**, 5586–5595 (1996).
293. A. Bochkarev, R. A. Pfuetzner, A. M. Edwards, and L. Frappier, *Nature* **385**, 176–181 (1997).
294. S. J. Brill and S. Bastin-Shanower, *Mol. Cell. Biol.* **18**, 7225–7234 (1998).
295. D. K. Kim, E. Stigger, and S. H. Lee, *J. Biol. Chem.* **271**, 15124–15129.
296. D. Philipova, J. Mullen, H. Maniar, J. Lu, C. Gu, and S. Brill, *Genes Dev.* **10**, 2222–2233.
297. E. Bochkareva, S. Korolev, and A. Bochkarev, *J. Biol. Chem.* **275**, 27332–27338 (2000).
298. Y. Lao, C. G. Lee, and M. S. Wold, *Biochemistry* **38**, 3974–3984 (1999).
299. S. J. Brill and B. Stillman, *Nature* **342**, 92–95 (1989).
300. S. J. Brill and B. Stillman, *Genes Dev.* **5**, 1589–1600 (1991).
301. A. Bochkarev, E. Bochkareva, L. Frappier, and A. M. Edwards, *EMBO J.* **18**, 4498–4504 (1999).
302. G. S. Brush, D. M. Morrow, P. Hieter, and T. J. Kelly, *Proc. Natl. Acad. Sci. USA* **93**, 15075–15080 (1996).
303. G. S. Brush and T. J. Kelly, *Nucleic Acids Res.* **28**, 3725–3732 (2000).
304. S.-U. Din, S. J. Brill, M. P. Fairman, and B. Stillman, *Genes Dev.* **4**, 968–977 (1990).
305. A. Dutta and B. Stillman, *EMBO J.* **11**, 2189–2199 (1992).
306. F. Fang and J. W. Newport, *J. Cell Sci.* **106**, 983–994 (1993).
307. R. Fotedar and J. M. Roberts, *EMBO J.* **11**, 2177–2187 (1992).
308. A. Georgaki and U. Hubscher, *Nucleic Acids Res.* **21**, 3659–3665 (1993).
309. Z. Q. Pan, C. H. Park, A. A. Amin, J. Hurwitz, and A. Sancar, *Proc. Natl. Acad. Sci. USA* **92**, 4636–4640 (1995).
310. A. Pellicioli, S. E. Lee, C. Lucca, M. Foiani, and J. E. Haber, *Mol. Cells* **7**, 293–300 (2001).
311. G. Rodrigo, S. Roumagnac, M. S. Wold, B. Salles, and P. Calsou, *Mol. Cell. Biol.* **20**, 2696–2705 (2000).
312. Y. Wang, X. Y. Zhou, H. Wang, M. S. Huq, and G. Iliakis, *J. Biol. Chem.* **274**, 22060–22064 (1999).
313. S. E. Morgan and M. B. Kastan, *Cancer Res.* **57**, 3386–3389 (1997).
314. P. M. Burgers, *Chromosoma* **107**, 218–227 (1998).
315. D. N. Frick and C. C. Richardson, *Annu. Rev. Biochem.* **70**, 39–80 (2001).
316. T. S.-F. Wang, in "DNA Replication in Eukaryotic Cells" (M. L. DePamphilis, ed.), pp. 461–493 Cold Spring Harbor Laboratory Press, Cold Spring Harbor Laboratory, New York, 1996.
317. A. Brückner, F. Stadlbauer, L. A. Guarino, A. Brunahl, C. Schneider, C. Rehfuess, C. Prives, E. Fanning, and H. P. Nasheuer, *Mol. Cell. Biol.* **15**, 1716–1724 (1995).
318. K. L. Conger, J. S. Liu, S. R. Kuo, L. T. Chow, and T. S. F. Wang, *J. Biol. Chem.* **274**, 2696–2705 (1999).
319. I. Dornreiter, A. Höss, A. K. Arthur, and E. Fanning, *EMBO J.* **9**, 3329–3336 (1990).
320. I. Dornreiter, W. C. Copeland, and T. S. Wang, *Mol. Cell. Biol.* **13**, 809–820 (1993).
321. A. Kautz, A. Schneider, K. Weisshart, C. Geiger, and H. P. Nasheuer, *J. Virol.* **75**, 1751–1760 (2001).
322. A. Kautz, K. Weisshart, A. Schneider, F. Grosse, and H. P. Nasheuer, *J. Virol.* **75**, 8569–8578 (2001).
323. P. J. Masterson, M. A. Stanley, A. P. Lewis, and M. A. Romanos, *J. Virol.* **72**, 7407–7419.

324. P. Park, W. Copeland, L. Yang, T. Wang, M. R. Botchan, and I. J. Mohr, *Proc. Natl. Acad. Sci. USA* **91**, 8700–8704 (1994).
325. C. Schneider, K. Weisshart, L.A. Guarino, I. Dornreiter, and E. Fanning, *Mol. Cell. Biol.* **14**, 3176–3185 (1994).
326. M. Takemura, T. Kitagawa, S. Izuta, J. Wasa, A. Takai, T. Akiyama, and S. Yoshida, *Oncogene* **15**, 2483–2492 (1997).
327. G. D'Urso, B. Grallert, and P. Nurse, *J. Cell Sci.* **108**, 3109–3118 (1995).
328. M. Foiani, G. Lucchini, and P. Plevani, *Trends Biochem. Sci.* **22**, 424–427 (1997).
329. M. P. Longhese, M. Foiani, M. Muzi-Falconi, G. Lucchini, and P. Plevani, *EMBO J.* **17**, 5525–5528 (1998).
330. F. Marini, A. Pellicioli, V. Paciotti, G. Lucchini, P. Plevani, D. F. Stern, and M. Foiani, *EMBO J.* **16**, 639–650 (1997).
331. A. M. Holmes and J. E. Haber, *Cell* **96**, 415–424 (1999).
332. J. E. Haber, *Trends Biochem. Sci.* **24**, 271–275 (1999).
333. F. Dantzer, H. P. Nasheuer, J. L. Vonesch, G. de Murcia, and J. Ménissier-de Murcia, *Nucleic Acids Res.* **26**, 1891–1898 (1998).
334. M. P. Longhese, R. Fraschini, P. Plevani, and G. Lucchini, *Mol. Cell. Biol.* **16**, 3235–3244 (1996).
335. S. Yoshida and C. M. Simbulan, *Mol. Cell. Biochem.* **138**, 39–44 (1994).
336. A. Adams Martin, I. Dionne, R. J. Wellinger, and C. Holm, *Mol. Cell. Biol.* **20**, 786–796 (2000).
337. M. J. Carson and L. Hartwell, *Cell* **42**, 249–257 (1985).
338. S. J. Diede and D. E. Gottschling, *Cell* **99**, 723–733 (1999).
339. H. Qi and V. A. Zakian, *Genes Dev.* **14**, 1777–1788 (2000).
340. K. Nozawa, M. Suzuki, M. Takemura, and S. Yoshida, *Nucleic Acids Res.* **28**, 3117–3124 (2000).
341. P. M. Reveal, K. M. Henkels, and J. J. Turchi, *J. Biol. Chem.* **272**, 11678–11681 (1997).
342. A. M. Zahler and D. M. Prescott, *Nucleic Acids Res.* **17**, 6299–6317 (1989).
343. K. Kuwahara, S. Tomiyasu, S. Fujimura, K. Nomura, Y. Xing, N. Nishiyama, M. Ogawa, S. Imajoh-Ohmi, S. Izuta, and N. Sakaguchi, *Proc. Natl. Acad. Sci. USA* **98**, 10279–10283 (2001).
344. M. A. Augustin, R. Huber, and J. T. Kaiser, *Nat. Struct. Biol.* **8**, 57–61 (2001).
345. C. J. Bakkenist and S. Cotterill, *J. Biol. Chem.* **269**, 26759–26766 (1994).
346. R. G. Brooke, R. Singhal, D. C. Hinkle, and L. B. Dumas, *J. Biol. Chem.* **266**, 3005–3015 (1991).
347. W. C. Copeland and T. S. Wang, *J. Biol. Chem.* **266**, 22739–22748 (1991).
348. W. C. Copeland and T. S. Wang, *J. Biol. Chem.* **268**, 26179–26189 (1993).
349. H.-P. Nasheuer and F. Grosse, *J. Biol. Chem.* **263**, 8981–8988 (1988).
350. C. Santocanale, M. Foiani, G. Lucchini, and P. Plevani, *J. Biol. Chem.* **268**, 1343–1348 (1993).
351. A. Schneider, R. W. P. Smith, A. R. Kautz, K. Weisshart, F. Grosse, and H. P. Nasheuer, *J. Biol. Chem.* **273**, 21608–21615 (1998).
352. F. Stadlbauer, A. Brueckner, C. Rehfuess, C. Eckerskorn, F. Lottspeich, V. Förster, B. Y. Tseng, and H. P. Nasheuer, *Eur. J. Biochem.* **222**, 781–793 (1994).
353. B. Arezi and R. D. Kuchta, *Trends Biochem. Sci.* **25**, 572–576 (2000).
354. T. Mizuno, N. Ito, M. Yokoi, A. Kobayashi, K. Tamai, H. Miyazawa, and F. Hanaoka, *Mol. Cell. Biol.* **18**, 3552–3562 (1998).
355. G. Desogus, S. Onesti, P. Brick, M. Rossi, and F. M. Pisani, *Nucleic Acids Res.* **27**, 4444–4450 (1999).
356. H. Miyazawa, M. Izumi, S. Tada, R. Takada, M. Masutani, M. Ui, and F. Hanaoka, *J. Biol. Chem.* **268**, 8111–8122 (1993).
357. C. E. Prussak, M. T. Almazan, and B. Y. Tseng, *J. Biol. Chem.* **264**, 4957–4963 (1989).
358. B. W. Kirk and R. D. Kuchta, *Biochemistry* **38**, 7727–7736 (1999).
359. X. Shao and N. V. Grishin, *Nucleic Acids Res.* **28**, 2643–2650 (2000).

360. J. L. Keck, D. D. Roche, A. S. Lynch, and J. M. Berger, *Science* **287**, 2482–2486 (2000).
361. J. L. Keck and J. M. Berger, *Nat. Struct. Biol.* **8**, 2–4 (2001).
362. H. Pan and D. B. Wigley, *Struct. Fold Des.* **8**, 231–239 (2000).
363. J. Ito and D. K. Braithwaite, *Nucleic Acids Res.* **19**, 4045–4057 (1991).
364. V. Damagnez, J. Tillit, A.-M. d. Recondo, and G. Baldacci, *Mol. Gen. Genet.* **226**, 182–189 (1991).
365. S. W. Wong, A. F. Wahl, P.-M. Yuan, N. Arai, B. E. Pearson, K.-I. Arai, D. Korn, M. W. Hunkapillar, and T. S.-F. Wang, *EMBO J.* **7**, 37–47 (1988).
366. W. C. Copeland and T. S.-F. Wang, *J. Biol. Chem.* **268**, 11028–11040 (1993).
367. Q. Dong, W. C. Copeland, and T. S. F. Wang, *J. Biol. Chem.* **268**, 24163–24174 (1993).
368. Q. Dong, W. C. Copeland, and T. S. F. Wang, *J. Biol. Chem.* **268**, 24175–24182 (1993).
369. H. P. Nasheuer, A. Moore, A. F. Wahl, and T. S. Wang, *J. Biol. Chem.* **266**, 7893–7903 (1991).
370. V. N. Podust, O. I. Lavrik, H. P. Nasheuer, and F. Grosse, *FEBS Lett.* **245**, 14–16 (1989).
371. M. Weinreich and B. Stillman, *EMBO J.* **18**, 5334–5346 (1999).
372. S. W. Wong, L. R. Paborsky, P. A. Fisher, T. S. Wang, and D. Korn, *J. Biol. Chem.* **261**, 7958–7968 (1986).
373. K. L. Collins, A. A. R. Russo, B. Y. Tseng, and T. J. Kelly, *EMBO J.* **12**, 4555–4566 (1993).
374. S. Cotterill, I. R. Lehman, and P. McLachlan, *Nucleic Acids Res.* **20**, 4325–4330 (1992).
375. M. Foiani, C. Santocanale, P. Plevani, and G. Lucchini, *Mol. Cell. Biol.* **15**, 883–891 (1994).
376. M. Foiani, G. Liberi, G. Lucchini, and P. Plevani, *Mol. Cell. Biol.* **15**, 883–891 (1995).
377. U. Hübscher, H. P. Nasheuer, and J. Syväoja, *Trends Biochem. Sci.* **25**, 143–147 (2000).
378. K. L. Collins and T. J. Kelly, *Mol. Cell. Biol.* **11**, 2108–2115 (1991).
379. Y. Ishimi, A. Claude, P. Bullock, and J. Hurwitz, *J. Biol. Chem.* **263**, 19723–19733 (1988).
380. S. Waga, G. Bauer, and B. Stillman, *J. Biol. Chem.* **269**, 10923–10934 (1994).
381. S. Waga and B. Stillman, *Nature* **369**, 207–212 (1994).
382. P. A. Fisher, T. S. Wang, and D. Korn, *J. Biol. Chem.* **254**, 6128–6137 (1979).
383. K. T. Hohn and F. Grosse, *Biochemistry* **26**, 2870–2878 (1987).
384. G. Maga, M. Stucki, S. Spadari, and U. Hübscher, *J. Mol. Biol.* **295**, 791–801 (2000).
385. R. Mossi, R. C. Keller, E. Ferrari, and U. Hübscher, *J. Mol. Biol.* **295**, 803–814 (2000).
386. A. Yuzhakov, Z. Kelman, J. Hurwitz, and M. O'Donnell, *EMBO J.* **18**, 6189–6199 (1999).
387. P. Bouvier, A. M. DeRocondo, and G. Baldacci, *Exp. Cell Res.* **207**, 41–47 (1993).
388. H. Park, R. Davis, and T. S. Wang, *Nucleic Acids Res.* **23**, 4337–4344 (1995).
389. C. Voitenleitner, E. Fanning, and H. P. Nasheuer, *Oncogene* **14**, 1611–1615 (1997).
390. C. Voitenleitner, C. Rehfuess, M. Hilmes, L. O'Rear, P. C. Liao, D. A. Gage, R. Ott, H. P. Nasheuer, and E. Fanning, *Mol. Cell. Biol.* **19**, 646–656 (1999).
391. J. J. Champoux, *Annu. Rev. Biochem.* **70**, 369–413 (2001).
392. A. Hangaard, C. Bendixen, and O. Westergaard, in "DNA Replication in Eukaryotic Cells" (M. L. DePamphilis, ed.), pp. 587–617 Cold Spring Harbor Laboratory Press, Cold Spring Harbor Laboratory.
393. J. C. Wang, *Annu. Rev. Biochem.* **65**, 635–692 (1996).
394. Y. Pommier, G. S. Laco, G. Kohlhagen, J. M. Sayer, H. Kroth, and D. M. Jerina, *Proc. Natl. Acad. Sci. USA* **97**, 10739–10744 (2000).
395. M. R. Redinbo, L. Stewart, P. Kuhn, J. J. Champoux, and W. G. Hol, *Science* **279**, 1504–1513 (1998).
396. M. R. Redinbo, L. Stewart, J. J. Champoux, and W. G. Hol, *J. Mol. Biol.* **292**, 685–696 (1999).
397. M. R. Redinbo, J. J. Champoux, and W. G. Hol, *Curr. Opin. Struct. Biol.* **9**, 29–36 (1999).
398. J. Sekiguchi and S. Shuman, *EMBO J.* **15**, 3448–3457 (1996).
399. Z. Yang and J. J. Champoux, *J. Biol. Chem.* **276**, 677–685 (2001).
400. L. Stewart, G. C. Ireton, and J. J. Champoux, *J. Biol. Chem.* **274**, 32950–32960 (1999).
401. J. L. Nitiss, *Biochim. Biophys. Acta* **1400**, 63–81 (1998).

402. R. M. Snapka, *NCI Monogr.* **4,** 55–60 (1987).
403. K. Soe, G. Dianov, H. P. Nasheuer, V. A. Bohr, F. Grosse, and T. Stevnsner, *Nucleic Acids Res.* **29,** 3195–3203 (2001).
404. A. F. Stewart and G. Schutz, *Cell* **50,** 1109–1117 (1987).
405. A. F. Stewart, R. E. Herrera, and A. Nordheim, *Cell* **60,** 141–149 (1990).
406. L. Yang, M. S. Wold, J. J. Li, T. J. Kelly, and L. F. Liu, *Proc. Natl. Acad. Sci. USA* **84,** 950–954 (1987).
407. D. T. Simmons, T. Melendy, D. Usher, and B. Stillman, *Virology* **222,** 365–374 (1996).
408. D. T. Simmons, P. W. Trowbridge, and R. Roy, *Virology* **242,** 435–443 (1998).
409. Y. Pommier, P. Pourquier, Y. Fan, and D. Strumberg, *Biochim. Biophys. Acta* **1400,** 83–105 (1998).
410. J. F. Diffley, *Genes Dev.* **10,** 2819–2830 (1996).
411. A. D. Donaldson and J. J. Blow, *Curr. Opin. Genet. Dev.* **9,** 62–68 (1999).
412. A. Dutta and S. P. Bell, *Annu. Rev. Cell Dev. Biol.* **13,** 293–332 (1997).
413. J. Leatherwood, *Curr. Opin. Cell Biol.* **10,** 742–748 (1998).
414. P. Hughes, I. Tratner, M. Ducoux, K. Piard, and G. Baldacci, *Nucleic Acids Res.* **27,** 2108–2114 (1999).
415. L. Liu, J. Mo, E. M. Rodriguez-Belmonte, and M. Y. Lee, *J. Biol. Chem.* **275,** 18739–18744 (2000).
416. M. Simon, L. Giot, and G. Faye, *EMBO J.* **10,** 2165–2170 (1991).
417. R. E. Goldsby, N. A. Lawrence, L. E. Hays, E. A. Olmsted, X. Chen, M. Singh, and B. D. Preston, *Nat. Med.* **7,** 638–639 (2001).
418. A. Morrison, J. B. Bell, T. A. Kunkel, and A. Sugino, *Proc. Natl. Acad. Sci. USA* **88,** 9473–9477 (1991).
419. A. Morrison and A. Sugino, *Mol. Gen. Genet.* **242,** 289–296 (1994).
420. T. A. Weinert and L. H. Hartwell, *Science* **241,** 317–322 (1988).
421. L. H. Hartwell and T. A. Weinert, *Science* **246,** 629–634 (1989).
422. A. G. Paulovich, R. U. Margulies, B. M. Garvik, and L. H. Hartwell, *Genetics* **145,** 45–62 (1997).
423. S. J. Elledge, *Science* **274,** 1664–1672 (1996).
424. A. Hirao, Y. Y. Kong, S. Matsuoka, A. Wakeham, J. Ruland, H. Yoshida, D. Liu, S. J. Elledge, and T. W. Mak, *Science* **287,** 1824–1827 (2000).
425. K. B. Ritchie, J. C. Mallory, and T. D. Petes, *Mol. Cell. Biol.* **19,** 6065–6075 (1999).
426. X. Wu, V. Ranganathan, D. S. Weisman, W. F. Heine, D. N. Ciccone, T. B. O'Neill, K. E. Crick, K. A. Pierce, W. S. Lane, G. Rathbun, *et al.*, *Nature* **405,** 477–482 (2000).
427. Y. Xu and D. Baltimore, *Genes Dev.* **10,** 2401–2410 (1996).
428. A. de Klein, M. Muijtjens, R. van Os, Y. Verhoeven, B. Smit, A. M. Carr, A. R. Lehmann, and J. H. Hoeijmakers, *Curr. Biol.* **10,** 479–482 (2000).
429. Q. Liu, S. Guntuku, X. S. Cui, S. Matsuoka, D. Cortez, K. Tamai, G. Luo, S. Carattini-Rivera, F. DeMayo, A. Bradley, *et al.*, *Genes Dev.* **14,** 1448–1459 (2000).
430. E. C. Friedberg, G. C. Walker, and W. Siede, "DNA Repair and Mutagenesis." ASM Press, Washington, D.C., 1995.
431. S. Griffin, *Curr. Biol.* **6,** 497–499 (1996).
432. P. Modrich, *J. Biol. Chem.* **272,** 24727–24730 (1997).
433. D. L. Croteau and V. A. Bohr, *J. Biol. Chem.* **272,** 25409–25412 (1997).
434. R. D. Wood, *J. Biol. Chem.* **272,** 23465–23468 (1997).
435. M. B. Kastan and D. S. Lim, *Nat. Rev. Mol. Cell Biol.* **1,** 179–186 (2000).
436. B. B. Zhou and S. J. Elledge, *Nature* **408,** 433–439 (2000).
437. T. Caspari, M. Dahlen, G. Kanter-Smoler, H. D. Lindsay, K. Hofmann, K. Papadimitriou, P. Sunnerhagen, and A. M. Carr, *Mol. Cell. Biol.* **20,** 1254–1262 (2000).

438. D. J. Griffiths, N. C. Barbet, S. McCready, A. R. Lehmann, and A. M. Carr, *EMBO J.* **14**, 5812–5823 (1995).
439. M. Shimada, D. Okuzaki, S. Tanaka, T. Tougan, K. K. Tamai, C. Shimoda, and H. Nojima, *Mol. Biol. Cell* **10**, 3991–4003 (1999).
440. V. N. Noskov, H. Araki, and A. Sugino, *Mol. Cell. Biol.* **18**, 4914–4923 (1998).
441. K. Sugimoto, T. Shimomura, K. Hashimoto, H. Araki, A. Sugino, and K. Matsumoto, *Proc. Natl. Acad. Sci. USA* **93**, 7048–7052 (1996).
442. A. E. Parker, I. Van de Weyer, M. C. Laus, I. Oostveen, J. Yon, P. Verhasselt, and W. H. Luyten, *J. Biol. Chem.* **273**, 18332–18339 (1998).
443. Y. Saka and M. Yanagida, *Cell* **74**, 383–393 (1993).
444. H. M. Verkade and M. J. O'Connell, *Mol. Gen. Genet.* **260**, 426–433 (1998).
445. R. J. Edwards, N. J. Bentley, and A. M. Carr, *Nat. Cell Biol.* **1**, 393–398 (1999).
446. J. Soulier and N. F. Lowndes, *Curr. Biol.* **9**, 551–554 (1999).
447. J. E. Vialard, C. S. Gilbert, C. M. Green, and N. F. Lowndes, *EMBO J.* **17**, 5679–5688 (1998).
448. F. Esashi and M. Yanagida, *Mol. Cells* **4**, 167–174 (1999).
449. X. Xu, Z. Weaver, S. P. Linke, C. Li, J. Gotay, X. W. Wang, C. C. Harris, T. Ried, and C. X. Deng, *Mol. Cells* **3**, 389–395 (1999).
450. V. Paciotti, M. Clerici, G. Lucchini, and M. P. Longhese, *Genes Dev.* **14**, 2046–2059 (2000).
451. R. G. Martinho, H. D. Lindsay, G. Flaggs, A. J. DeMaggio, M. F. Hoekstra, A. M. Carr, and N. J. Bentley, *EMBO J.* **17**, 7239–7249 (1998).
452. R. S. Weiss, T. Enoch, and P. Leder, *Genes Dev.* **14**, 1886–1898 (2000).
453. E. J. Brown and D. Baltimore, *Genes Dev.* **14**, 397–402 (2000).
454. R. S. Tibbetts, D. Cortez, K. M. Brumbaugh, R. Scully, D. Livingston, S. J. Elledge, and R. T. Abraham, *Genes Dev.* **14**, 2989–3002 (2000).
455. W. M. Michael, R. Ott, E. Fanning, and J. Newport, *Science* **289**, 2133–2137 (2000).
456. H. D. Lindsay, D. J. Griffiths, R. J. Edwards, P. U. Christensen, J. M. Murray, F. Osman, N. Walworth, and A. M. Carr, *Genes Dev.* **12**, 382–395 (1998).
457. N. C. Walworth and R. Bernards, *Science* **271**, 353–356 (1996).
458. P. U. Christensen, N. J. Bentley, R. G. Martinho, O. Nielsen, and A. M. Carr, *Proc. Natl. Acad. Sci. USA* **97**, 2579–2584 (2000).
459. M. A. de la Torre-Ruiz, C. M. Green, and N. F. Lowndes, *EMBO J.* **17**, 2687–2698 (1998).
460. Y. Sanchez, B. A. Desany, W. J. Jones, Q. Liu, B. Wang, and S. J. Elledge, *Science* **271**, 357–360 (1996).
461. Z. Sun, D. S. Fay, F. Marini, M. Foiani, and D. F. Stern, *Genes Dev.* **10**, 395–406 (1996).
462. J. M. Sidorova and L. L. Breeden, *Genes Dev.* **11**, 3032–3045 (1997).
463. T. A. Navas, Y. Sanchez, and S. J. Elledge, *Genes Dev.* **10**, 2632–2643 (1996).
464. M. Lopes, C. Cotta-Ramusino, A. Pellicioli, G. Liberi, P. Plevani, M. Muzi-Falconi, C. S. Newlon, and M. Foiani, *Nature* **412**, 557–561 (2001).
465. J. A. Tercero and J. F. Diffley, *Nature* **412**, 553–557 (2001).
466. N. Rhind and P. Russell, *J. Cell Sci.* **113**, 3889–3896 (2000).
467. M. N. Boddy, B. Furnari, O. Mondesert, and P. Russell, *Science* **280**, 909–912 (1998).
468. M. J. O'Connell, J. M. Raleigh, H. M. Verkade, and P. Nurse, *EMBO J.* **16**, 545–554 (1997).
469. Y. Sanchez, J. Bachant, H. Wang, F. Hu, D. Liu, M. Tetzlaff, and S. J. Elledge, *Science* **286**, 1166–1171 (1999).
470. J. D. Siliciano, C. E. Canman, Y. Taya, K. Sakaguchi, E. Appella, and M. B. Kastan, *Genes Dev.* **11**, 3471–3481 (1997).
471. N. Dumaz and D. W. Meek, *EMBO J.* **18**, 7002–7010 (1999).
472. P. K. Jackson, S. Chevalier, M. Philippe, and M. W. Kirschner, *J. Cell Biol.* **130**, 755–769 (1995).
473. A. Di Leonardo, S. P. Linke, K. Clarkin, and G. M. Wahl, *Genes Dev.* **8**, 2540–2551 (1994).

474. X. Xu, W. Qiao, S. P. Linke, L. Cao, W. M. Li, P. A. Furth, C. C. Harris, and C. X. Deng, *Nat. Genet.* **28**, 266–271 (2001).
475. S. Waga, G. J. Hannon, D. Beach, and B. Stillman, *Nature* **369**, 574–578 (1994).
476. H. Tanaka, H. Arakawa, T. Yamaguchi, K. Shiraishi, S. Fukuda, K. Matsui, Y. Takei, and Y. Nakamura, *Nature* **404**, 42–49 (2000).
477. D. Filatov, S. Bjorklund, E. Johansson, and L. Thelander, *J. Biol. Chem.* **271**, 23698–23704 (1996).
478. W. R. Taylor and G. R. Stark, *Oncogene* **20**, 1803–1815 (2001).
479. M. J. Waterman, E. S. Stavridi, J. L. Waterman, and T. D. Halazonetis, *Nat. Genet.* **19**, 175–178 (1998).
480. M. Huang, Z. Zhou, and S. J. Elledge, *Cell* **94**, 595–605 (1998).
481. R. B. Painter and B. R. Young, *Proc. Natl. Acad. Sci. USA* **77**, 7315–7317 (1980).
482. T. T. Paull, D. Cortez, B. Bowers, S. J. Elledge, and M. Gellert, *Proc. Natl. Acad. Sci. USA* **98**, 6086–6091 (2001).
483. T. T. Paull and M. Gellert, *Genes Dev.* **13**, 1276–1288 (1999).
484. Q. Zhong, C. F. Chen, S. Li, Y. Chen, C. C. Wang, J. Xiao, P. L. Chen, Z. D. Sharp, and W. H. Lee, *Science* **285**, 747–750 (1999).
485. Y. Wang, D. Cortez, P. Yazdi, N. Neff, S. J. Elledge, and J. Qin, *Genes Dev.* **14**, 927–939 (2000).
486. R. Baskaran, L. D. Wood, L. L. Whitaker, C. E. Canman, S. E. Morgan, Y. Xu, C. Barlow, D. Baltimore, A. Wynshaw-Boris, M. B. Kastan, *et al.*, *Nature* **387**, 516–519 (1997).
487. N. Rhind and P. Russell, *Curr. Biol.* **10**, R908–R911 (2000).

Deoxyribonucleotide Synthesis in Anaerobic Microorganisms: The Class III Ribonucleotide Reductase

MARC FONTECAVE,*
ETIENNE MULLIEZ,*
AND DEREK T. LOGAN[†]

*Laboratoire de Chimie et Biochimie des
Centres Rédox Biologiques
UMR CNRS/CEA/Université Joseph
Fourier No. 5047
DRDC-CB, CEA Grenoble
38054 Grenoble Cedex 9, France
†Department of Molecular Physics
University of Lund
Center for Chemistry and Chemical
Engineering
S-221 00 Lund, Sweden

I. Introduction... 96
II. The Anaerobic Ribonucleotide Reductase: A Multicomponent System..... 99
III. The nrdD Protein: The Reductase Component....................... 99
   A. A Glycyl Radical Enzyme 99
   B. Site-Directed Mutagenesis Studies 101
   C. The Three-Dimensional Structure............................. 103
   D. The Reduction of the Substrates: A Radical Mechanism 106
   E. Allosteric Regulation.. 108
IV. The nrdG Protein: The Activase................................... 114
   A. An Iron–Sulfur Protein 114
   B. The Interaction with the Reductase: The $\alpha_2\beta_2$ Complex 115
   C. The Activation Reaction: The Role of the Iron–Sulfur Center of
      Protein β in the Generation of the Glycyl Radical of Protein α 116
   D. The nrdG Protein as a Prototype for a Superfamily of Iron–Sulfur
      Proteins Requiring AdoMet for Activity 119
V. Gene Organization and Regulation 120
VI. The Anaerobic RNR: The Link between the RNA World and
   the DNA World... 120
   References... 125

For growth under oxygen-free atmosphere, some strict or facultative anaerobes depend on a class III ribonucleotide reductase for the synthesis of deoxyribonucleotides, the DNA precursors. Prototypes for this class of enzymes are ribonucleotide reductases from Escherichia coli and bacteriophage T4. This review article describes their structural and mechanistic properties as well as their complex allosteric regulation. Their evolutionnary relationship to class I and class II ribonucleotide reductases is also discussed. © 2002, Elsevier Science (USA).

I. Introduction

DNA synthesis depends on a balanced supply of the four deoxyribonucleotides (1). In all living organisms, with no exception so far, this is achieved by reduction of the corresponding ribonucleoside diphosphates (NDPs) or triphosphates (NTPs) (Fig. 1). This reaction is catalyzed by a family of allosterically regulated metalloenzymes named ribonucleotide reductases (RNRs), which have the remarkable property that they use free radical chemistry for this purpose (2, 3).

It is now generally accepted that life was first based on RNA and that the emergence of a ribonucleotide reductase was the key event that allowed the transition from the RNA to the DNA world. According to that concept we would expect to find only one type of enzyme with the same general structure and the same ribonucleotide reductase mechanism in all organisms. This is to a great extent the case, since all structurally characterized RNRs have, despite the frequent lack of significant primary sequence homology, structural homology in their active site. This site is exquisitely designed to convert a critical cysteine residue into a catalytically essential thiyl radical. It is thus very likely that all RNRs in contemporary metabolism are the products of divergent evolution from a common ancestor (4, 5).

On the other hand, whereas they have conserved a ribonucleotide reduction radical mechanism, RNRs have used different methods for introduction of that

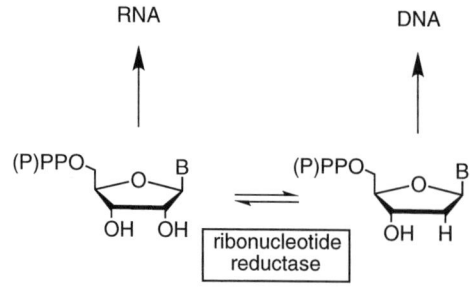

FIG. 1. The reaction catalyzed by ribonucleotide reductases.

radical. Consequently, RNRs have been divided into three classes on the basis of the ancillary metallocofactors and the chemical mechanisms used for the formation of the thiyl radical. The evolutionary relationship between these three mechanisms will be discussed later in this chapter.

Class I RNR is found in mammals and plants, a few bacteria, such as *Escherichia coli,* and bacteriophages. The protein component which holds the active site where the nucleoside diphosphate substrates bind and are reduced is a homodimer of 2 × (80–100) kDa, encoded by the *nrdA* gene and named protein R1 (6). The active site of protein R1 contains in addition to the critical thiyl radical a pair of cysteines, which form a redox-active dithiol providing the two electrons required for the reduction of a ribonucleotide into a deoxyribonucleotide, with concomitant formation of a disulfide bridge. A reduced dithiol is continuously regenerated by the action of the nicotinamide adenine dinucleotide phosphate (NADPH):thioredoxin reductase:thioredoxin system. Protein R1 also contains specific sites for allosteric effectors, which allow the enzyme to reduce each of the four substrate nucleotides in a balanced way. Protein R1 obtains its radical by radical transfer from a second homodimeric protein of 2 × 40 kDa, encoded by the *nrdB* gene and named protein R2. This protein contains a tyrosyl radical, which is derived from the one-electron oxidation of a specific tyrosine residue of the polypeptide chain by molecular oxygen. The reaction is catalyzed by an adjacent nonheme diiron center (7). It follows that RNR activation depends strictly on the presence of oxygen and, accordingly, class I RNRs are only found in aerobic organisms. A subdivision of class I into Ia and Ib has been proposed, with the *E. coli* enzyme being the prototype for Ia and that from *Salmonella typhimurium* that for Ib (2). A distinguishing feature of Ib enzymes is that they lack approximately 50 amino-terminal residues of the Ia enzyme and consequently have different allosteric regulation (2).

Class II RNR is found in bacteria and archaea, both aerobes and anaerobes (8). It occurs as a dimer (also as a monomer) of about 2 × 85 kDa, whose activity is also allosterically regulated. Its active site contains the three essential cysteines present in class I RNRs and the reducing equivalents are also provided by the NADPH/thioredoxin system. The thiyl radical is generated during reaction with adenosylcobalamin. The latter undergoes a homolytic cleavage of its Co—C bond and generates a 5′-deoxyadenosyl radical, which abstracts the hydrogen atom of the key cysteine residue of the active site.

Class III RNR is found in anaerobically growing bacteria and archaea, strict or facultative anaerobes (Table I) (8). Details of its structure, chemistry, and regulation will be given in this chapter. Briefly, it consists of a homodimeric protein, encoded by the *nrdD* gene and named protein α, which differs from

TABLE I
OCCURRENCE OF CLASS III RIBONUCLEOTIDE REDUCTASES
IN NATURE

Archaebacteria	
Methanococcus jannaschii	M_jan
Methanobacterium thermoautotrophicum	M_the
Pyrococcus horikoshii	P_hor
Eubacteria	
Escherichia coli	E_col
Salmonella typhimurium	S_typ
Shewanella putrefaciens	S_put
Vibrio cholerae	V_cho
Haemophilus influenzae	H_inf
Streptococcus pyogenes	S_pyo
Streptococcus mutans	S_mut
Streptococcus pneumoniae	S_pne
Enterococcus faecalis	E_fae
Lactococcus lactis	L_lac
Staphylococcus aureus	S_aur
Clostridium acetobutylicum	C_ace
Thermotoga maritima	T_mar
Clostridium difficile	C_dif
Pseudomonas aeruginosa	P_aer
Ralstonia eutropha	R_eut
Viruses	
Bacteriophage T4	BP_T4

class I and class II RNRS both by its requirement for formate as the reducing agent for ribonucleotide reduction and the absence of a redox dithiol in the active site, as well as by the mechanism of radical formation. Class III RNR uses a second protein, encoded by the *nrdG* gene and named protein β, for activation, as does class I RNR; however, β is an iron–sulfur protein which catalyzes the oxidation of a glycine residue of the active site of protein α to a stable, but oxygen-sensitive glycyl radical by S-adenosylmethionine. This glycyl radical plays the role of tyrosyl radical in class I and adenosylcobalamin in class II RNRs, namely the abstraction of the H atom of the critical cysteine in the active site.

In the following the class III RNR is described mainly on the basis of the results obtained with the *E. coli* enzyme. Detailed studies of the class III RNR from bacteriophage T4 by Britt-Marie Sjöberg's group or *Lactococcus lactis* by Peter Reichard's group are also available. They essentially show that the three enzymes behave similarly, except for minor differences related to allosteric regulation.

II. The Anaerobic Ribonucleotide Reductase: A Multicomponent System

In 1989, with P. Reichard at the Karolinska Institute (Stockholm, Sweden), we demonstrated in cell-free extracts from anaerobically grown *E. coli* the presence of an oxygen-sensitive enzymatic activity, different from class I and class II RNRs, that reduced NTPs to dNTPs (9). Standard fractionation and biochemical recombination techniques demonstrated the requirement for three separable protein fractions. One contained the reductase itself (protein α) in association with significant amounts of a small protein, the second the electron transfer protein NADPH:flavodoxin oxidoreductase, and the third flavodoxin, the substrate of the flavodoxin reductase (10–13). In addition the following low-molecular weight compounds were all found to be required for the reduction of the CTP substrate: K^+, S-adenosylmethionine (AdoMet), dithiothreitol (DTT), Mg^{2+}, ATP, and formate (10, 14, 15). Because of a tight interaction between protein α and the small protein, the purest preparations of protein α were always contaminated with that protein. These preparations were by themselves completely inactive, but could be activated during prolonged anaerobic preincubation with all of the other components before initiating the reaction with the substrates.

When the *nrdD* gene, 2136 bp long and coding for protein α, which contains 712 amino acid residues, was cloned it was realized that the small protein, with a polypeptide chain of 17.5 kDa, was encoded by a gene (*nrdG*) immediately downstream of the *nrdD* gene, within the same operon (11). Then standard molecular biology techniques were used to express each of the proteins separately in *E. coli* at a high level and purify them. It then became clear that the product of the *nrdG* gene, named protein β, was absolutely essential for the activation of protein α and was thus defined as an activating component or an activase.

III. The *nrdD* Protein: The Reductase Component

A. A Glycyl Radical Enzyme

The overexpressed reductase from *E. coli* could be purified to homogeneity in a few steps. Chromatography on an affinity dATP–Sepharose column proved to be extremely efficient as a consequence of the allosteric regulation of protein α by nucleotides and its ATP- and dATP-binding properties (10). The protein was characterized as an α_2 homodimer (2 × 80 kDa) with no bound cofactors and no activity.

Protein α was generally purified aerobically. However, it was recently realized that this resulted in the formation of a disulfide bridge probably located on the C-terminal end of the polypeptide containing conserved cysteines essential for activation (*16, 17*). We thus had an explanation for the requirement for DTT. Indeed, reduction of protein α with DTT or, even better, with the NADPH:thioredoxin reductase:thioredoxin system followed by purification resulted in a preparation whose activity was no longer dependent on DTT (*16*).

During anaerobic reaction with protein β, AdoMet, DTT, and a reducing system (the flavodoxin system, dithionite, or photoreduced deazaflavin), protein α is converted into an active enzyme. The latter contains a stable organic radical, characterized by an electron paramagnetic resonance (EPR) signal at $g = 2.0033$, showing a large doublet splitting (14 G) (*18, 19*).

A series of observations has unambiguously established that the organic radical is localized on the glycine residue Gly681 of the large protein and that catalysis of ribonucleotide reduction by formate is achieved only by the glycyl-radical-containing form of protein α. First, active enzymes selectively labeled with ^{13}C- or ^{2}H-glycine display significantly different EPR spectra (*19*). In particular the presence of a singlet signal in the ^{2}H-glycine-substituted enzyme demonstrates that the 14-G doublet splitting in the wild-type enzyme originates from the hyperfine coupling of the unpaired electron spin to the α-hydrogen nucleus of the glycyl radical center. Second, mutation of glycine 681 of protein α to alanine results in an enzyme which completely lacks a detectable organic radical as well as enzymatic activity. Third, the organic radical is extremely oxygen-sensitive and exposure of the active enzyme to air leads to truncation of the polypeptide chain at Gly681 accompanied by irreversible inactivation (*20*). Fourth, there is a good correlation between the amount of radical present in the large component and the specific activity of the holoenzyme. Fifth, as shown with the *L. lactis* protein α, which can be readily separated from its associated protein β, the latter is no longer required for activity once the radical is introduced into protein α (*21*).

The extreme sensitivity of the radical may explain why it was generally lost during purification and why the purified enzyme was inactive. However, with stricter adherence to anaerobic conditions during purification, it was possible to isolate an enzyme that maintained a sizable amount of its glycyl radical without any *in vitro* activation step. It was thus found that the reduction of the CTP substrate was dependent only on Mg-ATP and formate. All the other components are involved only in the activation of the enzyme. Following this realization, the activity of protein α is now assayed in two steps. In the first, protein α is activated by anaerobic incubation with protein β in the presence of AdoMet, DTT, NADPH, flavodoxin reductase and flavodoxin, and K$^+$. Dithionite or photoreduced deazaflavin can substitute for the flavodoxin system (*13, 22*). In the second step, an anaerobic solution containing the radioactive substrate CTP, formate, ATP, and Mg^{2+} is added to the activation mixture and dCTP formation is monitored as a function of time. Under these conditions, the maximal amount of glycyl radical that could be incorporated in the first step was one equivalent per homodimer (23). Whether this reflects an intrinsic property of the radicalization reaction or is due to an inefficient reaction under the *in vitro* conditions utilized is unknown.

The class III RNR from *E. coli* shares many properties with pyruvate formate lyase (PFL). PFL is a glycyl radical enzyme which converts pyruvate into formate and acetyl-CoA (24, 25). It is interesting to note that these two systems are essential enzymes of anaerobic metabolism and linked together by formate, the product of the PFL reaction and the substrate of the RNR. The radical is very sensitive to oxygen, and truncation of PFL occurs during air exposure of the active enzyme as well (26). PFL also acquires its radical during anaerobic reaction with AdoMet and an activating enzyme in the presence of the flavodoxin system (27).

B. Site-Directed Mutagenesis Studies

Comparison of the 20 available sequences of the *nrdD* gene product reveals a strongly conserved RXXGY sequence around G681 (corresponding to G580 in T4 RNR) in 18 of the sequences (Fig. 2A). Site-directed mutagenesis studies in the case of the *E. coli* enzyme have nevertheless shown that the adjacent conserved tyrosine is not absolutely required for radical formation and activity (19). The Y682F protein had about 10% activity and radical formation as compared to the wild-type enzyme. Highly homologous boxes have been identified in other glycyl radical enzymes, such as pyruvate formate lyase (PFL) and benzyl succinate synthase (BSS) (28). On this basis it is quite clear that methanogens such as *Methanococcus jannaschii* and *Methanobacterium thermoautotrophicum* and hyperthermophilic archaea such as *Pyrococcus horikoshii* depend on a glycyl radical-dependent class III RNR for growth, even though these enzymes have not been isolated yet.

Furthermore, among the 15–20 Cys generally present in these sequences, 6 (Cys79, 290, 543, 546, 561, 564) are completely conserved in 16 RNRs (Figs. 2A–2C). Site-directed mutagenesis studies on the enzyme from bacteriophage T4 have demonstrated that 2 of the 6 invariant cysteines are directly involved in ribonucleotide reaction, in agreement with the presence of these 2 cysteines (Cys79 and Cys290 in the T4 enzyme) in the active site as revealed by the three-dimensional structure (see below) (17). The 4 other invariants are part of two neighboring CXXC boxes (not visible in the three-dimensional structure) located near the C-terminus of the polypeptide, just upstream of the radical site (Fig. 2A). Each of these cysteines was found to be essential for the formation of the protein radical, as shown by site-directed mutagenesis studies (17).

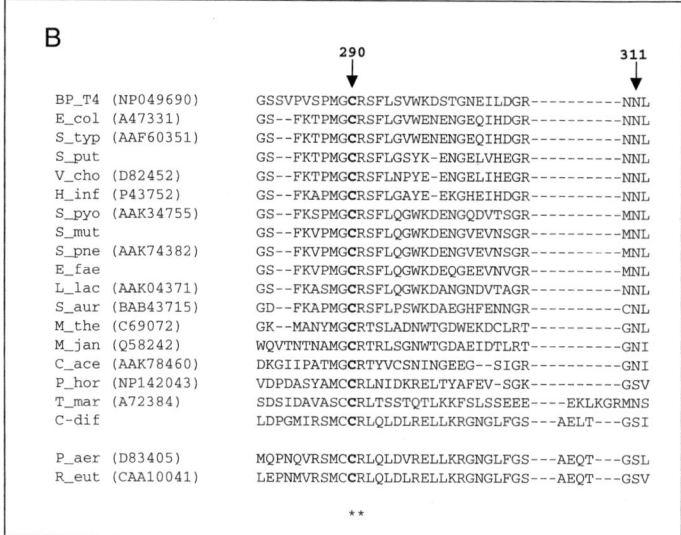

FIG. 2. Amino acid sequence alignments for the *nrdD* gene product.

The RNRs from *T. maritima* and *C. difficile* lack the cysteine corresponding to Cys546 (Fig. 2A). Site-directed mutagenesis studies on the RNR from *Ralstonia eutrophus* demonstrate that the radical site is located upstream of the conserved cysteines of the C-terminus (Fig. 2A) and furthermore that one cysteine of the CXXC boxes is missing (29). The enzyme from

```
                      C                          79
                                                 ↓
        BP_T4  (NP049690)    IDYSPALPFTNCCLVDLKGMLENGFKLGNA
        E_col  (A47331)      LDYSPFFPMFNCMLIDLKGMLTQGFKMGNA
        S_typ  (AAF60351)    LDYSPFFPMFNCMLIDLKGMLTQGFKMGNA
        S_put                LDYSPFFPMFNCMLIDLAGMLTHGFKMGNA
        V_cho  (D82452)      LDYAPFFPMFNCMLIDLKGMLTHGFKMGNA
        H_inf  (P43752)      LDYAPFFPMFNCMLVDLEGMLSRGFKMGNA
        S_pyo  (AAK34755)    LDYSPYTPMTNCCLIDFKGMLANGFKIGNA
        S_mut                LDYSPYTPMTNCCLIDFKGMLHSGFKIGNA
        S_pne  (AAK74382)    LDYSPYTPMTNCCLIDFKGMLENGFKIGNA
        E_fae                LDYHPYTPMTNCCLIDFKGMLNNGFKIGNA
        L_lac  (AAK04371)    LDYSPFTTMANCCLIDFKNMFENGFKLGNA
        S_aur  (BAB43715)    LDYHPFQPLTNCCLIDAKNMLHNGFEIGNA
        M_the  (C69072)      LEFFAARPL-NCLQHDLRLFIRHGLRVDGT
        M_jan  (Q58242)      LEYAATRPV--CLQHDLRPFFKYGLKVDGT
        C_ace  (AAK78460)    LD--SYNLTTNCLHIPTKEVLQSGFNTGYG
        P_hor  (NP142043)    LPYSLYIP--YCTGHSIARLLEKGLKTPTI
        T_mar  (A72384)      RHHAALMP--YCFAYTLKPIVEKGLPFIKT
        C_dif                LD-- MLSG--YCAGWSLRQLLEEGFGGVPG

        P_aer  (D83405)      LD--MLAG--YCAGWSLRTLLNEGLNGVPG
        R_eut  (CAA10041)    LD--MLSG--YCAGWSLRTLLHEGFNGVPG
                                      *          *
```

FIG. 2. *(continued)*

P. aeruginosa looks more similar to that from *R. eutrophus* than to the others.

C. The Three-Dimensional Structure

The only three-dimensional structure of a class III ribonucleotide reductase available is that from bacteriophage T4 shown in Fig. 3 (30). Because the enzyme was expressed and isolated as a complex of the reductase with the activase protein, a Gly580Ala mutant protein (corresponding to Gly681 in the *E. coli* sequence) was used so as to avoid complications with possible cleavage and denaturation of the protein due to the oxygen sensitivity of the radical site. The current refined model is at a resolution of around 2.5 Å, with an R factor of 22%, partly due to the presence of several disordered or poorly ordered regions.

Despite the lack of homology between class I protein R1 and class III protein α, a very similar structural organization is observed for the two systems, in particular as far as the active site is concerned (30, 31). Although it has not been possible to obtain a class III enzyme–substrate complex, a good idea of where the substrate binding occurs has been obtained by modeling the coordinates of GDP as observed in the class I enzyme (32) into the T4 ARR active site (Fig. 4). A crystal structure of a class II reductase confirms both the general organization and the conservation of the key structural features associated with the catalytic and regulatory machinery of class I and class III RNRs (66).

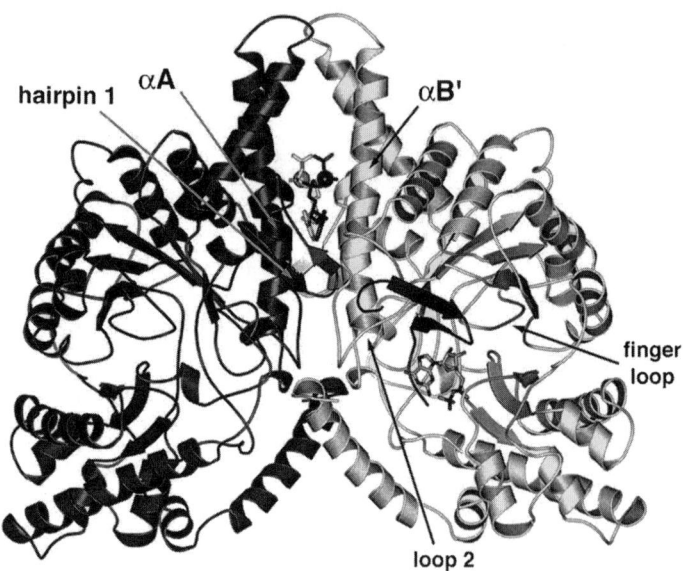

FIG. 3. View of the *nrdD* protein dimer as seen in the crystal structure. Monomer A is on the right and shown in light gray; monomer B is on the left and shown in dark gray. The dimer axis is vertical. The C-terminal finger of monomer A, containing the glycyl radical site at its tip, is drawn in black. Mg-dATP as observed in the crystal structure is seen close to the dimer interface; in this orientation the symmetry-related molecule of Mg-dATP lies behind. A model for substrate binding is given by the position of GDP in the active site of monomer 1. Note, however, that substrates of RNR are triphosphates. Secondary structure elements relevant for the discussion of allosteric regulation are labeled.

The T4 enzyme is shaped like a barrel made up of two parallel, five-stranded β sheets arranged in antiparallel manner and linked by a loop protruding through the inside of the barrel ("the finger loop"). The barrel itself is flanked by helices, which connect the strands. The major differences between the structures of class I and class III RNRs lie at the dimer interface and at the C-terminus. In class III RNR each monomer contibutes two extended helices, which pack against each other in an antiparallel fashion at the twofold axis. Moreover, the two finger loops inside each barrel point in opposite directions. In contrast, in class I RNR the two barrels point approximately 90° from each other (33). As a result the dimer axis is almost at 90° with respect to that observed in the class III RNR.

These structural homologies help in identifying the key residues of the class III RNR, as they are found at the same positions as essential residues in the class I RNR. The active site contains two conserved cysteines, which can be superimposed on two of the three essential cysteines present in the structure

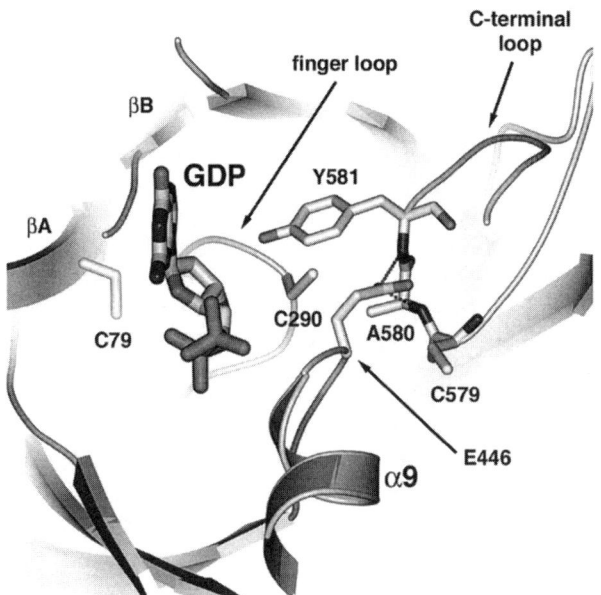

FIG. 4. The C-terminal loop containing the glycyl radical site and its interactions with the active site. The view is looking directly down into the α–β barrel. The finger loop protrudes up through the center of the barrel and the C-terminal loop points down into the barrel from the right. The tips of these loops meet such that the distance between the sulfur atom of Cys290 and the alpha carbon of Ala580 (in the G580A mutant used to solve the crystal structure) is 5.2 Å. However, this distance may be shorter in the native structure. Main-chain and side-chain atoms are shown for the residues flanking Ala580. Note the hydrogen bonds between the main-chain nitrogens of the radical site and the conserved residue Glu446 on helix α9.

of class I R1 (Fig. 4). These cysteines are respectively Cys290 (equivalent to Cys439 of R1), placed at the extremity of the finger loop, and Cys79 (equivalent to Cys225 of R1) on strand βA (the numbers refer to the T4 enzyme). This suggests strongly that Cys290 is the thiyl radical site required for substrate activation and is consistent with the site-directed mutagenesis experiments already mentioned, as well as studies by photoinduced radiolabeling (17, 34). In place of the third cysteine (Cys462 of R1), Asn311 is found, which is also conserved in 16 of 20 $nrdD$ protein sequences (in the others it is replaced by Ser). In class I RNA, Cys462 and Cys225 provide the redox-active dithiol, which serves as the two-electron donor. The absence of a dithiol in class III RNR is in line with formate being the hydrogen donor (see below). Asn311 has been shown to be important for activity and proposed to participate in correct positioning of formate in the active site (35). Two further important side chains of class I RNR are absent in the $nrdD$ protein: Glu441 and Asn437 are replaced by Met288 and Ser292, respectively. These two residues are oriented in such a way that they cannot interact with the substrate. In particular, Met288 is buried in a hydrophobic pocket and this orientation is probably essential for stabilization of the finger loop.

The glycyl radical site, an alanine in the structure, is located near the C-terminus (at position 580 in T4 RNR) and is a relatively ordered segment of a region which otherwise shows high flexibility. The barrel structure ends at around residue 543 and, of the C-terminal residues, only 571–586 (of a total of 605) could be located in the structure. These form a loop directed right into the barrel where the tip containing the mutated glycyl radical site Ala580 meets the tip of the "finger loop" in the active site (Fig. 4). This places the radical site very close (5.2 Å) to Cys290, the putative thiyl-radical-abstracting species, and suggests that the thiyl radical is formed by direct H-atom abstraction by the glycyl radical. The side chain of Cys290, with the sulfur radical at the extremity, thus almost certainly swings around in order to react with the substrate. Although the crystallographic unit contained the activase protein, the latter did not appear in the structure, probably because of a low occupancy and/or disorder. Therefore at present we can only speculate on the way the β protein binds to the α protein.

D. The Reduction of the Substrates: A Radical Mechanism

Little is known on the mechanism of ribonucleotide reduction by class III enzymes subsequent to cysteinyl radical formation. However, it is very likely that the reaction proceeds by a radical mechanism similar to that of class I and class II enzymes, which in both cases has been delineated, mainly from J. Stubbe's outstanding studies (36). Accordingly, mechanism-based inhibitors of class I and class II RNRs, such as nucleotide analogs carrying azido, chloro, or fluoro groups at the 2' position, are also excellent inhibitors of class III enzymes (37).

By analogy with Stubbe's mechanism we propose that the reaction is initiated by the abstraction of the 3'-H atom of the ribose to generate a substrate radical from which 2'C—OH bond heterolysis is facilitated. The abstracting species is likely to be a thiyl radical, from Cys290 (for the T4 enzyme). This is deduced from the active-site three-dimensional structure and site-directed mutagenesis studies (see above). When the reaction was conducted in deuterated buffer a small amount (1–2%) of deuterium was found in the 3' position of the deoxyribonucleotide product (38). This is consistent with the hypothesis that the abstracted 3'-H atom is reintroduced at the same position of the deoxyribose at the end of the cycle and that the protein radical site which abstracts the 3'-H atom is exchangeable with the solvent. It thus cannot be the glycyl radical itself, since no exchange with water could be observed at that residue, and fits better with a thiyl radical.

CLASS III RIBONUCLEOTIDE REDUCTASE

Whereas class I and class II RNRs use a pair of redox cysteines as the primary electron donor and an enzymatic reducing system to keep them reduced, class III RNR uses formate as the external reducing agent. This has been clearly shown by using ^{14}C-labeled formate and correlating the amount of [^{14}C]CO$_2$ formed with that of the dCTP product (15). Moreover, using [^{3}H]-labeled formate, the isotope was recovered in water exclusively, demonstrating that the hydrogen from formate is not transferred directly to the 2' position of the substrate, but instead indirectly through exchangeable sites. This is in agreement with the observation of a stereospecific incorporation of deuterium at the 2' position (with retention of configuration with regard to the 2'-OH group) when the reaction was performed in D$_2$O and the reaction product analyzed by nuclear magnetic resonance spectroscopy (38).

From these results we proposed the following mechanism (Fig. 5) (8, 31). After abstraction of the 3'-H of the ribose by the Cys290 radical on one face of the ribose moiety, the 2'-OH is protonated by the second cysteine, Cys79,

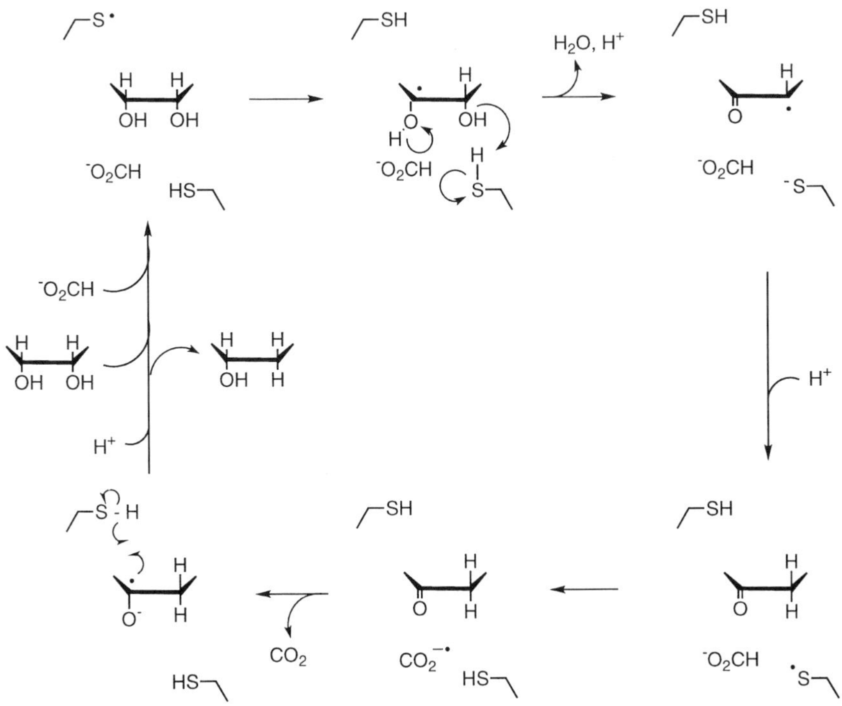

FIG. 5. Proposed mechanism for ribonucleotide reduction by formate catalyzed by class III ribonucleotide reductases.

on the other face (Fig. 4). Loss of water at this position and deprotonation of the 3'-OH generates a 3'-keto radical intermediate. A one-electron reduction of that intermediate by the anionic Cys79 would then yield on one hand the 3'-keto nucleotide common to previously proposed mechanisms for all classes of RNRs (36) and on the other hand a Cys79 radical, which is proposed to abstract a hydrogen atom from formate. The resulting CO_2^- formyl radical is an excellent reducing agent, capable of reducing ketones into ketyl radicals. We propose that the intermediate 3'-ketyl radical finally yields the deoxyribonucleotide by recovering the 3'-H (radical abstraction) from Cys290 and the 3'-OH group (protonation). In this hypothesis formate is thus viewed as a source of CO_2^- formyl radicals. A recent report based on quantum mechanical calculations supports a mechanism in which formate plays two roles (39). First, it would assist in the deprotonation of the 3'-OH group, thus facilitating the release of a water molecule from the 2'-C—OH bond heterolysis. In class I RNR this role is fullfilled by a glutamate side chain, Glu441. Second, it would be oxidized by the Cys79 thiyl radical intermediate to generate the highly reducing CO_2^- formyl radical.

E. Allosteric Regulation

Class III RNR is an allosterically regulated enzyme. In this case allosteric mechanisms allow the same enzyme to reduce the four common ribonucleoside triphosphates in a regulated manner so that the cell is supplied with a balanced production of deoxyribonucleotides, which is a requirement for DNA replication and repair (1). By definition an allosteric enzyme contains a specific binding site for an effector distinct from the substrate-binding site. Binding of the effector leads to a conformational change in the protein affecting the substrate binding and the catalytic activity. The allosteric regulation of class III RNR has been extensively investigated in the case of the *E. coli*, *L. lactis*, and bacteriophage T4 enzymes (40–42). It has been shown that the allosteric effectors are deoxyribonucleoside triphosphates dTTP, dATP, and dGTP, the products of the reaction, and ATP. There is no effect of dCTP.

In the *E. coli* and *L. lactis* enzymes comparable results were obtained, which could be interpreted in terms of the presence of two effector sites on the same polypeptide chain (Fig. 6). One of the sites binds one molecule of either dATP or ATP. Binding of ATP stimulates the reduction of the two pyrimidine ribonucleotides CTP and UTP, whereas dATP binding shuts off the enzyme. This site has been named the pyrimidine site. The second effector site, named the purine site, since it regulates the reduction of ATP and GTP, binds one molecule of dATP, dGTP, or dTTP. With dGTP bound, the enzyme reduces ATP, whereas with dTTP bound it reduces GTP. Again, binding of dATP results in an inactive enzyme. Consequently, with ATP bound to the pyrimidine site the enzyme

CLASS III RIBONUCLEOTIDE REDUCTASE

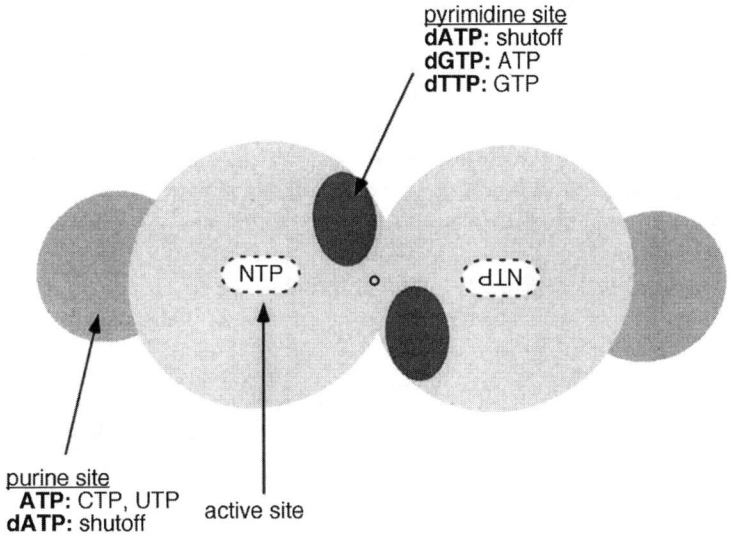

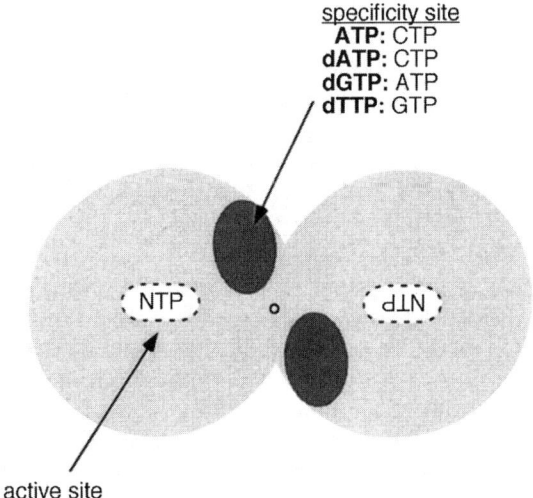

FIG. 6. Schematic overview of allosteric regulation in the class III RNRs.

produces dTTP, which, upon binding to the purine site, turns off UTP and CTP reduction and favors the production of dGTP. Binding of dGTP then results in the accumulation of dATP, which finally, by binding to both effector sites, shuts off the whole process. K_m values have been determined in the case of the E. coli enzyme (40). For substrates they are in the millimolar range, whereas apparent K_m values for modulators are between 10 and 100 μM, the K_m values for the adenosine nucleotides (4 mM for substrate ATP and 100 μM for effector dATP) being the largest and those for guanosine nucleotides (0.4 mM for substrate GTP and 4 μM for effector dGTP) the smallest. These values are consistent with the physiological cellular concentrations of ribo- and deoxyribonucleotides.

The allosteric regulation is slightly different in the case of the bacteriophage T4 enzyme (Fig. 6), which contains only one effector site, which shares a number of properties with the purine site of the E. coli RNR (41). dATP and high concentrations of ATP are positive effectors for CTP reduction, dGTP is a positive effector for ATP reduction, and dTTP is a positive effector for GTP reduction. This is closer to the classical pattern of allosteric regulation in ribonucleotide reductases as exemplified by the class I aerobic enzymes. The major difference is that there is no general negative allosteric effector. This latter feature might be advantageous for a phage, as it allows a continuous supply of dNTPs for efficient replication. K_d values for the effectors are of the same order of magnitude as for the E. coli enzyme (41).

One difference in primary structure between the E. coli and the bacteriophage T4 ARR is that the latter lacks approximately 100 residues present at the N-terminal end of the E. coli enzyme. It is thus tempting to suggest that the second effector site (the pyrimidine site) of the E. coli RNR is present in its N-terminal end. This is supported by the finding of limited but significant sequence homology between this N-terminal end and that of the E. coli aerobic RNR, which was shown by X-ray crystallography to carry one of the two effector sites (32). Also, unlike the E. coli and L. lactis enzymes, the T4 RNR is unable to reduce UTP, but phage T4 is assumed to synthesize its precursors for the dTTP pool via deamination of dCMP (41).

That a nucleotide-binding site, distinct from the substrate-binding site, is present on the class III RNR was unambiguously demonstrated by X-ray crystallographic analysis of the T4 enzyme. Crystal structures have been determined in complex with dATP, dTTP, dGTP, and dCTP (43). The dNTP effectors bind to one and the same site close to the dimer interface, along the length of two α helices that make up an important part of the dimer interaction area (Figs. 3 and 7). One of these helices comes from each monomer. Also involved is a β hairpin (hairpin 1) that spans the dimer interface. About half of the effector is buried in interactions with each monomer. From the enzyme–GDP model we

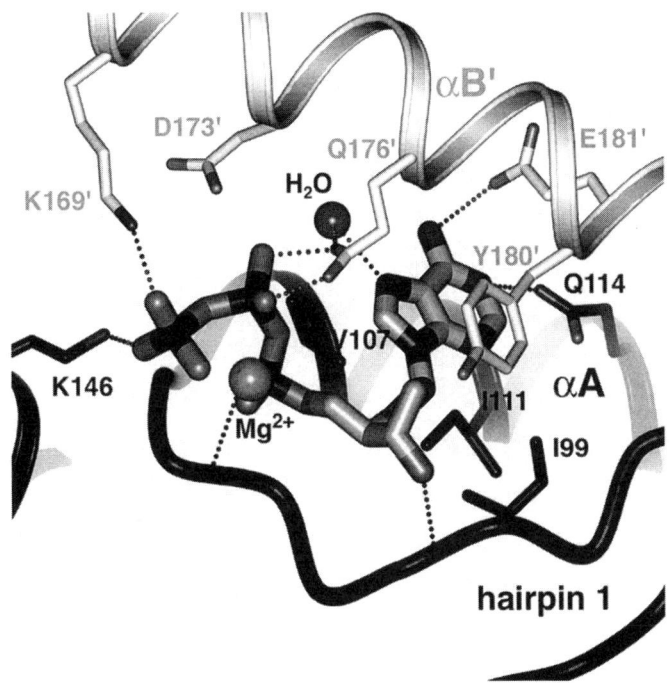

FIG. 7. Closeup view of allosteric specificity of effector binding. dATP has been taken as a representative example. The view is approximately a 90° anticlockwise rotation with respect to Fig. 1. Helix αA and hairpin 1 from monomer 1 are drawn in black; helix αB from monomer 2 is drawn in gray. Important residues in the vicinity of the effector are labeled. A prime symbol indicates that the residue comes from monomer 2. Hydrogen bonds between the effector and the surrounding residues are represented by dotted lines.

can calculate that the distance between the specificity site and either of the two active sites is approximately 25 Å (Fig. 3).

The recognition of the ribose and phosphate groups of the effectors, and to a certain extent the bases, is essentially identical in all the complexes (Fig. 7). The negative charges of the phosphate groups are neutralized by a number of lysine residues in the vicinity. In addition, the electron density appears to show an Mg^{2+} ion coordinated between the three phosphate groups, although this has not been directly confirmed experimentally. The β-phosphate moieties make hydrogen-bonding interactions with the highly conserved side-chain Gln176. In some of the structures it has been possible to identify bound water molecules, which in some cases may be metal ions. The ribose moiety of the nucleotide makes a single hydrogen bond, through the 3'-OH atom, to the main chain of

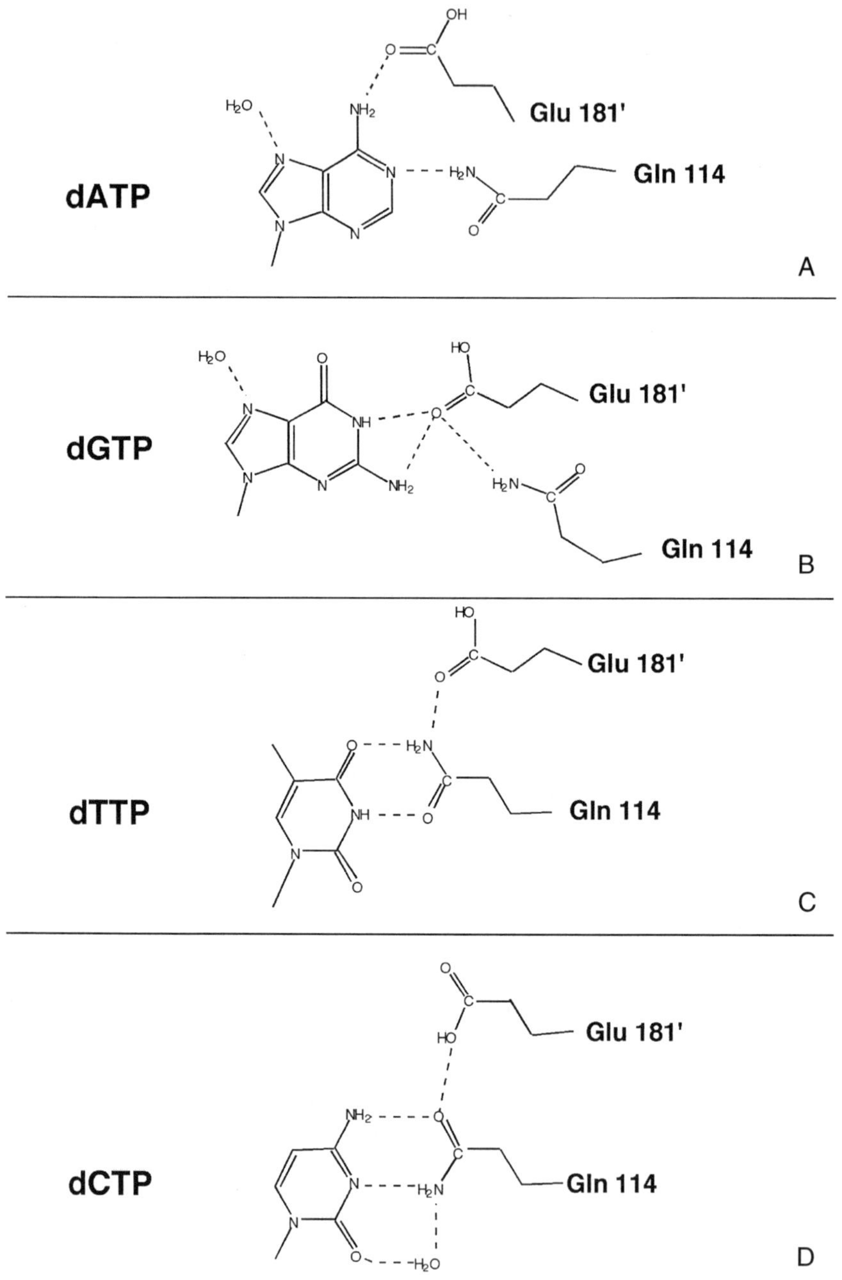

dATP

Glu 181'
Gln 114

A

dGTP

Glu 181'
Gln 114

B

dTTP

Glu 181'
Gln 114

C

dCTP

Glu 181'
Gln 114

D

residue 100 in hairpin 1; another H bond to hairpin 1 is made by one of the α-phosphate oxygens. The base is in all complexes sandwiched between Tyr180′ and Ile111 (where the prime indicates a residue from the second monomer) (Fig. 7).

The discrimination between the different specificity effectors rests on recognition of the hydrogen-bonding groups on the base moiety. This is achieved by two amino acid side chains, one from each monomer: Gln114 and Glu181′. These residues adopt different conformational patterns in each complex that are optimally matched to the hydrogen-bonding capacity of the base. The pattern for dATP is shown in Fig. 7. Figure 8 concentrates on the hydrogen-bonding interactions with the four dNTP bases. The question is how such subtle conformational changes at the specificity site affect the relative affinities for the substrates at the active site. When the three complexes are compared, conformational changes can be observed that stretch all the way between the two sites. In the case of the two purine effectors, dATP and dGTP, the essential difference between the two bases is the presence of the extracyclic 2-NH_2 group in dGTP. This affects Gln 114, which is unable to hydrogen-bond to the base, as it does in the other complexes. Conformational changes in both Gln114 and Glu181′ propagate, through a cascade of side-chain movements in αB′ and the following loop (loop 2, see Fig. 3), to Phe 194′, which interacts with the substrate base through stacking in our current model for substrate binding (not shown in Fig. 4). In contrast to this simple side-chain switching, the change from purine to pyrimidine effector binding, as exemplified by the dATP and dTTP complexes, causes more dramatic conformational changes, involving the large-scale refolding of loop 2 and a shift in hairpin 1. However, the net effect is still that Phe 194′ changes side-chain conformation.

In the absence of structures of substrate complexes we can only speculate on how these conformational changes affect substrate binding. However, the mechanism would appear to be somewhat different from that employed at the effector site, in that the only active-site side chain which undergoes a significant conformational change, Phe194, has no hydrogen-bonding capacity. Although the relative amounts of given dNTP products are only raised by factors ranging from 2 to 60 by interaction of the enzyme with appropriate effectors, alteration of the stacking interaction may not in itself suffice to allow it to discriminate

---

FIG. 8. Schematic view of the interactions of the side chains of Gln114 and Glu181′ with the bases of four allosteric effectors: (A) dATP, (B) dGTP, (C) dTTP, and (D) dCTP. In each case, the hydrogen-bonding pattern of the side chains is perfectly matched to the donor and acceptor groups available on the six-membered ring of the base. Note that dCTP has no effect on RNR at physiological concentrations. The complex was obtained by soaking crystals in 10 mM dCTP. Nevertheless, the results show that the specificity site is primed for recognition of dCTP in a similar manner to dTTP.

between different substrate bases. The most likely scenario is that Phe194 affects the substrate conformation in such a way that it interacts in different ways with fixed hydrogen-bonding groups in the active site. Main-chain atoms near the base may participate in these interactions (43).

In the approximately 20 currently known *nrdD* protein sequences, the conservation of residues involved in the substrate specificity site is very good. On this basis we can consider the structure of the specificity site of the T4 enzyme to be representative for all class III RNRs and any mechanism derived from it to be general for this class. The nature of the pyrimidine site in the *E. coli* and *L. lactis* enzymes is an unresolved question which awaits more structural data.

IV. The *nrdG* Protein: The Activase

As mentioned above, the β protein is absolutely required for the introduction of the glycyl radical into the α polypeptide and is consequently defined as the activating component. The fact that this protein acts catalytically, that is, one molecule can activate several molecules of the reductase, allows us to define it as an activase (44). Accordingly, a radical-containing protein α no longer requires protein β for activity, as shown with the *L. lactis* enzyme (21).

A. An Iron–Sulfur Protein

Overexpression of the *nrdG* protein in *E. coli* BL21(DE3) leads to red-colored protein extracts from which protein β can be purified. An ultraviolet-visible analysis of the pure protein showed a broad absorption band at around 420 nm characteristic for S-to-Fe charge transfer (11, 18, 45). Metal and sulfur quantitation confirmed the presence of roughly equal and substoichiometric amounts of iron and sulfide, thus suggesting the presence of iron–sulfur centers in the protein.

During incubation with an excess of ferrous iron and sodium sulfide under strict anaerobiosis, the β apopolypeptide takes up to four Fe and four S to form a $[4Fe-4S]^{2+}$ center, as unambiguously shown by Mössbauer spectroscopy (44). This center is redox-active and can reach the EPR-active $S = 1/2$ $[4Fe-4S]^+$ state during anaerobic reduction with strong reductants such as dithionite or photoreduced deazaflavin (46). Reduction is never quantitative, as the yield of the reaction is not more than 50–80%, depending on the reductant and protein concentration. This is in agreement with the very low redox potential of the cluster, which was determined from direct electrochemistry of protein β adsorbed on a gold electrode in the notable absence of additives and/or mediators (23). The β protein responds amazingly well to the electrode under these conditions, suggesting a direct interaction between the center and the electrode. In the absence of DTT, the protein displays a pseudo-reversible wave at $E'^0 = -500$ mV (vs. NHE) which is shifted to slightly more negative value (-550 mV) in the presence of 10 mM DTT.

The iron–sulfur center proved to be very sensitive to oxygen, in the presence of which it is converted into one stable $[2Fe-2S]^{2+}$ cluster per monomer (47, 48). The reaction can be reversed, since under reducing conditions the iron and sulfide of the $[2Fe-2S]^{2+}$ centers can be remobilized and assembled back into a $[4Fe-4S]^{2+}$ center in half of the polypeptides. Thus the Fe–S center has the ability to undergo a $[4Fe-4S]^{2+}$-to-$[2Fe-2S]^{2+}$ interconversion. The physiological significance of this reaction is unknown.

The protein sequence shows the presence of a highly conserved box of three cysteines in a CXXXCXXC motif (the "cysteine triad") located in the N-terminal part of the polypeptide. The *E. coli* β protein has two other nonconserved cysteines. Site-directed mutagenesis of the five cysteines has led to the following conclusions (49): (1) The iron center is bound to the protein by the three cysteines of the triad only, in agreement with Mössbauer properties of the cluster of the wild-type protein; (2) the triad mutants retain the ability to form a [4Fe–4S] center, presumably bound by only two cysteines, also in agreement with Mössbauer data; and (3) the fourth ligand, still unidentified, might be another side chain of a residue of the polypeptide chain or a water molecule. This coordination site seems to be occupied by DTT when the reduced protein is incubated with DTT. The iron–DTT interaction is indicated by the strong effect of DTT on the shape and the microwave power saturation properties of the EPR signal of the $[4Fe-4S]^+$ center (46). The latter is initially rhombic with g values at 2.03, 1.92, and 1.86, and, in the presence of 10 mM DTT, becomes axial with g values at 2.03 and 1.92. The effect of DTT on the redox potential of the cluster (see above) is also consistent with an iron–DTT interaction. Taken together, these results thus demonstrate that the Fe–S center of the β protein has an exchangeable site.

It is interesting to note that the cysteine triad is also present in the PFL activase and that the latter is also an iron–sulfur protein carrying a [4Fe–4S] center chelated by the three cysteines of the triad, as shown by site-directed mutagenesis (50). As shown below, this triad defines a superfamily of enzymes.

B. The Interaction with the Reductase: The $\alpha_2\beta_2$ Complex

In the case of the *E. coli* RNR, the interaction between the two proteins α and β proved so strong that it was not possible to separate them by chromatography. Generally, a purified complex can be obtained using a dATP–Sepharose affinity column specific for the α protein. Sucrose gradient sedimentation experiments with such preparations showed that the unactivated α protein forms a tight 1:1 complex with the β protein, whether the latter contains an [Fe–S] center or not. Since protein α is in the form of a homodimer, the system is thus better defined as a $\alpha_2\beta_2$ complex. It should be noted that, in solution, isolated protein β occurs as a mixture of monomer, dimer, and polymers in equilibrium. Binding to the homodimeric reductase thus shifts the equilibrium to the dimer form. No structure of the complex is available and the molecular details of the α/β interface are thus unknown. It should be noted that the physical properties of the iron cluster within the $\alpha_2\beta_2$ complex are grossly similar to those of that in the β protein alone. However, the redox potential of the complex is found at -550 mV and is not shifted by the addition of DTT.

The strong affinity of the unactivated reductase for the activase is apparently inconsistent with the activase playing a catalytic role during the activation. Whereas a tight complex is likely to create an environment which protects the radical transfer from one protein to another, dissociation is required to allow protein β to activate other protein α molecules. The critical interactions which are transiently broken after radical introduction to allow this dissociation are unknown. A much less stable complex occurs in the class III enzyme from L. lactis, but it is too early to understand the structural basis for this difference (21).

C. The Activation Reaction: The Role of the Iron–Sulfur Center of Protein β in the Generation of the Glycyl Radical of Protein α

The introduction of the glycyl radical in protein α requires the participation of the following components working under strict anaerobiosis: the activating β protein, an electron source such as flavodoxin, and S-adenosylmethionine (AdoMet).

The simplest working hypothesis for the mechanism of radical formation is shown in Fig. 9. More complicated versions of this mechanism have appeared in recent review articles (25, 51–53). In the first step, the Fe–S center of protein β is reduced by one electron. Then AdoMet binds to the protein to form a complex in which it lies in close proximity to the cluster from which it receives one electron to generate a sulfuranyl radical. Finally, the latter is cleaved into methionine and a 5′-deoxyadenosyl radical that generates the glycyl radical by abstracting a hydrogen atom from the α-carbon of glycine 681.

The following results strongly support this mechanism.

First, the evidence for an intermediate AdoMet–β protein complex comes from the strong effect of AdoMet on the EPR signal of the reduced Fe–S center, with the low-field feature at $g = 2.02$ becoming high-field-shifted and

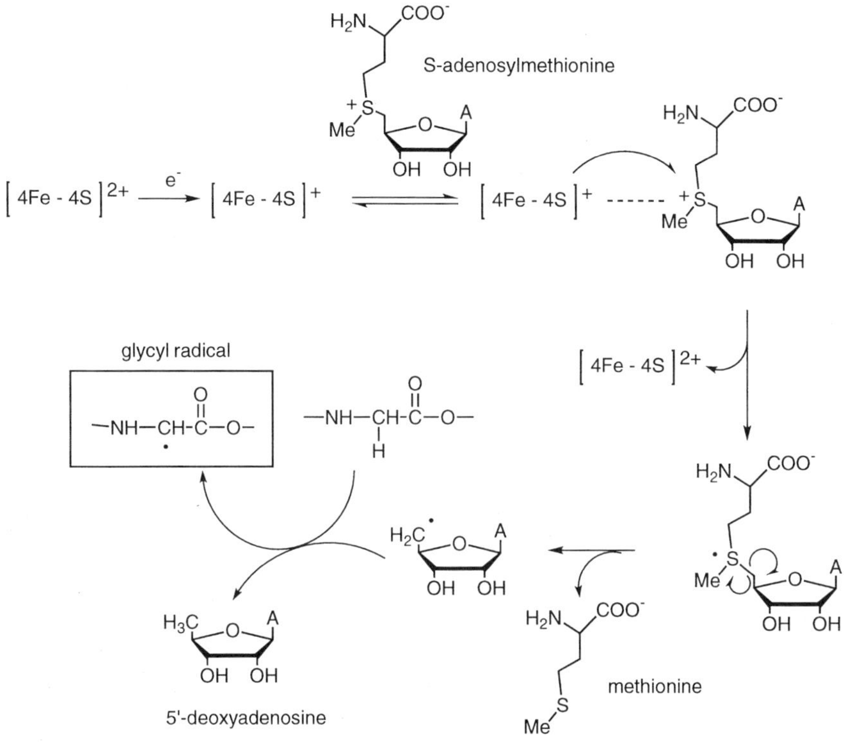

FIG. 9. Proposed mechanism for the generation of the glycyl radical. The iron–sulfur cluster and S-adenosylmethionine are on protein β and the glycyl radical is on protein α.

the high-field feature being greatly broadened (22, 46). The new signal also displays different microwave power saturation properties. Experiments using *methyl*-^{3}H AdoMet and a filter binding assay demonstrated that binding occurs only under anaerobic conditions when the enzyme is in the reduced state and in the presence of K^+. A K_d value of 10 μM for AdoMet was determined (22). From the amino acid primary sequence, no obvious AdoMet-binding peptide can be identified.

Second, during reaction of reduced $\alpha_2\beta_2$ complex, containing $[4Fe-4S]^+$ centers, with an excess of AdoMet, the cluster is rapidly and fully oxidized, AdoMet is cleaved into methione and 5′-deoxyadenosine, and the glycyl radical is formed (22, 46). The three products are formed in stoichiometric amounts, corresponding to approximately one equivalent with regard to the initial amount of reduced cluster. No reaction occurs when all the clusters are in the $[4Fe-4S]^{2+}$ state.

Third, deuterium was incorporated into 5'-deoxyadenosine when the reaction was carried out with ^{2}H-glycine-substituted protein α preparations, in agreement with an intermediate 5'-deoxyadenosyl radical abstracting a hydrogen atom from Gly681 of protein α. Deuterium was incorporated into 5'-deoxyadenosine when the reaction was carried out with ^{2}H-glycine-substituted protein α preparations. Almost no deuterium incorporation occurred when the wild-type enzyme was reacted in D$_2$O. It should be noted that the reduced cluster and AdoMet are on the activase, where the 5'-deoxyadenosyl radical is generated, whereas the glycine residue is on the reductase. How communication between the two proteins occurs is still unknown. The glycyl radical site lies quite near the surface of protein α. However, the cystal structure of protein α shows that the methyl group of Ala580, corresponding to the abstracted hydrogen atom of Gly580, is directed toward the protein interior and is inaccessible to solvent (30). This is mainly due to a conserved side chain, Glu446, which helps anchor the Gly radical site by hydrogen bonds to surrounding main-chain atoms (Fig. 4). It is thus very likely that the formation of Gly° is dependent on some conformational rearrangement of the proteins.

The reduction of the cluster by the one- and two-electron reduced forms (SQ and HQ, respectively) of flavodoxin (fldx) is thermodynamically unfavorable. Indeed, the redox potential of the fldx/SQ couple lies at -250 mV and that of the SQ/HQ couple at -440 mV (pH = 7.5, vs. NHE), both potentials more positive than that of the Fe–S center of the β protein (23). Accordingly, neither SQ or HQ could reduce the cluster of protein β alone to a detectable extent. The fact that AdoMet is nevertheless reductively cleaved and the glycyl radical formed with either reduced form of flavodoxin during activation of protein α by protein β is the result of coupling the thermodynamically unfavorable reduction of the cluster and the reduction of AdoMet by the reduced cluster to the thermodynamically favorable oxidation of the glycine residue by the 5'-deoxyadenosyl radical. Indeed there is a large difference between the C—H bond dissociation energies of 5'-deoxyadenosine (100 kcal/mole) and glycine (79 kcal/mole). This is consistent with the observation that no AdoMet cleavage occurs during activation of a protein α mutant in which Gly681 has been changed into an alanine (23).

Altogether, these results fit with the mechanism shown in Fig. 9, which assigns a key role to the reduced [4Fe–4S]$^+$ center of protein β during radical formation as a one-electron reductant for AdoMet cleavage. However, it is also possible that this cluster is not the actual donor, but instead the precursor for an active species, as shown from recent experiments in our laboratory described below.

Reduced protein β, containing [4Fe–4S]$^+$ centers, is unable to reduce and cleave AdoMet in the absence of protein α. This in all probability is so for thermodynamic reasons, considering the very negative redox potential of

CLASS III RIBONUCLEOTIDE REDUCTASE 119

sulfonium ions in general. However, when DTT is added in place of protein α, the reaction takes place, albeit at a much reduced rate compared to the reaction in the presence of protein α. In this reaction, two methionine and two 5′-deoxyadenosine equivalents *and not one* were formed per [4Fe–4S]$^+$ center (*46*). This suggests that, in the presence of DTT, the [4Fe–4S]$^+$ center has unexpectedly gained the ability to react as a two-electron donor. Since DTT binds in the vicinity of the cluster (EPR and ENDOR, unpublished results), it is tempting to suggest that DTT transiently dissociates the [4Fe–4S]$^+$ center into an Fe(DTT) complex and a [3Fe–4S]$^{2-}$ hyperreduced center, proposed to be the active species. Hyperreduced [3Fe–4S]$^{2-}$ centers have been characterized in a few [3Fe–4S] and [7Fe–8S] ferredoxins by F. Armstrong and colleagues using electrochemistry as the analytical tool (*54, 55*). Interestingly, these studies have shown that oxidation of these centers occurs in a two-electron electrochemical wave in the −600- to −800-mV range. In the case of the Fe–S center of protein β, the intermediate formation of a hyperreduced [3Fe–4S]$^{2-}$ center would thus explain why two molecules of AdoMet can be reduced. This novel mechanism remains to be further experimentally substantiated. So far the only support for it, in addition to the unexpected stoichiometry, is the detection of intermediate [3Fe–4S]^0 centers during the reaction by Mössbauer spectroscopy (*46*). Whether this DTT reaction is a model for the reductive cleavage of AdoMet during activation of protein α and whether the latter also involves [3Fe–4S] intermediate species remain to be studied.

D. The *nrdG* Protein as a Prototype for a Superfamily of Iron–Sulfur Proteins Requiring AdoMet for Activity

Using powerful bioinformatics and information vizualization methods, a recent study demonstrated that more than 600 proteins contain the unusual conserved CXXXCXXC sequence, which was shown in the case of the activases of RNR and PFL to be involved in the chelation of a unique iron–sulfur center designed for AdoMet reductive cleavage and controlled radical formation (*56*). It thus seems that a radical mechanism based on the combination of an iron–sulfur center and AdoMet is used in a large number of biological reactions involved in DNA precursor, vitamin, cofactor, antibiotic and herbicide biosynthesis, biodegradation pathways, and DNA repair. Some enzymes from that list have already been confirmed to carry such a cluster (*25*). This is the case for lysine aminomutase, biotin synthase, lipoate synthase, spore photoproduct lyase, and an enzyme involved in methylthiolation of tRNA (M. Atta *et al.*, unpublished results). It is likely that the mechanism for radical generation in RNR described in this chapter in its main lines applies to all these systems.

V. Gene Organization and Regulation

The anaerobic RNR from *E. coli* is encoded by the *nrdD* and *nrdG* genes belonging to the same operon located at around 96 min on the *E. coli* genomic map (*11, 57*). The operon is probably regulated by the Arc/Fnr system, since a binding site for the Fnr protein, the anaerobic transcriptional activator, is present in the promoter region, 228 bp upstream of the initiator ATG. Thus synthesis of *nrdD* and *nrdG* protein is repressed under oxic conditions and induced by a shift from aerobic to anaerobic growth. Knockout mutants of either *nrdD* or *nrdG* cannot grow under strict anaerobiosis (*58*). That anaerobic deoxyribonucleotide biosynthesis is dependent on a glycyl radical on the *nrdD* protein was shown from the inability of an *nrdD*-deficient *Ralstonia eutrophus* mutant strain in which only the critical glycine residue was changed to an alanine to grow under anoxic conditions (*29*).

Even though the details of the transcriptional regulation remain to be established, *nrdD* and *nrdG* genes appear to be cotranscribed from a common promoter located upstream of *nrdD*. However, the expression of *nrdD* is much larger than that of *nrdG*, in agreement with the *nrdG* protein playing a catalytic role in the activation of the *nrdD* protein (*29*).

VI. The Anaerobic RNR: The Link between the RNA World and the DNA World

There is today a general agreement that an RNA world, or most probably an RNA–protein world, preceded the present DNA world. The observation that in all living organisms deoxyribonucleotides are synthesized by enzymatic reduction of ribonucleotides fits well with such a concept and furthermore suggests that the advent of an RNR at the surface of the earth was a key event for the conversion from the RNA to the DNA world (*4, 5, 59*). Since the reaction requires complex proteins, cofactors, and mechanisms, it is very likely that DNA could evolve only after a considerable complexification of the protein world had been achieved. Thus RNR did not appear in the prebiotic soup or the RNA world, as sometimes discussed, but instead in a complex and late RNA–protein world.

The hypothesis of a primitive RNR from which the three classes of RNR present in contemporary metabolism evolved by divergent evolution is supported by the great similarities between them. These similarities extend from the catalytic mechanisms, with the common requirement for a conserved cysteinyl radical as the initiating species and a similar complex radical-based chemistry, to the sophisticated allosteric regulation, with basically similar responses in the enzyme to a given allosteric effector. With such a conservation of essential mechanisms, it is not surprising that the three classes display similar three-dimensional structures, despite very limited primary sequence homology (30, 31, 34; J. Stubbe and C. Drennan, personal communication). Whether the major driving force in the evolution of RNRs resides in the radical chemistry or the allosteric regulation is a matter of discussion (4, 5, 59).

In each class of RNR, the active site is in the center of a 10-stranded α/β barrel with a finger loop that projects up though the interior with the cysteine becoming the essential thiyl radical residing at the tip of the finger loop, the sulfur atom being at 3–4 Å from the 3'-H of the ribose substrate, in a suitable position to initiate the reaction by H abstraction. The relative position of the radical precursor (a tyrosyl radical, AdoCbl, and a glycyl radical in classes I, II, and III RNR, respectively) with regard to the thiyl radical is remarkably conserved.

Furthermore, the architecture of the allosteric sites and their communication to the active sites in classes I and II RNR are remarkably similar. In the class I RNR structure, with dTTP as an effector and GDP as a substrate, dTTP binds at the edge of the dimer interface which is formed from the association of two helices from each subunit forming a four-helix bundle (32). In this configuration dTTP is closest to the GDP in the active site of the opposite monomer, 13 Å away (instead of 27 Å from the GDP of the same monomer). In addition, three loops become ordered on binding of dTTP, two of these coming from one monomer and one from the other monomer. These loops are in contact with each other. Taken together this strongly supports the idea of an intersubunit communication between allosteric and activity sites.

Class II RNRs can be divided into dimeric and monomeric enzymes. The former have not been structurally characterized. However, they appear to have conserved many of the residues of the class I allosteric specificity site and thus most likely utilize similar structures and mechanisms for allosteric regulation. A recently determined crystal structure of the monomeric enzyme from *Lactobacillus leichmanii* shows the presence of a four-helix bundle, which, on the basis of its strong structural homology to the effector specificity site of class I RNR, is likely to play the same role (J. Stubbe and C. Drennan, personal communication). This structural conservation of effector sites is remarkable considering that in dimeric class I RNR the bundle is made from the association of two helices from each monomer at the dimer interface, whereas in monomeric class II RNR it corresponds to a single large sequence insertion, within the same polypeptide, relative to the dimeric enzymes. This insertion mimics the topology of the other half of the dimer interface in class I RNR. Furthermore, the distance between the allosteric site in one subunit and the substrate site in the second one in class I

RNR is comparable to that between the two sites in the monomeric class II RNR, further supporting the notion of an intersubunit communication in the class I RNR dimer. It is tempting to envisage a continuous evolutionary process leading from a primitive monomeric reductase whose allosteric regulation was entirely intramonomeric, through a dimeric system where regulation is mainly within one monomer, but with compensatory rearrangements of the other (as in class III RNR), to the situation where regulation is highly intermonomeric and the effector and active sites have also moved closer to one another (as in class I RNR).

In the structure of the class III RNR from bacteriophage T4, with dATP, dGTP, or dTTP bound at the unique effector site, the dimer interface is significantly different and rather more complex, with three helices from each subunit contributing to that interface. Nevertheless deoxyribonucleotides also bind at the interface, but along the length of the helices rather than across their ends (30, 43). As a consequence, the allosteric site is at comparable distances from both active sites of the dimer (25 Å). It is thus possible that in this case communication is intrasubunit. In addition, the details of effector recognition are quite different from class RNR, although only one class I dNTP–enzyme complex is structurally known (32). No base-specific hydrogen bonds are made in class I RNR, whereas effector recognition in class III RNR is acutely dependent on matching side chains to the hydrogen-bonding pattern of the base. Despite these differences in effector recognition, the allosteric signal appears to be transmitted to the active site via the same loop (loop 2).

In fact, the three classes of RNRs are not best defined on the basis of the protein component that carries the ribonucleotide-binding site where the catalysis takes place and the allosteric sites, but rather on the basis of the mechanisms of activation of this component and the metal–radical cofactor required for the activation (2, 3). In contrast to the mechanism of nucleotide reduction and to the structure of the active site, the mechanism of activation has not been conserved and appears to be an adaptational response to the environment. In class I RNR the formation of the cysteinyl radical in the active site is achieved by intermolecular radical transfer through the action of a second component, a protein carrying a stable tyrosyl radical. This radical is generated during the one-electron oxidation of a specific tyrosine residue by molecular oxygen catalyzed by a nonheme dinuclear iron center. In class II RNR the cysteinyl radical is formed by abstraction of the hydrogen atom of the corresponding cysteine by a 5'-deoxyadenosyl radical generated from adenosylcobalamin (AdoCbl). Finally, in class III RNR the cysteinyl radical derives from an intramolecular radical transfer from an adjacent oxygen-sensitive glycyl radical, generated by the action of a second component, an iron–sulfur protein, in combination with S-adenosylmethionine, as described above. It is striking that there is absolutely no sequence homology between the activating components of the three classes. In fact AdoCbl is not even a protein, and the activating proteins of the class I and III enzymes are totally different both in terms of sequence and metallocofactor. Therefore it looks like these systems are not evolutionarily related and nature has independently evolved three different activating mechanisms. It is thus interesting to discuss which came first, or rather which modern class of RNR best resembles the ancestral enzyme and what kind of environmental selective pressure has been at the origin of the advent of new mechanisms.

It seems quite reasonable to suggest that the class I RNR activating system is a latecomer which arose only after photosynthesis and the emergence of an O_2 atmosphere. Dependent on its being activated, molecular oxygen is a suitable oxidant for radical generation. The activating catalyst that has been selected is a protein carrying a μ-oxo-bridged diferric nonheme center, an "iron–oxide" species which is functional for a large number of O_2-dependent oxygenases of the oxygen world, the prototype of which is methane monooxygenase (60). Accordingly, mammals and plants exclusively use class I RNRs for deoxyribonucleotide synthesis.

It is a fair assumption that DNA existed before oxygen appeared on our planet. Thus the primitive RNR was an anaerobic enzyme such as class II and class III RNRs. A distinction between these two classes is not trivial. Indeed, they are found both in eubacteria and archaea and thus probably existed before the two kingdoms diverged. However, there are several arguments which support the suggestion that the class III RNR activating mechanism is the closest to the primitive one.

First, whereas AdoCbl provides the simplest and most beautiful mechanism for radical generation, that is the homolytic cleavage of the Co—C bond of the metallocofactor, its availability in quantities sufficient to be recruited by an enzyme is questionable. Indeed, the organic part is of an extreme complexity and the biosynthetic pathway in today's organisms requires about 30 different enzymes (61, 62). As we hypothesize that the living world was full of proteins and enzymes, the presence of AdoCbl in early metabolism cannot be completely excluded, but is not so likely. Furthermore, cobalt is very scarce at the surface of the earth (63). It is also possible that the simplest cobalamin versions of AdoCbl were available at that time and active, but we have no molecular fossils at our disposal to support such a suggestion. The source of 5′-deoxyadenosyl radicals in class III RNR is S-adenosylmethionine, a smaller, simpler, much more synthetically accessible molecule of the RNA–protein world. The fact that AdoMet generates the same organic radical as AdoCbl has led to the suggestion that AdoMet was a primitive version of the

AdoCbl cofactor (64). It is important to note that the AdoMet-dependent mechanism is also used by pyruvate–formate lyase, which was supposed to be essential for anaerobic energy metabolism very early during evolution. Nevertheless, conversion of AdoMet into a radical in addition requires a reducing agent. Today's organisms utilize a complex electron transfer chain, involving NADPH, flavodoxin reductase, and flavodoxin. The complexity of this chain is against AdoMet being a simple radical source; however, it is possible that simpler and equally potent reducing agents were available very early to support AdoMet reduction. For example, dithionite and irradiated deazaflavin are active *in vitro*.

Second, the dependence on an Fe–S cluster for catalysis of electron transfer and radical generation is a further argument in favor of the primitive nature of the class III RNR activating mechanism, since minerals of sulfur-linked iron were abundant at the surface of the early earth, in much larger amounts than cobalt (63).

Third, the electron source during the reduction of ribonucleotides in class II RNR is a complex protein machinery involving NADPH, thioredoxin, and thioredoxin reductase, whereas class III RNR gets its electrons from formate, one of the simplest organic reductants present in large quantities very early at the surface of the earth.

In conclusion it is reasonable to think that the primitive RNR used AdoMet, with the help of an iron–sulfur center, as the radical source and formate for the reduction of ribonucleotides. Recruitment of AdoCbl as the activase occurred later, when the cofactor became available. When oxygen appeared in the atmosphere air-sensitive organisms depending on class III RNRs had to migrate to completely anaerobic habitats where one can find them now or, instead, to recruit a new oxygen-tolerant activase (and become a facultative anaerobe). In that context the organisms that contained a class II RNR had a great advantage, since they could continue to produce deoxyribonucleotides and survive the oxygenated conditions. The oxygen solution came with the invention of nonheme diferric proteins and the utilization of the powerful oxidizing chemistry of the iron–O_2 combination. Reductases which recruited such a protein in place of AdoCbl or the iron–sulfur protein became a class I RNR. Such a conclusion has been recently challenged on the basis of a critical reexamination of several of the above assumptions (65). Nevertheless, the adaptation of the reductase protein to the various activating mechanisms was accompanied by only minor structural variations, as shown from the striking conservation of the active-site architecture. This shows that once the "thiyl radical" solution to the difficult chemical problem of the conversion of ribo- to deoxyribonucleotides was discovered, it proved to be so satisfactory that it has been conserved throughout evolution up to the present.

Acknowledgment

We are grateful to Mohammed Atta for his contribution to the sequence analysis and helpful comments.

References

1. P. Reichard, *Annu. Rev. Biochem.* **57**, 349 (1988).
2. A. Jordan and P. Reichard, *Annu. Rev. Biochem.* **67**, 71 (1998).
3. B.-M. Sjöberg, *Struct. Bonding* **88**, 139 (1997).
4. P. Reichard, *Trends Biochem. Sci.* **22**, 81 (1997).
5. J. Stubbe, *Curr. Opin. Struct. Biol.* **10**, 731 (2000).
6. M. Sahlin and B.-M. Sjöberg, *Subcell. Biochem.* **35**, 405 (2000).
7. J. Stubbe and P. Riggs-Gelasco, *Trends Biochem. Sci.* **23**, 438 (1998).
8. M. Fontecave and E. Mulliez, in "Chemistry and Biochemistry of B12" (R. Banerjee, ed.), pp. 731–756. Wiley, New York, 1999.
9. M. Fontecave, R. Eliasson, and P. Reichard, *Proc. Natl. Acad. Sci. USA* **86**, 2147 (1989).
10. R. Eliasson, E. Pontis, M. Fontecave, C. Gerez, J. Harder, H. Jörnvall, M. Krook, and P. Reichard, *J. Biol. Chem.* **267**, 25541 (1992).
11. X. Sun, R. Eliasson, E. Pontis, J. Anderson, G. Buist, B.-M. Sjöberg, and P. Reichard, *J. Biol. Chem.* **270**, 2443 (1995).
12. V. Bianchi, P. Reichard, R. Eliasson, E. Pontis, M. Krook, H. Jörnvall, and E. Haggard-Ljungquist, *J. Bacteriol.* **175**, 1590 (1993).
13. V. Bianchi, R. Eliasson, M. Fontecave, E. Mulliez, D. M. Hoover, R. G. Matthews, and P. Reichard, *Biochem. Biophys. Res. Commun.* **197**, 792 (1995).
14. R. Eliasson, M. Fontecave, H. Jörnvall, M. Krook, E. Pontis, and P. Reichard, *Proc. Natl. Acad. Sci. USA* **87**, 3314 (1990).
15. E. Mulliez, S. Ollagnier, M. Fontecave, R. Eliasson, and P. Reichard, *Proc. Natl. Acad. Sci. USA* **92**, 8759 (1995).
16. D. Padovani, E. Mulliez, and M. Fontecave, *J. Biol. Chem.* **276**, 9587 (2001).
17. J. Andersson, M. Westman, M. Sahlin, and B.-M. Sjöberg, *J. Biol. Chem.* **275**, 19449 (2000).
18. E. Mulliez, M. Fontecave, J. Gaillard, and P. Reichard, *J. Biol. Chem.* **268**, 2296 (1993).
19. X. Sun, S. Ollagnier, P. P. Schmidt, M. Atta, E. Mulliez, L. Lepape, R. Eliasson, A. Gräslund, M. Fontecave, P. Reichard, and B.-M. Sjöberg, *J. Biol. Chem.* **271**, 6827 (1996).
20. D. S. King and P. Reichard, *Biochem. Biophys. Res. Commun.* **206**, 731 (1995).
21. E. Torrents, R. Eliasson, H. Wolpher, A. Gräslund, and P. Reichard, *J. Biol. Chem.* **276**, 33488 (2001).
22. S. Ollagnier, E. Mulliez, P. P. Schmidt, R. Eliasson, J. Gaillard, C. Deronzier, T. Bergman, A. Gräslund, P. Reichard, and M. Fontecave, *J. Biol. Chem.* **272**, 24216 (1997).
23. E. Mulliez, D. Padovani, M. Atta, C. Alcouffe, and M. Fontecave, *Biochemistry* **40**, 3730 (2001).
24. A. F. V. Wagner, M. Frey, F. A. Neugebauer, W. Schäfer, and J. Knappe, *Proc. Natl. Acad. Sci. USA* **89**, 996 (1992).
25. J. Cheek and J. Broderick, *J. Biol. Inorg. Chem.* **6**, 209 (2001).
26. S. G. Reddy, K. K. Wong, C. V. Parast, J. Peisach, R. S. Magliozzo, and J. W. Kozarich, *Biochemistry* **37**, 558 (1998).
27. H. P. Blaschkowski, G. Neuer, M. Ludwig-Festl, and J. Knappe, *Eur. J. Biochem.* **123**, 563 (1982).

28. B. Leuthner, C. Leutwein, H. Schulz, P. Horth, W. Haehnel, E. Schiltz, H. Schagger, and J. Heider, *Mol. Microbiol.* **28,** 615 (1998).
29. A. Siedow, R. Cramm, R. A. Siddiqui, and B. Friedrich, *J. Bacteriol.* **181,** 4919 (1999).
30. D. T. Logan, J. Andersson, B.-M. Sjöberg, and P. Nordlund, *Science* **283,** 1499 (1999).
31. H. Eklund and M. Fontecave, *Structure* **7,** R257 (1999).
32. M. Eriksson, U. Uhlin, S. Ramaswamy, M. Ekberg, K. Regnström, B.-M. Sjöberg, and H. Eklund, *Structure* **5,** 1077 (1997).
33. U. Uhlin and H. Eklund, *Nature* **370,** 533 (1996).
34. M. C. Olcott, J. Andersson, and B.-M. Sjöberg, *J. Biol. Chem.* **273,** 24853 (1998).
35. J. Andersson, S. Bodevin, M. Westman, M. Sahlin, and B.-M. Sjöberg, *J. Biol. Chem.* **276,** 40457 (2001).
36. J. Stubbe and W. A. van der Donk, *Chem. Rev.* **98,** 705 (1998).
37. R. Eliasson, E. Pontis, F. Eckstein, and P. Reichard, *J. Biol. Chem.* **269,** 26116 (1994).
38. R. Eliasson, P. Reichard, E. Mulliez, S. Ollagnier, M. Fontecave, E. Liepinsh, and G. Otting, *Biochem. Biophys. Res. Commun.* **214,** 28 (1995).
39. K. Cho, F. Himo, A. Gräslund, and P. E. M. Siegbahn, *J. Phys. Chem. B* **105,** 6445 (2001).
40. R. Eliasson, E. Pontis, X. Sun, and P. Reichard, *J. Biol. Chem.* **269,** 26052 (1994).
41. J. Andersson, M. Westman, A. Hofer, and B.-M. Sjöberg, *J. Biol. Chem.* **275,** 19443 (2000).
42. E. Torrents, G. Buist, A. Liu, R. Eliasson, J. Kok, I. Gibert, A. Gräslund, and P. Reichard, *J. Biol. Chem.* **275,** 2643 (2000).
43. K.-M. Larsson, J. Andersson, B.-M. Sjöberg, P. Nordlund, and D. T. Logan, *Structure* **9,** 739 (2001).
44. Tamarit, E. Mulliez, C. Meier, A. Trautwein, and M. Fontecave, *J. Biol. Chem.* **274,** 31291 (1999).
45. S. Ollagnier, E. Mulliez, J. Gaillard, R. Eliasson, M. Fontecave, and P. Reichard, *J. Biol. Chem.* **271,** 9410 (1996).
46. D. Padovani, F. Thomas, A. X. Trautwein, E. Mulliez, and M. Fontecave, *Biochemistry* **40,** 6713, (2001).
47. S. Ollagnier, C. Meier, E. Mulliez, J. Gaillard, V. Schuenemann, A. Trautwein, T. Mattioli, M. Lutz, and M. Fontecave, *J. Am. Chem. Soc.* **121,** 6344 (1999).
48. E. Mulliez, S. Ollagnier de Choudens, C. Meier, M. Cremonini, C. Luchinat, A. X. Trautwein, and M. Fontecave, *J. Biol. Inorg. Chem.* **4,** 614 (1999).
49. J. Tamarit, C. Gerez, C. Meier, E. Mulliez, A. Trautwein, and M. Fontecave, *J. Biol. Chem.* **275,** 15669 (2000).
50. R. Külzer, T. Pils, R. Kappl, J. Hüttermann, and J. Knappe, *J. Biol. Chem.* **273,** 4897 (1998).
51. P. A. Frey and S. Booker, *Adv. Free Radical Chem.* **2,** 1 (1999).
52. P. A. Frey, *Curr. Opin. Chem. Biol.* **1,** 347 (1997).
53. M. Fontecave, E. Mulliez, and S. Ollagnier-de-Choudens, *Curr. Opin. Chem. Biol.* **5,** 506 (2001).
54. J. L. Duff, J. L. Breton, J. N. Butt, F. A. Armstrong, and A. J. Thomson, *J. Am. Chem. Soc.* **118,** 8593 (1996).
55. J. Hirst, G. N. L. Jameson, J. W. A. Allen, and F. A. Armstrong, *J. Am. Chem. Soc.* **120,** 11994 (1998).
56. H. J. Sofia, G. Chen, B. G. Hetzler, J. F. Reyes-Spindola, and N. E. Miller, *Nucleic Acids Res.* **29,** 1097 (2001).
57. X. Sun, J. Harder, M. Krook, H. Jörnvall, B.-M. Sjöberg, and P. Reichard, *Proc. Natl. Acad. Sci. USA* **90,** 577 (1993).
58. Garriga, R. Eliasson, E. Torrents, A. Jordan, J. Barbé, I. Gibert, and P. Reichard, *Biochem. Biophys. Res. Commun.* **229,** 189 (1996).
59. J. Stubbe, J. Ge, and C. S. Yee, *Trends Biochem. Sci.* **26,** 93 (2001).
60. B. J. Wallar and J. D. Lipscomb, *Chem. Rev.* **96,** 2625 (1996).

61. A. R. Battersby and F. J. Leeper, in "Chemistry and Biochemistry of B12" (R. Banerjee, ed.), pp. 507–536. Wiley, New York, 1999.
62. A. I. Scott, C. A. Roessner, and P. J. Santander, in "Chemistry and Biochemistry of B12" (R. Banerjee, ed.), pp. 537–556. Wiley, New York, 1999.
63. R. J. P. Williams, *Coord. Chem. Rev.* **216**, 583 (2001).
64. P. A. Frey, *FASEB J.* **7**, 662 (1993).
65. A. M. Poole, D. T. Logan, and B.-M. Sjöberg, *J. Mol. Evol.* (2002) in press.
66. M. D. Sintchak, G. Arjara, B. A. Kellog, J. Stubbe, and C. L. Drennan, *Nature Struct. Biol.* **9**, 293 (2002).

Regulation of Pathways of mRNA Destabilization and Stabilization

ROBIN E. DODSON
AND DAVID J. SHAPIRO

*Department of Biochemistry
University of Illinois at Urbana-Champaign
Urbana, Illinois 61801*

I. Life and Half-Life of mRNA. 130
II. Degradation of mRNA through the General Pathway;
Deadenylation-Dependent mRNA Decay. 131
III. Special Pathways for Regulating the Stability of mRNAs 133
   A. Regulating Decay of Unstable mRNAs 134
   B. Removal of Error-Prone RNA: RNA Surveillance or
      Nonsense-Mediated Decay. 135
   C. Stabilization and Destabilization of Endonucleolytically Cleaved RNAs . 136
IV. Regulation of Vitellogenin and Albumin mRNA Stability. 142
   A. Background and Early Work. 142
   B. Overview of Regulation of Vitellogenin Gene Expression. 143
   C. Basic Features of Vitellogenin mRNA Stabilization 144
   D. The Estrogen-Inducible Vitellogenin mRNA 3′-UTR-Binding Protein
      Is Vigilin. 145
   E. Vigilin and Vitellogenin mRNA Stabilization. 154
   F. Future Prospects. 156
   References. 159

The level of an mRNA in the cytoplasm represents a balance between the rate at which the mRNA precursor is synthesized in the nucleus and the rates of nuclear RNA processing and export and cytoplasmic mRNA degradation. Although most studies of gene expression have focused on gene transcription and

Abbreviations: 2-5A, 2′–5′-linked oligoadenylate; ARE, AU-rich element; αCP, poly(C)-binding protein; CRDBP, coding region response element-binding protein; DDP1, dodeca-satellite protein 1; ELAV, embryonic lethal abnormal vision; GMC-SF, granulocyte macrophage colony stimulating factor; hnRNP, heteronuclear ribonucleoprotein; INF, interferon; Ire1, endoplasmic reticulum-associated type 1 transmembrane protein kinase; KH, K homology; mCRD, major coding region instability determinant; PABP, poly(A)-binding protein; PARN, poly(A) ribonuclease; PMR-1, polysomal ribonuclease 1; RRM, RNA recognition motif; SCP160, 160-kDa *Saccharomyces cerevisiae* control of ploidy; TNFα, tumor necrosis factor α; TfR, transferrin receptor; 3′-UTR, 3′ untranslated region.

its control, a great deal of information indicates that mRNA degradation and its regulation are major control mechanisms that help govern cellular mRNA levels. The objective of this chapter is not to exhaustively review our present knowledge in the area of eukaryotic mRNA degradation, but to provide a short general discussion of the importance of mRNA degradation and its regulation and a brief overview of recent findings and present knowledge. The overview is followed by a more in-depth discussion of one of the several pathways for mRNA degradation. We concentrate on the pathway for regulated mRNA degradation mediated by mRNA-binding proteins and endonucleases that cleave within the body of mRNAs. As a potential example of this type of control, we focus on the regulated degradation of the egg yolk precursor protein vitellogenin on the mRNA-binding protein vigilin and the mRNA endonuclease polysomal ribonuclease 1 (PMR-1).
© 2002, Elsevier Science (USA).

I. Life and Half-Life of mRNA

The amount of protein that can be translated from each molecule of an mRNA is determined by the amount of time the mRNA persists in the cytoplasm (1). An mRNA with a long half-life will usually undergo a great many rounds of translation, producing very large quantities of protein, thereby amplifying effects based on regulation of gene transcription. For example, the long half-life of vitellogenin mRNA allows for production of vast amounts of the egg yolk proteins needed during reproduction (2). Conversely, an mRNA with a short half-life changes its level quickly in response to alterations in transcription rate. For these unstable mRNAs, changes in transcription rate are therefore rapidly reflected by a change in the rate of protein production. Transcription factors which are needed transiently are typically encoded by mRNAs with short half-lives (3–5). This allows for a brief period of translation and a transient pulse of protein synthesis, rather than a prolonged period of continuous protein production.

The half-lives of many RNAs are an intrinsic property determined by their rate of decay through the general degradation pathway (5–7; see below). There are mechanisms for changing the half-life of an mRNA, making the rate of decay a dynamic property. This allows another level of control in protein production. Under one set of conditions, a message can be stabilized and protein production increased, but with a change in conditions, the mRNA can be destabilized, shutting down protein production more rapidly (1). Regulation of the half-life of an mRNA is not only important in protein production, but in other cellular functions of mRNA as well, including the storage of maternal RNAs (8) and establishment of protein gradients in *Drosophila* embryos (9). The importance of regulation of mRNA half-life is reflected in the number of systems in which it is found and by the many aspects of mRNA metabolism that are coupled to the rate of mRNA turnover.

Changes in rates of mRNA turnover accompany a wide range of physiological and clinical events including development and differentiation, reproduction, immune responses, and viral infections. Additionally, changes in mRNA decay rates occur during the cell cycle and in response to changes in cell environment, for example, during hypoxia, inflammation, and other stresses, and changes in levels of glucose, iron, and calcium. The mechanisms activating mRNA decay or stabilization have not been completely worked out, but often involve a number of cell signaling cascades including the mitogen and stress-activated protein kinase pathways, steroid hormone receptor-mediated events, and membrane receptor-mediated signal transduction (for reviews see Refs. *1, 10–12*; also see below).

There are several mechanisms by which an RNA can be degraded. The general decay pathway, which is considered to be deadenylation-dependent RNA decay, will be discussed first. The general pathway can be altered to regulate mRNA levels; however, there are also a number of specialized pathways which exist for regulating the turnover of subsets of mRNAs or of individual mRNAs. These include pathways to degrade damaged or mutated mRNAs and viral RNAs and to regulate the turnover of unstable mRNAs and of other mRNAs whose stability is regulated in response to changes in cellular conditions.

II. Degradation of mRNA through the General Pathway; Deadenylation-Dependent mRNA Decay

Degradation of mRNAs by deadenylation-dependent decapping is thought to be the major mechanism by which mRNAs are destroyed (reviewed in Refs. *5, 6*). This occurs by removal of the poly(A) tail, which is added to the 3′-end of most RNAs, followed by removal of the 5′ CAP structure, which is a 5′–5′-linked methyl guanosine added to the 5′-end of the message. Both structures are thought to protect the mRNA from exonucleolytic attack, and are also important in mRNA translation through their interaction with other proteins (*1*). The poly(A) tail binds poly(A) binding protein (PABP) (*13*), or (Pab)1, as it is known in yeast (*14*). PABP is an RNA-binding protein containing four RNA recognition motifs (RRMs) and binds to a stretch of A's approximately 12 nucleotide in length (reviewed in Ref. *15*). The CAP is bound by the subunit of the eIF4E initiation factor, which is one of three components of the CAP binding complex, eIF4F (*16*). PABP binds a second component of eIF4F called eIF4G (*17–19*). This interaction leads to a circularization of the mRNA, and facilitates translation initiation (*20*). A role for this complex in stabilization is possible, but remains to be clearly established.

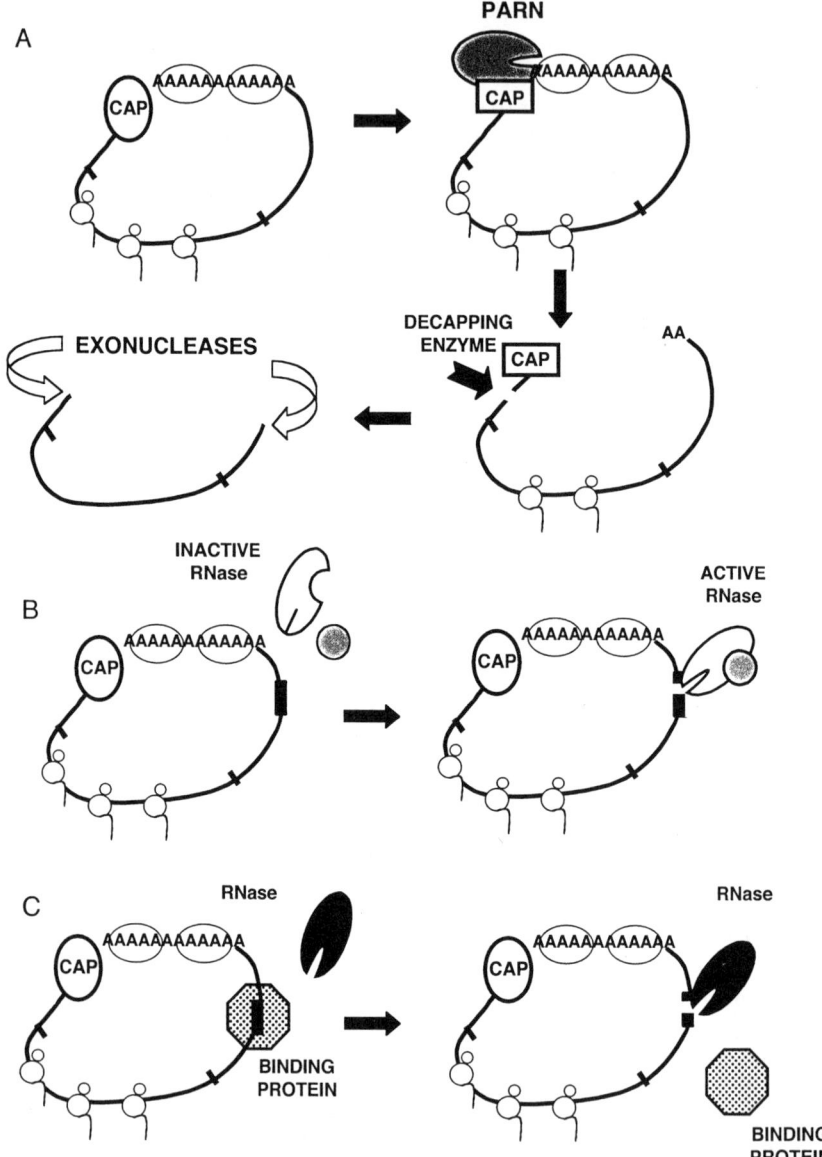

FIG. 1. Mechanism of endonucleolytic cleavage and RNA stabilization. The pathways for poly(A)-dependent mRNA decay and endonucleolytic cleavage are shown in schematic form. (A) Model of deadenylation-dependent decapping. Top left: Transcribed mRNA containing CAP-binding proteins and PABP. Top right: CAP-binding proteins are removed and PARN can now bind to the CAP. PARN digests the poly(A) tail and then dissociates. Bottom right: The decapping enzyme

Deadenylation-dependent decapping is best defined in yeast (reviewed in Refs. 6, 7, 21) and is initiated by removal of the poly(A) tail by poly(A) nuclease (22). After the poly(A) tail is removed, a decapping enzyme, Dcp1 (23; reviewed in Ref. 24), removes the CAP structure and the decapped mRNA is subject to rapid destruction by a 5'-to-3' exonuclease called Xrn1 (25, 26). Several other proteins play essential roles in the process, but their exact contributions have yet to be described (reviewed in Refs. 7, 27).

The deadenylation-dependent mRNA degradative pathway has not been fully characterized in vertebrate cells; however, most of the enzymes or activities in the yeast pathway correlate with proteins or activities in vertebrate systems. A poly(A) ribonuclease called PARN has been identified, decapping activities have been demonstrated in cell extracts, and a 5'-to-3' exonuclease homolog to Xrn1 has been found (28–30). However, some differences do exist; for example, degradation of vertebrate mRNAs may occur primarily in a 3'-to-5' direction (1, 31). Characterization of the precise degradation mechanism is ongoing, but some important interactions have been identified. PARN has a higher affinity for binding to the 5' CAP structure than to eIF4E; however, eIF4E in the presence of eIF4G and PABP can interfere with PARN binding (32). When PARN does bind to the CAP, it stimulates deadenylase activity and competes with decapping activity (33). A working hypothesis (reviewed in Ref. 7) (see Fig. 1A) is that instability of the CAP binding complex allows PARN to bind the CAP structure leading to deadenylation of the mRNA. After removal of the poly(A) tail, PARN falls off and leaves the 5' CAP structure available for cleavage by the decapping enzyme.

III. Special Pathways for Regulating the Stability of mRNAs

Passage of mRNAs through the general degradation pathway determines the intrinsic half-life of many mRNAs whose turnover is not typically regulated. The stability of many important mRNAs, however, is regulated, and their stabilization and destabilization is critically important for cellular function. To ensure

---

can now act to remove the CAP. Bottom left: Exonuclease rapidly destroys the remaining message. (B) Model of regulated endonuclease activity. Transcribed mRNA with an endonucleolytic cleavage site (black box on mRNA) is stable, since the endonuclease (white oval) is inactive. Binding of a factor (shaded circle) activates the endonuclease and it cleaves at the specific site on the RNA. (C) Model of an RNA-binding protein protecting an mRNA from an endonuclease. The translated mRNA contains an endonucleolytic cleavage site (black box). An RNA-binding protein (stippled hexagon) protects the site from an endonuclease (black oval). When the RNA-binding protein dissociates from the RNA, the endonuclease cuts at the cleavage site.

that levels of these mRNAs are tightly regulated, several pathways exist for altering the stability of individual mRNAs or of groups of mRNAs. The mechanisms discussed below include regulation of mRNA turnover through the degradation machinery used in the general pathway and through unique regulatory mechanisms that are independent of deadenylation and decapping.

A. Regulating Decay of Unstable mRNAs

Short-lived RNAs, including many cytokines, growth factors, and transcription factors, are ubiquitous factors required for regulation of the cell cycle, growth, and gene regulation. Aberrant expression of many of these factors leads to uncontrolled cell growth and cancer, or may, conversely, result in cell death (34). The transient nature of these signals requires that expression of these proteins be precisely controlled. To closely couple protein production and gene transcription, these factors are often encoded by short-lived mRNAs (3–5). One of the mechanisms for regulating the levels of these factors is the presence of *cis* destabilizing elements which can bind *trans*-acting factors thought to accelerate or decelerate entry into the general decay pathway (30, 35). Two elements have been defined that interact with this pathway, the AU-rich element (ARE), found in a number of labile mRNAs, and the major coding region determinants found in c-*fos* mRNA.

1. AU-RICH ELEMENTS AND ARE-BINDING PROTEINS

AU-rich elements are destabilizing *cis* elements found in a wide variety of short-lived mRNAs (reviewed in Refs. 3, 4). Several classes of AREs, found in different sets of mRNAs, determine the fate of a given message in a particular environment (4, 36). Some AREs consist of the pentamer AUUUA or sometimes multiple or overlapping series of this sequence. Other AREs do not conform to this sequence element, but are AU- or U-rich. These different AREs can have variable effects on the stability of a given mRNA depending on the cellular context. For example, the 3'-UTRs of c-*fos*, c-*myc*, and granulocyte macrophage colony stimulating factor (GM-CSF) all contain AREs. The 3' untranslated region (3'-UTR) of c-*fos* or c-*myc* can destabilize an mRNA in monocytes, whereas the 3'-UTR of GM-CSF does not (37). However, in monocytes treated with lipopolysaccharide, the ARE-containing mRNAs for inteleulein-2 (IL-2), IL-3, and tumor necrosis factor α (TNFα) are destabilized, but there is no effect on the stability of c-*fos* or c-*myc* mRNAs (37–39). In another cell type, activated T cells, several cytokine mRNAs are stabilized, but c-*fos* and c-*myc* mRNAs are not affected (40). This variable effect of different AREs is presumably brought about by interaction of the AREs with a variety of *trans*-acting factors. The abundance and activity of these proteins will vary widely in different cell contexts.

A number of ARE-binding proteins have been identified (reviewed in Ref. 7). AUF1/hnRNPD is an ARE-binding protein that has two RRM motifs and an RGG box and exists in four isoforms. It usually acts as a destabilizing factor, but can stabilize under some circumstances (*3, 41*). Tristetraproline is another factor involved in the destabilization of TNFα, IL-3, and GM-CSF RNAs (*42–44*). It contains two CCCH-type zinc fingers and is also called TISII and Nup475. Members of the embryonic lethal abnormal visual (ELAV) family, of which the ubiquitous protein HuR is the most prominent, are generally involved in the stabilization of mRNAs with AREs (*45*). HuR contains three RRM RNA-binding motifs. The cytoplasmic availability of HuR appears to be regulated by shuttling it in and out of the nucleus. HuR does not effect deadenlyation and is thought to inhibit subsequent steps in RNA decay. HuR interacts with a subset of mRNAs containing AREs, including glucose transporter-1, c-*fos*, and GM-CSF (*46–48*; also see Ref. *45*).

The levels and activity of ARE-binding proteins are regulated by a variety of factors including members of the steroid receptor family, stress-activated kinases, and membrane receptors (*49–51*). Levels of some of the ARE-binding proteins vary during development, cell senescence, and stress (*48, 52–57*). This list is by no means comprehensive, but illustrates that these diverse regulatory inputs further increase the potential of these proteins to specifically control the decay of subgroups of ARE-containing mRNAs.

2. THE C-*fos* MAJOR CODING REGION DETERMINANT

A deadenylation-dependent mechanism provides one site for control of the degradation of c-*fos* mRNA (*58*). In addition to an ARE in the 3'-UTR, c-*fos* has a second element called the major coding region instability element (mCRD1), which also regulates its rate of decay (*35, 59*). This region is a purine-rich segment whose distance from the poly(A) tail is functionally important (*60*). Five proteins have been shown to complex with this region: Unr (a purine-rich RNA-binding protein), PABP, poly(A)-binding interacting protein, AUF-1, and NSAIP1 (a heteronuclear ribonucleoprotein-like protein) (*60*). A model that has been suggested is that the bound proteins form a bridge between the poly(A) tail and the mCRD1 (*60*). In the absence of translation, the mRNA may be stable, but translating ribosomes may disrupt or reorganize the complex and lead to rapid decay of the message. Although the precise mechanics remain to be resolved, it seems likely that interactions between the various instability determinants, their specific binding proteins, and the deadenylation machinery will control the rate of mRNA decay.

B. Removal of Error-Prone RNA: RNA Surveillance or Nonsense-Mediated Decay

Nonsense-mediated decay or mRNA surveillance is a mechanism present in eukaryotic organisms that targets mRNAs containing errors for destruction, while allowing normal mRNAs to proceed through the checkpoint. mRNAs containing errors such as nonsense mutations, improperly spliced introns, or early-termination codons are generally destroyed through this pathway. In yeast, nonsense-mediated decay proceeds through decapping, independent of deadenylation; however, in higher eukaryotes the exact mechanism of RNA decay has not been determined and may be somewhat different than the yeast mechanism. There are several excellent recent reviews of this degradative pathway (61–63).

In yeast, nonsense-mediated decay proceeds through deadenylation-independent decapping (reviewed in Refs. 5, 6). A downstream sequence element is thought to interact with a protein called Hrp1 (64). If the downstream sequence element is in the coding region, translating ribosomes displace Hrp1 from the mRNA and the message is perceived as normal. If a premature stop codon leaves the downstream sequence element in the 3′-UTR, Hrp1 remains bound to this sequence. A complex of proteins which includes several Upf proteins scans the message after the stop codon, and if Hrp1 is still associated with the downstream sequence element, this somehow activates decapping of the mRNA via Dcp1 (23).

In vertebrate cells the mechanism is less clear. Nonsense-mediated decay is dependent on translation of the message and is associated with the first round of translation (65). Although vertebrate cells do not appear to have downstream sequence elements, other signals, such as an exon–exon junction in the 3′-UTR, are thought to activate this response (62, 63). Splicing junctions remain complexed to some nuclear proteins and since introns do not normally occur in the 3′-UTR of mRNAs, the presence of these proteins, which include Upf proteins, Y14, and RNPS1, in the 3′-UTR activates an as-yet-undescribed degradation mechanism (66–68).

C. Stabilization and Destabilization of Endonucleolytically Cleaved RNAs

Decay pathways involving deadenylation and decapping are not the only means by which RNAs can be destroyed. Endonucleolytic cleavage of mRNAs represents another important pathway by which mRNAs are degraded. These pathways rely on endonucleases to cleave within the body of the RNA, followed by exonucleolytic decay of the remaining message. Endonucleolytic pathways provide a mechanism for fine tuning levels of specific mRNAs or groups of RNAs or to differentially up- and downregulate RNAs. Control of this class of pathway can be elicited by regulating the activity of the endonuclease, controlling the availability of the cleavage site, or providing for regulation at both levels.

1. Turning Endonucleases on and off: Destruction of RNAs during Cell Stress

Systems which act to destabilize RNAs through endonucleolytic activity are triggered in response to infection or cellular crisis. By regulating the activity of the riboendonuclease, these systems can be kept ready, but inactive until they are turned on in response to cellular signals (Fig. 1B). Two pathways are activated in response to double-stranded RNA; the first is the interferon- or $2'$–$5'$-linked oligoadenylate (2-5A)-mediated response (69) and the second is interference RNA (70). A third pathway involving endonucleolytic cleavage is activated by unfolded protein in the endoplasmic reticulum (71).

Viral infection often leads to the appearance of double-stranded RNA in the cytoplasm. The cytoplasmic double-stranded RNA triggers activation of 2-5A synthase, leading to synthesis of $2'$–$5'$-linked oligoadenylate (2-5A) (reviewed in Ref. 69). The 2-5A together with double-stranded RNA activates RNAse L (see Table I), which in turn dimerizes to form an active endoribonuclease. RNase L cleaves single-stranded RNA with a preference for UA and UU residues. This system clearly targets foreign RNAs for degradation, but endogenous substrates, such as ISG43, a 43-kDa ubiquitin-specific protease, may also be targeted (72). Additionally, regulation of 2-5A has been seen in oviduct after estrogen administration and in mammary gland after lactation (69), suggesting that responding to viral infection is not the sole function of this system.

A distinct RNA degradation system called interference RNA is also triggered by double-stranded RNA and targets specific endonucleolytic cleavage of target RNA (reviewed in Refs. 70, 73). The specific mechanism and endonuclease involved in this process have not yet been well defined. However, it has been suggested that in response to double-stranded RNA, an RNase III-like activity cleaves double-stranded RNA into 21–23 nucleotide fragments. A recognition system which requires ATP is thought to unwind the fragments and use them as a template to target specific destruction of RNAs containing homologous sequences. The recent use of this system to selectively target destruction of vertebrate mRNAs will certainly further our understanding of this pathway (74).

Ire1, an endoplasmic reticulum-associated type 1 transmembrane protein kinase (see Table I), is a unique RNase containing both protein kinase and endoribonuclease domains. The function of this endonuclease is best characterized in yeast, but it may have a broader function in mammalian cells including down-regulation of mRNA levels. In yeast, the presence of unfolded protein, which occurs during various cellular stresses, triggers the transautophosphorylation and oligomerization of Ire1 (reviewed in Ref. 75). The activated endoribonuclease excises an intron from the HAC1 mRNA and a tRNA ligase splices the cleavage products. The encoded HAC1 protein is a transcription factor, which leads to transcription of genes required for the unfolded protein response. In mammals, Ire1 is present (76) and has a similar structure consisting of protein kinase and endoribonuclease domains. Mammalian Ire1 can autophosphorylate in *trans* and is implicated in the unfolded protein response (77, 78). Mammalian Ire1 can also cleave yeast HAC1 mRNA; however, HAC1 mRNA is not present in mammals, and a different set of transcription factors is necessary for the unfolded protein response. Interestingly, Ire1 can cleave its own mRNA and levels of endogenous Ire1 mRNA are downregulated in the unfolded protein response (71). The cleavage sites in HAC1 and Ire1 mRNAs are different, suggesting a wider range of mammalian Ire1 target sites. It is unclear whether mammalian Ire1 functions to excise introns or degrade mRNA, or perhaps has a role in both processes.

2. REGULATING ENDONUCLEOLYTIC CLEAVAGE SITES IN mRNAs

Regulation of availability of the endonucleolytic cleavage site, rather than the endonuclease itself, represents an important regulatory mechanism controlling mRNA degradation. This pathway is illustrated schematically in Fig. 1C. The basic concept is that the mRNA contains *cis* elements that are cleavage sites for sequence- or structure-specific endonucleases. In the absence of other factors, the presence of the active nuclease and target sequence in the mRNA leads to endonucleolytic cleavage. Since mRNA cleavage produces RNAs with unprotected 5′ and 3′ ends, the cleavage products are rapidly degraded by exonucleases. RNA-binding proteins are thought to control the availability of the mRNA by protecting the cleavage site from attack by endonuclease. Controlling the level, availability, or activity of the RNA-binding proteins provides an important mechanism of regulation. Working in concert, the presence or absence of *cis* elements and control at the level of the RNA-binding protein may allow targeting of specific mRNAs for cleavage, while sparing others.

Regulation of transferrin receptor (TfR) mRNA stability is the prototype for this mechanism. In response to low levels of iron, transferrin receptor levels are increased predominately by stabilizing TfR mRNA (reviewed in Refs. 79, 80). The TfR mRNA contains a series of stem–loop structures in its 3′-UTR, called iron response elements, that bind to a protein termed the iron response element-binding protein (IREBP) (81–83). Binding of the IREBP to the iron response element increases in the presence of low levels of iron and stabilizes TfR mRNA. The IREBP:iron response element complex dissociates in the presence of iron, destabilizing the mRNA. An endonuclease cleavage site was mapped to a region adjacent to one of the stem–loop structures, but the endonuclease responsible for cleavage at this site has yet to be identified (84). Interestingly, using several posttranscriptional mechanisms including mRNA stability, regulation of the activity of the IRE-binding proteins by phosphorylation via signal transduction pathways can coordinate levels of several proteins involved in iron and oxygen metabolism (reviewed in Ref. 85).

3. COORDINATE REGULATION OF mRNA BY ENDONUCLEASE CLEAVAGE AND DEADENYLATION-DEPENDENT DEGRADATION

The endonuclease cleavage pathways and the deadenylation-dependent decay pathway combine to regulate the stability of some mRNAs. Two examples of the intersection of these pathways are regulation of the stability of the highly unstable c-*myc* mRNA and the extremely stable α-globin mRNA. Both mRNAs contain *cis* elements which bind proteins and act as stability determinants by protecting the mRNAs from endonuclease cleavage (see Table I) and are destabilized by deadenylation-dependent mechanisms.

c-*myc* is a transcription factor important in cell growth. Elevated levels of c-*myc* are implicated in cellular transformation and formation of tumors (86). c-*myc* mRNA levels are tightly regulated and the mRNA is very short-lived. Two major elements play a role in the instability of c-*myc* mRNA (reviewed in Ref. 3). The first is an ARE in the 3'-UTR, which is thought to mediate decay through deadenylation. Additionally, there is a coding region instability determinant (CRD), which binds a protein called the CRD-binding protein (CRDBP) (87, 88). The CRDBP has an RNA-binding domain with multiple subunits consisting of four K-homology (KH) domains, two RNP motifs, and one RGG box (89). The increase in CRDBP levels during development, in cancers, and in transformed cell lines correlates with increases in c-*myc* RNA levels (89). CRDBP can also protect from an endonucleolytic activity that cleaves c-*myc* RNA specifically within the CRDBP-binding site (90). This is consistent with the presence of tissue-specific systems for regulating mRNA degradation.

α-Globin is a developmentally regulated protein encoded by an extremely stable mRNA (91). Destabilization of α-globin mRNA leads to the clinical condition α-thalassemia. The 3'-UTR of α-globin mRNA contains a stabilizing element which can bind several proteins, including the KH domain proteins αCP1 and 2 (also known as C-binding proteins or hnRNPE) (92) and AUF1 (93). The association of the proteins on the mRNA form what is called the α-complex. This region of the 3'-UTR also contains a cleavage site for an endonucleolytic activity that is enriched in erythryocytes called ErEN. Both the binding of the α-complex and the activity of ErEN are influenced by PABP and the poly(A) tail (94, 95). A unique model for degradation of α-globin mRNA has been proposed which can involve two possible routes (96). The first mechanism is the one described in this section, that is, the α-complex binding to the 3'-UTR protects the message from cleavage by ErEN. A second mechanism is initiated by deadenylation of the α-globin mRNA. In the absence of PABP and the poly(A) tail, αCP binding is destabilized and ErEN is activated, resulting in enhanced cleavage of the mRNA.

These two systems provide interesting, novel mechanisms by which the classic mRNA degradation machinery combined with endonucleolytic cleavage can be used to either destabilize or stabilize specific mRNAs.

4. Other Endonucleolytic Cleavage Systems

Endonucleolytic cleavage of mRNA in vertebrate cells is beginning to be identified, as shown by the examples described above. Evidence for regulation of these processes is also apparent, but a cohesive picture of the importance of these mechanisms in controlling mRNA turnover is only beginning to emerge. A summary of this information is given in Table I. Endonucleolytic cleavage sites have been identified in mRNAs from a variety of organisms including yeast, *Xenopus,* chicken, and mammals. The cleavage sites are quite heterogeneous in both sequence and structure. Cleavage sites include A-rich (c-*myc* and Tfr mRNAs), AU-rich (apolipoprotein II and gro-α mRNA), ACU-rich (Xlhbox2B, Hep B, and 9E3 mRNAs), U-rich (p27kip1), C-rich (α-globin mRNA), and G-rich (Igf II and PGK RNAs) regions. Whereas most cleavage sites are in single-stranded regions of the RNAs, some are also dependent on mRNA secondary structure (for example, the cleavage sites in IGF II and apolipoprotein mRNAs), and cleavage sites are found in all parts of the mRNA (Table I). The instability of the degradation intermediates has made identification of endonucleolytic cleavage sites difficult. Consequently, some sites have been identified *in vivo,* whereas others have only been characterized *in vitro.*

The isolation and characterization of the endonucleases which target specific mRNAs for decay has been equally elusive. Endonucleolytic activities are apparent in a variety of systems, but only a few endonucleases have been cloned and further characterized. Some of the cloned endonucleases and endonuclease activities are listed in Table I. As with the cleavage sites, putative endonucleases are highly heterogeneous, ranging in size from 13 to 120 kDa and they are found associated at a variety of sites including the nucleus, the cytosol, and on polysomes. Some show endonuclease activity only when activated, others are thought to be constitutively active. Interestingly, several of the proteins identified as endonucleases have either no homology with known nucleases, for example, G3BP (*97*), and some, such as PMR-1 (*98*), belong to other families of proteins with no previously known members that contain endonuclease activity. Nucleases with similar catalytic activities may diverge widely at the amino acid sequence level. For example, the nuclease called activator of RNA decay (ARD) has functional homology to RNase E (*99, 100*), but does not cross react with antibodies to RNase E. It will be of interest to see if these endonucleases exhibit a broader range of RNA targets or cleave a narrow group of substrates, and whether this is a large group of proteins or is limited to a few proteins.

There are a variety of RNA-binding motifs (*101*) and a large number of RNA-binding proteins interacting with RNAs that are both stabilized and destabilized. Some of the better characterized RNA-binding proteins that protect against endonuclease cleavage have already been mentioned and are included in Table I. Other proteins with established functions may also be found to integrate with this system of stabilization. For example, HuR is thought to stabilize mRNAs by interfering with the general degradation pathway. However, it may also act via a second mechanism by blocking an endonuclease

TABLE I
ENDONUCLEASES AND ENDONUCLEASE CLEAVAGE SITES OF mRNAs[a]

mRNA	Cleavage site	Endonuclease Name	Size (kDa)	Location	Activator	Binding protein	Ref.
c-myc	CAAUGAAA; coding		39	Polysomes	—	CRDBP	90
Undefined[b]	In c-myc 3'-UTR	G3BP	120	Cytoplasm	PO4	—	175
Albumin (X.l.)	APryUGA; 3'-UTR	PMR-1	60	Polysomes	Estrogen	—	176
IL-2	Coding region		60–70	Nonribosomal	—	—	177
α-Globin	CCCUCCUUGCACC	ErEN		Nonpolysomal	Erythroid-enriched	αCP	95
Undefined[b]	>24 nt, 3'-UTR		—	—	—	—	91
	AU-rich, ssRNA	ARD1	13.3	Polysomes	—	—	99, 100
	RNase E substrates		—	—	—	—	100
Xlhbox2B	CACCUACCUACCCAACUA		120	Polysomes	—	Endonuclease inhibitor	178
HepB	CCAUACU; ssRNA		26	Nuclear	INF, TNFα	La antigen	104
Apo-lipo II	AAUAA, AAUn(0–?)UAA, 3'-UTR; ssRNA in loop		—	—	Estrogen	—	179
TfR	GAACAAG, 3'-UTR; ssRNA adjacent to G-rich duplex structure, ssRNA, 3'-UTR		—	—	Iron	IREBP	84
IGF II	UUCGGUUUGUUUUUU; 5'-UTR		—	—	—	—	180
p27kip1		Endonuclease in HUR-binding site	—	—	—	HUR	102
Albumin (mouse)	CCAN(1–3)CUGN(0–1)UGAU		—	—	—	—	181
PGK1	GGUG, coding region		—	—	—	—	182
9E3	AUUCCUCCU, 3'-UTR		—	—	Serum	—	183
Gro-α	AAUAU, 3'-UTR		—	—	IL-1	—	184
Stress-regulated endonucleases							185
Viral RNA	AA and AU ssRNA	RNase L	40, 80	ER	INF, 2-5A	—	175
Ire1	CUGCAG, AAAACUA, AGUGAA	IRE1	110	ER	Unfolded Protein	—	71

[a]ss, Single-stranded; ER, endoplasmic reticulum; X.l., Xenopus laevis.
[b]Endogenous substrate unknown.

cleavage site (*102*). Another protein, the La antigen, which is involved in tRNA end processing (reviewed in Ref. *103*), may also interfere with degradation of viral RNA (*104*) and also binds to the 3′-UTR of histone mRNA (*105*). The levels or activity of a number of the RNA-binding proteins listed in Table I and their interactions with RNA-binding sites are altered by changes in conditions known to influence mRNA stability. This implies that protection of an endonuclease cleavage site by an mRNA-binding protein may regulate mRNA stability in a variety of circumstances affecting cells. An experimental system we have focused on in which a candidate RNA-binding protein and mRNase have been identified is the estrogen-mediated regulation of the stability of the mRNA encoding the *Xenopus* egg yolk precursor protein vitellogenin.

IV. Regulation of Vitellogenin and Albumin mRNA Stability

A. Background and Early Work

In oviparous vertebrates, seasonal reproduction requires a high level of production of oocytes. To produce the eggs, large quantities of the egg yolk precursor protein vitellogenin are synthesized in the liver (*106*, *107*), a process which is induced by estrogen. Newly synthesized vitellogenin is secreted into the blood system, and then taken up by developing oocytes by receptor-mediated endocytosis. In the developing oocytes vitellogenin is cleaved into its final products, the yolk proteins. To maintain osmolality of the serum and to offset the production of large quantities of serum proteins, as vitellogenin production is increased there is a concomitant decrease in liver production of albumin and other serum proteins. Although numerous systems are brought into play to coordinate increased vitellogenin synthesis and decreased serum protein production, one of the regulatory systems that we and others have focused on is the estrogen-induced stabilization of vitellogenin mRNA combined with the estrogen-induced destabilization of albumin mRNA.

In *Xenopus laevis*, the administration of estrogen to male animals results in the hepatic synthesis and secretion of massive amounts of vitellogenin (*108*, *109*). Induction of vitellogenin in the male has been chosen as the preferred model since it removes the complicating factor of endogenous estrogen, which is present in females. In the absence of exogenous estrogen, hepatocytes from male *Xenopus* do not contain detectable vitellogenin mRNA (*110*). After prolonged administration of estrogen, vitellogenin mRNA represents as much as half of the hepatocyte's mRNA (*109*, *111*). To identify the regulatory mechanisms by which estrogen achieves this impressive induction of vitellogenin mRNA, we carried out quantitative measurements of the flow of vitellogenin mRNA from its site of transcription to its site of translation on cytoplasmic polysomes.

B. Overview of Regulation of Vitellogenin Gene Expression

1. TRANSCRIPTION AND mRNA STABILIZATION ARE THE PRIMARY SITES FOR REGULATION OF VITELLOGENIN SYNTHESIS

To follow vitellogenin mRNA from the gene to the polysome, we determined the absolute rate of vitellogenin mRNA synthesis (112), the rate of vitellogenin mRNA degradation (111), and the efficiency of nuclear processing of vitellogenin mRNA precursor (111, 113). The absolute rate of vitellogenin mRNA synthesis was determined using pulse labeling and hybridization to immobilized vitellogenin cDNA clones, coupled with measurements of the specific radioactivity of nucleoside precursor pools (111). Estrogen increases the rate of vitellogenin gene transcription by >20,000-fold, from less than 1 molecule of vitellogenin mRNA precursor synthesized per cell per day to 20 molecules of vitellogenin mRNA precursor synthesized per cell per minute (Table II). The induction of specific vitellogenin gene transcription is accompanied by an increase of ~20-fold in the total rate of nuclear RNA synthesis.

To determine the rate of vitellogenin mRNA degradation in the presence and absence of estrogen, we pulse-labeled vitellogenin mRNA in liver fragment cultures exposed to estrogen, chased the label, and then maintained the cultures in medium either containing or lacking estrogen (111). The decay of the pulse-labeled vitellogenin mRNA was monitored by hybridization to immobilized vitellogenin cDNA clones. The efficiency of the chase was determined by measuring the specific radioactivity of nucleoside precursor pools. When estrogen was present, vitellogenin mRNA was largely exempt from mRNA degradation and exhibited an extremely long half-life of ~500 h. When estrogen was removed from the medium, vitellogenin mRNA was degraded about 30 times more rapidly and exhibited a half-life of 16 h at 25°C (and 30 h at 20°C) (111, 114).

TABLE II
ESTROGEN REGULATION OF VITELLOGENIN mRNA LEVELS[a]

Treatment	Nuclear RNA synthesis	Vitellogenin RNA synthesis	$t_{1/2}$ of vitellogenin mRNA (h)
−Estrogen	1	<1 molecule/cell day	16
+Estrogen	20	20 molecules/cell/min	~500
Fold change	20	>20,000	~30

[a]Data compiled from Brock and Shapiro (111, 112).

2. Processing of Vitellogenin mRNA Precursor Is Highly Efficient

Cytoplasmic vitellogenin mRNA levels represent a balance among nuclear vitellogenin gene transcription, splicing of the large vitellogenin mRNA precursor, polyadenylation, and nuclear export. For an mRNA with a very long half-life, such as vitellogenin, the nonsense-mediated decay system may be considered as an additional step between transcription and mRNA accumulation in the cytosol. To establish whether or not nuclear mRNA processing and transport and nonsense-mediated decay were major regulatory sites in this system, we measured the absolute rate of vitellogenin mRNA synthesis and the rate at which vitellogenin mRNA accumulates in the cytosol. These studies provided a rare measurement of the efficiency with which introns are excised from a nuclear pre-mRNA. We found that nuclear transcription produces 20 molecules of vitellogenin mRNA precursor per minute, whereas only 13 molecules appear and accumulate in the cytosol. This suggests that ∼35% of newly synthesized vitellogenin mRNA is subject to intranuclear degradation or nonsense-mediated decay. The efficiency with which each of the 32 introns in the vitellogenin mRNA precursor is excised therefore cannot average less than ∼99% (111, 113).

Since nuclear vitellogenin gene transcription was undetectable in the absence of estrogen and the steps between transcription and cytoplasmic vitellogenin mRNA accumulation were reasonably efficient, the massive estrogen induction of cytoplasmic vitellogenin mRNA is achieved through a combination of estrogen-induced transcription of the vitellogenin genes and estrogen-mediated stabilization of vitellogenin mRNA. The long half-life of vitellogenin mRNA resulting from estrogen-mediated stabilization provides a prolonged window for translation of each vitellogenin mRNA molecule. A very rough estimate based on apparent elongation rates for vitellogenin (115) and other mRNAs is that with a half-life of ∼500 h in the presence of estrogen, each vitellogenin mRNA can be translated at least 40,000 times.

C. Basic Features of Vitellogenin mRNA Stabilization

Vitellogenin mRNAs contain rather short, <20-nucleotide 5′-untranslated regions (116). The 6.3-kb vitellogenin mRNAs encode proteins of ∼180 kDa and are translated efficiently on large, circular, membrane-bound polysomes containing ∼30 ribosomes (114, 117, 118).

The ∼30-fold estrogen-mediated stabilization of vitellogenin mRNA is a specific effect of estrogen on vitellogenin mRNA stability. The half-life of total cell poly(A) mRNA is unchanged in the presence or absence of estrogen (111). The mRNA encoding the major liver-secreted protein albumin is destabilized by estrogen, perhaps by part of the same regulatory system implicated in vitellogenin mRNA degradation (see below). Thus, the regulation of mRNA stability by estrogen in *Xenopus* hepatocytes is a versatile regulatory mechanism, stabilizing some mRNAs, destabilizing other mRNAs in the same cells, and having no effect on the stability of many other mRNAs. Our early studies established several basic features of this system.

1. The stabilization of vitellogenin mRNA requires estrogen and estrogen receptor, and is a positive effect of the estrogen–estrogen receptor complex (*119*).
2. Since pulse-labeled cytoplasmic vitellogenin mRNA that is undergoing degradation can be restabilized on addition of estrogen to the culture medium, estrogen-mediated stabilization of vitellogenin mRNA is a reversible cytoplasmic effect (*111*).
3. The stabilization of vitellogenin mRNA requires that polysomes be associated with the mRNA, but does not require ongoing protein synthesis (*114*).
4. Transient transfection of a *Xenopus* liver-derived cell line indicated that >90% of the protein-coding region of vitellogenin mRNA could be deleted without loss of estrogen-mediated stabilization (*119*). However, our efforts to identify a short sequence motif in the vitellogenin mRNA 3′-UTR responsible for estrogen-mediated stabilization were unsuccessful.

Stabilization of the message by 30-fold, the involvement of the well-characterized estrogen receptor system, and our finding that this was a cytoplasmic effect likely mediated by sequences in the relatively short 3′-UTR made this an attractive model for more detailed studies.

D. The Estrogen-Inducible Vitellogenin mRNA 3′-UTR-Binding Protein Is Vigilin

To search for proteins which might be involved in the estrogen-mediated stabilization of vitellogenin mRNA, we carried out RNA gel mobility shift assays of sequences encompassing the entire 3′-UTR. We identified a protein in salt extracts from *Xenopus* liver polysomes that binds preferentially to the vitellogenin mRNA 3′-UTR. By analyzing binding of the protein in polysome salt extracts to different segments of the vitellogenin mRNA 3′-UTR, we identified a long, ~90-nucleotide fragment adjacent to the poly(A) tract which exhibited high-affinity binding to the protein. An adjacent segment of the 3′-UTR that exhibited a high degree of RNA secondary structure, when modeled using an RNA folding program, was unable to bind the protein (*120*).

Consistent with a potential role for the binding protein in estrogen-mediated stabilization of vitellogenin mRNA, binding of the protein to the vitellogenin mRNA 3′-UTR was induced four- to fivefold after estrogen administration (Fig. 2)

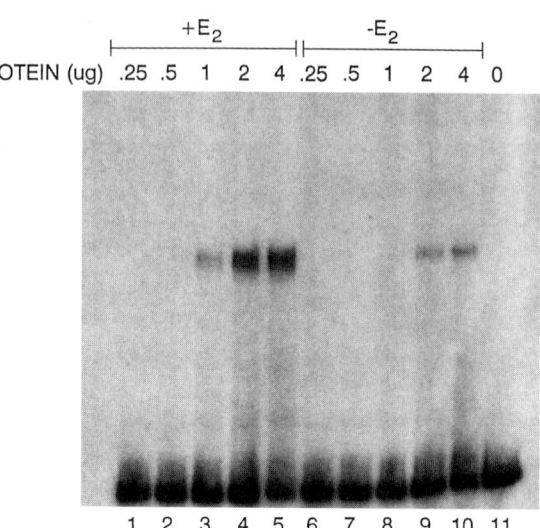

FIG. 2. Estrogen induces binding of the estrogen-inducible vitellogenin mRNA 3'-UTR-binding protein to the vitellogenin mRNA 3'-UTR. Representative autoradiogram of an RNA gel mobility shift assay using a ^{32}P-labeled, ~120-nucleotide segment of the vitellogenin 3'-UTR as a probe. The labeled RNA probe was incubated with increasing concentrations of protein extract (0.25–4 μg) from estrogen-treated (+E, lanes 1–6) or untreated (−E, lanes 7–10) *Xenopus laevis* liver. Probe alone is shown in lane 11. From Ref. *120*.

(*120*). Further analysis showed vitellogenin mRNA 3'-UTR-binding activity was present in several *Xenopus* tissues and was induced by estrogen and by testosterone in various tissues. A similar binding activity was detected in MCF-7 human breast cancer cells (*121*).

Size fractionation using fast protein liquid chromatography showed that the size of the vitellogenin mRNA 3'-UTR-binding protein was in the range of 130–180 kDa (*120*). We used affinity purification to identify the vitellogenin mRNA 3'-UTR-binding protein (*122*). A vitellogenin mRNA 3'-UTR-binding segment containing a short poly(A) tail was immobilized by binding to poly(U) agarose, and proteins salt-extracted from polysomes, obtained from estrogen-treated *Xenopus* liver, were passed over the column. A 150- to 155-kDa protein bound specifically to the vitellogenin 3'-UTR column and not to a column containing nonspecific RNA. Microsequencing identified the vitellogenin 3'-UTR-binding protein as vigilin, a large, 155-kDa protein which contains 15 related, but nonidentical, KH RNA-binding domains (*122*). The identity of the vitellogenin 3'-UTR-binding protein as vigilin was confirmed in electrophoretic mobility shift assays. Antibody to vigilin (*123*) specifically upshifted the vitellogenin mRNA 3'-UTR-binding protein:RNA complex (*122*).

The presence of 15 RNA-binding domains further confirmed that vigilin was the protein binding to vitellogenin mRNA. The KH domain was originally described as the RNA-binding domain of hnRNP-K, the Kth protein isolated in an extract of all heteronuclear RNA-binding proteins (*124, 125*). The KH domain has a variable consensus sequence, which is composed of about 70 amino acids, and has been identified in a wide variety of proteins that bind RNA (*126, 127*). The core motif centers around the sequence (I/L/V)IGxxGxx(I/L/V) (*127*) and the name vigilin derives from the first three amino acids in one of the KH domains, VIG (*128*).

1. VIGILIN IS A UBIQUITOUS, CONSERVED NUCLEIC ACID-BINDING PROTEIN

Vigilin is a ubiquitous nucleic acid-binding protein, having been identified and studied in yeast, *Drosophila*, chicken, *Xenopus*, and human cells (*122, 128–131*). Vigilin was first identified independently by two groups looking for very different proteins. Muller and colleagues (*128, 132*) identified a protein which was increased developmentally during chondrocyte differentiation. A second group identified vigilin as the high-density-lipoprotein-binding protein (*130*), however, this function is less likely, given a number of vigilin's properties. Others, including ourselves, identified vigilin in a variety of different screens from different organisms. In yeast, a partial clone called the HX protein (*133*) was later identified as the 160-kDa protein involved in *Saccharomyces cerevisiae* control of ploidy (SCP160) (*129*). In *Drosophila*, vigilin was identified as the dodecasatellite-binding protein (DDP1) (*131*). A vigilin is also apparent in the sequence of the *Caenorhabditis elegans* genome.

Vigilin is a unique protein, since virtually its entire length is taken up by 14 complete and one degenerate KH binding domains and by spacer regions between the KH domains. Vigilin has maintained a high level of amino acid sequence homology in organisms from yeast to humans (*126*). The high level of conservation of vigilin is reflected in the ability of *Drosophila* DDP1 to complement a knockout of yeast SCP160 (*131*). Analysis of the yeast and human genomes does not reveal any closely related family members or other proteins containing >10 KH domains. We speculate that vigilins may not need to be a family of proteins, because the presence of 15 related, but nonidentical KH domains imparts to vigilin a great deal of functional flexibility. With so many KH domains, point mutations which impair the function of a single KH domain may not have much impact on the function of the entire vigilin molecule.

2. LOCALIZATION AND REGULATION OF VIGILIN

Vigilin has been identified in a variety of tissues from different organisms. In *Xenopus*, we found vigilin in all tissues examined including muscle and reproductive organs (*121*). In mammals, vigilin has also been identified in pancreas (*134*), ovary (*135*), and other tissues (*136*). Most cell types that have been examined have also been found to contain vigilin including all cell culture lines (*121, 136*) and other cells such as adipocytes (*137*), chondrocytes (*132*), fibroblasts, and lymphocytes (*136*). Intracellularly, vigilin has been localized using both immunocytochemical and biochemical techniques and has been found in the cytosol predominantly associated with polysomes and in the nucleus associated with heterochromatin (*123, 138–140*). Vigilin can be salt-extracted from *Xenopus* liver polysomes (*120*), yeast vigilin coisolates with polysomes (*139, 140*), and a complex of vigilin with components of the translation apparatus has been reported (*141*).

Whereas vigilin is ubiquitous, there is considerable evidence that vigilin's levels are regulated within cells. Elevated levels of vigilin in vertebrate cells are often associated with either rapid cell proliferation or high levels of protein secretion. We found that estrogen and testosterone would increase or decrease vigilin levels depending on the specific tissue examined (*121*). For example, testosterone increases vigilin levels in muscle, where the exogenous hormone generally has anabolic activity, and decreases vigilin levels in testes, where the steroid leads to regression of the tissue. Vigilin levels vary in a cyclical manner in the ovary of rats (*135*) and levels are elevated in pancreas at times when trypsin secretion is high (*134*). Mitogens can cause elevation of vigilin levels in peripheral blood lymphocytes (*136*). In cell lines, vigilin appears to be constitutively expressed (*121, 136*), perhaps reflecting the high metabolic activity or rapid production of new protein in these cells. The human and chicken promoters have been isolated (*142–144*), but are not well characterized and it is not yet clear how the diverse regulatory mechanisms mentioned above play out at the level of regulated vigilin gene transcription.

3. Binding of Vigilin to the Vitellogenin mRNA 3′-UTR

The wide tissue and cellular distribution and intracellular localization of vigilin in both nucleus and cytosol suggest a variety of potential interactions with nucleic acids. In addition to vigilin, KH domains are found in a diverse set of prokaryotic and eukaryotic RNA-binding proteins (*126, 127, 145*). When we began these studies the ability of several proteins containing KH domains to interact with RNA homopolymers raised the question of the extent to which KH domain-containing proteins could preferentially recognize and bind to specific RNA sequences. One clinically important KH domain protein, Nova 1, was shown to recognize specific RNA sequences and to bind to glycine receptor $\alpha 2$ mRNA (*146*). Since vigilin is found in diverse cell types and organisms, it seemed clear that there were other intracellular RNA-binding sites than the region of the vitellogenin mRNA 3′-UTR we initially identified. Using *in vitro* genetic selection and biochemical studies, we analyzed the interaction of vigilin with RNAs (*147*). Computer prediction of RNA secondary structures indicated that strong vigilin-binding sites were relatively unstructured. Subsequently, similar requirements for unstructured templates were identified for a single-stranded-DNA-binding site (138).

The genetically selected binding sites were enriched in $(A)_n CU$ and $UC(A)_n$ sequences. Random selection of mutant RNAs with reduced ability to bind vigilin showed that mutations in the $(A)_n CU$ motifs and mutations likely to enhance the stability of RNA secondary structures were effective in reducing binding of vigilin to RNAs. Consistent with the view that RNAs bind to multiple vigilin KH domains, our genetic selection did not reveal any mutants in which a single nucleotide change resulted in greatly reduced binding of an RNA to vigilin (147).

To determine the minimum length of a high-affinity vigilin-binding site, we created RNA ladders using either 5'-end-labeled or 3'-end-labeled RNAs, isolated the vigilin–RNA complexes on nitrocellulose filters, and analyzed them by gel electrophoresis. Since binding declined sharply when sequences at either the 3'-end or the 5'-end of the RNA were deleted, we concluded that there was a sequence-independent minimum length requirement for vigilin binding to RNA. With full-length vigilin, RNAs >70 nucleotides in length were required for efficient binding and binding was negligible when the RNA length was <55 nucleotides. Perhaps because it contains so many KH domains, these sequence and structural requirements for vigilin binding to RNA differ from those of many other RNA-binding proteins, which recognize short sequence motifs (4, 148–151).

Using this information on synthetic vigilin-binding sites, we scanned segments of the human genome database and predicted and subsequently identified a strong vigilin-binding site at the 3'-end of dystrophin mRNA (147). The tight *in vitro* binding of vigilin to the 3'-end of dystrophin mRNA, the presence of vigilin in muscle, and the regulation of vigilin's mRNA-binding activity by testosterone in *Xenopus* muscle raise the intriguing possibility that binding of vigilin to dystrophin mRNA is a biologically significant interaction. Whereas the Becker form of muscular dystrophy is characterized by reduced levels of wild-type dystrophin, there is as yet no direct evidence for an *in vivo* role for vigilin in the metabolism of dystrophin mRNA.

Our analysis of the interaction of vigilin with sequences in several mRNAs suggests that vigilin has some capacity to bind to a variety of mRNA sequences, but exhibits strong preferences for binding to specific sequences. The presence of 15 related, but nonidentical RNA-binding domains on vigilin raised the possibility that different subsets of vigilin KH domains are responsible for binding to different RNAs and single-stranded DNAs. The possibility that a specific KH domain or domains represented a nucleation site that initiated RNA binding also required testing. Using recombinant vigilin segments containing 6 KH domains each, we examined the interaction of vigilin with diverse RNAs and single-stranded DNA. These studies indicated that discrete subregions of vigilin KH

A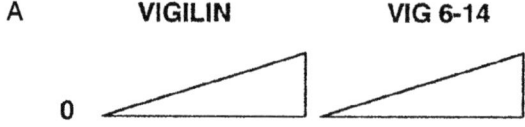

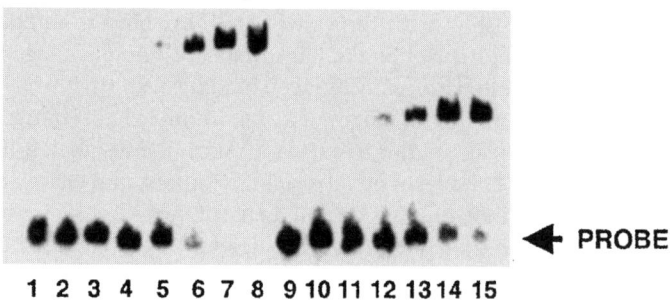

B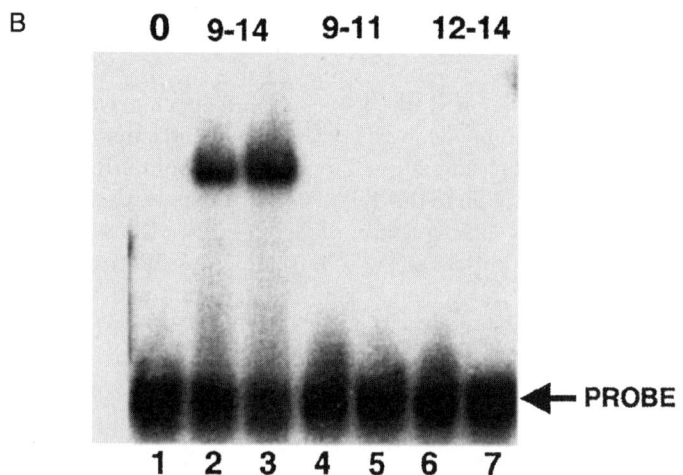

FIG. 3. Contribution of the number of KH domains in vigilin to binding to the vitellogenin mRNA 3'-UTR segment. Flag epitope-tagged full-length vigilin (vigilin) or vigilin truncations containing nine KH domains (KH domains 6–14; VIG 6–14), six KH domains (KH domains 9–14, VIG 9–14), or three KH domains (KH domains 9–11 or 12–14, VIG 9–11 and VIG 12–14, respectively) were expressed using a baculovirus expression system and purified over a flag antibody affinity column. Proteins were used in an RNA gel mobility shift assay using a portion of the vitellogenin mRNA 3'-UTR as probe. (A) Increasing amounts (10^{-9} to $10^{-6}M$) of full-length vigilin (lanes 2–8) or of VIG 6–14 (lanes 9–15) were added to the probe with 5 μg each of tRNA and heparin. (B) Binding

domains are not responsible for vigilin's ability to bind to diverse nucleic acids, but some regions of vigilin appear to have a higher affinity for certain areas of the vitellogenin 3′-UTR (R. Dodson and D. Shapiro, manuscript in preparation). Additionally, we find that binding to the vitellogenin mRNA 3′-UTR required several KH domains. Several different recombinant proteins containing 3 vigilin KH domains failed to bind to the RNA, whereas vigilin segments containing 6, 9, or all 15 KH domains bound progressively better to the vitellogenin 3′-UTR (see Fig. 3) (unpublished data).

4. MODEL FOR INTERACTION OF VIGILIN WITH SINGLE-STRANDED NUCLEIC ACIDS

The above data are most consistent with a model in which initial binding of a subset of vigilin's KH domains to a relatively high-affinity site in an RNA target is followed by a "zippering," or processive binding, process in which additional regions of the highly flexible nucleic acid template make weak contacts with additional vigilin KH domains. By summing up the individually weak interactions of all of its many KH domains with RNA, vigilin achieves high-affinity binding to nucleic acids. This unusual mode of RNA interaction is illustrated schematically in the model shown in Fig. 4.

Although the structure of vigilin KH domains bound to RNA has not been solved, the structures of a few KH domains containing bound RNAs have recently been described (153, 154). KH domains consist of globular, compact, three-stranded antiparallel β sheets with two or three α helices behind the β sheets (127, 152). The structure of the third KH domain of Nova-1 bound to RNA is quite different than the β-sheet interaction system seen for RNA-binding proteins containing RRM motifs. In Nova-1, single-stranded RNA binds to a platform created by two α helices and one edge of a β sheet. The RNA is clamped between two segments of the protein, with one side of the clamp consisting of the conserved Gly–X–X–Gly motif characteristic of KH domains, and the other side of the clamp containing the relatively unstructured variable loop. Recently, the structure of NusA bound to RNA was reported (154). NusA contains two KH domains and two additional RNA-binding domains. The RNA interaction motifs form a relatively continuous extended RNA-binding surface. A great deal of flexibility is required for the bound RNA to loop and wind around the protein. Whereas the RRM motif interacts with RNAs through a β sheet and KH domains contain a different composite α/β aliphatic platform, comparison of KH domain:RNA interaction with recently described structures of RRM

---

of VIG 9–14 (200 ng, lane 2; and 400 ng, lane 3), VIG 9–11 (290 ng, lane 4; and 580 ng, lane 5) and VIG 12–14 (796 ng, lane 6; and 1.6 μg, lane 7) to labeled vitellogenin 3′-UTR RNA probe in the presence of 0.05 μg of heparin. Probe alone is indicated as 0 (lane 1 in both A and B).

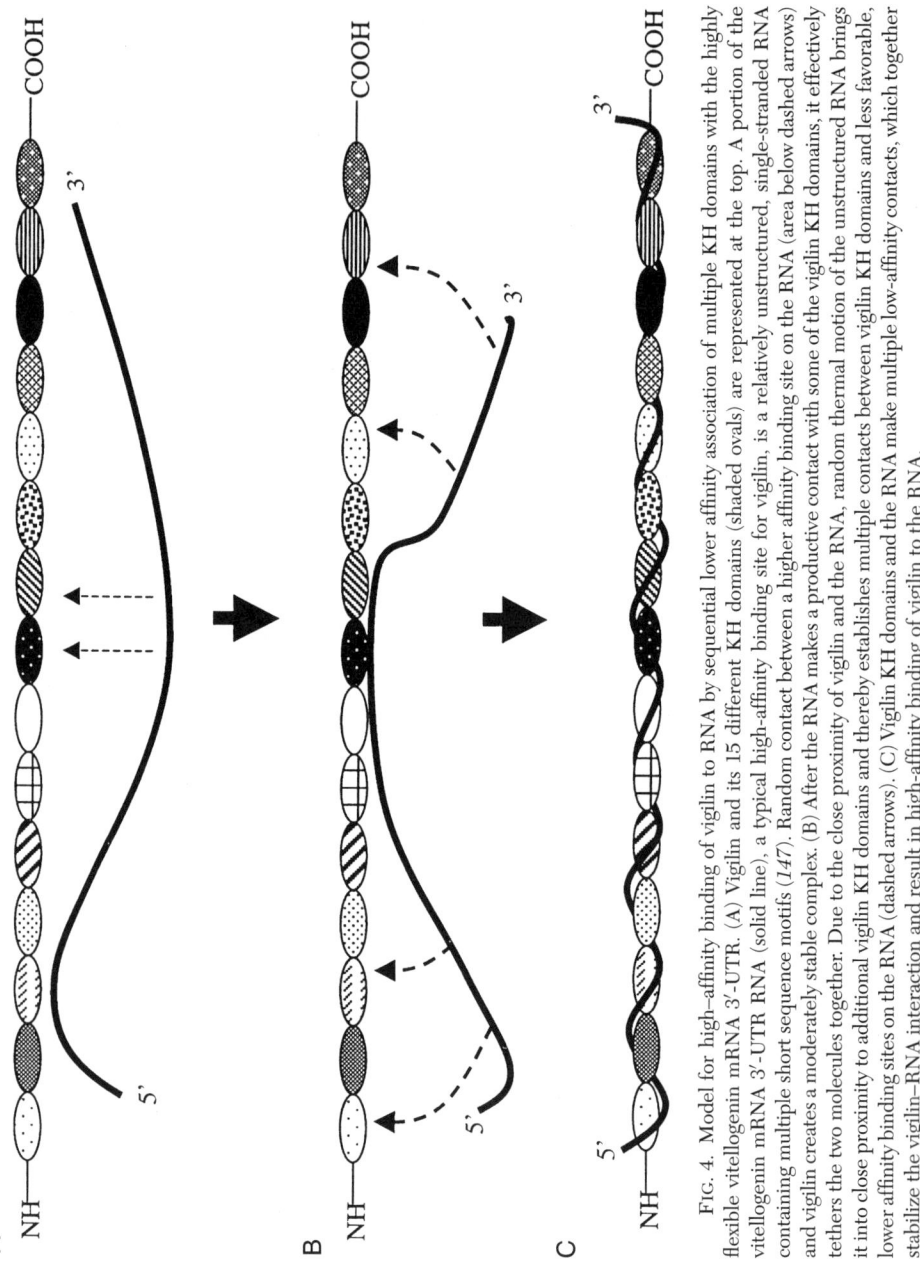

FIG. 4. Model for high-affinity binding of vigilin to RNA by sequential lower affinity association of multiple KH domains with the highly flexible vitellogenin mRNA 3'-UTR. (A) Vigilin and its 15 different KH domains (shaded ovals) are represented at the top. A portion of the vitellogenin mRNA 3'-UTR RNA (solid line), a typical high-affinity binding site for vigilin, is a relatively unstructured, single-stranded RNA containing multiple short sequence motifs (147). Random contact between a higher affinity binding site on the RNA (area below dashed arrows) and vigilin creates a moderately stable complex. (B) After the RNA makes a productive contact with some of the vigilin KH domains, it effectively tethers the two molecules together. Due to the close proximity of vigilin and the RNA, random thermal motion of the unstructured RNA brings it into close proximity to additional vigilin KH domains and thereby establishes multiple contacts between vigilin KH domains and less favorable, lower affinity binding sites on the RNA (dashed arrows). (C) Vigilin KH domains and the RNA make multiple low-affinity contacts, which together stabilize the vigilin–RNA interaction and result in high-affinity binding of vigilin to the RNA.

proteins bound to single-stranded RNAs (155–158) is of interest. For the RRM-containing proteins sex lethal and poly(A) binding protein (PABP) (155, 156), whose structures bound to RNAs have been solved, and for Nova-1 and Nova-2, a single RNA-binding motif interacts poorly or not at all with its RNA target. The presence of a second RNA-binding motif results in a large increase in binding (15, 146, 159). Similarly for vigilin, we find that multiple KH domains are required for detectable binding to RNA and to single-stranded DNA.

When an RNA-binding protein interacts with a structure such as a stem-and-loop, the structure of the RNA is usually largely determined before binding. In contrast, the interaction of single-stranded-RNA-binding proteins with RNAs can profoundly alter the structure of the RNA. Interestingly, the RNA segments bound to each of the four RNA interaction modules in NusA have quite different conformations, as do the RNAs bound to the two RRM motifs in poly(A)-binding proteins (156), which are themselves completely different than the conformation of the RNAs bound to the two RRM motifs in sex lethal U1A and the U2B:U2A complex (155, 160, 161). These data are consistent with our proposal that interaction of a long, >70-nucleotide unstructured RNA-binding site with vigilin involves substantial conformational changes in the RNA, enabling it to make appropriate contacts with multiple vigilin KH domains (145; R. E. Dodson and D. J. Shapiro, manuscript in preparation). In this model, highly structured RNAs may be unable to create stable complexes with vigilin's KH domains, because of the larger energy input required to deform their structures.

The third KH domain of Nova-1 binds to RNA using a combination of van der Waals contacts, stacking interactions, and hydrogen bonds. Although contacts are made with the backbone and the bases, in apparent contrast to the RRM proteins, sex lethal, PABP, and TRAP, hydrogen bonds between $2'$-OH groups of the sugar and amino acid side chains do not appear to play important roles in interactions of KH domains with their RNA targets (153, 156, 159, 162). Because interactions with the $2'$-OH of the ribose are important for RRM proteins, they show little or no ability to bind to single-stranded DNA. In contrast, vigilin, which contains KH domains, binds effectively to unstructured single-stranded DNA. These binding data also suggest that $2'$-OH groups in the RNA are not important in stabilizing binding of vigilin to RNAs. Of course, the requirement for conformational flexibility in the interaction of KH domains with nucleic acids makes it unlikely that relatively rigid double-stranded DNAs are high-affinity in vivo binding sites for vigilin and other multi-KH-domain proteins.

5. PROPOSED FUNCTIONS OF VIGILIN

Vigilin clearly binds specific single-stranded RNAs and DNAs, but the exact functions of vigilin have not been determined. However, some roles are suggested. Targeted disruption of yeast SCP160 leads to abnormal partioning of chromosomes at mitosis, which is referred to as control of ploidy (129). It is not clear whether the observed effects of vigilin/SCP160 on nuclear segregation reflect direct effects or, as recently suggested (*163*), reflect downstream effects due to altered expression of other gene products when vigilin is absent. Consistent with a possible direct effect of vigilin at the chromosome, *Drosophila* DDP1/vigilin binds to single-stranded dodeca-satellite DNA located in the pericentric heterochromatin and is proposed to play a role in its organization (*131*). Additionally, it is found in other areas of heterochromatin colocalizing with heterochromatin protein 1. The presence of 15 RNA-binding domains, the localization of vigilin on polysomes, the correlation of increased vigilin levels with increased protein production or cell proliferation, and our findings that vigilin binds to RNA suggest a role in RNA metabolism, and our data suggest a role in the estrogen-mediated stabilization of vitellogenin mRNA.

E. Vigilin and Vitellogenin mRNA Stabilization

The most straightforward role for vigilin in the estrogen regulation of vitellogenin mRNA stability would be for vigilin to bind to a segment of the vitellogenin mRNA 3'-UTR and to protect this segment of the mRNA against the action of a degradative enzyme or system. To accomplish the selective estrogen regulation of vitellogenin mRNA stability, several criteria probably must be met. First, in response to estrogen, the binding ability of vigilin would be expected to increase. Second, vigilin should show a binding preference for vitellogenin mRNA and reduced binding to albumin mRNA, which is destabilized by estrogen in the same hepatocytes in which vitellogenin mRNA is stabilized. Third, a targeted degradation mechanism should be present whose activity can be blocked by the RNA-binding protein. These points are discussed below.

1. Vigilin would be expected to exhibit estrogen-induced binding to the vitellogenin mRNA 3'-UTR. Consistent with a possible role for vigilin in vitellogenin mRNA stabilization, both vigilin mRNA and vigilin's vitellogenin mRNA 3'-UTR-binding activity are induced by estrogen in *Xenopus* liver (Fig. 5) (*122*).

2. A second prediction of a model in which vigilin plays a role in vitellogenin mRNA stabilization is that vigilin should show preferential binding to the vitellogenin mRNA 3'-UTR. Specifically, we anticipate that vigilin should show higher affinity binding to the vitellogenin mRNA 3'-UTR-binding site than to the albumin mRNA 3'-UTR. We tested binding of vigilin to vitellogenin and albumin mRNA-binding sites (*164*). To simulate the presence of other cellular nucleic acids, we used significant levels of nonspecific competitors as well as the test RNAs. Using both recombinant human vigilin and vigilin in extracts from *Xenopus* liver polysomes, we found that vigilin exhibited far higher affinity binding to the vitellogenin mRNA 3'-UTR site than to the albumin mRNA site. These data are consistent with the idea that estrogen-mediated stabilization might involve estrogen induction of additional vigilin, which then exhibits

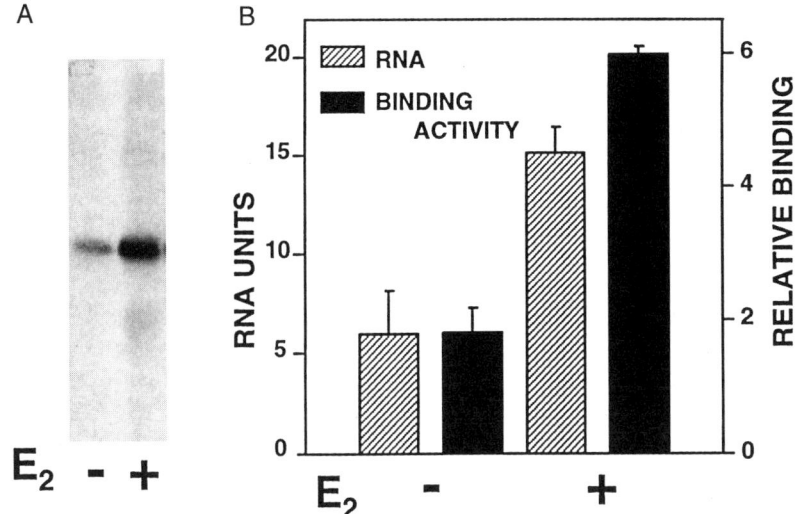

FIG. 5. Estrogen induces vigilin mRNA and RNA binding activity. (A) Northern blot analysis of total RNA from livers of *Xenopus* treated with vehicle (−) or with estradiol (+) and probed with vigilin cDNA. (B) Comparison of the induction of vigilin mRNA and vitellogenin mRNA 3′-UTR-binding activity from livers of *Xenopus* treated with vehicle (−) or estrogen (+). RNA levels (hatched bars) were determined by Northern analysis. Relative binding activity of vigilin in liver extracts (solid bars) was determined in RNA gel mobility shift assays. Gel-shifted bands were quantified using a PhosphorImager. The data represent the average from three animals ± SE. From Ref. *122*.

preferential binding to the vitellogenin mRNA 3′-UTR and thereby protects vitellogenin mRNA against endonuclease cleavage.

3. An estrogen-regulated stabilization and degradation system for vitellogenin mRNA should involve both mRNA-binding proteins and a degradative enzyme or apparatus. Recently, a candidate mRNA-degrading enzyme in *Xenopus* liver polysomes whose properties are consistent with a role in vitellogenin mRNA degradation has been identified (*165*).

1. THE PMR-1 NUCLEASE SYSTEM

Administration of estrogen reprograms the *Xenopus* hepatocyte's spectrum of secreted proteins. Estrogen induces the coordinate stabilization of the egg yolk precursor protein vitellogenin mRNA and the destabilization of the mRNAs encoding albumin and other serum protein mRNAs (*166, 167*). Schoenberg and coworkers identified and characterized a polysomal riboendonuclease, PMR-1, likely to be involved in this process (*98, 168*). PMR-1 is a 62/64-kDa protein belonging to the peroxidase family. Although PMR-1 does not exhibit peroxidase activity, it has a 57% sequence homology with myeloperoxidase (98). PMR-1 exists in a latent form on polysomes and its activity is induced >20-fold by estrogen (165). The potent estrogen activation of polysomal PMR-1 activity suggests a mechanism for estrogen-mediated destabilization of the mRNAs encoding albumin and other secretory proteins. Vitellogenin mRNA, which is stabilized by estrogen, would have to be protected against PMR-1 cleavage.

2. VIGILIN BINDING PROTECTS THE VITELLOGENIN mRNA 3'-UTR SITE FROM PMR-1 CLEAVAGE *IN VITRO*

PMR-1 preferentially cuts RNA within a single-stranded consensus sequence APyrUGA and at some nonconsensus sites in albumin mRNA. Cleavage of albumin mRNA at these sites has been demonstrated both *in vivo* and *in vitro* in estrogen-induced *Xenopus* liver (169, 170). Interestingly, two copies of the APyrUGA pentamer are located in the vitellogenin B1 mRNA 3'-UTR region, which is an effective vigilin-binding site (164). These data naturally lead to the hypothesis that the high-affinity binding of vigilin to the vitellogenin mRNA 3'-UTR binding site protects this region of the mRNA from cleavage by PMR-1.

When no protein is bound, the free vitellogenin mRNA 3'-UTR region is efficiently cleaved by PMR-1 at the APyrUGA sites. Vigilin largely binds to and protects the vitellogenin mRNA 3'-UTR site from PMR-1 cleavage. In contrast, under the same binding conditions, because of vigilin's much lower affinity for the albumin mRNA site, it is unable to bind to the albumin mRNA, which is effectively cleaved by PMR-1 (Fig. 6) (164).

3. MODEL FOR THE ESTROGEN-MEDIATED STABILIZATION OF VITELLOGENIN mRNA

The presence of vigilin and PMR-1 in polysomes (120, 165), the *in vitro* data showing that vigilin binding interferes with cleavage by PMR-1 (164), the data showing that estrogen treatment results in the induction of vigilin (122), and analogy to other systems, such as the regulated degradation of α-globin and TfR mRNA (79, 95), are consistent with the conceptual model for estrogen-mediated stabilization of vitellogenin mRNA shown in Fig. 7. Because the half-life of vitellogenin mRNA in the presence of estrogen is very long ($\sim$500 h), prospects for development of an *in vitro* mRNA decay system mimicking estrogen action are poor. Intracellular studies of the actions of vigilin and PMR-1 will therefore be required to directly test this model under physiological conditions.

F. Future Prospects

Although this model accounts for many observations, important questions remain. Although these observations provide a plausible model for how vigilin binding could protect vitellogenin mRNA from the estrogen-induced degradative system implicated in the accelerated degradation of other serum protein

REGULATION OF mRNA STABILITY

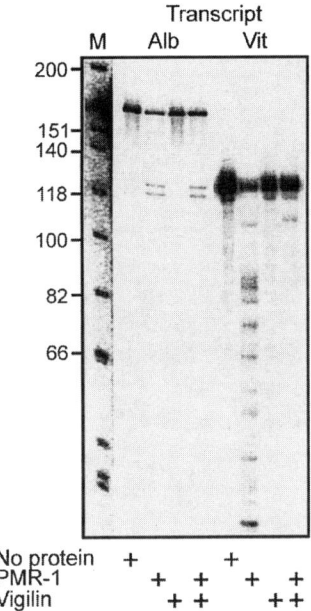

FIG. 6. Vigilin binding selectively inhibits cleavage of the vitellogenin mRNA 3'-UTR by PMR-1. A portion of the albumin mRNA 5'-UTR containing PMR-1 cleavage sites (Alb) or vitellogenin mRNA 3'-UTR was uniformly labeled and incubated with buffer alone (No protein), 40 units of PMR-1, 500 ng of vigilin, or both PMR-1 and vigilin. Samples were recovered from each reaction and were electrophoresed on a 6% polyacrylamide–urea gel and visualized using a PhophorImager. Lane 1 contains a size marker of Hinf1-cut ϕX174 DNA. From Ref. 164.

mRNAs, they shed no light on the mechanism by which vitellogenin mRNA evades degradation by the standard mRNA degradative pathway initiated by shortening of the poly(A) tail. With its half-life in the presence of estrogen of ~500 h, in metabolically active liver cells, vitellogenin mRNA in estrogen-treated cells appears to be exempt from normal intracellular pathways of mRNA degradation. Whether this is due to intrinsic properties of polysomal vitellogenin mRNA or represents another aspect of estrogen action in this system remains to be established. It is interesting that early electron microscopy data suggested that vitellogenin may be translated on large, circular, membrane-bound polysomes in which the 3'- and 5'-ends of the mRNA are in contact (117, 118). This is consistent with recent observations in other systems demonstrating interaction between the 3'- and 5'-ends of mRNAs (20). One intriguing and potentially testable possibility is that contact between the poly(A) tail and its bound PABP and proteins bound at the 5'-end of vitellogenin mRNA somehow protect the poly(A) tail against exonuclease digestion.

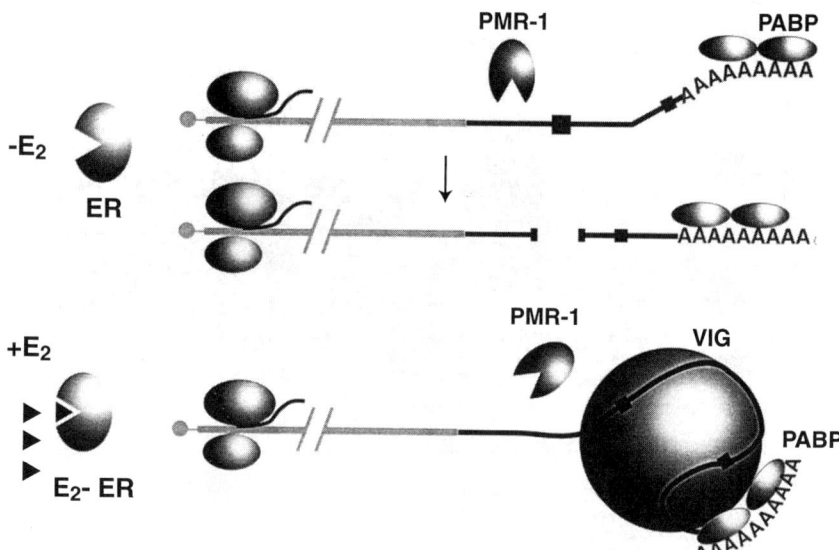

FIG. 7. A model for the stabilization and degradation of vitellogenin mRNA by vigilin and PMR-1. In the absence of estrogen ($-E_2$, top), the estrogen receptor (ER) is unliganded, vigilin is not induced, and the vitellogenin mRNA cleavage sites (indicated by squares) are unoccupied by vigilin and therefore susceptible to attack by PMR-1 RNase. Internal cleavage by PMR-1 exposes unprotected 3'- and 5'-ends leaving the mRNA susceptible to destruction by exonucleases. In the presence of estrogen, the estrogen receptor is activated by the presence of bound ligand, and, by an as-yet-undetermined mechanism, increases levels of vigilin (VIG). Vigilin binds to a high-affinity site in the vitellogenin mRNA 3'-UTR and occludes the RNase cleavage site, stabilizing the mRNA.

Recent studies suggest that several proteins which influence the cytoplasmic stability of mRNAs initially associate with pre-mRNA in the nucleus. Also, a number of RNA-binding proteins can function in both the nucleus and cytoplasm (reviewed in Ref. *171*). For example, HuR, which stabilizes ARE-containing mRNAs, is predominantly nuclear, and shuttles between the nucleus and cytoplasm using a novel shuttling sequence (reviewed in Ref. *45*). HuR crosslinks *in vivo* to both nuclear and cytoplasmic RNAs (*97, 172, 173*). Additionally, agents which induce mRNA stabilization at AREs elicit increased cytoplasmic localization of HU proteins. Vigilin can bind to unstructured single-stranded DNAs and to RNAs and is found both in the nucleus associated with chromatin and in the cytoplasm associated with polysomes (*123, 131, 139, 140*). It is therefore tempting to suggest that under conditions in which requirements for new proteins are reduced, vigilin is largely bound to nuclear DNA, not to cytoplasmic RNAs. Increased demand for protein production might result in movement of vigilin into the cytoplasm. A nuclear localization signal has been identified between the second and third KH domains of vigilin (*123*); however, candidate nuclear export signals in vigilin remain to be identified.

The absence of experimentally tractable yeast systems in which an RNA-binding protein protects against endonuclease cleavage makes it difficult to bring the power of genetic analysis to bear. In this context, the recent description of interference RNA for selective ablation of mRNAs encoding individual vertebrate proteins exhibits great promise (*74, 174*). Insights from structural biology and biochemistry as well as cell biology approaches will be needed to achieve a molecular understanding of how these proteins achieve specificity in their interactions with specific mRNAs. An important issue in these systems is the relation between the binding protein:endonuclease degradative system and the deadenylation pathway for mRNA decay. The binding protein:endonuclease cleavage systems suggest the simple and testable model that binding of the protein to the RNA physically occludes or renders inaccessible the recognition elements of the mRNase cleavage site. In the deadenylation-based mRNA decay pathway the mechanism(s) by which the binding of various proteins to sequences such as AREs actually modulates degradation remains an important and unresolved issue. The coupling of most mRNA degradative pathways to cell division and translation remains poorly understood.

Although mRNA degradation is now widely accepted as an important regulatory site and the machinery for some pathways of mRNA degradation is emerging, deciphering the process by which this control plays itself out at the level of most individual mRNAs remains an ongoing challenge. Developing an understanding of the integration of regulation of mRNA degradation into the spectrum of controls governing the life and death of cells represents a major long-term research challenge that we are only beginning to approach.

ACKNOWLEDGMENT

The research from our laboratory described in this article was supported by NIH grant DK50080.

REFERENCES

1. J. Ross, *Microbiol. Rev.* **59**, 423 (1995).
2. D. J. Shapiro, R. E. Dodson, M. R. Acena, V. A. DiPippo, and H. Kanamori, *in* "Hormones and Growth Factors in Development and Neoplasia" (R. B. Dickson and D. S. Salomon, eds.), p. 79. Wiley-Liss, New York, 1998.
3. G. M. Wilson and G. Brewer, *Prog. Nucleic Acid Res. Mol. Biol.* **62**, 257 (1999).
4. C. Y. Chen and A. B. Shyu, *Trends Biochem. Sci.* **20**, 465 (1995).
5. A. Jacobson and S. W. Peltz, *Annu. Rev. Biochem.* **65**, 693 (1996).

6. C. A. Beelman and R. Parker, *Cell* **81,** 179 (1995).
7. C. J. Wilusz, M. Wormington, and S. W. Peltz, *Nat. Rev. Mol. Cell. Biol.* **2,** 237 (2001).
8. R. F. Bachvarova, *Cell* **69,** 895 (1992).
9. A. Bashirullah, R. L. Cooperstock, and H. D. Lipshitz, *Proc. Natl. Acad. Sci. USA* **98,** 7025 (2001).
10. W. R. Paulding and M. F. Czyzyk-Krzeska, *Adv. Exp. Med. Biol.* **475,** 111 (2000).
11. J. M. Staton, A. M. Thomson, and P. J. Leedman, *J. Mol. Endocrinol.* **25,** 17 (2000).
12. J. Guhaniyogi and G. Brewer, *Gene* **265,** 11 (2001).
13. G. Blobel, *Proc. Natl. Acad. Sci. USA* **70,** 924 (1973).
14. A. B. Sachs, M. W. Bond, and R. D. Kornberg, *Cell* **45,** 827 (1986).
15. A. A. Antson, *Curr. Opin. Struct. Biol.* **10,** 87 (2000).
16. A. C. Gingras, B. Raught, and N. Sonenberg, *Annu. Rev. Biochem.* **68,** 913 (1999).
17. A. B. Sachs, P. Sarnow, and M. W. Hentze, *Cell* **89,** 831 (1997).
18. S. Z. Tarun, Jr. and A. B. Sachs, *EMBO J.* **15,** 7168 (1996).
19. H. Imataka, A. Gradi, and N. Sonenberg, *EMBO J.* **17,** 7480 (1998).
20. S. E. Wells, P. E. Hillner, R. D. Vale, and A. B. Sachs, *Mol. Cells* **2,** 135 (1998).
21. G. Caponigro and R. Parker, *Microbiol. Rev.* **60,** 233 (1996).
22. C. E. Brown and A. B. Sachs, *Mol. Cell. Biol.* **18,** 6548 (1998).
23. D. Muhlrad and R. Parker, *Nature* **340,** 578 (1994).
24. M. Tucker and R. Parker, *Annu. Rev. Biochem.* **69,** 571 (2000).
25. D. Muhlrad, C. J. Decker, and R. Parker, *Genes Dev.* **8,** 855 (1994).
26. C. L. Hsu and A. Stevens, *Mol. Cell. Biol.* **13,** 4826 (1993).
27. W. He and R. Parker, *Curr. Opin. Cell Biol.* **12,** 346 (2000).
28. V. I. Bashkirov, H. Scherthan, J. A. Solinger, J. M. Buerstedde, and W. D. Heyer, *J. Cell Biol.* **136,** 761 (1997).
29. C. G. Korner, M. Wormington, M. Muckenthaler, S. Schneider, E. Dehlin, and E. Wahle, *EMBO J.* **17,** 5427 (1998).
30. M. Gao, C. J. Wilusz, S. W. Peltz, and J. Wilusz, *EMBO J.* **20,** 1134 (2001).
31. A. van Hoof and R. Parker, *Cell* **99,** 347 (1999).
32. E. Dehlin, M. Wormington, C. G. Korner, and E. Wahle, *EMBO J.* **19,** 1079 (2000).
33. M. Gao, D. T. Fritz, L. P. Ford, and J. Wilusz, *Mol. Cells* **5,** 479 (2000).
34. A. O. Perantoni, in "The Biological Basis of Cancer" (R. G. McKinnel, R. E. Parchment, A. O. Perantoni, and G. B. Pierce, eds.), p. 133. Cambridge University Press, Cambridge, 1998.
35. A. B. Shyu, M. E. Greenberg, and J. Belasco, *Genes Dev.* **3,** 60 (1989).
36. S.-P. Peng, C.-Y. Chen, N. Xu, and A.-B. Shyu, *EMBO J.* **17,** 3461 (1998).
37. G. D. Schuler and M. D. Cole, *Cell* **55,** 1115 (1988).
38. T. J. Ernst, A. R. Ritchie, G. D. Demetri, and J. D. Griffin, *J. Biol. Chem.* **264,** 5700 (1989).
39. G. Shaw and R. Kamen, *Cell* **46,** 659 (1986).
40. T. Lindsten, C. H. June, G. Ledbetter, G. Stella, and C. B. Thompson, *Science* **244,** 339 (1989).
41. N. Xu, C.-Y. A. Chen, and A.-B. Shyu, *Mol. Cell. Biol.* **21,** 6960 (2001).
42. G. Stoecklin, X. F. Ming, R. Looser, and C. Moroni, *Mol. Cell. Biol.* **20,** 3753 (2000).
43. W. S. Lai, E. Carballo, J. R. Strum, E. A. Kennington, R. S. Phillips, and P. J. Blackshear, *Mol. Cell. Biol.* **19,** 4311 (1999).
44. E. Carballo, W. S. Lai, and P. J. Blackshear, *Science* **281,** 1001 (1998).
45. C. M. Brennan and J. A. Steitz, *Cell. Mol. Life Sci.* **58,** 266 (2001).
46. W. J. Ma, S. Chung, and H. Furneaux, *Nucleic Acids Res.* **25,** 3564 (1997).
47. R. G. Jain, L. G. Andrews, K. M. McGowan, P. H. Tekala, and J. D. Keene, *Mol. Cell. Biol.* **17,** 954 (1997).
48. W. Wang, M. C. Caldwell, S. Lin, H. Furneaux, and M. Gorospe, *EMBO J.* **19,** 2340 (2000).
49. K. J. Busam, A. G. Geiser, A. B. Roberts, and M. B. Sporn, *Oncogene* **8,** 2267 (1993).
50. A. Sela-Brown, J. Silver, G. Brewer, and T. Naveh-Many, *J. Biol. Chem.* **275,** 7424 (2000).

51. L. G. Sheflin, W. Zhang, and S. W. Spaulding, *Endocrinology* **142**, 2361 (2001).
52. W. Wang, X. Yang, V. J. Cristofalo, N. J. Holbrook, and M. Gorospe, *Mol. Cell. Biol.* **21**, 5889 (2001).
53. W. Wang, H. Furneaux, H. Cheng, M. C. Caldwell, D. Hutter, Y. Liu, N. Holbrook, and M. Gorospe, *Mol. Cell. Biol.* **20**, 760 (2000).
54. A. Gouble and D. Morello, *Oncogene* **19**, 5377 (2000).
55. C. Blattner, P. Kannouche, M. Litfin, K. Bender, H. J. Rahmsdorf, J. F. Angulo, and P. Herrlich, *Mol. Cell. Biol.* **20**, 3616 (2000).
56. J. S. Buzby, G. Brewer, and D. J. Nugent, *J. Biol. Chem.* **274**, 33973 (1999).
57. G. Laroia, R. Cuesta, G. Brewer, and R. J. Schneider, *Science* **284**, 499 (1999).
58. M. E. Greenberg, A. B. Shyu, and J. G. Belasco, *Enzyme* **44**, 181 (1990).
59. S. C. Schiavi, C. L. Wellington, A. B. Shyu, C. Y. Chen, M. E. Greenberg, and J. G. Belasco, *J. Biol. Chem.* **269**, 3441 (1994).
60. C. Grosset, C. Y. Chen, N. Xu, N. Sonenberg, H. Jacquemin-Sablon, and A. B. Shyu, *Cell* **103**, 29 (2000).
61. M. W. Hentze and A. E. Kulozik, *Cell* **96**, 307 (1999).
62. L. E. Maquat and G. G. Carmichael, *Cell* **104**, 173 (2001).
63. P. Mitchell and D. Tollervey, *Curr. Opin. Cell Biol.* **13**, 320 (2001).
64. C. I. Gonzalez, M. J. Ruiz-Echevarria, S. Vasudevan, M. F. Henry, and S. W. Peltz, *Mol. Cells* **5**, 489 (2000).
65. Y. Ishigake, X. L. Li, G. Serin, and L. E. Maquat, *Cell* **106**, 607 (2001).
66. J. Lykke-Andersen, M. D. Shu, and J. A. Steitz, *Cell* **103**, 1121 (2000).
67. J. Lykke-Andersen, M. D. Shu, and F. A. Steitz, *Science* **293**, 1836 (2001).
68. V. N. Kim, N. Kataoka, and G. Dreyfuss, *Science* **293**, 1832 (2001).
69. G. R. Stark, I. M. Kerr, B. R. Williams, R. H. Silverman, and R. D. Schreiber, *Annu. Rev. Biochem.* **67**, 227 (1998).
70. P. A. Sharp, *Genes Dev.* **15**, 485 (2001).
71. W. Tirasophon, K. Lee, B. Callaghan, A. Welihinda, and R. J. Kaufman, *Genes Dev.* **14**, 2725 (2000).
72. X. L. Li, J. A. Blackford, C. S. Judge, M. Liu, W. Xaio, D. V. Kalvakolanu, and B. A. Hassel, *J. Biol. Chem.* **275**, 8880 (2000).
73. B. L. Bass, *Cell* **101**, 235 (2000).
74. S. M. Elbashir, J. Harborth, W. Lendeckel, A. Yalcin, K. Wever, and T. Tuschl, *Nature* **411**, 494 (2001).
75. C. Patil and P. Walter, *Curr. Opin. Cell Biol.* **13**, 349 (2001).
76. X. Z. Wang, H. P. Harding, Y. Zhang, E. M. Jolicoeur, M. Kuroda, and D. Ron, *EMBO J.* **17**, 5708 (1998).
77. W. Tirasophon, A. Welihinda, and R. J. Kaufman, *Genes Dev.* **12**, 1812 (1998).
78. M. Niwa, C. Sidrauski, R. J. Kaufman, and P. Walter, *Cell* **99**, 691 (1999).
79. J. B. Harford, in "Control of mRNA Stability" (J. Belasco and G. Brawerman, eds.), p. 239. Academic Press, San Diego, California, 1993.
80. M. W. Hentze and L. C. Kuhn, *Proc. Natl. Acad. Sci. USA* **93**, 8175 (1996).
81. E. W. Mullner, B. Neupert, and L. C. Kuhn, *Cell* **58**, 373 (1989).
82. T. A. Rouault, C. D. Stout, S. Kaptain, J. B. Harford, and R. D. Klausner, *Cell* **64**, 881 (1991).
83. R. D. Klausner and J. B. Harford, *Science* **246**, 870 (1989).
84. R. Binder, J. A. Horowitz, J. P. Bassilion, D. M. Koeller, R. D. Klausner, and J. B. Harford, *EMBO J.* **13**, 1969 (1994).
85. E. C. Theil and R. S. Eisenstein, *J. Biol. Chem.* **275**, 40659 (2000).
86. C. Grandori, S. M. Cowley, L. P. James, and R. N. Eisenman, *Annu. Rev. Cell Dev. Biol.* **16**, 653 (2000).
87. R. Wisdom and W. Lee, *Genes Dev.* **5**, 232 (1991).

88. R. D. Prokipcak, D. J. Herrick, and J. Ross, *J. Biol. Chem.* **269,** 9261 (1994).
89. G. A. Doyle, N. A. Betz, P. F. Leeds, A. J. Fleisig, R. D. Prokipcak, and J. Ross, *Nucleic Acids Res.* **26,** 5036 (1998).
90. C. H. Lee, P. Leeds, and J. Ross, *J. Biol. Chem.* **273,** 25261 (1998).
91. J. E. Russell, J. Morales, and S. A. Liebhaber, *Prog. Nucleic Acid Res. Mol. Biol.* **57,** 249 (1997).
92. M. Kiledjian, X. Wang, and S. A. Liebhaber, *EMBO J.* **14,** 4357 (1995).
93. M. Kiledjian, C. T. DeMaria, G. Brewer, and K. Novick, *Mol. Cell. Biol.* **17,** 4870 (1997).
94. Z. Wang, N. Day, P. Trifillis, and M. Kiledjian, *Mol. Cell. Biol.* **19,** 4552 (1999).
95. Z. Wang and M. Kiledjian, *EMBO J.* **19,** 295 (2000).
96. Z. Wang and M. Kiledjian, *Mol. Cell. Biol.* **20,** 6334 (2000).
97. I. E. Gallouzi, C. M. Brennan, M. G. Stenberg, M. S. Swanson, A. Eversole, M. Maizels, and J. A. Steitz, *Proc. Natl. Acad. Sci. USA* **97,** 3073 (2000).
98. E. Chernokalskaya, A. N. Dubel, K. S. Cunningham, M. N. Hanson, R. E. Dompenciel, and D. R. Schoenberg, *RNA* **4,** 1537 (1998).
99. A. Wennborg, B. Sohlberg, D. Angerer, G. Klein, and A. von Gabain, *Proc. Natl. Acad. Sci. USA* **92,** 7322 (1995).
100. F. Claverie-Martin, M. Wang, and S. N. Cohen, *J. Biol. Chem.* **272,** 13823 (1997).
101. H. Siomi and G. Dreyfuss, *Curr. Opin. Gen. Dev.* **7,** 345 (1997).
102. Z. Zhao, F. C. Chang, and H. M. Furneaux, *Nucleic Acids Res.* **28,** 2695 (2000).
103. R. J. Maraia and R. V. Intine, *Mol. Cell. Biol.* **21,** 367 (2001).
104. T. Heise, L. G. Guidotti, and F. V. Chisari, *J. Virol.* **75,** 6874 (2001).
105. R. S. McLaren, N. Caruccio, and J. Ross, *Mol. Cell. Biol.* **17,** 3028 (1997).
106. R. A. Wallace, in "Developmental Biology" (L. W. Browder, ed.), p. 127. Plenum Press, New York, 1985.
107. D. J. Shapiro, M. C. Barton, D. M. McKearin, T. C. Chang, D. Lew, J. Blume, D. A. Nielson, and L. Gould, *Rec. Prog. Hormone Res.* **45,** 29 (1989).
108. D. J. Shapiro and H. J. Baker, in "Ontogeny of Receptors and Reproductive Hormone Action" (T. H. Hamilton, J. H. Clark, and W. A. Sadler, eds.), p. 309. Raven Press, New York, 1979.
109. H. J. Baker and D. J. Shapiro, *J. Biol. Chem.* **252,** 8428 (1977).
110. H. J. Baker and D. J. Shapiro, *J. Biol. Chem.* **253,** 4521 (1978).
111. M. L. Brock and D. J. Shapiro, *Cell* **34,** 207 (1983).
112. M. L. Brock and D. J. Shapiro, *J. Biol. Chem.* **9,** 5449 (1983).
113. D. J. Shapiro, D. A. Nielsen, J. Blume, and D. M. McKearin, in "Steroid and Sterol Hormone Action" (T. C. Spelsberg and R. Kumar, eds.), p. 117. Nijhoff, Boston, 1987.
114. J. E. Blume and D. J. Shapiro, *Nucleic Acids Res.* **17,** 9003 (1989).
115. D. J. Shapiro, H. J. Baker, and D. T. Stitt, *J. Biol. Chem.* **251,** 3105 (1976).
116. S. Gerber-Huber, D. Nardelli, J.-A. Haefliger, D. N. Cooper, F. Givel, J. E. Germond, J. Engel, N. M. Green, and W. Wahli, *Nucleic Acids Res.* **15,** 4737 (1987).
117. B. K. Follett, T. J. Nicholls, and M. R. Redshaw, *J. Cell. Physiol.* **72**(Suppl.), 91 (1968).
118. M. V. Berridge, S. R. Farmer, G. C. D., E. C. Henshaw, and J. R. Tata, *Eur. J. Biochem.* **62,** 161 (1976).
119. D. A. Nielsen and D. J. Shapiro, *Mol. Cell. Biol.* **10,** 371 (1990).
120. R. E. Dodson and D. J. Shapiro, *Mol. Cell. Biol.* **14,** 3130 (1994).
121. R. E. Dodson, M. R. Acena, and D. J. Shapiro, *J. Steroid Biochem. Mol. Biol.* **52,** 505 (1995).
122. R. E. Dodson and D. J. Shapiro, *J. Biol. Chem.* **272,** 12249 (1997).
123. S. Kugler, A. Grunweller, C. Probst, M. Klinger, P. K. Muller, and C. Kruse, *FEBS Lett.* **382,** 330 (1996).
124. G. Dreyfuss, M. S. Swanson, and S. Pinol-Roma, *Trends Biochem. Sci.* **13,** 86 (1988).
125. H. Siomi, M. J. Matunis, W. M. Michael, and G. Dreyfuss, *Nucleic Acids Res.* **21,** 1193 (1993).

126. G. Musco, G. Stier, C. Hoseph, M. A. Castiglione Morelli, M. Nilges, T. J. Gibson, and A. Pastore, *Cell* **85**, 237 (1996).
127. S. Adinolfi, C. Bagni, M. A. Castiglione Morelli, F. Fraternali, G. Musco, and A. Pastore, *Biopolymers* **51**, 153 (1999).
128. C. Schmidt, B. Henkel, E. Poschl, H. Zorbas, W. G. Purschke, T. R. Gloe, and P. K. Muller, *Eur. J. Biochem.* **206**, 625 (1992).
129. U. Wintersberger, C. Kuhne, and A. Karwan, *Yeast* **11**, 929 (1995).
130. G. L. McKnight, K. O. Sundquist, B. Hokland, P. A. McKernan, J. Champagne, C. J. Johnson, M. C. Bailey, R. Holly, P. J. O'Hara, and J. F. Oram, *J. Biol. Chem.* **267**, 12131 (1992).
131. A. Cortes, D. Huertas, L. Fanti, S. Pimpinelli, F. X. Marsellach, B. Pina, and F. Azorin, *EMBO J.* **18**, 3820 (1999).
132. G. Plenz, Y. Gan, H. M. Raabe, and P. K. Muller, *Cell Tiss. Res.* **273**, 381 (1993).
133. A. Delahodde, A. M. Becam, J. Perea, and C. Jacq, *Nucleic Acids Res.* **14**, 9213 (1986).
134. C. Kruse, M. H. Klinger, A. Grunweller, J. Pohwedel, H. J. Krammer, W. Kuhnel, and P. K. Muller, *Exp. Cell Res.* **239**, 111 (1998).
135. E. Rumpel, C. Kruse, P. K. Muller, and W. Kuhnel, *Anat. Anz.* **178**, 337 (1996).
136. G. Neu-Yilik, H. Zorbas, T. R. Gloe, H. M. Raabe, T. A. Hopp-Christensen, and P. K. Muller, *Eur. J. Biochem.* **213**, 727 (1993).
137. S. Boeuf, M. Klingenspor, N. L. van Hal, T. Schneider, J. Keifer, and S. Klaus, *Physiol. Genom.* **7**, 15 (2001).
138. A. Cortes and F. Azorin, *Mol. Cell. Biol.* **20**, 3860 (2000).
139. S. Frey, M. Pool, and M. Seedorf, *J. Biol. Chem.* **276**, 15905 (2001).
140. B. D. Lang and J. L. Fridovich-Keil, *Nucleic Acids Res.* **28**, 1576 (2000).
141. C. Kruse, A. Grunweller, H. Notbohm, S. Kugler, W. G. Purschke, and P. K. Muller, *Biochem. J.* **320**, 247 (1996).
142. A. Grunweller, W. G. Purschke, S. Kugler, C. Kruse, and P. K. Muller, *Biochem. J.* **326**, 601 (1997).
143. S. Kugler, G. Plenz, and P. K. Muller, *Eur. J. Biochem.* **238**, 410 (1996).
144. G. Plenz, S. Kugler, S. Schnittger, H. Rieder, C. Fonatsch, and P. K. Muller, *Hum. Gen.* **93**, 575 (1994).
145. F. Fraternali, P. Amodeo, G. Musco, M. Nilges, and A. Pastore, *Proteins* **34**, 484 (1999).
146. R. J. Buckanovich and R. B. Darnell, *Mol. Cell. Biol.* **17**, 3194 (1997).
147. H. Kanamori, R. E. Dodson, and D. J. Shapiro, *Mol. Cell. Biol.* **18**, 3991 (1998).
148. C. G. Burd and G. Dreyfuss, *EMBO J.* **13**, 1197 (1994).
149. M. Gorlach, C. G. Burd, and G. Dreyfuss, *J. Biol. Chem.* **269**, 23074 (1994).
150. M. S. Swanson and G. Dreyfuss, *Mol. Cell. Biol.* **8**, 2237 (1988).
151. D. E. Tsai, D. S. Harper, and J. D. Keene, *Nucleic Acids Res.* **19**, 4931 (1991).
152. H. A. Lewis, H. Chen, C. Edo, R. J. Buckanovich, Y. Y. Yang, K. Musunuru, R. Zhong, R. B. Darnell, and S. K. Burley, *Struct. Folding Design* **7**, 191 (1999).
153. H. A. Lewis, K. Musunuru, K. B. Hensen, C. Edo, H. Chen, R. B. Darnell, and S. K. Burley, *Cell* **100**, 323 (2000).
154. M. Worbs, G. P. Bourenkov, H. D. Bartunik, R. Huber, and M. C. Wahl, *Mol. Cells* **7**, 1177 (2001).
155. N. Handa, O. Nureki, K. Kurimoto, I. Kim, H. Sakamoto, Y. Shimura, Y. Muto, and S. Yokoyama, *Nature* **398**, 579 (1999).
156. R. C. Deo, J. B. Bonanno, N. Sonenberg, and S. K. Burley, *Cell* **98**, 835 (1999).
157. C. E. Bogden, D. Fass, N. Bergman, M. D. Nichols, and J. M. Berger, *Mol. Cells* **3**, 487 (1999).
158. A. A. Antson, E. J. Dodson, G. Dodson, R. B. Greaves, X. Chen, and P. Gollnick, *Nature* **401**, 235 (1999).
159. R. Kanaar, A. L. Lee, D. Z. Rudner, D. E. Wemmer, and D. C. Rio, *EMBO J.* **14**, 4530 (1995).

160. C. Oubridge, N. Ito, P. R. Evans, C. H. Teo, and K. Nagai, *Nature* **372**, 432 (1994).
161. S. R. Price, P. R. Evans, and K. Nagai, *Nature* **394**, 645 (1998).
162. M. B. Elliott, P. A. Gottlieb, and P. Gollnick, *RNA* **5**, 1277 (1999).
163. B. D. Lang, A. Li, H. D. Black-Brewster, and J. L. Fridovich-Keil, *Nucleic Acids Res.* **29**, 2567 (2001).
164. K. S. Cunningham, R. E. Dodson, M. A. Nagel, D. J. Shapiro, and D. R. Schoenberg, *Proc. Natl. Acad. Sci. USA* **97**, 12498 (2000).
165. K. S. Cunningham, M. N. Hanson, and D. R. Schoenberg, *Nucleic Acids Res.* **29**, 1156 (2001).
166. D. R. Schoenberg, J. E. Moskaitis, L. H. Smith, and R. L. Pastori, *Mol. Endocrinol.* **3**, 805 (1989).
167. A. T. Riegel, M. B. Martin, and D. R. Schoenberg, *Mol. Cell. Endocrinol.* **44**, 201 (1986).
168. R. E. Dompenciel, V. R. Garnepudi, and D. R. Schoenberg, *J. Biol. Chem.* **270**, 6108 (1995).
169. E. Chernokalskaya, R. Dompenciel, and D. R. Schoenberg, *Nucleic Acids Res.* **25**, 735 (1997).
170. M. N. Hanson and D. R. Schoenberg, *J. Biol. Chem.* **276**, 12331 (2001).
171. A. B. Shyu and M. F. Wilkinson, *Cell* **102**, 135 (2000).
172. X. C. Fan and F. A. Steitz, *EMBO J.* **17**, 3448 (1998).
173. X. C. Fan and J. A. Steitz, *Proc. Natl. Acad. Sci. USA* **95**, 15293 (1998).
174. N. J. Caplen, S. Parrish, F. Imani, A. Fire, and R. A. Morgan, *Proc. Natl. Acad. Sci. USA* **98**, 9742 (2001).
175. I. E. Gallouzi, F. Parker, K. Chebli, F. Maurier, E. Labourier, I. Barlat, J. P. Capony, B. Tocque, and J. Tazi, *Mol. Cell. Biol.* **18**, 3956 (1998).
176. K. S. Cunningham, M. N. Hansen, and D. R. Schoenberg, *Methods. Enzymol.* **342**, 28 (2001).
177. J. Hua, R. Garner, and V. Paetkau, *Nucleic Acids Res.* **21**, 155 (1993).
178. B. D. Brown, I. D. Zipkin, and R. M. Harland, *Genes Dev.* **7**, 1620 (1993).
179. R. Binder, S.-P. L. Hwang, R. Ratnasabapathy, and D. L. Williams, *J. Biol. Chem.* **264**, 16910 (1989).
180. F. C. Nielsen and J. Christiansen, *J. Biol. Chem.* **267**, 19404 (1992).
181. D. Meinsma, W. Scheper, P. E. Holthuizen, J. L. Van den Brande, and J. S. Sussenbach, *Nucleic Acids Res.* **20**, 5003 (1992).
182. S. Tharun and Sirdeshmukh, *Nucleic Acids Res.* **23**, 641 (1995).
183. P. Vreken and H. A. Raue, *Mol. Cell. Biol.* **12**, 2986 (1992).
184. M. Y. Stoeckle, *Nucleic Acids Res.* **20**, 1123 (1992).
185. M. Y. Stoeckle and H. Hanafusa, *Mol. Cell. Biol.* **9**, 4738 (1989).

Jasmonates and Octadecanoids: Signals in Plant Stress Responses and Development

CLAUS WASTERNACK
AND BETTINA HAUSE

Institute of Plant Biochemistry
Weinberg 3
D-06120 Halle, Germany

I.	Introduction	166
II.	Occurrence of Jasmonates and Octadecanoids	168
III.	Biosynthesis of Jasmonates and Octadecanoids	171
	A. The LOX Pathway	171
	B. The AOS (Jasmonate) Branch within the LOX Pathway	173
IV.	Jasmonate-Induced Gene Expression	180
	A. Downregulation	181
	B. Upregulation	181
	C. Jasmonate-Responsive Promoters	184
V.	Jasmonates in Stress Response and Signal Transduction Pathways	185
	A. The Wound Response Pathway	185
	B. Plant/Insect Interactions	190
	C. Plant/Plant Interactions	193
	D. Mycorrhiza	193
	E. Plant/Pathogen Interactions	194
VI.	Jasmonates and Octadecanoids in Plant Development	196
	A. Germination and Seedling Development	196
	B. Flower Development	198
	C. Tuberization	200
	D. Tendril Coiling	202
	E. Senescence	202
VII.	Concluding Remarks	205
	References	206

Plants are sessile organisms. Consequently they have to adapt constantly to fluctuations in the environment. Some of these changes involve essential factors such as nutrients, light, and water. Plants have evolved independent systems to sense nutrients such as phosphate and nitrogen. However, many of the environmental factors may reach levels which represent stress for the plant. The fluctuations can range between moderate and unfavorable, and the factors can be of biotic or abiotic origin. Among the biotic factors influencing plant life are pathogens and herbivores. In case of bacteria and fungi, symbiotic interactions such as nitrogen-fixating nodules and mycorrhiza, respectively, may be established. In case of insects, a tritrophic interaction of herbivores, carnivores, and plants may occur mutualistically or parasitically. Among the numerous abiotic factors are low temperature, frost, heat, high light conditions, ultraviolet light, darkness, oxidation stress, hypoxia, wind, touch, nutrient imbalance, salt stress, osmotic adjustment, water deficit, and desiccation.

In the last decade jasmonates were recognized as being signals in plant responses to most of these biotic and abiotic factors. Signaling via jasmonates was found to occur intracellularly, intercellularly, and systemically as well as interorganismically. Jasmonates are a group of ubiquitously occurring plant growth regulators originally found as the major constituents in the etheric oil of jasmine, and were first suggested to play a role in senescence due to a strong senescence-promoting effect. Subsequently, numerous developmental processes were described in which jasmonates exhibited hormone-like properties. Recent knowledge is reviewed here on jasmonates and their precursors, the octadecanoids. After discussing occurrence and biosynthesis, emphasis is placed upon the signal transduction pathways in plant stress responses in which jasmonates act as a signal. Finally, examples are described on the role of jasmonates in developmental processes. © 2002, Elsevier Science (USA).

I. Introduction

In addition to ethylene, gibberellins, auxins, cytokinins, and abscisic acid (ABA), further plant hormones were identified and characterized in the last two decades, for example, brassinosteroids and jasmonates (for reviews see Refs. *1–8*). Both of them were found to occur ubiquitously in higher plants. They are plant-specific signaling compounds which have animal counterparts, the steroids and prostaglandins. There are remarkable similarities between jasmonates and prostaglandins with respect to their structure and function in plants and animals. Jasmonates exhibit similar properties in the wound response of leaves to effects of the antiinflammatory prostaglandins in animals.

Jasmonic acid (JA) was first identified in the etheric oil of jasmin in 1962 (*9*). In the 1980s, numerous physiological effects of jasmonates, such as root growth inhibition and promotion of senescence, were observed (*10, 11*). In addition, the biosynthetic pathway was elucidated (*12*). In the last decade, jasmonic acid and related compounds were found to function as a "master switch" in many plant responses to biotic or abiotic factors (*13*). In the subsequent analysis of the expression of numerous genes, some showed jasmonate-induced upregulation, for example, genes coding for plant defense proteins such as proteinase inhibitors or enzymes of phytoalexin synthesis. Other genes showed jasmonate-specific downregulation, such as those coding for housekeeping proteins.

External stimuli such as elicitors or osmotic stress cause an endogenous rise in jasmonate levels (*14–16*) (Fig. 1). This observation led to increased

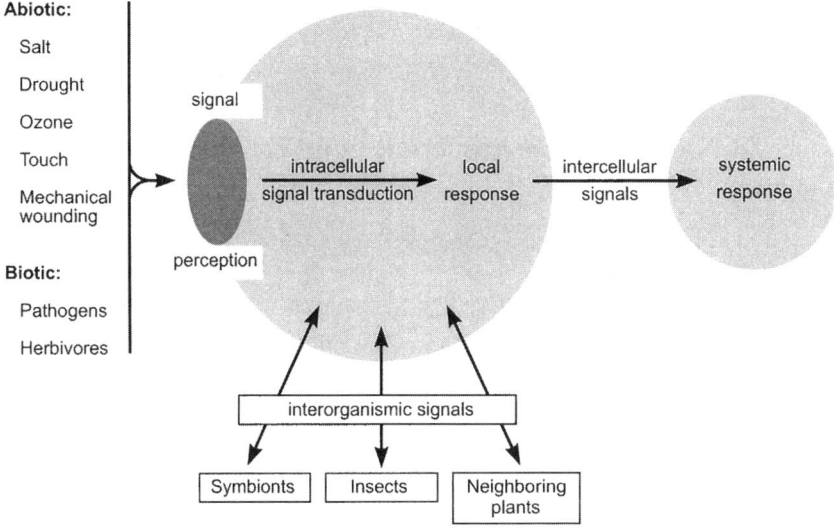

FIG. 1. Scheme illustrating signal transduction in plant interactions with the biotic and abiotic environment. Adapted from Ref. 161.

interest in jasmonate biosynthesis and its regulation. Octadecanoids and jasmonates originate from the polyunsaturated fatty acid (PUFA) α-linolenic acid (α-LeA). Upon oxygenation by lipoxygenase (LOX) at carbon atom 13, the resulting (13S)-hydroperoxide is converted by the concerted action of an allene oxide synthase (AOS) and an allene oxide cyclase (AOC) into cis(+)-12-oxo-phytodienoic acid (OPDA). Upon reduction of the double bond within the cyclopentenone ring and β-oxidation of the side chain, (+)-7-iso-jasmonic acid is formed. Except for the β-oxidation steps, all enzymes specific for JA biosynthesis have been cloned, and functional analysis has been started by reverse genetics.

A remarkable amount of information has accumulated in recent years on signal transduction pathways carrying JA as an essential element. Among the well-established signaling pathways are those of the wound response and the response to pathogens. Here, protein phosphatase/kinase cycles could be identified upstream and downstream of the signal JA, including a role for Ca^{2+}. Some mutants characterized for JA biosynthesis and JA signaling allowed the identification of distinct signaling pathways as well as cross-talk among them (17, 18). First hints are available on jasmonate-specific transcription factors, which belong to a subclass of the large family of AP2 domain proteins (19). Although still poorly understood in respect to the mechanism, the role of JA in distinct developmental processes is becoming clear. Here, we will discuss recent knowledge on occurrence, biosynthesis, and function of jasmonates and octadecanoids in plant stress responses and development.

II. Occurrence of Jasmonates and Octadecanoids

JA is a cyclopentanone compound carrying a pentenyl side chain and a carboxylic acid side chain (Fig. 2). Among the four possible stereoisomers, the (−)-JA enantiomer having the absolute configuration (3R, 7R) and (+)-7-*iso*-JA (3R, 7S) are native, with a preferential appearance of the thermodynamically more stable *trans*-(3R, 7R) enantiomer. The methyl ester (JAME), as well as the hydroxyderivatives 11-hydroxy-JA (11-OH-JA) and 12-hydroxy-JA (12-OH-JA), the amino acid conjugates of JA and the O-glycosides of 12-OH-JA, are naturally occurring derivatives, which are collectively named jasmonates. All of them originate from OPDA, which occurs also as methyl ester (OPDAME). These compounds as well as OPDA-containing derivatives are collectively named octadecanoids. In analogy to the mammalian eicosanoids, jasmonates and octadecanoids are members of a large group of so-called "oxylipins", which includes metabolites formed by oxidation of PUFAs.

Some of the naturally occurring JA derivatives are formed during JA biosynthesis, for example, 9,10-dihydro JA (20), whereas most of them are metabolic products originating from JA. Among them are methyl, glucosyl, and gentobiosyl esters, O-glucosylates of 11- and 12-OH-JA, O-glucolytes of cucurbic acid, and conjugates of amino acids (21). In many plants, amino acid conjugates of JA are permanent constituents besides JA (22). Among them, the isoleucine conjugate occurs predominantly and accumulates upon stress (22, 23). In leaves, its level can reach up to 10% of the amount of JA, whereas in flowers, its level can exceed that of JA remarkably (24, 25) (Section VI.B). Higher plants can release jasmonates as volatiles. *Jasminum grandiflorum* and *Artemisia vulgaris* are the most prominent examples. Many plants release a specific blend of jasmonates, which are part of direct or indirect defense mechanisms mainly against insects (Section V.B) or part of allelochemical interaction between different plants (Section V.C). Among the volatile defense compounds of the jasmonate type, *cis*-jasmone was recently identified (26). Fungi grown in suspension are known to synthesize numerous jasmonates, which are predominantly released into the culture filtrate. In *Aspergillus nidulans*, 22 different JA-like compounds were detected (27). Also, *Botryodiplodia* and *Gibberella* species form numerous stereochemically pure jasmonate compounds (28).

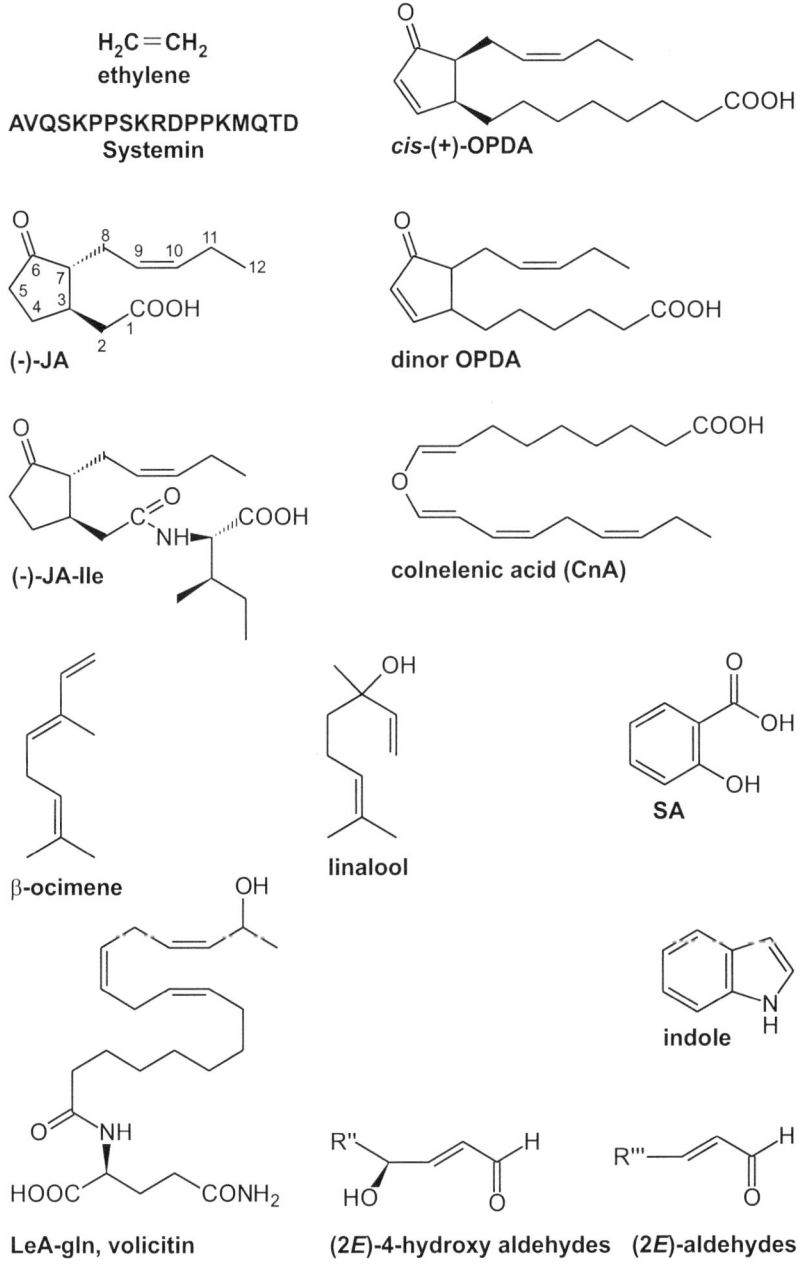

FIG. 2. Structures of important jasmonates and octadecanoids, as well as of compounds which function in JA signaling or are formed in response to JA treatment.

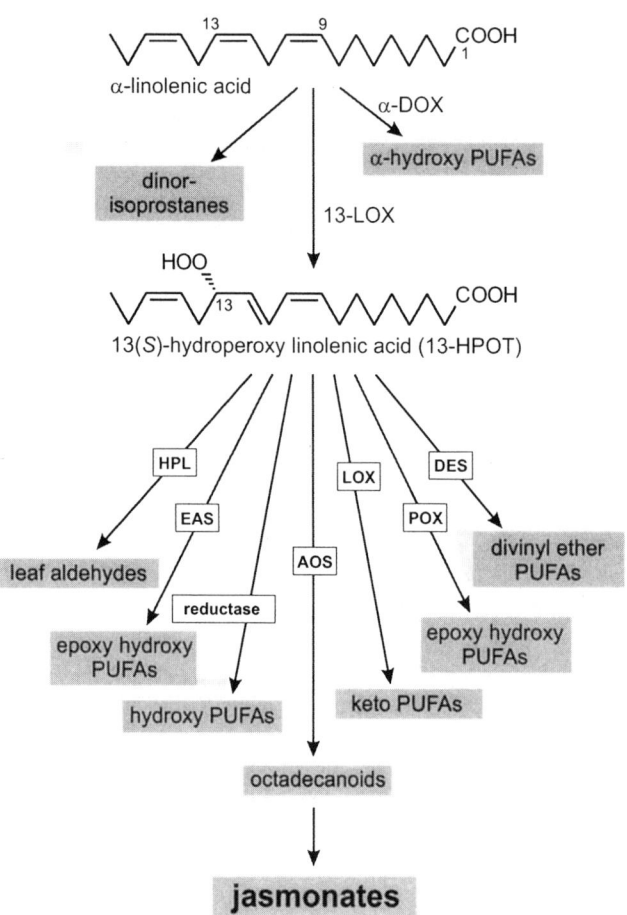

FIG. 3. The different branches in the LOX pathway. AOS, Allene oxide synthase; DES, divinyl ether synthase; α-DOX, α-dioxygenase; EAS, epoxyalcohol synthase; HPL, hydroperoxide lyase; POX, peroxygenase; LOX, lipoxygenase; PUFAs, polyunsaturated fatty acids. Adapted from Ref. 1.

In recent years, octadecanoids have been detected in addition to the jasmonates in plant tissues due to improved analytical techniques such as combined gas chromatography–mass spectroscopy (GC-MS) and related techniques. In particular, the free acid OPDA and its methyl ester were identified as well (20, 23, 29–32). Recently, in leaves of *Arabidopsis thaliana* OPDA was detected as an esterified derivative, the (12-oxophytodienoyl)-monogalactosyl diglyceride, from which OPDA can be released by sn-1-specific lipases (33). Although the biosynthetic origin of the esterified OPDA is not known, stress-induced release of OPDA can be discussed as a new type of signal generation, with a possible function for esterified OPDA as a storage compound.

Another important new aspect introduced by improved analytical techniques was the discovery of dinor-phytodienoic acid as a product of a (7Z, 10Z, 13Z)-hexadecatrienoic acid occurring alongside OPDA, which is the product of an 18-carbon fatty acid, α-LeA (*34*). Like other oxylipins, the level of dinor-phytodienoic acid fluctuates characteristically depending on different stress conditions. Consequently, an altered ratio of the various oxylipins, the so-called "oxylipin signature," has been suggested to be a signal in plant stress responses (*34*) (Section V.A).

The basic level of jasmonates and octadecanoids varies remarkably between different plant species, but is usually in the picomolar range per gram fresh weight of leaf tissue and can dramatically increase upon external stimuli. Some organs and tissues exhibit up to 10-fold higher levels of jasmonates and octadecanoids than leaves. There are hints that such elevated levels of jasmonates and octadecanoids have distinct functions in developmentally regulated processes (Section VI.B).

Apart from the enzymatically formed jasmonates and octadecanoids, other octadecanoids have been recently identified, including the phytoprostanes (Fig. 3) (*35*). They are formed by free radical-catalyzed oxidation reactions of α-LeA, analogous to reactions in animals in which prostaglandin-like compounds are generated nonenzymatically from arachidonic acid. It will be interesting to see whether these phytoprostanes have specific physiological functions or whether they are only indicative of oxidative stress in plants.

III. Biosynthesis of Jasmonates and Octadecanoids

A. The LOX Pathway

Phospholipid membranes are the source of different groups of signaling compounds. Inositol phosphate, diacylglycerol, and phosphatidic acid are generated mainly by the action of phospholipases C and D (*36*). In addition, PUFAs esterified to the glycerol backbone in cellular membranes and in storage lipids function not only in mobilization of energy, but also in the generation of numerous signaling compounds. Among them, jasmonates and octadecanoids are most prominent. Release of α-LeA, presumably by a phospholipase A (PLA) activity, might be in many cases the initial reaction for the generation of metabolites from the so-called LOX pathway. Increase in PLA activity was found to precede a rise in endogenous oxylipins as well as jasmonates after elicitation of various cells grown in suspension (*15, 37, 38*), upon wounding of tomato leaves (*39*), and in tobacco mosaic virus-infected tobacco leaves (*40*).

However, due to the exclusive location of the JA biosynthetic enzymes 13-LOX, AOS, and AOC within chloroplast (Section III.B), α-LeA, the substrate of JA biosynthesis, should be generated within chloroplasts, but not at the plasma membrane, where the PLA_2 is assumed to be located. Another α-LeA-releasing activity might be phospholipase D (PLD). This enzyme generates phosphatidic acid (41), an activator of an acyl hydrolase, which is also assumed to be located at the plasma membrane. The recently cloned wound-inducible PLD was shown to mediate systemic activation of the 13-LOX *AtLOX2* and an *AOS* in *A. thaliana* (42), suggesting its involvement in wound signaling. However, most recently a chloroplast-located phospholipase A1 was detected which hydrolyzes phospholipids in an *sn*-1-specific manner. The mutant *dad1* defect in this enzyme is JA-deficient, suggesting that α-LeA of chloroplast origin and released by this enzyme is the substrate of JA (43). This observation is in accordance with our knowledge on chloroplast-specific initiation of jasmonate synthesis.

Enzymatic lipid peroxidation of PUFAs such as LeA occurs by LOXs (EC 1.13.11.12) or by the recently identified α-dioxygenase, the plant homolog of the cyclooxygenase. LOX catalyzes a stereospecific insertion of molecular oxygen either at carbon atom C-9 (9-LOX) or C-13 (13-LOX). The resulting hydroperoxy fatty acid isomers can be further converted into fatty acid derivatives harboring divinyl ethers, ketols, cyclopentanones, aldehydes, ω-oxo groups, epoxy alcohols, and polyhydroxides as functional groups by enzymes of the so-called LOX pathway. However, jasmonates are specifically formed by the action of 13-LOXs.

LOXs constitute a family of nonheme iron-containing dioxygenases, which are ubiquitously distributed in plants and animals (44–46). From α-LeA, 13-LOXs generate (13S, 9Z, 11E, 15Z)-13-hydroperoxy-9,11,15-octadecatrienoic acid (13-HPOT). In most plants analyzed so far, several 13-LOX forms have been identified, and corresponding cDNAs have been isolated from potato (38), tomato (47), soybean (48), *Arabidopsis* (49), barley (50, 51), wheat (52), and rice (53). Additional LOX cDNAs and a phylogenetic tree analysis for more than 50 LOXs have been recently discussed (1). Some of the LOXs belong to the so-called *type 2* LOXs, which are characterized by the presence of a putative chloroplast transit peptide sequence and a moderate overall sequence homology. They are transcriptionally upregulated following wounding (47), pathogen attack (54), or treatment with jasmonates (38, 47, 51) or salicylate (55). Although multiple *type 2* LOX isoforms in plants exist, to date there is no proof that one of these 13-LOX forms is specifically responsible for jasmonate biosynthesis. Moreover, chloroplast-located 13-LOX forms exhibited different biochemical properties (38). Therefore, it is unclear how the specificity for JA biosynthesis is given *in vivo*, where the LOX-product 13-HPOT is used in different branches of the LOX pathway. Apart from the enzyme family of AOSs, there are five other enzyme families which may use 13-HPOT as the substrate (Fig. 3) (56): (1) LOX itself, catalyzing formation of keto PUFAs, (2) divinyl ether synthases (DESs), which form divinyl ether-containing PUFAs, (3) peroxygenases (POXs), which catalyze formation of epoxy hydroxy-PUFAs, (4) hydroperoxy lyases (HPLs), which cleave 13-HPOT into ω-oxo fatty acids and aldehydes, and (5) epoxy alcohol synthases (EASs), catalyzing formation of epoxy hydroxy PUFAs. Moreover, a reductase reaction leads to hydroxy PUFAs. This may be catalyzed by the endogenous level of glutathione (1).

There are indications of changes in the ratio of activities of the different branches of the LOX pathway. Upon external stimuli such as osmotic stress, jasmonate, or salicylate treatment, 13-HPOT is shifted into the AOS branch, the HPL branch, or the reductase branch, respectively, as shown by metabolic profiling (16, 57, 58). Most enzymes directing the substrate 13-HPOT into the various branches have been cloned. Therefore, reverse genetics is now being used to address questions on regulation of these branches of the LOX pathway as well as cross-talk between them. These aspects of the LOX pathway have been recently reviewed (1).

B. The AOS (Jasmonate) Branch within the LOX Pathway

1. THE ENZYME AOS

A first scheme for biosynthesis of jasmonates was proposed by Vick and Zimmerman (59) (Fig. 4). Here, 13-LOX-generated 13-HPOT was found to be the substrate. Initially, a single enzyme called hydroperoxide cyclase was assumed for the conversion of the hydroperoxide into cis(+)-OPDA. Subsequently, Hamberg and Hughes (60) identified two different enzyme activities, AOS (EC 4.2.1.92), which forms the unstable epoxide (9Z, 11E, 15Z, 12R)-12, 13-epoxy-9, 11, 15-octadecatrienoic acid, and AOC (EC 5.99.6), which converts the epoxide into the enantiomerically pure (9S, 13S)-OPDA [cis(+)-OPDA] (61). Nonenzymatic hydrolysis of the allene oxide may occur spontaneously, yielding α- and γ-ketols or racemic OPDA. Among the various AOSs analyzed to date, the majority of enzymes have a high specificity for 13-HPOT. However, a recombinant AOS of barley is able to convert not only the (13S)-hydroperoxides, but also (9S)-hydroperoxides of linoleic and linolenic acid (62). Similar properties were found for the AOS activity of the 100,000 × g membrane pellet of corn seed homogenate (63). Formation of metabolites derived from (9S)-hydroperoxides may be facilitated via a conversion by such an AOS activity. However, the clear preference of all chloroplast LOXs for formation of 13-HPOT may suggest a preferential function of AOS in JA biosynthesis.

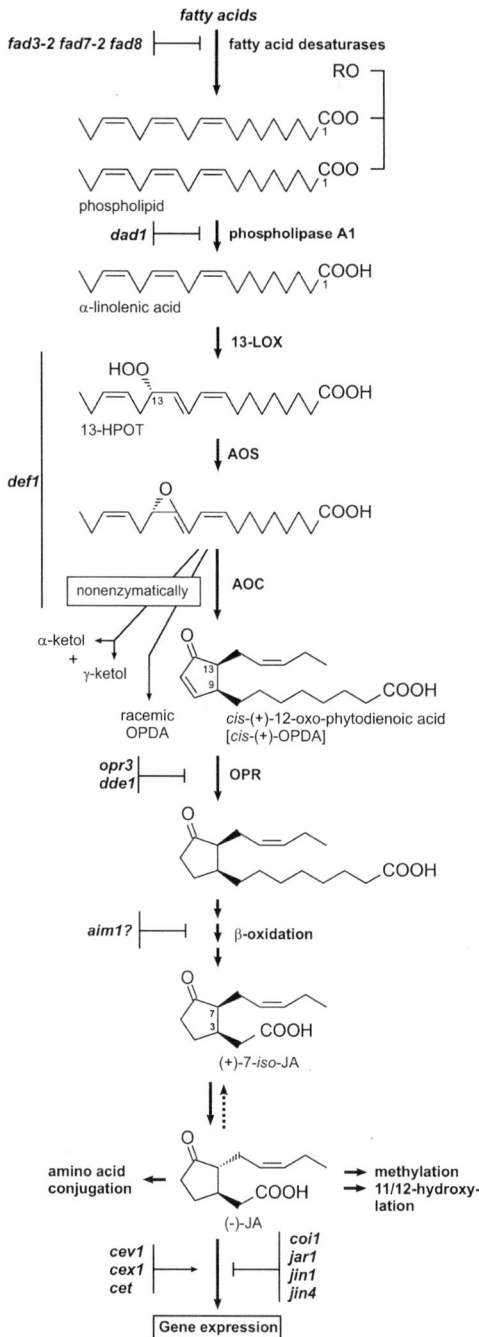

FIG. 4. The biosynthesis of jasmonates and mutants in JA biosynthesis and JA signaling isolated to date. For further details on mutants see Table I.

The first purification and cloning of AOS was done with flax seeds (64, 65). AOS cDNAs were characterized from guayule (66), Arabidopsis (67), tomato (68, 69), barley (62), and other plants, and using phylogenetic tree analysis, different subgroups with distinct biochemical properties were found (1). Moreover, similarities and differences of AOSs, HPLs, and DESs indicated their phylogenetic relation to each other (1). All of them are cytochrome P450-containing enzymes belonging to one family. Due to some special characteristics (independence of cosubstrates such as NADPH, oxygen, and P450 reductases, lack of transmembrane domains, low affinity to CO, and clear differences in the heme-binding domain) (70) compared to other P450 enzymes, all of them are grouped into one subfamily, CYP74. Whereas AOSs are grouped as CYP74A, the CYP74B/C group comprises the HPLs, and the DESs belong to the CYP74D group (1, 3, 71, 72).

AOSs from flax (65), tomato (68, 69), as well as Arabidopsis (67) contain a chloroplast transit peptide. These observations are in accordance with AOS activity data measured in isolated chloroplasts of spinach (73, 74). AOSs from guayule (66) and barley (62) lack a transit peptide for chloroplast targeting. However, for barley immunocytochemical localization and detection of AOS protein in isolated chloroplasts revealed that it also occurs in this compartment (62). Recent import studies for AOS from tomato (75) substantiated biochemical evidence for the occurrence of the enzyme in the inner envelope membrane (73).

2. AOC

AOC was detected as a separate enzyme activity, because it was absent in AOS-containing membrane fractions of corn seeds. Here, addition of a supernatant fraction led to the formation of $cis(+)$-OPDA from 13-HPOT, instead of formation of α- and γ-ketols (61). Whereas AOSs appear to be trimeric or tetrameric proteins (72, 76), homogeneous AOC purified from dry corn seeds is a dimeric protein of 47 kDa (77). AOC catalyzes the enzymatic step in which the correct enantiomeric form of naturally occurring octadecanoids and jasmonates is established. AOC converts preferentially fatty acid derivatives carrying an epoxide group in the n-6,7 position and a double bound in the n-3 position (63). This specificity supports its function in jasmonate biosynthesis and argues against a parallel pathway originating from linoleic acid, whereby AOS is able to convert hydroperoxy fatty acids derived from linoleic and linolenic acid. A predominant formation of JA from α-LeA is also supported by the fact that the recently isolated JA-deficient mutant *dad1* can only be normalized by α-LeA, but not by other PUFAs (43).

Recently, an AOC-encoding cDNA was cloned for the first time from tomato (78). In this plant, a single gene located on chromosome 2 codes for a chloroplast-located protein of 26 kDa. Only a truncated version of the recombinant tomato

AOC corresponding to the mature protein exhibited activity which led to exclusive formation of the (9S,13S) enantiomer. This corresponds to the exclusive formation of (9S,13S)-OPDA measured *in vivo* with wounded tomato leaves (78). In addition, four different cDNAs coding for proteins with AOC activity were isolated from *A. thaliana* and designated as *AOC1*, *AOC2*, *AOC3*, and *AOC4* (79). All of them carry a putative chloroplast targeting sequences. *AOC1–AOC3* map on chromosome 3, whereas *AOC4* is located on chromosome 1.

3. OPR AND β-OXIDATION

OPDA can exist in four different enantiomeric structures. However, only the (9S,13S)-enantiomeric structure of OPDA [*cis*-(+)-OPDA) formed by AOC occurs in the naturally occurring JA. Therefore, in the subsequent enzymatic step catalyzed by OPDA reductase (OPR), the structure should be retained. A cDNA clone (*OPR1*) coding for a reductase was isolated which exhibited sequence similarity and similar properties to Warburg's old yellow enzymes (OYEs) from yeast (80), which catalyze the reduction of α,β-olefinic ketones. The *OPR1* cloned from *Arabidopsis* and tomato (81, 82) showed preferential activity with (9R,13R)-OPDA, which excludes a function in JA biosynthesis. The recent X-ray structure analysis with OPR1 from tomato revealed a monomer which folds into a $(\beta\alpha)_8$ barrel with an overall structure similar to OYE proteins, and substrate-binding specificity for the (9R,13R)-OPDA structure (83). This suggests that OPR1 could function *in vivo* possibly in removing (9R,13R)-OPDA for retaining preferential occurrence of the JA precursor (9S,13S)-OPDA (83). A second reductase, OPR2, cloned from *Arabidopsis*, exhibited only a partial activity with the essential substrate (9S,13S)-OPDA (81, 84). Finally, a third OPR was cloned, exhibiting the correct substrate specificity (85–88). OPR3 contains a peroxisomal targeting sequence, suggesting its location in peroxisomes, where enzymes of β-oxidation are also located. This would suggest a direct transport of OPDA from chloroplasts into the peroxisomes. To date, there are no specific transport systems known for this group of compounds. The final steps of JA biosynthesis are three cycles of β-oxidation of the carboxylic side chain (59). That β-oxidation is crucial is substantiated by the fact that only derivatives carrying an odd-numbered carboxylic acid side chain are precursors of JA (89).

4. REGULATION OF JA BIOSYNTHESIS

Most of the enzymes involved in JA biosynthesis are transcriptionally upregulated by treatment with jasmonates or octadecanoids. Within 1 h of treatment, leaves accumulate, transiently or steadily, several mRNAs, including LOX from tomato (47), barley (51), or potato (38), AOS from *Arabidopsis* (90), tomato (68, 69), or barley (62), AOC from tomato (91), and OPR3 from *Arabidopsis* (86). Also, upon exposure to numerous biotic and abiotic factors leading to an endogenous rise in octadecanoids and jasmonates, mRNAs of biosynthetic enzymes accumulate. These data were taken to be indicative for a feedforward regulation in JA biosynthesis (31, 68, 90). However, the following facts are not in line with this type of control in short times: (1) Despite a transcriptional upregulation of AOS by JA or salicylate (SA) in barley leaves, its substrate, 13-HPOT, was shifted into branches other than the AOS branch of the LOX pathway by these treatments (57), and also HPL mRNA accumulated in response to JA treatment (76). (2) Although 13-LOX, AOS, AOC, or OPR mRNAs accumulate following JA treatment, isotopic dilution analysis performed with barley and tomato leaves revealed that endogenous formation of jasmonates was not increased within 24 h (23, 89). (3) Transient AOC mRNA accumulation upon wounding of tomato leaves did not result in elevated levels of AOC protein, despite a preceding rise in the JA level (91). (4) The (9Z,12R,13S)-12, 13-epoxy-9-octadecanoic acid, which can be formed in the peroxygenase branch of the LOX pathway (92), inhibits the AOC (63), and the linoleic acid-derived product 13-HPOD was found to irreversibly inhibit AOS activity of corn (93).

More recently, strong arguments were proposed for control of activity by substrate availability in short term responses: (1) Wounding, which generates the AOS substrate 13-HPOT, elevates JA levels in AOS-overexpression lines of tobacco (94, 95). (2) Treatment with 13-HPOT elevates JA levels in both AOC-overexpression lines and wild-type tomato, but not in AOC *antisense* plants (91). (3) Despite the fact that AOC-overexpression lines of tomato contain high amounts of AOC protein in all leaf tissues, generating 10-fold higher *in vitro* enzyme activity, JA levels rose only upon wounding or other treatments generating 13-HPOT (91). (4) Nontransgenic leaves of *A. thaliana* have constitutively high levels of 13-LOX, AOS, and AOC protein, and JA is formed upon wounding, which generates free α-LeA and 13-HPOT (79).

These data suggest that generation, accumulation, and action of jasmonate might be regulated by the following factors: (1) Substrate generation following external stimuli, initiating biosynthesis of jasmonates (91), (2) intracellular sequestration of accumulated jasmonates, precluding its action (96), (3) tissue-specific generation and accumulation (25, 91), (4) temporally different pattern of accumulation of JA by concerted action of transcriptional and posttranslational control of JA biosynthetic enzymes (23, 91), (5) metabolic transformation of JA leading to spatial and temporal differences in the removal of the jasmonate signal (97), and (6) positive feed back by JA in leaf development (79).

5. METABOLISM OF JASMONIC ACID

Metabolic transformation of JA has been known for a long time based on the identification of various metabolites. These can be formed by 11-hydroxylation, 12-hydroxylation, O-glycosylation, methylation, or addition of amino acids (21, 98) (Figs. 2, 4). Although an involvement of enzymes in the metabolism of JA is clearly indicated by the stereospecificity of the metabolites formed, only very few reactions have been characterized at the biochemical and the molecular level.

a. Hydroxylation. 12-OH-JA, also known as tuberonic acid, and 11-OH-JA are constituents in *Solanum*. The tuber-inducing activity of the *O*-glycoside of 12-OH-JA was also demonstrated (*99*) (Section VI.C). No specific hydroxylase gene has been cloned. However, the recently cloned sulfotransferase of *A. thaliana* acting specifically with 12-OH-JA (*100*) sheds some light on the signaling properties of 12-OH-JA. Decreased levels of 12-OH-JA, detected following overexpression of the sulfotransferase, were accompanied by a delayed onset of flowering time (*100*). This indicates that both developmental processes, flowering and tuberization, have a common signaling compound. Recent inspection of several plants revealed 11-OH-JA and 12-OH-JA as permanent constituents of leaves following stress-induced or transgenic elevation of JA (O. Miersch, unpublished).

b. Methylation. Besides JA, JAME is present permanently in plant tissues, suggesting the existence of an equilibrium between the free acid and its methyl ester. Recently, a JA-specific methyl transferase of *A. thaliana* was cloned (*97*), which is expressed locally and systemically in response to wounding or other stresses. The preferential accumulation of JAME compared to JA under these conditions was taken as an indication for signaling properties of JAME.

c. Other Metabolic Transformations. Cucurbic acid, a cyclopentanol derivative, *O*-glucosides, and L-amino acid conjugates are further metabolites of JA. Their synthesis and regulation is less well understood due to the lack of data for corresponding enzymes and cDNAs, respectively. In case of amino acid conjugates, the isoleucine conjugate of JA was the predominant compound. It accumulates upon stress (*22, 23*). As new type of JA conjugate, the (–)-JA-N-tyramine was identified (*101*). Since JA, its amino acid conjugates, and other JA derivatives differ in their biological activity (*102–104*), cloning and characterization of the corresponding enzymes in jasmonate metabolism will improve our understanding of the signaling properties of these compounds. This includes the question of how *cis*-jasmone is formed to function as a flower volatile in plant–insect interaction (Section V.B).

6. MUTANTS IN JASMONATE BIOSYNTHESIS

A number of mutants have been identified which are altered in JA biosynthesis (Table I, Fig. 4). In *A. thaliana*, the critical requirement for jasmonate in plant reproduction was found by characterization of the triple mutant *fad3–2fad7–2fad8*, which is unable to form the precursors of the JA pathway,

TABLE I
MUTANTS OF *ARABIDOPSIS THALIANA*[a]

	Aspect affected	Morphological phenotype	Ref.
JA-deficient mutants			
dad	Phospholipase A1	Male-sterile	43
fad3-2 fad7-2 fad8	Biosynthesis of α-LeA	Normal vegetative phenotype, male-sterile	105
opr3	OPR3	Male-sterile	88
dde1	OPR3	Male-sterile	87
JA-insensitive mutants			
jar1	Unknown	Similar to wild type	243
jin1, jin4	Unknown	Similar to wild type	245
coi1	F-box protein, targeting of repressors to ubiquitination	Male-sterile, deficient in wound response	110, 254
Constitutive JA-response mutants			
cev1	Unknown	Small, stunted roots, anthocyanin accumulation	111
cex1	Unknown	Stunted growth, stunted roots	112
cet1–cet9	Unknown	Stunted growth, callus- and blister-like structures on leaves, spontaneous lesions	113

[a]Abbreviations: *dad*, defective in anther dehiscence; *dde1*, delayed dehiscence 1; *jar1*, jasmonate-resistant; *jin1, jin4*, jasmonate-insensitive; *coi1*, coronatine-insensitive; *cev1*, constitutive expression of VSPs; *cex1*, constitutive expression; *cet1–cet9*, constitutive expression of thionins.

α-LeA (18:3) and hexatrienoic acid (16:3). This mutant is male-sterile (*105*), and is more sensitive to plant pathogens (*106*). Three other male-sterile mutants, *defective in anther dehiscence1* (*dad1*), *delayed dehiscence1* (*dde1*), and *opr3*, exhibit an altered timing of pollen release and code for a defective phospholipase A1 and an OPR3, the *cis*(+)-OPDA metabolizing enzyme, respectively (*43, 87*). In *dad1*, OPDA and JA deficiency occurred (*43*), whereas in the *opr3* mutants, elevated levels of OPDA were found upon wounding or pathogen infection, accompanied by expression of JA-responsive genes as well as a resistant phenotype (*107*). Thus, signaling qualities can be distinguished now between octadecanoids and jasmonates by use of *opr3*. This could also hold true for the recently isolated male-sterile *abnormal inflorescence mutant1* (*aim1*) mutant (*108*). This mutant is defective in acyl-CoA hydratase, an enzyme in β-oxidation, thus possibly leading to an altered capacity for JA biosynthesis. For tomato, a mutant, *defenseless 1* (*def1*), was isolated with diminished response to wounding, including less wound-induced JA accumulation. The mutation might be related to regulatory elements of JA biosynthesis distinct from the LOX, AOS, and AOC steps, since these enzymes are formed as active proteins in *def1* plants (*69, 91, 109*).

Altered JA levels may occur not only by a defect in JA biosynthesis, but also by JA accumulation due to a defect in JA signaling. Signaling mutants may have (1) a loss of the JA response, as occurring in *coi1* (*110*) (Section VI.B), or (2) a constitutive JA response. Constitutive expression of JA-responsive genes was recently found in three new groups of mutants of *A. thaliana: cev* (*111*), *cex* (*112*), and *cet* (*113*) (Table I). Only for some *cet* mutants have elevated levels of jasmonates and octadecanoids been identified, and may be responsible for the constitutive JA response.

7. Transgenic Manipulation of JA Biosynthesis

In addition to the isolation of mutants, the generation of transgenic plants has been used to modulate JA levels as reported for several plant hormones (*114*). Only a few examples of transgenic plants showing up- or downregulation of jasmonates have been found. In potato, constitutive expression of flax AOS led to elevated JA levels, which were, however, functionally inactive, possibly due to intracellular sequestration (*96*). Constitutive expression of AOS in tobacco (*94, 95*) and AOC in tomato (*91*) led to high levels of active protein. However, JA increased only upon wounding or other conditions which generated the LOX product 13-HPOT. Obviously, this compound or its precursor, α-LeA, is limiting as the substrate of the JA pathway, suggesting regulation of JA biosynthesis by substrate availability (*91*) (Section III.B). Accordingly, diminishing levels of α-LeA by *antisense* expression of ω-3-desaturase (*115*) or cosuppression of *LOX2* from *A. thaliana* abolished wound-induced JA formation without alteration of the basal level of JA (*116*). Several studies failed to limit JA formation via *antisense* expression of individual LOX forms, presumably due to a compensatory effect among the LOX forms. In addition to overexpression of biosynthetic enzymes, overexpression of the JA methyl transferase, a metabolic enzyme, altered JA responses in *A. thaliana*. These plants exhibited elevated JAME levels with constant JA levels, activation of JA-responsive gene expression, and increased resistance against the fungal pathogen *Botrytis cineria* (*97*).

IV. Jasmonate-Induced Gene Expression

Jasmonate treatment leads to dramatic alterations in gene expression in most plant tissues analyzed. Two different responses can be distinguished: (1) downregulation of housekeeping genes such as those coding for proteins of the phytosynthetic apparatus, and (2) upregulation of JA-responsive genes (JRGs) leading to the synthesis of JA-induced proteins (JIPs).

A. Downregulation

The JA-induced decrease in expression of nuclear- and plastid-encoded proteins, mainly involved in photosynthesis (*117*), may contribute to the senescence-promoting effect of jasmonates (*10*). The decrease was shown to be translationally regulated in the case of the large subunit of ribulose bisphosphate carboxylase/oxygenase (Rubisco) and of the chlorophyll a/b-binding protein (*118*) and represents an early response to JA. A late JA-induced alteration of gene expression takes places in the case of some plastid-located proteins such as the light harvesting complex II, small subunit of Rubisco, D1, and early light-induced proteins (ELIPs) through a transcriptional control (*117–119*). Interestingly, the most abundant JA-induced protein, of 23 kDa, from barley leaves (JIP23; see below) exerts a similar effect if overexpressed in tobacco (*120*). Possibly, this cumulative effect of downregulation of photosynthetic genes by JA and JIP23 may explain the strong promotion of senescence in a monocotyledonous plant such as barley.

The effect of elevated levels of JA can be discussed in terms of JA-specific downregulation of photosynthetic genes (*121*) in photosynthetic-inactive tissues (*62, 122*) (Section VI.A). Here, JA may inhibit premature accumulation of the photosynthetic proteins at times of carbon flow from storage sources and may also minimize photochemical damage in tissues lacking radical scavengers like chlorophyll. Interestingly, a rise in sugar content in sucrose-transporter *antisense* plants upregulates JA-responsive genes such as PIN2, but downregulates those of Rubisco (O. Herde, unpublished). Such a coordinated regulation corresponds to the presence of both sugar- and JA-responsive elements within promoters of JA-responsive genes (Section IV.C).

B. Upregulation

In the late 1980s, the first JIPs were described for barley leaf segments treated with JA (*123*). Induction of many genes by JA has been reported for most plants analyzed. These genes code for proteins with functions in a wide range of cellular responses (Table II), illustrating the pleiotropic effects of jasmonates. JA-inducible proteins can be grouped as follows:

1. Enzymes of JA biosynthesis and metabolism (*85, 124*)
2. Enzymes of secondary metabolism and phytoalexin synthesis (*19*)
3. Antinutritional proteins such as proteinase inhibitor proteins (PINs) and polyphenoloxidases (*4, 125*)
4. Pathogenesis-related (PR) proteins (*17, 126*)
5. Defense proteins such as thionins and defensins (*127, 128*)
6. Proteinases such as leucine aminopeptidase (LAP) (*5, 129*)

TABLE II
PROTEINS ENCODED BY GENES EXHIBITING ALTERED EXPRESSION IN RESPONSE TO JASMONATE[a]

Upregulation
  Enzymes of JA biosynthesis
    LOX
    AOS
    AOC
    OPR3
  Enzymes in secondary metabolism
    Strictosidine synthase
    Chalcone synthase
    Phenylalanine-ammonia lyase
    Berberin bridge enzyme
    Myrosinase
  Antinutritional proteins
    PIN1, PIN2
    Cystatin (cys proteinase inhibitor)
    Polyphenoloxidase
    Putrescine-N-methyltransferase
      (nicotine formation)
    Cysteine endopeptidase
  Pathogenesis-related (PR) proteins
    Chitinases
    Glucanases
    PR-10
    Defensins
    Thionins
  Proteinases
    LAP
    Carboxypeptidase
  Proteins of cell wall formation
    Glycerin-rich proteins
    Hydroxyproline-rich proteins

  Stress-protective proteins
    Osmotin
    JIP23
    Threonine deaminase
    Glutathione S transferase
  Vegetative storage proteins
    VspA
    VspB
    Soybean LOX
  Seed storage proteins
    Napin
    Cruciferin
  Proteins of signal transduction
    Calmodulin
    Prosystemin
    Polygalacturonase
    Phospholipase A1
    ACC oxidase (ethylene)
  Proteins with function in generation of ROS
    NADPH oxidase
    Peroxidases

Downregulation
  Chlorophyll a/b-binding protein
  Rubisco
  D1 of photosynthetic apparatus
  Light harvesting complex II
  ELIPs

[a] Based on data from Refs. 1, 4, 5, 7, 13, 19, 29, 121, 164.

7. Proteins of cell wall formation (121)
8. Stress-protective proteins such as osmotin (121)
9. Vegetative and seed storage proteins such as vegetative storage proteins (VSPs) and napin (130)
10. Proteins involved in signal transduction such as calmodulin (4)

Although originally identified in JA-treated tissues, most JA-inducible proteins can be found following exposure to external stimuli leading to endogenous rises of JA, for example, wounding (Section V.A), or in distinct stages of development that exhibit an endogenous rise of JA (Section VI).

The diverse functions of JIPs (Table II) illustrate the role of JA as a "master switch" (13). This is can be especially documented with JIPs of barley. The abundant appearance of JIPs of molecular masses of 6, 23, 37, and 60 kDa was observed upon treatment of barley leaf segments with JA (123, 131). These proteins were designated as JIP6, JIP23, JIP37, and JIP60, respectively. Cloning revealed that JIP6 is a thionin (132), a member of a large, multigene family in barley with antifungal properties (127, 133). Although cDNAs coding for JIP23 lack homology to database sequences (132), numerous correlative data argue for a protective function during environmental stress (16) and during development (122, 134). JIP37 is also stress-induced (16, 135). Although there is homology to a phytase, so far no activity of the recombinant JIP37 has been detected. JIP60 was found as a stress-induced protein (136) and identified subsequently as a ribosome-inactivating protein (RIP) (137, 138). RIPs belong to a large family of ubiquitous proteins with antiviral and antifungal properties (139). Many of them contain a lectin domain, and all of them exhibit an N-glycosidase activity which site-specifically cleaves ribosomal RNA. For JIP60 such an N-glycosidase activity was shown in vitro (137) and in vivo (140). The function of JIP60 as a RIP was also demonstrated by its transgenic expression in tobacco, leading to dramatic alteration of the translational machinery and of the phenotype (141). Most RIPs can be targeted into the cell wall, the vacuole, the protein bodies, or the apoplast. Therefore, they cannot affect conspecific ribosomes, but they can, if present extracellularly, inhibit protein synthesis of invading pathogens. However, JIP60 lacks a lectin domain and is retained intracellularly (141). Possibly, the JA-induced RIP formation in barley supports downregulation of the cellular machinery during stress.

Apart from the abundantly accumulating JIPs of 6, 23, 37, and 60 kDa, several cDNAs of other JA-inducible proteins were cloned from barley. Among these were enzymes of the LOX pathway and JA biosynthesis such as a *type2* 13-LOX (51), two AOSs (62), and an AOC (I. Stenzel, unpublished). Another set of JA-inducible proteins from barley belongs to the group of light stress-inducible proteins. Under high light conditions, two proteins of about 32 kDa accumulate preferentially in the hypocotyl (142). These proteins are upregulated by JA and light stress. In contrast, ELIPs and thionins are antagonistically regulated by both factors (119). Thionins are upregulated by JA and downregulated by light stress, whereas ELIPs are upregulated by light stress and downregulated by JA. There is a synergistic effect of JA and ABA on the induction of the expression of some genes. The JIPs mentioned above are synthesized upon JA and ABA treatment (16), whereas another set of genes, including those encoding a caffeoyl *O*-methyl transferase and a jacalin-like protein, were exclusively expressed in response to JA (143). This illustrates that different sets of genes respond to JA and additional factors. Even JA itself, depending on its origin, can act differentially: The stress-induced endogenous rise in jasmonates and octadecanoids (23) leads to expression of a set of genes different from that expressed upon JA treatment, indicating the existence of different signaling pathways. The consequences of upregulation by JA in barley have different functions in osmotic stress (JIP23) (122, 134), plant defense (JIP6) (127, 133, 144), growth inhibition via decreased translational activity (JIP60) (140, 141), light stress (JIP32) (119), phytoalexin synthesis, and cell wall strengthening (caffeoyl O-methyl transferase) (145).

C. Jasmonate-Responsive Promoters

In the 1990s, a few JA-responsive promoters were analyzed. All of them contain a G-box (CACGTG), which is necessary, but not sufficient for JA-inducible gene expression. Palindromic sequences were found to be critical for the JA response of the *LOX1* promoter of barley (146) and were also described for the *nos* promoter (147), the promoter of the cathepsin D inhibitor (148), the *vspB* promoter (149), and the *PIN2* promoter of potato (150). In addition, these cis-elements were identified to be responsive to sugars (149, 150), developmental changes (151, 152), and wounding (151). The repeatedly seen coordinate activation of an *as-1*-type element in JA-responsive promoters points to a convergence in the different responses to osmotic stress, cold stress, or pathogen attack. This was nicely documented by recent isolation and characterization of the first transcription factors for JA-responsive gene expression by J. Memelink's group (19).

Cells grown in suspension are responsive to elicitation, leading to an endogenous rise of jasmonate and to alkaloid synthesis (14), or can form alkaloids upon jasmonate treatment (20, 153). This is a common property of cell suspension cultures (154). In case of cell suspension cultures of *Catharanthus roseus* abundant accumulation of monoterpenoid indole alkaloids (TIA) occurs (155). Taking advantage of the strong JA-inducibility of enzymes in TIA biosynthesis, cis- and trans-acting factors were identified (156).

In the promoter of the gene coding for strictosidine synthase, which forms an intermediate of the terminal alkaloid product vinblastine, two functionally significant regions were identified. An elicitor-responsive BA region is activated in a Ca^{2+}-dependent and JA-independent manner via MYB-like transcription factors (157). Apart from BA regions, JA-and elicitor-responsive elements (JEREs) were identified (156, 158). It was shown that JERE responds exclusively to JA and its precursor, OPDA. Therefore, subsequently identified transcription factors were called ORCAs (octadecanoid-responsive *Catharanthus* AP2-domain proteins). Those characterized to date (ORCA2, ORCA3) are members of the AP2/ERF domain family of transcription factors (156, 158, 159). JA has dual functions. It activates preexisting ORCA proteins and induces *ORCA* gene expression. ORCA3 was shown by overexpression studies to function as a master switch in JA-regulated alkaloid synthesis via posttranslational regulation (160).

Most striking is the wide range of occurrence of AP2/ERF-domain transcription factors. Whereas ORCAs may have a role in the switch from primary to secondary metabolism, others are known to function in drought stress (DREB2), low temperature (CBF/DREB-1), osmotic stress (MYC/MYB; bZip), pathogen attack (TGA), or ethylene response (EREBF/ERF) (19). In this way a modular principle of interaction is possible leading to specificity and flexibility in the terminal part of the signal transduction of stress responses (161). It will be interesting to see how the chemical signal jasmonate is mechanistically linked to the signal transduction and the function of transcription factors.

V. Jasmonates in Stress Response and Signal Transduction Pathways

A. The Wound Response Pathway

Herbivory is an intricate mechanism of insects and other animals for taking nutrients from the environment. Plants protect themselves by chemical defense mechanisms. At least three types of mechanisms, all of them containing jasmonates as a signaling compound, can be distinguished: (1) Wounding may result in the production of proteineous defense compounds such as PINs, which reduce the nutritional quality of consumed proteins by inhibiting digestive proteases of the herbivore (4); (2) phytoalexins such as the alkaloid nicotine might be formed upon wounding or challenge with other environmental factors, thereby avoiding further attack of herbivores by direct defense (162) (Section V.B); (3) wounding or other factors may generate volatiles, which in turn can attract insect predators that are part of a tritrophic interaction (162). Wounding by herbivores is different from mechanical wounding in terms of expressed genes (163). This might be caused by the fact that herbivore attack is the sum of mechanical damage and release of compounds (regurgitants) from the herbivore (Section V.B).

For wound-induced formation of defense proteins, a local and a systemic response pathway were proposed by Ryan and coworkers, using tomato as a model plant (4, 125, 164). Both responses are characterized by the activation of individual sets of genes in a spatial and temporal manner (Fig. 5). In the most recent version of the model (4), early gene activation, mainly of signal pathway genes in vascular bundles, is separated from late gene activation, preferentially of defense genes, in mesophyll cells. At least two parts of wound-induced signal transduction may be related to jasmonates or related compounds: (1) local generation of signals which function locally and systemically and (2) defense gene expression upon activation in the systemic tissue by a systemic signal. Both will be discussed primarily with data from tomato.

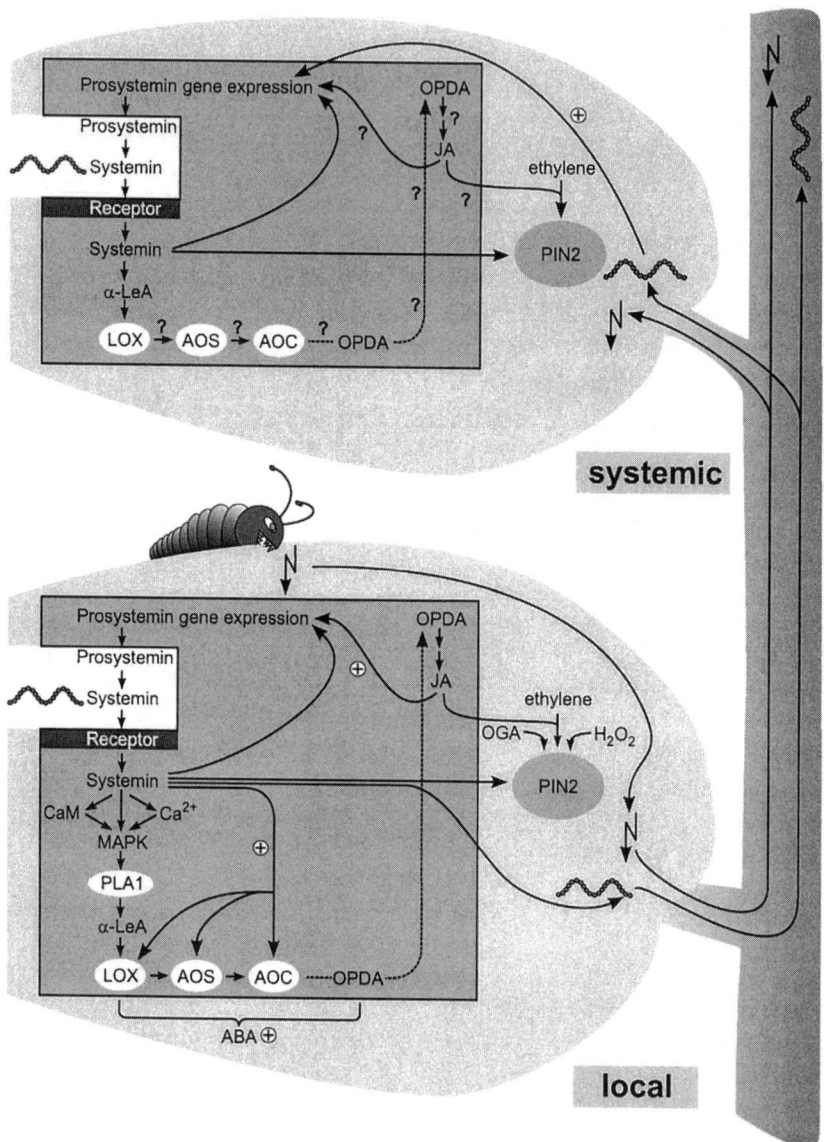

FIG. 5. The wound response pathway. ABA, Abscisic acid; AOC, allene oxide cyclase; AOS, allene oxide synthase; CaM, calmodulin; jagged arrow, electric signal; JA, jasmonic acid; LOX, 13-lipoxygenase; MAPK, mitogen-activated protein kinase; OGA, oligogalacturonic acid; OPDA, 12-oxo-phytodienoic acid; PLA1, phospholipase A1; unknown reactions are indicated by question marks; ⊕ indicates positive regulation; the white area indicates the apoplast; the vascular bundles are framed as a rectangle; a mesophyll cell is indicated by an elliptic frame. Data are summarized in Refs. 2, 4, 7.

1. LOCAL GENERATION OF WOUND SIGNALS SUCH AS SYSTEMIN AND AMPLIFICATION BY JASMONATES

Leaf damage by chewing insects or mechanical wounding is accompanied by the generation of several putative signaling compounds: (1) oligogalacturonic acids (OGAs) released from the damaged cell wall, (2) systemin, a peptide (165), (3) reactive oxygen species (ROSs) such as superoxide anion and H_2O_2 (166, 167), (4) ABA (168), (5) ethylene (169), and (6) jasmonates (170). Based on recent data, these signals can be grouped into a hypothetical model (Fig. 5). A central element is the generation of a systemic signal, most presumably the peptide systemin (4). However, participation of other signals such as a hydraulic wave (171), electric pulses (172), or action and variation potentials (173) cannot be ruled out unequivocally. Convincing data show that systemin might function as a systemic signal upon generation at the wound site. In tomato, systemin is a 18-amino-acid peptide processed from prosystemin, a 200-amino-acid precursor protein (174), and is active in the range of femtomoles per plant (165). Prosystemin and its truncated versions can already activate the wound response if the intact systemin-containing N-terminal part is present (175, 176). Generation of systemin requires processing by so-far-unknown proteolytic enzymes (4, 164). Prosystemin gene expression occurs tissue-specifically in vascular bundles (177), and unwounded tomato plants exhibit constitutive expression of prosystemin at a low level (165). Obviously, compartmentalization of prosystemin in vascular bundles protects the leaf tissues from a permanent wound response, but keeps the tissue in an activated state which allows an immediate response upon wounding. Interestingly, in tomato the crucial enzyme in JA synthesis, AOC (Section III.B), is constitutively expressed in the same tissue. AOC occurs in vascular bundles of main veins and parenchymatic cells of small veins and the AOC gene can be transcriptionally activated in a systemin-dependent manner (25, 91). As a consequence, systemin-induced AOC expression and the jasmonate-induced prosystemin expression (178) may amplify the generation of systemin upon wounding. According to the apparent regulation of jasmonate biosynthesis by substrate availability, elevation of JA levels occur only upon wounding. In tomato leaves, this rise in JA levels is about 10-fold and occurs specifically in AOC-containing tissues such as the main veins (91). The common tissue-specific expression of genes coding for prosystemin and AOC in vascular bundles keeps both proteins separate from the target cells and may lead to the above-mentioned preactivated state of this tissue.

Only a few systemin-like peptides have been found (179). Homologs of the tomato systemin were identified in potato, nightshade, and pepper (180). In contrast, tobacco contains two systemins, which are structurally unrelated to tomato systemin and are processed from each end of a common 165-amino-acid precursor (181). Although different in structure, tobacco systemins function like tomato systemin. This principle of signaling—divergent generation of the signals, but convergent transduction—has not yet been found in animals or yeast.

Local defense gene activation is known to occur in mesophyll cells (182). Systemin, generated in the vascular tissues upon wounding, may activate the surrounding parenchymatic cells via binding to a receptor. This is supported by identification of a systemin-binding protein from the plasma membrane of leaf tissues (183) and tomato cells cultured in suspension (184, 185). In tomato cells, suramin, an inhibitor of tumor necrosis factor α-mediated signal perception in animal cells, was shown to inhibit binding of systemin to its putative receptor, suggesting an extracellular binding side (186).

2. Intracellular Signal Transduction in Local Action of Systemin via Jasmonate

Several downstream events of systemin perception were found. OGAs and ROSs are generated at the wound site (166) and systemin can potentiate the oxidative burst (187). Increases in Ca^{2+} concentration, released from an intracellular (vacuolar) source (188) and from an extracellular source (189), have been shown to occur within minutes. This is accompanied by depolarization of plasma membranes, increases in extracellular proton concentration (164, 184), and inactivation of a plasma membrane H^+-ATPase (164). Subsequently, activation of a mitogen-activated protein (MAP) kinase occurs (190). Data from tobacco and *Arabidopsis* revealed the regulatory role of protein kinases in wound-induced JA-dependent and JA-independent signaling (reviewed in Ref. 7). Activation of phospholipase A_2 (39) and transcriptional activation of calmodulin (191, 192) also occur within the first 30 min after wounding. Release of α-LeA, the initial precursor of JA, is assumed as a downstream event. Indeed, wound-induced rise in α-LeA (193) and JA (32, 91, 109, 169) is well documented for tomato leaves.

For the local wound response, the following scenario can be given as a summary: Wounding leads to the generation of systemin, since preexisting prosystemin is reached by proteolytic enzymes. Systemin perceived in cells of vascular tissues may activate AOC, which leads to the synthesis of jasmonate in parenchyma cells of vascular bundles (91, 167). Jasmonate may there amplify the generation of systemin by prosystemin expression. Both signals are known to induce polygalacturonase located in these tissues (191) and lead to the release of OGA. The nonmobile OGA generates highly diffusible H_2O_2 in the vascular tissue, and H_2O_2 may diffuse to the surrounding mesophyll cells, where *PIN2* expression takes place (167). This scenario is nicely supported by the spatial and temporal pattern of expression and generation of all components: prosystemin/systemin, AOC/JA, polygalacturonase/OGA, and H_2O_2.

3. THE SYSTEMIC WOUND RESPONSE

Local wounding of tomato leaves by herbivores leads to the expression of defense genes such as *PINs* in systemic leaves, thereby contributing to plant immunity against an additional herbivore attack. This systemic defense mechanism and the role of systemin were clearly shown in transgenic tomato plants carrying the prosystemin encoding cDNA in *sense* and *antisense* orientation. 35S::*prosystemin* plants (*178*) exhibited a constitutive *PIN* expression, whereas the defense reactions against attacking insect larvae were compromised in 35S::*prosystemin antisense* plants (*174, 194*). Systemin generated locally can be loaded into the phloem to be transported into a systemic leaf as shown by ^{14}C-labeled systemin (*182*) and by grafting experiments with wild-type and 35S::*prosystemin*-overexpression lines (*178*).

Signal transduction within the systemic leaf is partially understood. A role of systemin is clearly indicated (*4*) and phosphatidic acid may participate (*41*). Systemic induction of AOS (*69*) and AOC (*91*) suggests that *PIN* expression in the systemic leaf is JA-dependent. On the other hand, systemin, but not JA, failed to induce *PIN* gene expression in a *spr* mutant (suppressors of prosystemin-mediated responses) (*195*). Furthermore, preliminary data on grafting of 35S-AOC *antisense* plants on wild-type plants revealed a systemic *PIN2* expression upon local wounding, suggesting JA-independent action of systemin (C. Kutter and C. Wasternack, unpublished).

4. ADDITIONAL SIGNALS ACTING IN THE WOUND RESPONSE (ETHYLENE, ABA, SA)

Apart from OGA and H_2O_2, other chemical signals were shown to act synergistically in the wound response. Initially, ethylene was identified as a signal of the jasmonate-mediated wound response of tomato leaves (*169*). JA and ethylene, applied together, induced plant defense gene expression in tobacco more rapidly than when applied individually (*196*). This synergism was frequently observed in plant–pathogen interactions (Section V.E). However, antagonistic effects also exist. Attack of the specialist herbivore *Manduca sexta* on its natural host *Nicotiana attenuata* induces ethylene formation, which in turn decreases jasmonate-induced nicotine formation. Thereby, direct defense (nicotine accumulation) is shifted into an indirect defense (volatile emission) (*197*) (Section V.B). Another example of negative cross-talk between ethylene and JA is the wound-induced expression of tobacco peroxidases (*198*).

In *Arabidopsis*, ethylene may influence the ratio of local and systemic responses (*2*). There is a simultaneously occurring JA-dependent and JA-independent expression of genes upon wounding. OGAs and ethylene, both generated at the wound site, repress JA-dependent signaling locally, but not systemically (*199*). In contrast to the wound-response pathway, ethylene and JA can act synergistically upon pathogen infection (*200, 201*) (Section V.E).

ABA was discussed in relation to its function in the wound-response pathway of tomato upstream of JA, since an ABA-deficient mutant was unable to express *PINs* upon systemin treatment, but could do so after addition of exogenous ABA (*168*). The loss of wound-induced *PIN* expression in transgenic tomato by expressing the mutant allele of *Arabidopsis abi1-1* clearly demonstrates that ABA perception is part of the wound-response pathway in this plant (*202*). However, further data indicate that ABA cannot be a primary signal (*203*). ABA is possibly necessary for a maximal response (*204*).

In the wound response, SA was found to inhibit wound-induced JA formation (*168*). However, inhibitory effects were found also downstream of JA (*205*). Such different cross-talk between JA and related signals points to the combinatorial nature of its signaling pathways. Signaling is organized like a network which combines independent inputs and outputs (*206, 207*). Synergistic and antagonistic effects of signals such as JA and SA or ethylene may sustain their individual responses at different levels, thereby contributing, together with temporal and spatial regulation, to the adaptiveness of the plant.

B. Plant/Insect Interactions

Among the environmental conditions which have to be tolerated by plants, insects play a dominant role due to their high number (60% of all species on earth) and diversity. Plants use this diversity to discriminate among the insects for optimizing their response between cost and benefit for growth, defense, and reproduction (*162*). Physical or chemical principles are used in these optimizing processes. A response is defined as an *induced defense* if the fitness of the responding plant is increased (*208*). The induced defense can be *direct* or *indirect*. In both types of defense mechanisms jasmonates function as important signals (Fig. 6).

1. Direct Defense

Physical barriers such as a thick cuticle or chemicals such as toxins and secondary metabolites like nicotine, glucosinolates, and flavonoids may contribute to a direct defense. Furthermore, many defense-related proteins (Section V.A) are part of a direct defense of plants. PINs can inhibit elastases in the larval midguts, and polyphenol oxidases (PPOs) crosslink proteins or form reactive quinones, which may then polymerize. Synthesis of most of these low molecular weight or macromolecular compounds is inducible by jasmonates and was shown to be part of numerous plant–insect interactions (*6*). The most prominent example of a plant defense mechanism against insects is the response against herbivores. Here, plants may immunize themselves by local and systemic formation of long-lived PINs (Fig. 5). However, plants and herbivores exhibit diversity in their weapons: Plants may have up to 200 different PIN genes divided into serine-, cystein-, aspartic-, and metallo-PINs (*209*), whereas herbivores can counter the

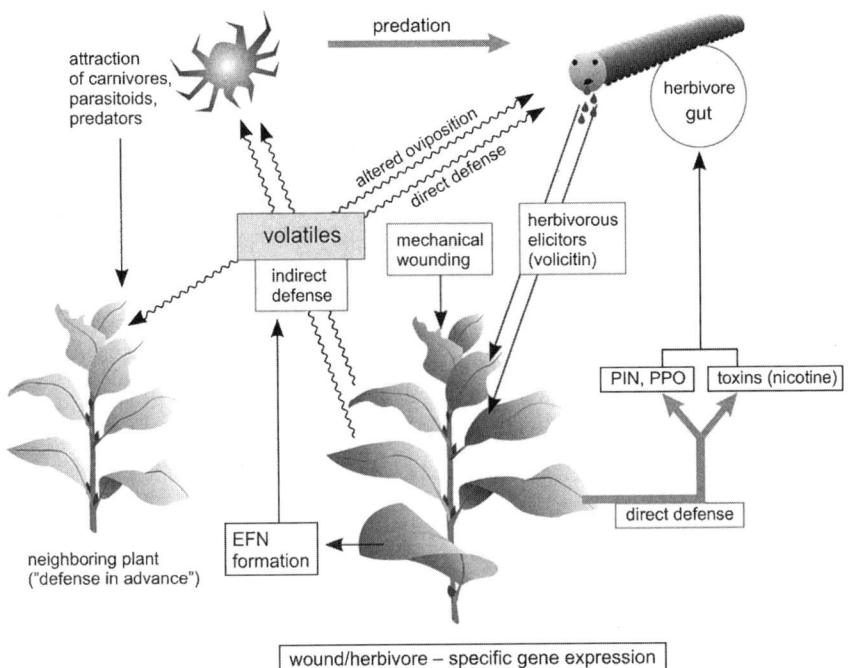

FIG. 6. Plant/insect interactions via jasmonates. EFN, Extrafloral nectar; PIN, proteinase inhibitor; PPO, polyphenol oxidase. Adapted from Refs. 6, 221.

deterrent role of PINs by altering expression of digestive proteinases. In addition, herbivores can produce enzymes that degrade toxins of plant origin (210).

Secondary metabolites, inducible upon insect attack, can exert direct defensive properties against feeding insects. The most prominent example is the formation of nicotine in the interaction of Manduca sexta and Nicotiana attenuata, as analyzed by I. T. Baldwin's group. Upon attack of M. sexta, nicotine is synthesized by a jasmonate-dependent activation of putrescine methyl transferase in roots. Subsequently, nicotine is transported via xylem and can accumulate 10-fold in the leaves (211). These levels are lethal for many herbivores (210). Recent data suggest an even more complex scenario: The rapidly activated JA-dependent direct defense by nicotine formation has a negative effect on the indirect defense (see below). The plant can distinguish between different insects and may respond specifically by use of ethylene and jasmonate. Larval feeding leads to an ethylene burst. As a consequence, the direct defense, mediated by JA and leading to nitrogen-intensive synthesis of nicotine, is suppressed (197), whereas there is no influence on the indirect defense (212).

Nonchewing insects differ remarkably in their action from herbivores (5). Insects which feed by piercing/sucking the phloem, for example, whitefly or aphids, activate apoplastic chitinases, β-1,3-glucanases, or peroxidases. Due to minute tissue damage, wound-induced genes such as *PIN*s or *LAP*s are not activated. Responses to whitefly are systemin-independent. Among the specifically induced genes are those coding for a M206 peptidase, a glucosyl hydrolase, and a gp91-phox homolog, a subunit of the NADPH oxidase, indicating activation of the ROS pathway (5, 213). The whitefly activates gene expression in a JA- and ethylene-dependent manner both locally and systemically.

2. INDIRECT DEFENSE

Attack of herbivores, for example, arthropods, on plant leaves induces immediately a release of volatiles, which leads to attraction of carnivores (Fig. 6). Plant–caterpillar–parasitoid interactions are one of the best-studied examples (214). Among the volatiles emitted are LOX pathway-derived compounds such as (3Z)-hexenal and *cis*-(3Z)-hexenol, essentially formed by HPL activity. These so-called green leaf aldehydes/alcohols may also act in a direct defense (215). Other volatile compounds formed *de novo* are indoles, methyl salicylate, and mono-, homo-, and sesquiterpenes.

Although most volatile formation occurs by wound-induced activation of the octadecanoid/jasmonate pathway within the plant (5, 216), the oral secretion (regurgitants) of insects such as caterpillars mimick JA. Regurgitants are necessary and sufficient for a herbivore-specific plant response (217). The main active compound of regurgitants is volicitin, identified as conjugated 17-hydroxylated linolenic acid (Fig. 2) (214, 218). Volicitin can be formed by microbes within the insect gut (219). The plants respond to a herbivore attack by a rapid burst of jasmonates, regurgitant-specific synthesis of growth- and defense-related mRNAs, and finally a systemic release of volatiles such as mono- and sesquiterpenes. Although volicitin contains the JA precursor α-LeA, the JA burst is not formed from the α-LeA moiety of volicitin (217).

Plants release a specific volatile blend which attracts predators including parasitoids and carnivores. Due to the specific blend, parasitoids can recognize these plants to find their host (220). The volatile emission influences not only the predation, but also the oviposition rate of herbivores (221). Plants can increase the amount of volatiles under field conditions leading up to 90% reduction of the number of herbivores (222). Despite the cost of volatile synthesis, there is benefit for the plant (162). Volatiles released by herbivore-infested plants attract predators to the infested plant, but induce also a jasmonate-responsive set of defense genes in neighboring plants (223). This response represents a "defense in advance."

These examples illustrate the central role of JA in orchestrating the different interactions within interorganismic communication. In terms of evolution, different levels of specificity and diversity, most of them directed by the plant octadecanoid pathway, can be distinguished in plant–insect–interactions (Fig. 6): (1) specific composition of the microbially formed regurgitants, (2) plant-specific gene activation, leading to synthesis of directly or indirectly acting defense compounds, (3) specific blends of volatiles, which attract predators, can increase predation of herbivore eggs, can alter oviposition of the infesting herbivores, and/or may act on neighboring plants, and (4) behavioral memory of predators for plant-specific blends leading to host specificity.

C. Plant/Plant Interactions

The volatile nature of methyl jasmonate led to its discovery in the flower scent of *J. grandiflorum* (9). Thirty years later, it was discovered as a signal in plant–plant interactions. Airborne methyl jasmonate was shown to act by interspecific signaling in *PIN* expression of tomato leaves following its release from a neighboring sagebrush plant, both kept in a closed chamber (224). Recently, the ecophysiological relevance was demonstrated under field conditions, where tobacco became significantly resistant against natural herbivores if leaf-damaged sagebrush was nearby (within 15 cm). The biologically active enantiomer of JAME could be trapped from the air up to a distance of 3 m from the sagebrush (225).

Volatile signaling between different individuals of the same species (intraspecific signaling) was shown for black alder trees. Here, an intact plant was less damaged by herbivores if a neighboring alder tree was manually defoliated. The universality of intraspecific signaling is unclear, but may occur in dense swards of grass (226). Regarding the high physiological activity of jasmonates on growth and numerous developmental processes, it cannot be ruled out that individual plants compete for common resources such as nutrients via volatiles.

D. Mycorrhiza

The term mycorrhiza refers to the mutualistic interaction between plant roots and fungi. This interaction varies widely in structure and function, resulting in different types of mycorrhizas (227). Among them, arbuscular mycorrhiza (AM) is the most common type and is formed between roots of more than 80% of all terrestrial plant species and zygomycete fungi from the order Glomales. In contrast to plant–pathogen interactions, the AM association is a relatively nonspecific, highly compatible, and long-lasting mutualism from which both partners derive benefit. The plant supplies the fungus with carbon, whereas the fungus assists the plant with the acquisition of phosphate and other mineral nutrients from the soil. In addition, the association may also enhance the plant resistance to biotic and abiotic stresses (228). These beneficial effects of the AM symbiosis result from a complex molecular dialogue between the two symbiotic partners.

Some processes occurring in this dialogue are suggested to be mediated by JA on the plant side, but the precise role of JA in the interaction is unknown. Although JA applied exogenously promotes colonization and development of mycorrhizal structures (229), an endogenous rise of jasmonates correlating with mycorrhization of barley, wheat, and tomato is more indicative for a role in AM (B. Hause, unpublished). The endogenous content of JA and its amino acid conjugates increases parallel to the expression of *AOS* as well as of jasmonate-induced genes such as that coding for JIP23 in barley. Since JA levels increased *after* the initial step of the plant–fungal interaction, the development of mycorrhiza rather than the recognition of the interacting partners may cause expression of JA-biosynthetic genes and finally elevate JA levels. Among the JA-inducible genes expressed in mycorrhizal roots are those which code for proteins with a function in cell wall crosslinking or phytoalexin formation. As a consequence, mycorrhizal roots may be more resistant against secondary infection and/or other stresses.

E. Plant/Pathogen Interactions

In the diverse range of enemies which influence the plant life cycle, microbial pathogens are among the most prominent factors. Besides a resistance (R) gene-mediated defense acting through recognition of a pathogen-encoded avirulence (Avr) gene, non-host-pathogen interaction occurs. A role for JA in the overall response to pathogens was unambiguously shown for non-host specific interactions. Jasmonates were identified in two SA-independent defense mechanisms: (1) the recently discovered induced systemic resistance (ISR) and (2) the infection by necrotic pathogens, leading to expression of genes encoding toxic proteins such as defensins and thionins. In contrast, the SA-dependent systemic acquired resistance (SAR), which confers a systemic immunity against a broad spectrum of pathogens by expression of PR proteins, is clearly JA-independent.

The ISR is ethylene- and JA-dependent and occurs upon colonization of roots by certain rhizosphere bacteria. Since the JA-insensitive mutant *jar1* can form ISR upon ethylene treatment, ethylene functions downstream of JA. ISR is linked to a basal resistance (230) and its JA dependence is not inhibited by SA (231). Both SAR and ISR require the *NPR1* gene product (nonexpressor of PR-1), illustrating its central role in plant–pathogen interactions (Fig. 7). A putative ankyrin domain identified within NPR1 suggests involvement of protein–protein interactions. Indeed, a nuclear-localized NPR1 was shown to interact with TGA factors. Some of these transcription factors bind specifically to SA-responsive promoter elements such *as-1* occurring in the *PR-1* promoter (Fig. 7). The *as-1* elements are also identified in promoters activated by abiotic stress, and many of these stress- and pathogen-related transcription factors have a common domain, the AP2 domain (Section IV.C). Consequently, expression

JASMONATES AND OCTADECANOIDS

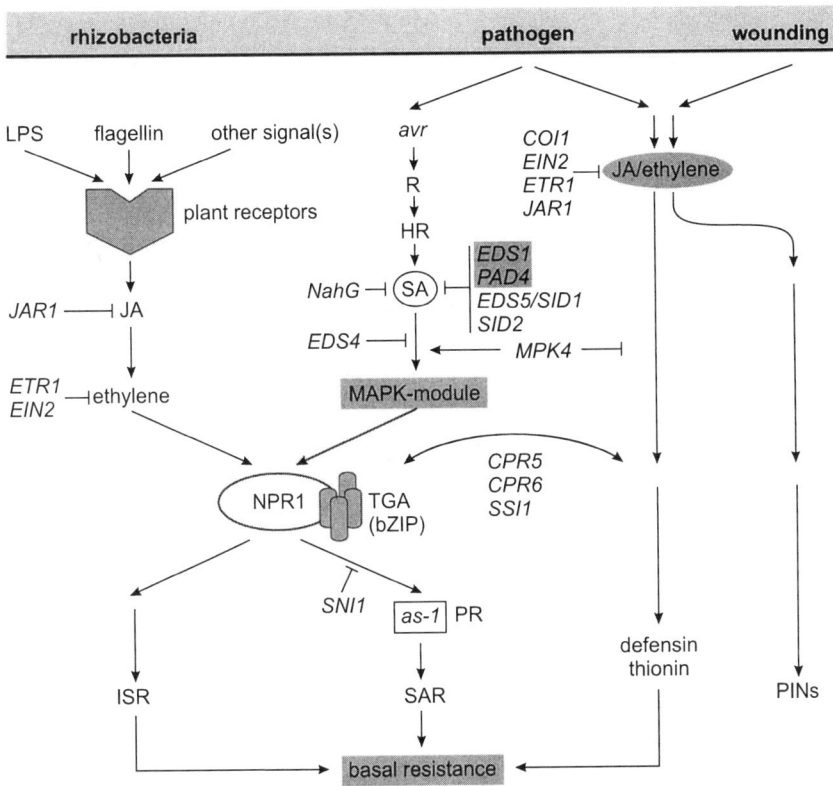

FIG. 7. Plant/pathogen interactions via jasmonates. avr, Avirulence; HR, hypersensitive response; ISR, induced systemic resistance; JA, jasmonic acid; LPS, lipopolysaccharide; PR, pathogenesis-related protein; PINs, proteinase inhibitor proteins; R, receptor; SA, salicylic acid; SAR, systemic acquired resistance; TGA, transcription factor subfamily. For further details see Section V.D, Table I, and Refs. 17, 126. Scheme adapted from Refs. 17, 126.

of a cDNA encoding an AP2/EREBP-type transcription factor increased resistance to pathogens and tolerance to salt stress, illustrating how biotic and abiotic stress signaling can converge at the level of transcription factors (232). It will be interesting to see how the modular principle of having various transcription factors contributes to the overlap of gene expression in response to different environmental stimuli.

The other JA-dependent defense mechanism is the expression of defensins (200, 201) and thionins (233, 234). Here, JA and ethylene act simultaneously, as shown by JA- and ethylene-insensitive mutants (17, 126). Recent data on the JA-deficient *opr3* mutant (Table I) revealed that OPDA, but not JA, is the important regulator in establishing resistance to necrotrophic fungi (*107*). The JA-dependent pathway of defensin/thionin expression is inhibited by SA, and the SA-dependent PR expression is inhibited by JA (*126*). However, numerous mutants indicate cross-talk between both pathways (Fig. 7), for example, in the mutant *mpk4*, the JA-dependent defensin expression is blocked by lack of MAP kinase4 (MPK4) activity and SA signaling is constitutively activated (*235*). Other examples of cross-talk between SA-dependent and JA-dependent signaling pathways are the *cpr* mutants, the *eds1* and *eds5* mutants, and the *pad4* mutant (*17, 126*). Such cross-talk may occur by physical interaction of pathway-specific proteins in a time-dependent manner as shown recently for EDS1 and PAD4 (*236*). For details on the roles of JA in plant–pathogen interactions, which cover a tremendous amount of data due to an increasing number of characterized mutants, the available reviews are recommended (*17, 126, 128, 237–239*).

VI. Jasmonates and Octadecanoids in Plant Development

Jasmonates and octadecanoids can modulate various developmental processes in plants. The most prominent examples, found by application experiments, are the inhibition of germination and seedling development, the stimulation of tendril coiling, the production of viable pollen (flower development), and the promotion of leaf senescence. Expression of genes coding for enzymes of JA biosynthesis, the elevation of JA levels, and the expression of JA-regulated genes was found to be correlated temporally in distinct developmental processes or plant organogenesis. Further insights into function of jasmonates were provided by characterization of mutants which are JA-insensitive or JA-deficient (Table I). Only few data obtained by reverse genetic approaches exist.

A. Germination and Seedling Development

Although of limited relevance in analyzing the role of JA in germination and seedling development, the effects of exogenously applied JA were tested. Jasmonates exert both inhibitory and promoting effects on seed germination. Jasmonates inhibit germination of nondormant seeds, whereas germination of dormant seeds is stimulated (*11*). However, the levels of endogenous jasmonates exhibit characteristic changes during embryogenesis, seedling development, and germination. In soybean, JA levels in late stages of seed development are remarkably higher than in developing seeds shortly after anthesis (*240*). After imbibition, even the relatively high level of jasmonates in dry seeds increases, but declines during later stages of seedling development. The increase of JA levels in early developmental stages of soybean seedlings correlates with seed reserve mobilization, suggesting that JA synthesis is a consequence, rather than a trigger, of germination (121).

In all seedlings analyzed so far, elevated JA levels and concomitant gene expression took place in tissues which are known to facilitate transport of metabolizable sugars or act as sink tissues. Accordingly, during germination of barley seeds, elevated levels of JA occur preferentially within the scutellar node and the base of the primary leaf. Both tissues are photosynthetically inactive and represent sink tissues. Increased levels of JA are also indicated by increased transcript accumulation of JA-inducible genes such as that coding for JIP23 (122). Interestingly, AOS is upregulated in these tissues (62). Another example of high levels of jasmonates correlating with upregulation of JA-responsive genes is the hypocotyl hook of seedlings of Arabidopsis and soybean (130). This organ is a highly active sink for carbon and nitrogen (240). Furthermore, in the ovary of tomato flower buds, there is an upregulation of AOC by glucose accompanied by elevated OPDA/JA levels (25). This suggests that sink tissues may use jasmonates/octadecanoids to sustain coordinate expression of genes related to sink/source relationships (25). A characteristic feature of sink tissues is the expression of sink-specific as well as defense-related genes, whereas the expression of photosynthetic genes is downregulated (241). All these data fit into the following scenario: Sugars formed by degradation of carbohydrates during germination or formed via photosynthesis in source tissues are translocated into sink tissues. Due to putative osmotic stress, this may lead to expression of genes coding for enzymes of JA biosynthesis. JA biosynthetic genes are stress/glucose-inducible. As a consequence, a rise of JA may lead to downregulation of genes coding for proteins of the photosynthetic apparatus and/or an upregulation of defense-related genes (121) (Sections IV.A, IV.B). This may contribute to an increased defense status as well as to protection against oxidative stress in those tissues lacking radical scavengers like chlorophyll.

Another obvious effect of JA on seedling development is the inhibition of root growth (242). This feature was used to isolate JA-insensitive mutants by screening for plants with less sensitivity to JAME in comparison to the wild type (243–245). JA-responsive gene expression is lacking or reduced in leaves of these mutants. JA-insensitive mutants were used to dissect common signaling pathways for jasmonates and ethylene. In this way, it was shown that inhibition of root growth by JA is not mediated by ethylene. JA and ethylene are independent signals in inhibiting root growth (111, 245). However, the JA-insensitive mutants are not altered in root morphology. In contrast, mutants exhibiting constitutive expression of JA-responsive genes (see Table I) have stunted roots with an excess of root hairs and are smaller than the wild type. This altered root morphology and the gene expression pattern in those mutants can be phenocopied by treatment of wild-type plants with JAME.

B. Flower Development

Flower development is the best-studied developmental process in which an essential role of JA and octadecanoids has been shown. For a long time a role of JA was assumed in flower development, because of the relatively high JA levels in developing plant reproductive tissues (*121*). Jasmonates have been shown also to occur in mature pollen (*101, 246*). In pine pollen, a high amount of JA amino acid conjugates was detected, but the function is unknown (*24*). In tomato flowers, OPDA, JA, and their amino acid conjugates and methyl esters accumulate to high levels exceeding that of leaves remarkably (*25*). In flowers, specific JA derivatives such as [(−)-jasmonoyl]-tyramine conjugate or 12-OH-JA are formed (*101*).

In addition, OPDA, JA, and their derivatives occur in a characteristic ratio within distinct flower organs (Fig. 8). Such a tissue-specific oxylipin signature may result in differential gene expression. Indeed, a number of JA-induced genes are specifically expressed within flowers. Among them are those coding

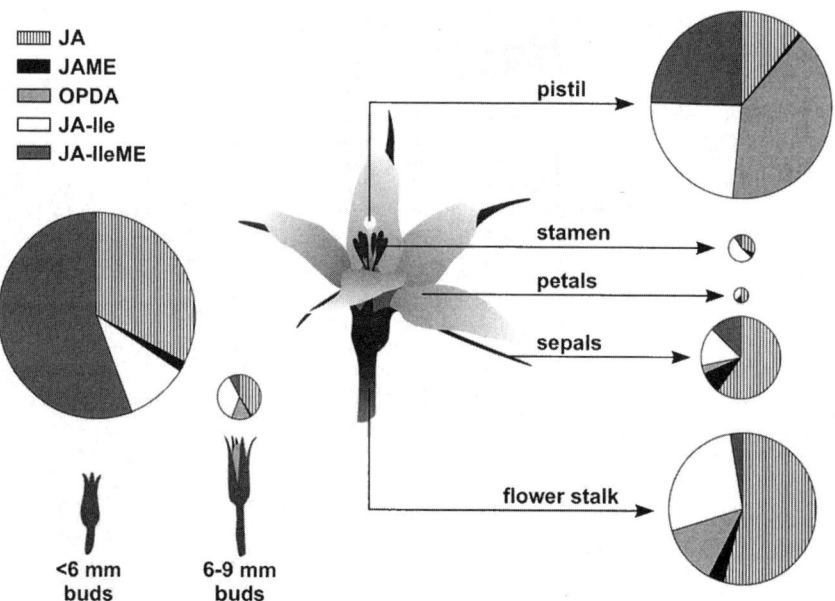

FIG. 8. Jasmonate and octadecanoids in flower buds and different flower organs of tomato. The sum of all compounds is given by the diameter, for example, buds of <6 mm contain 20.5 nmole per g fresh weight. The percentage of each compound is indicated by sectors. JA, Jasmonic acid; JAME, JA methyl ester; JA-Ile, JA–L-isoleucine conjugate; JA-IleME, JA–L-isoleucine conjugate methyl ester; OPDA, 12-oxo-phytodienoic acid. Data from Ref. 25.

for VSPs (244), histones (247), defense-related proteins such as PIN2 (248), PR1 (249, 250), a dioxygenase (251), the leucine amino peptidase A (129), a myrosinase (252), and threonine deaminase (253). Most of these genes are preferentially expressed within the female reproductive organ. This sink tissue is an attractive target for pathogens. The increased expression of defense-related genes might be of biological significance if it leads to constitutive defense against pathogens/insects. This may occur by formation of deterrant alkaloids (251), formation of glucosinolates due to activity of myrosinase (252), or, in the case of threonine deaminase, which is expressed in parenchyma cells of the flower (253), formation of α-aminobutyric acid, a compound known to function in defense reactions against pathogens.

Another putative role of jasmonates in flowers is related to male fertility, as shown by mutant analysis with *A. thaliana*. The coronatine- and jasmonate-insensitive mutant *coi1* cannot produce viable pollen. As a consequence, *coi1* is male-sterile and cannot be restored by application of JA (254). Another group of mutants exhibits deficiency in JA biosynthesis, and again they are also defective in pollen development (Table I): This includes the triple mutant *fad3-2 fad7-2 fad8* (105), the *dad1* mutant (43), the *opr3* mutant (88), and the *delayed dehiscence1* mutant (*dde1*) (87). All four mutants exhibit identical characteristics in the male-sterile phenotype: (1) Floral organs develop normally within closed buds, but the anther filaments do not elongate sufficiently to position the locules above the stigma at anthesis. (2) The anthers lack the proper dehiscence of the stomium at the time of flower opening. Although the tapetum is correctly broken down, pollen cannot be released from locules. (3) The viable, tricellular pollen is produced in smaller amount than in the wild type and does not germinate. The triple-*fad* mutant can be restored by application of α-LeA, OPDA, and JA. However, *opr3* and *dde1* ultimately need JA for recovery, whereas OPDA is inactive. This is in accordance with the fact that *opr3* and *dde1* have mutations in the gene encoding OPR3, which catalyzes a reaction downstream of OPDA. Three distinct steps can be distinguished in which JA may function as a signal: (1) elongation of anther filaments, (2) maturation of viable pollen, and (3) control of timing of anther dehiscence. In the latter case, JA-regulated changes in the cell wall structure of the stomium and in the cellular water relations within the endothecium may result in successful dehiscence of the anthers.

The requirement of JA for pollen development seems to be in contrast to the tissue-specific expression of JA biosynthetic genes in flowers: All JA biosynthetic enzymes cloned have been shown to be expressed within female flower organs of various plants. Although not unequivocally shown to be related to JA biosynthesis, *LOX* expression is associated with carpel development in pea (255) and rose (256). *AOS* is expressed in early stages of carpel development in *Arabidopsis* (257). In tomato flowers, the AOC protein preferentially accumulates in ovules and in the transmission tissue of style (25). Also, in flowers of *Arabidopsis*, the AOC accumulates preferentially within sepals, petals, filaments, and the ovary (79). This parallels the expression of *OPR3*, which was shown to occur in all *Arabidopsis* flower organs at early stages of floral development, and preferentially in pistil, filaments, and petals at later stages (87, 258). These data strongly suggest that all floral tissues except the anthers may be effective sources of JA, which in turn contribute to the regulation of pollen development. If this is true, JA formed in sepals and other organs of the flower should be targeted to the stomium. It is not yet clear how transport of jasmonates occurs between different flower tissues. It could be mediated by simple diffusion of JA or after its conversion into JAME, which is a volatile compound. The enzyme catalyzing this conversion, the jasmonic acid carboxy methyltransferase (JMT), is expressed in developing flowers of *Arabidopsis*. Its highest expression occurs during opening of flowers and decreases after opening (97). In conclusion, JA or its conjugates synthesized in the female and other tissues of a flower could act as a signal mediating interorgan plant communications, thus regulating pollen development.

Another effect of jasmonates in flower development is related to the onset of flowering. A delayed onset is caused by decreased levels of 12-OH-JA, as shown with 12-OH-JA sulfotransferase-overexpression lines (*100*) (Section III.B). Also, the nonenzymatically formed byproducts of the AOS reactions (Fig. 4), α-ketols such as 12Z,15Z-9-hydroxy-10-oxo-12,15-octadecadienoic acid, may have flower-inducing capacity (259).

C. Tuberization

Tuber formation in potatoes is a complex developmental process that requires the interaction of environmental, biochemical, and genetic factors. Under conditions of a short-day photoperiod and cool night temperature, *Solanum* species are able to form tubers. Tuberization is initiated by cell division and cell expansion and by change in the orientation of cell growth at the subapical region of the stolon tip (260). Tuberization is first indicated by cessation of stolon growth followed by swelling at the subapical region of the stolon (261). Subsequently, vigorous thickening growth occurs, and starch accumulates. Tuberization is completed when the apical meristem has sunk into the swelling tuber and has formed eyes.

Little is known about the morphological control of tuberization, although a prominent role of phytohormones is obvious (99, 262). There are combined actions of several hormones: (1) Decreased gibberellin levels are a prerequisite for tuberization (263). (2) Elevated jasmonate levels within stolons (264) may cause cessation of stolon elongation and may induce slight swelling at the subapical region of the stolon. JA-induced expansion of cells in potato tubers is thereby mediated by changing cell wall architecture and reorganization of cortical microtubules (265). (3) Cytokinins appear to induce vigorous thickening growth due to cell division (266). (4) The ABA content was shown to be increased during tuberization, thereby inducing dormancy of the apical meristem (261).

The existence of a tuber-inducing stimulus, generated in leaves under short-day conditions, was suggested based on results of grafting experiments (267). Subsequently, a JA-related compound, tuberonic acid glucoside (TAG), was identified as a compound stimulating tuberization (268). The aglycone of this glycoside is 12-OH-JA, which was named tuberonic acid (TA). TA and TAG do not exhibit effects known for JA, such as promotion of leaf senescence and inhibition of seedling growth (269), suggesting that both compounds have functions different from that of JA. Independently of day length, TAG is synthesized within the leaves and is transported throughout the whole plant, as shown by application of $[2-^{14}C]$-JA, which is converted completely into TAG within 10 days (270). However, the photoperiod affects the transport of TAG. TAG was found preferentially in stolons at short days, but accumulated preferentially in flowers at long days (270). This correlates with the effect of the photoperiod on hydroxylation of JA. Under short-day conditions, 12-OH-JA was found in the leaves of *Solanum demissum* that had formed tubers, whereas those plants grown in long-day conditions did not form tubers and 12-OH-JA was undetectable (271). Therefore, 12-OH-JA is assumed to promote tuberization, whereas TAG may represent its transport form. This glucoside is capable of long-distance transport throughout the plant.

The question arises, Which compound represents the trigger of potato tuberization? All putative compounds, TA, TAG, JA, and JAME, show strong tuber-inducing activity in an *in vitro* assay (272). In contrast to JA and JAME, neither TA and TAG causes cell expansion, which is essential for tuberization, and is one of the characteristic features of the action of JA and JAME (273). The following model summarizes most of the data (99): TA and TAG are signals for tuberization and are synthesized within the leaves. Upon long-distance transport throughout the plant, they reach the stolons. Here, an increase in the level of JA may be induced, which triggers tuberization by inducing cell expansion. This model is supported by the fact that the accumulation of a tuber-specific LOX mRNA (*POTLX-1*) occurred in apical and subapical regions of developing tubers, stolons, and roots (274). The accumulation of POTLX-1 mRNA increased with tuber initiation and growth. However, POTLX-1 was characterized as a 9-LOX, thus being obviously not involved in JA biosynthesis. Nevertheless, suppression of *POTLX-1* expression by an antisense approach led to the disruption of tuber development and to the formation of malformed tubers. The question remains, How can the suppression of a tuber-specific 9-LOX affect the tuber-forming activity? Since jasmonates can be generated only via the activity of a

13-LOX, tuberization should depend not only on JA-related compounds, but also on 9-LOX activity. The tuber 9-LOX could contribute to the formation of cyclopentanone derivatives derived from a (9S)-hydroperoxide, as described recently for potato tubers (275).

D. Tendril Coiling

Other examples of morphological changes apparently mediated by jasmonates and/or octadecanoids are tendril coiling in Cucurbitaceae and the thigmomorphogenetic response of bean internodes. Both processes result from the so-called mechanotransduction following stimulation by mechanical forces (touch, wind). In *Bryonia dioica*, touch causes a release of Ca^{2+} from the endoplasmic reticulum (ER) lumen into the cytosol through an ER-located Ca^{2+} channel (276). This is followed by immediate changes in protein phosphorylation (277) and results finally in tendril coiling. This process can be elicited also by airborne JAME without any mechanical contact (278). In this way, the chemically induced reaction is morphologically and biochemically indistinguishable from the mechanically induced process. Interestingly, in tendril coiling, the endogenous signal is OPDA, but not JA (279). The OPDA level rises transiently and correlates with the degree of coiling, whereas JA levels remain low and constant (30). Furthermore, structure–activity tests with numerous structural derivatives of altered octadecanoids and jasmonates revealed clear preference for octadecanoid-like compounds as inducers of tendril coiling (280). JA is active only indirectly by increasing the endogenous level of OPDA (31). Apparently, OPDA coordinates the growth reactions of tendrils downstream of the calcium signal upon touch of the blebs (281). In conclusion, the perception of a mechanical stimulus results in different processes such as rapid ion flux, protein phosphorylation, and octadecanoid signaling, which all converge in a common response of the organ (7).

E. Senescence

Senescence is the last stage of development and leads to the death of a single cell, a tissue, an organ, or a whole organism. In plants, senescence is a structurally, physiologically, and genetically orchestrated process whereby cellular organelles and their constituents are sequentially broken down. The released nutrients are recycled to actively growing organs such as young leaves, developing seeds, and fruits (282). Three phases can be distinguished in leaf senescence (283): (1) The initiation phase or the onset of senescence is marked by alterations of photosynthetic membranes, decrease in number of ribosomes and polysomes, and higher activity of ubiquitin-mediated protein degradation. (2) The reorganization or progression phase is manifested by a thylakoid breakdown, mainly indicated by loss of chlorophyll (284, 285). This process is accompanied by an increase in the number and size of osmiophilic globuli within the plastids (286), whereas other cell organelles such as mitochondria, nuclei, and vacuoles persist with only slight alterations (284). (3) The terminal phase is marked by changes of the plasma membrane leading to a collapse of cellular homeostasis and ultimately resulting in cell death. In all phases of leaf senescence, gene expression is dramatically altered. Genes which are active in nonsenescing leaves are downregulated, as shown for photosynthetic genes, while a subset of genes is upregulated (for review see Ref. 287).

Leaf senescence is a developmentally regulated program, but is superimposed by numerous internal and environmental factors which can influence onset, progression, and termination of senescence. Among these are darkness and dehydration as well as hormones, such as ABA, jasmonates, and ethylene. Most data on altered gene expression in senescence were obtained with detached leaf tissues, upon dark-induced senescence, or by application of senescence-promoting compounds. However, treatments that cause loss of chlorophyll do not necessarily induce the complete program of nutrient salvage that is characteristic of the age-induced senescence (287). Therefore, many genes identified in a senescence-related screen were not senescence-specific, but were stress-specific (136). Despite these discrepancies between the age-induced "normal" senescence and exogenously induced "artificial" senescence, the identification of genes induced by darkness or upon hormone treatment has led to some insights into the regulation of senescence (see below). Another problem in analyzing the genetic program underlying leaf senescence is the occurrence of nonenzymatic events. As shown recently, occurrence of stereo-random lipid peroxidation products indicates a dominant role for nonenzymatic lipid peroxidation in late stages of leaf development (288).

The most prominent feature of leaf senescence, the rapid loss of chlorophyll, is also caused by exogenously applied jasmonates (10). Therefore, JA was referred to as a senescence-promoting hormone. Cytokinins counteract this effect (123). In barley leaf segments, the senescence-promoting effect of jasmonates might be the consequence of the two types of JA-inducible downregulation of gene expression (Section IV.A). The capacity of jasmonates to induce downregulation of photosynthetic genes might be one determinant in the onset of senescence. In this respect it is interesting to note that barley and *Arabidopsis* plants exhibit high JA levels in tissues undergoing age-dependent senescence (O. Miersch and I. Feussner, personal communication).

The senescence-promoting feature of jasmonate is also supported by identification of two cDNAs coding for the initial enzyme of chlorophyll breakdown, the chlorophyllase, from *Arabidopsis* (289). One of them is highly inducible by JAME. However, instead of an expected chloroplast transit peptide, this cDNA exhibits a vacuolar sorting determinant (289). It is still unclear how a vacuolar-located chlorophyllase could perform the first step of the chlorophyll breakdown, the three subsequent steps of which are known to be located within the chloroplast (*290*). Therefore, it will be interesting to analyze whether the recently described senescence-related secretion of chlorophyll into vacuoles (*291*) is functionally linked to chlorophyllase activity and formation of chlorophyll catabolites.

In order to monitor alterations in gene expression, about 50 different cDNAs were identified as being expressed during leaf senescence. They were collectively named senescence-associated genes (*SAGs*; *292–295*) and clustered into the following groups (*283*): (1) Genes encoding proteins which function in proteolytic processes. Among these are cystein proteases and ubiquitin. (2) Genes encoding proteins involved in plant defense such as proteins with antifungal properties, PR proteins, chitinases, and metallothionins. The latter could be involved in defense against metal ion-mediated oxidative damage or in ion storage and transport. (3) Genes encoding proteins related to gluconeogenesis.

However, some of these genes are active at later stages of senescence and might be expressed rather as a consequence, than a cause, of senescence (*286*). Studies on expression of different *SAGs* revealed a diverse range of gene activation patterns during senescence, indicating the existence of multiple regulatory pathways (*287*). In order to understand the molecular basis of leaf senescence, it is necessary to identify genes whose products are components of such pathways. Therefore, new strategies were used to identify genes upregulated during leaf senescence (*296*). More than 140 new putative senescence-regulated genes were identified by using *Arabidopsis* enhancer trap lines. Most of these genes exhibited different expression patterns in age-dependent and induced senescence. This led to the proposal of a regulatory network, which includes, apart from the developmental regulation, different senescence-promoting factors such as ethylene, JA, ABA, dehydration, and darkness (*296*). In this respect, a link between jasmonate and senescence was shown by recent characterization of the *ore9* mutant of *Arabidopsis* (*297*). This mutant exhibits increased longevity during age-dependent natural senescence of leaves by delaying the onset of senescence. It also displays delayed senescence symptoms during hormone-modulated senescence. Therefore, ORE9 may be linked to a common step of age-dependent senescence and senescence induced by ABA, ethylene, and jasmonates. The gene *ORE9* codes for an F-box protein characterized by an F-box motif interacting with SKP1, a component of the plant SCF complex (*297*). The SCF complex is one type of ubiquitin-protein ligase (E3) functioning in ubiquitin-dependent protein degradation (*298*). Thus, ORE9 may be involved in ubiquitin-dependent proteolysis of negative regulators of senescence, such as transcriptional repressors of *SAGs*. Interestingly, by analyzing the JA-insensitive mutant *coi1* (Section VI.B), another F-box protein, COI1, was identified which may function in ubiquitin-dependent removal of a negative regulator of jasmonate-induced gene expression (*110*). It is tempting to speculate that age-dependent and jasmonate-induced senescence converge into a common pathway by combinatorial action of F-box proteins in the SCF complex. Thus, ORE9 could function as a positive regulator of leaf senescence and COI1 could switch on jasmonate-dependent gene expression.

VII. Concluding Remarks

Twenty-five years after the identification of JA as a constituent of flower-released odor, the role of JA in two physiological processes was established: (1) a senescence-promoting effect with newly formed proteins (*123, 131*) and (2) abundant PIN accumulation in tomato leaves by air-borne JAME (*224*). These observations initiated an exponentially growing interest in jasmonates. Besides elucidation of the biosynthetic pathway of JA in the 1980s (*59, 60*), its roles in the wound response upon herbivore attack (*125, 170*), phytoalexin synthesis upon elicitation of plant cell cultures (*14*), pollen development and male sterility (*105*), and plant–pathogen interactions (*200*) became landmarks in the elucidation of function. One of the most important breakthroughs was the identification and characterization of the JA-insensitive mutant *coi1* (*110, 254*). In this mutant an F-box protein, a member of the ubiquitin-dependent protein degradation machinary, is defective. This principle of regulation of gene expression was also supported by isolation of corresponding F-box proteins for auxin response (TIR1), light response (COP9), circadian clock (FKF1, ZTL), patterning and growth of floral meristem (UFO), and senescence (ORE9) (*297–299*).

Analysis of target proteins of COI1 will be an important issue in elucidating the mechanism of JA-induced gene expression. Possibly, such target proteins form a link to JA-responsive transcription factors such as the ORCAs. The basic structure of these ORCAs as AP2/EREB proteins suggests a modular principle of action. ORCAs seem to be a switch between primary and secondary metabolism, and their mode of action highlights how both processes might be orchestrated by JA. Analysis of the mode of action of JA-dependent transcription factors is expected to be a major goal for the near future, and will give insights into how the different environmental and developmental signals are integrated and transduced into a distinct pattern of gene expression.

Upstream events in JA-dependent signal transduction pathways will be of increasing interest in the future. Some examples in which the roles of MAP kinases and Ca^{2+} were identified as positive or negative regulators in JA signaling have already illustrated the potential of these steps of signal transduction for JA specificity and cross-talk with other signaling pathways.

Nothing is known about the putative JA receptors. Here, we can expect a new breakthrough in understanding JA perception and signaling. The observed tissue-specific generation and function of JA has led to numerous new questions on intercellular signaling, long-distance transport, and developmentally regulated processes. Here, we will obtain insight into how plants perform permanent adjustment to diverse and variable environments and how a distinct developmental program is sustained via a specific signature of compounds. Finally, exciting information can be expected on interorganismic communication in an ecosystem by JA-induced or JA-mediated events. Here, the tritrophic interactions among herbivores, carnivores, and plants via the versatile signal jasmonate hint at adaptive mechanisms and evolutionary advantage. It will be interesting to see how the chemical diversity of plants is integrated into specific adaptive responses.

ACKNOWLEDGMENTS

We apologize to scientists whose work we overlooked or were not able to include because of space limitations. We thank Dr. S. Rosahl, Dr. J. Rudd, and Dr. I. Feussner for critical reading of the manuscript. We are grateful to C. Kutter for aid in designing Fig. 5, to C. Kaufmann for help in preparing all the figures, and to C. Dietel for the large amount of work in typing the manuscript.

REFERENCES

1. I. Feussner and C. Wasternack, The lipoxygenase pathway. *Annu. Rev. Plant Physiol.*, **53**, 275 (2002).
2. J. Leon, E. Rojo, and J. J. Sanchez-Serrano, Wound signalling in plants. *J. Exp. Bot.* **52**, 1 (2001).
3. F. Schaller, Enzymes of the biosynthesis of octadecanoid-derived signalling molecules. *J. Exp. Bot.* **52**, 11 (2001).
4. C. A. Ryan, The systemin signaling pathway: differential activation of plant defensive genes. *Biochim. Biophys. Acta* **1477**, 112 (2000).
5. L. L. Walling, The myriad plant responses to herbivores. *J. Plant Growth Regul.* **19**, 195 (2000).
6. M. Dicke and R. M. P. van Poecke, *in* "Plant Signal Transduction" (D. Scheel and C. Wasternack, eds.), in press. Oxford University Press, Oxford, 2002.
7. F. Schaller and E. W. Weiler, *in* "Plant Signal Transduction" (D. Scheel and C. Wasternack, eds.), in press. Oxford University Press, Oxford, 2002.
8. L. Xiong and J.-K. Zhu, *in* "Plant Signal Transduction" (D. Scheel and C. Wasternack, eds.), in press. Oxford University Press, Oxford, 2002.
9. E. Demole, E. Lederer, and D. Mercier, Isolement et détermination de la structure du jasmonate de méthyle, constituant odorant charactéristique de lèssence de jasmin. *Helv. Chim. Acta* **45**, 675 (1962).
10. B. Parthier, Jasmonates, new regulators of plant growth and development: many facts and few hypotheses on their actions. *Bot. Acta* **104**, 446 (1991).
11. B. Parthier, C. Brückner, W. Dathe, B. Hause, G. Herrmann, H.-D. Knöfel, H.-M. Kramell, R. Kramell, J. Lehmann, O. Miersch, S. Reinbothe, G. Sembdner, C. Wasternack, and U. zur Nieden, *in* "Progress in Plant Growth Regulation" (C. M. Karssen, L. C. van Loon, and D. Vreugdenhil, eds.), p. 276. Kluwer, Netherlands, 1992.

12. B. A. Vick and D. C. Zimmerman, The biosynthesis of jasmonic acid: a physiological role for plant lipoxygenase. *Biochem. Biophys. Res. Commun.* **111**, 470 (1983).
13. C. Wasternack and B. Parthier, Jasmonate-signalled plant gene expression. *Trends Plant Sci.* **2**, 302 (1997).
14. H. Gundlach, M. J. Müller, T. M. Kutchan, and M. H. Zenk, Jasmonic acid is a signal transducer in elicitor-induced plant cell cultures. *Proc. Natl. Acad. Sci. USA* **90**, 2389 (1992).
15. M. J. Mueller, W. Brodschelm, E. Spannagl, and M. H. Zenk, Signaling in the elicitation process is mediated through the octadecanoid pathway leading to jasmonic acid. *Proc. Natl. Acad. Sci. USA* **90**, 7490 (1993).
16. J. Lehmann, R. Atzorn, C. Brückner, S. Reinbothe, J. Leopold, C. Wasternack, and B. Parthier, Accumulation of jasmonate, abscisic acid, specific transcripts and proteins in osmotically stressed barley leaf segments. *Planta* **197**, 156 (1995).
17. B. J. Feys and J. E. Parker, Interplay of signaling pathways in plant disease resistance. *Trends Genet.* **16**, 449 (2000).
18. S. Berger, Jasmonate-related mutants of *Arabidopsis* as tools for studying stress signaling. *Planta*, **214**, 497 (2002).
19. J. Memelink, R. Verpoorte, and J. W. Kijne, ORCAnization of jasmonate-responsive gene expression in alkaloid metabolism. *Trends Plant Sci.* **6**, 212 (2001).
20. H. Gundlach and M. H. Zenk, Biological activity and biosynthesis of pentacyclic oxylipins: the linoleic acid pathway. *Phytochemistry* **47**, 527 (1998).
21. G. Sembdner, R. Atzorn, and G. Schneider, Plant hormone conjugation. *Plant Mol. Biol.* **26**, 1459 (1994).
22. R. Kramell, R. Atzorn, G. Schneider, O. Miersch, C. Brückner, J. Schmidt, G. Sembdner, and B. Parthier, Occurrence and identification of jasmonic acid and its amino acid conjugates induced by osmotic stress in barley leaf tissue. *J. Plant Growth Regul.* **14**, 29 (1995).
23. R. Kramell, O. Miersch, R. Atzorn, B. Parthier, and C. Wasternack, Octadecanoid-derived alteration of gene expression and the 'oxylipin signature' in stressed barley leaves—implications for different signalling pathways. *Plant Physiol.* **123**, 177 (2000).
24. H.-D. Knöfel and G. Sembdner, Jasmonates from pine pollen. *Phytochemistry* **38**, 569 (1995).
25. B. Hause, I. Stenzel, O. Miersch, H. Maucher, R. Kramell, J. Ziegler, and C. Wasternack, Tissue-specific oxylipin signature of tomato flowers—allene oxide cyclase is highly expressed in distinct flower organs and vascular bundles. *Plant J.* **24**, 113 (2000).
26. M. A. Birkett, C. A. M. Campbell, K. Chamberlain, E. Guerrieri, A. J. Hick, J. L. Martin, M. Matthes, J. A. Napier, J. Pettersson, J. A. Pickett, G. M. Poppy, E. M. Pow, B. J. Pye, L. E. Smart, G. H. Wadhams, L. J. Wadhams, and C. M. Woodcock, New roles for cis-jasmone as an insect semiochemical and in plant defense. *Proc. Natl. Acad. Sci. USA* **97**, 9329 (2000).
27. O. Miersch, A. Porzel, and C. Wasternack, Microbial conversion of jasmonates—hydroxylations by *Aspergillus niger*. *Phytochemistry* **50**, 1147 (1999).
28. O. Miersch, A. Preiss, G. Sembdner, and K. Schreiber, (+)-7-*iso*-Jasmonic acid and related compounds from *Botryodiplodia theobromae*. *Phytochemistry* **26**, 1037 (1987).
29. E. W. Weiler, Octadecanoid-mediated signal transduction in higher plants. *Naturwissenschaften* **84**, 340 (1997).
30. B. A. Stelmach, A. Müller, P. Hennig, D. Laudert, L. Andert, and E. W. Weiler, Quantitation of the octadecanoid 12-oxo-phytodienoic acid, a signalling compound in plant mechanotransduction. *Phytochemistry* **47**, 539 (1998).
31. B. A. Stelmach, A. Müller, and E. W. Weiler, 12-oxo-phytodienoic acid and indole-3-acetic acid in jasmonic acid-treated tendrils of *Bryonia dioica*. *Phytochemistry* **51**, 187 (1999).
32. S. Parchmann, H. Gundlach, and M. J. Mueller, Induction of 12-oxo-phytodienoic acid in wounded plants and elicited plant cell cultures. *Plant Physiol.* **115**, 1057 (1997).

33. B. A. Stelmach, A. Müller, P. Hennig, S. Gebhardt, M. Schubert-Zsilavecz, and E. W. Weiler, A novel class of oxylipins, sn1-O-(12-oxophytodienoyl)-sn2-O-(hexadecatrienoyl)-monogalactosyl diglyceride, from Arabidopsis thaliana. J. Biol. Chem. **276,** 1282 (2001).
34. H. Weber, B. A. Vick, and E. E. Farmer, Dinor-oxo-phytodienoic acid: A new hexadecanoid signal in the jasmonate family. Proc. Natl. Acad. Sci. USA **94,** 10473 (1997).
35. R. Imbusch and M. J. Müller, Analysis of oxidative stress and wound-inducible dinor isoprostanes F1 (phytoprostanes F1) in plants. Plant Physiol. **124,** 1293 (2000).
36. T. Munnick, Phosphatidic acid: an emerging plant lipid second messenger. Trends Plant Sci. **6,** 227 (2001).
37. C. Göbel, I. Feussner, A. Schmidt, D. Scheel, J. J. Sanchez-Serrano, and S. Rosahl, Oxylipin profiling reveals the preferential stimulation of the 9-lipoxygenase pathway in elicitor-treated potato cells. J. Biol. Chem. **276,** 6267 (2001).
38. J. Royo, G. Vancanneyt, A. G. Perez, C. Sanz, K. Störmann, S. Rosahl, and J. J. Sanchez-Serrano, Characterization of three potato lipoxygenases with distinct enzymatic activities and different organ-specific and wound-regulated expression patterns. J. Biol. Chem. **271,** 21012 (1996).
39. J. Narváez-Vásquez, J. Florin-Christensen, and C. A. Ryan, Positional specificity of a phospholipase A2 activity induced by wounding, systemin, and oligosaccharide elicitors in tomato leaves. Plant Cell **11,** 2249 (1999).
40. S. Dhondt, P. Geoffroy, B. A. Stelmach, M. Legrand, and T. Heitz, Soluble phospholipase A2 activity is induced before oxylipin accumulation in tobacco mosaic virus-infected tobacco leaves and is contributed by patatin-like enzymes. Plant J. **23,** 431 (2000).
41. S. Lee, S. Suh, S. Kim, R. C. Crain, J. M. Kwak, H. G. Nam, and Y. Lee, Systemic elevation of phosphatidic acid and lysophospholipid levels in wounded plants. Plant J. **12,** 547 (1997).
42. C. Wang, C. A. Ziehn, M. Afitlhile, R. Welti, D. F. Hildebrand, and X. Wang, Involvement of phospholipase D in wound-induced accumulation of jasmonic acid in Arabidopsis. Plant Cell **12,** 2237 (2000).
43. S. Ishiguro, A. Kawai-Oda, J. Ueda, I. Nishida, and K. Okada, The defective in anther dehiscence 1 gene encodes a novel phospholipase A1 catalyzing the initial step of jasmonic acid biosynthesis, which synchronyzes pollen maturiaton, anther dehiscence, and flower opening in Arabidopsis. Plant Cell **13,** 2191 (2001).
44. A. R. Brash, Lipoxygenases: Occurrence, functions, catalysis, and acquisition of substrate. J. Biol. Chem. **274,** 23679 (1999).
45. S. Rosahl, Lipoxygenases in plants—their role in development and stress response. Z. Naturforsch. **51c,** 123 (1996).
46. J. N. Siedow, Lipoxygenases. Annu. Rev. Plant Physiol. Plant Mol. Biol. **42,** 145 (1991).
47. T. Heitz, D. R. Bergey, and C. A. Ryan, A gene encoding a chloroplast-targeted lipoxygenase in tomato leaves is transiently induced by wounding, systemin, and methyl jasmonate. Plant Physiol. **114,** 1085 (1997).
48. T. W. Bunker, D. S. Koetje, L. C. Stephenson, R. A. Creelman, J. E. Mullet, and H. D. Grimes, Sink limitation induces the expression of multiple soybean vegetative lipoxygenase mRNAs while the endogenous jasmonic acid level remains low. Plant Cell **7,** 1319 (1995).
49. E. Bell and J. E. Mullet, Characterization of an Arabidopsis lipoxygenase gene responsive to methyl jasmonate and wounding. Plant Physiol. **103,** 1133 (1993).
50. J. R. Van Mechelen, M. Smits, A. C. Douma, J. Rouster, V. Cameron-Mills, F. Heidekamp, and B. E. Valk, Primary structure of a lipoxygenase from barley grain as deduced from its cDNA sequence. Biochim. Biophys. Acta **1254,** 221 (1995).
51. K. Vörös, I. Feussner, H. Kühn, J. Lee, A. Graner, M. Löbler, B. Parthier, and C. Wasternack, Characterization of methyljasmonate-inducible lipoxygenase from barley (Hordeum vulgare cv. Salome) leaves. Eur. J. Biochem. **251,** 36 (1998).

52. U. Schaffrath, F. Zabbai, and R. Dudler, Characterization of RCI-1, a chloroplastic rice lipoxygenase whose synthesis is induced by chemical plant resistance activators. *Eur. J. Biochem.* **267**, 5935 (2000).
53. Y. L. Peng, Y. Shirano, H. Ohta, T. Hibino, K. Tanaka, and D. J. Shibata, A novel lipoxygenase from rice. Primary structure and specific expression upon incompatible infection with rice blast fungus. *J. Biol. Chem.* **269**, 3755 (1994).
54. I. Rancé, J. Fournier, and M.-T. Esquerré-Tugayé, The incompatible interaction between *Phytophthora parasitica* var. *nicotiana* race 0 and tobacco is suppressed in transgenic plants expressing antisense lipoxygenase sequences. *Proc. Natl. Acad. Sci. USA* **95**, 6554 (1998).
55. B. Hause, K. Vörös, K.-H. Kogel, K. Besser, and C. Wasternack, A jasmonate responsive lipoxygenase of barley leaves is induced by plant activators but not by pathogens. *J. Plant Physiol.* **154**, 459 (1999).
56. I. Feussner, H. Kühn, and C. Wasternack, Lipoxygenase dependent degradation of storage lipids. *Trends Plant Sci.* **6**, 368 (2001).
57. H. Weichert, I. Stenzel, E. Berndt, C. Wasternack, and I. Feussner, Metabolic profiling of oxylipins upon salicylate treatments in barley leaves—preferential induction of the reductase pathway by salicylate. *FEBS Lett.* **464**, 133 (1999).
58. M. Kohlmann, A. Bachmann, H. Weichert, A. Kolbe, T. Balkenhohl, C. Wasternack, and I. Feussner, Formation of lipoxygenase-pathway-derived aldehydes in barley leaves upon methyl jasmonate treatment. *Eur. J. Biochem.* **260**, 885 (1999).
59. B. A. Vick and D. C. Zimmerman, Biosynthesis of jasmonic acid by several plant species. *Plant Physiol.* **75**, 458 (1984).
60. M. Hamberg and M. Hughes, Fatty acid allene oxides. III. Albumin-induced cyclization of 12,13(S)-epoxy-9(Z),11-octadecadienoic acid. *Lipids* **23**, 469 (1988).
61. M. Hamberg and P. Fahlstadius, Allene oxide cyclase: a new enzyme in plant lipid metabolism. *Arch. Biochem. Biophys.* **276**, 518 (1990).
62. H. Maucher, B. Hause, I. Feussner, J. Ziegler, and C. Wasternack, Allene oxide synthases of barley (*Hordeum vulgare* cv. Salome)—tissue specific regulation in seedling development. *Plant J.* **21**, 199 (2000).
63. J. Ziegler, C. Wasternack, and M. Hamberg, On the specificity of allene oxide cyclase. *Lipids* **34**, 1005 (1999).
64. W. C. Song and A. R. Brash, Purification of an allene oxide-synthase and identification of the enzyme as a cytochrome P-450. *Science* **253**, 781 (1991).
65. W. C. Song, C. D. Funk, and A. R. Brash, Molecular cloning of an allene oxide synthase. A cytochrome P-450 specialized for metabolism of fatty acid hydroperoxides. *Proc. Natl. Acad. Sci. USA* **90**, 8519 (1993).
66. Z. Pan, F. Durst, D. Werck-Reichhart, H. W. Gardner, B. Camara, K. Cornish, and R. A. Backhaus, The major protein of guayule rubber particles is a cytochrome P450. *J. Biol. Chem.* **270**, 8487 (1995).
67. D. Laudert, U. Pfannschmidt, F. Lottspeich, H. Holländer-Czytko, and E. W. Weiler, Cloning, molecular and functional characterization of *Arabidopsis thaliana* allene oxide synthase (CYP74), the first enzyme of the octadecanoid pathway to jasmonates. *Plant Mol. Biol.* **31**, 323 (1996).
68. S. Sivasankar, B. Sheldrick, and S. J. Rothstein, Expression of allene oxide synthase determines defense gene activation in tomato. *Plant Physiol.* **122**, 1335 (2000).
69. G. A. Howe, G. I. Lee, A. Itoh, L. Li, and A. E. DeRocher, Cytochrome P450-dependent metabolism of oxylipins in tomato. Cloning and expression of allene oxide synthase and fatty acid hydroperoxide lyase. *Plant Physiol.* **123**, 711 (2000).

70. S.-M. C. Lau, P. A. Harder, and D. P. O'Keefe, Low carbon monoxide affinity allene oxide synthase is the predominant cytochrome P450 in many plant tissues. *Biochemistry* **32**, 1945 (1993).
71. C. Chapple, Molecular-genetic analysis of plant cytochrome P450-dependent monooxygenases. *Annu. Rev. Plant Physiol. Plant Mol. Biol.* **49**, 311 (1998).
72. A. Itoh and G. A. Howe, Molecular cloning of a divinyl ether synthase. *J. Biol. Chem.* **276**, 3620 (2001).
73. E. Blée and J. Joyard, Envelope membranes from spinach chloroplasts are a site of the metabolism of fatty acid hydroperoxides. *Plant Physiol.* **110**, 445 (1996).
74. B. A. Vick and D. C. Zimmerman, Pathways of fatty acid hydroperoxide metabolism in spinach leaf chloroplasts. *Plant Physiol.* **85**, 1073 (1987).
75. J. E. Froehlich, A. Itoh, and G. A. Howe, Tomato allene oxide synthase and fatty acid hydroperoxide lyase, two cytochrome P450s involved in oxylipin metabolism, are targeted to different membranes of chloroplast envelope. *Plant Physiol.* **125**, 306 (2001).
76. K. Matsui, J. Wilkinson, B. Hiatt, V. Knauf, and T. Kajiwara, Molecular cloning and expression of *Arabidopsis* fatty acid hydroperoxide lyase. *Plant Cell Physiol.* **40**, 477 (1999).
77. J. Ziegler, M. Hamberg, O. Miersch, and B. Parthier, Purification and characterization of allene oxide cyclase from dry corn seeds. *Plant Physiol.* **114**, 565 (1997).
78. J. Ziegler, I. Stenzel, B. Hause, H. Maucher, O. Miersch, M. Hamberg, M. Grimm, M. Ganal, and C. Wasternack, Molecular cloning of allene oxide cyclase: The enzyme establishing the stereochemistry of octadecanoids and jasmonates. *J. Biol. Chem.* **275**, 19132 (2000).
79. I. Stenzel, B. Hause, O. Miersch, T. Kurz, H. Maucher, H. Weichert, J. Ziegler, I. Feussner, and C. Wasternack, Jasmonate biosynthesis and the allene oxide cyclase family of Arabidopsis thaliana. (2002 submitted).
80. F. Schaller and E. W. Weiler, Molecular cloning and characterization of 12-oxophytodienoate reductase, an enzyme of the octadecanoid signaling pathway from *Arabidopsis thaliana*. *J. Biol. Chem.* **272**, 28066 (1997).
81. F. Schaller, P. Hennig, and E. W. Weiler, 12-Oxophytodienoate-19,11-reductase: Occurrence of two isoenzymes of different specificity against stereoisomers of 12-oxophytodienoic acid. *Plant Physiol.* **188**, 1345 (1998).
82. J. Strassner, A. Fürholz, P. Macheroux, N. Amrhein, and A. Schaller, A homolog of old yellow enzyme in tomato. Spectral properties and substrate specificity of the recombinant protein. *J. Biol. Chem.* **274**, 35067 (1999).
83. C. Breithaupt, J. Strasser, U. Breitinger, R. Huber, P. Macheroux, and A. Schaller, X-Ray structure of 12-oxophytodienoate reductase 1 provides structural insight into substrate binding and specificity within the family of OYE. *Structure* **9**, 419 (2001).
84. C. Biesgen and E. W. Weiler, Structure and regulation of *OPR1* and *OPR2*, two closely related genes encoding 12-oxophytodienoic acid-10,11-reductases from *Arabidopsis thaliana*. *Planta* **208**, 155 (1999).
85. F. Schaller, C. Biesgen, C. Müssig, T. Altmann, and E. W. Weiler, 12-Oxophytodienoate reductase 3 (OPR3) is the isoenzyme involved in jasmonate biosynthesis. *Planta* **210**, 979 (2000).
86. C. Müssig, C. Biesgen, J. Lisso, U. Uwer, E. W. Weiler, and T. Altmann, A novel stress-inducible 12-oxophytodienoate reductase from *Arabidopsis thaliana* provides a potential link between brassinosteroid action and jasmonic acid synthesis. *J. Plant Physiol.* **157**, 155 (2000).
87. P. M. Sander, P. Y. Lee, C. Biesgen, J. D. Boone, T. P. Beals, E. W. Weiler, and R. B. Goldberg, The *Arabidopsis DELAYED DEHISCENCE1* gene encodes an enzyme in the jasmonic acid synthesis pathway. *Plant Cell* **12**, 1041 (2000).
88. A. Stintzi and J. Browse, The *Arabidopsis* male-sterile mutant, *opr3*, lacks the 12-oxophytodienoic acid reductase required for jasmonate synthesis. *Proc. Natl. Acad. Sci. USA* **97**, 10625 (2000).

89. O. Miersch and C. Wasternack, Octadecanoid and jasmonate signalling in tomato leaves (*Lycopersicon esculentum* Mill.)—endogenous jasmonates do not induce jasmonate biosynthesis. *Biol. Chem.* **381**, 715 (2000).
90. D. Laudert and E. W. Weiler, Allene oxide synthase: a major control point in *Arabidopsis thaliana* octadecanoid signalling. *Plant J.* **15**, 675 (1998).
91. I. Stenzel, B. Hause, M. Maucher, A. Pitzschke, O. Miersch, R. Kramell, J. Ziegler, C. A. Ryan, and C. Wasternack, Allene oxide cyclase transgenes amplify jasmonate biosynthesis and the wound-response of tomato leaves—Vascular bundle-specific generation of jasmonates. *Plant J.* submitted (2002).
92. E. Blée, Phytooxylipins and plant defense reactions. *Lipid Res.* **37**, 33 (1998).
93. Y. Utsunomiya, T. Nakayama, H. Ohira, R. Hirota, T. Mori, F. Kawai, and T. Ueda, Purification and inactivation by substrate of an allene oxide synthase (CYP74) from corn (*Zea mays* L.) seeds. *Phytochemistry* **53**, 319 (2000).
94. C. Wang, S. Avdiushko, and D. F. Hildebrand, Overexpression of a cytoplasm-localized allene oxide synthase promotes the wound-induced accumulation of jasmonic acid in transgenic tobacco. *Plant Mol. Biol.* **40**, 783 (1999).
95. D. Laudert, A. Schaller, and E. W. Weiler, Transgenic *Nicotiana tabacum* and *Arabidopsis thaliana* plants overexpressing allene oxide synthase. *Planta* **211**, 163 (2000).
96. K. Harms, R. Atzorn, A. R. Brash, H. Kühn, C. Wasternack, L. Willmitzer, and H. Peña-Cortés, Expression of a flax allene oxide synthase cDNA leads to increased endogenous jasmonic acid (JA) levels in transgenic potato plants but not to a corresponding activation of JA-responding genes. *Plant Cell* **7**, 1645 (1995).
97. H. S. Seo, J. T. Song, J.-J. Cheong, Y.-H. Lee, Y.-W. Lee, I. Hwang, J. S. Lee, and Y. D. Choi, Jasmonic acid carboxyl methyl transferase: A key enzyme for jasmonate-regulated plant responses. *Proc. Natl. Acad. Sci. USA* **98**, 4788 (2001).
98. G. Sembdner and B. Parthier, The biochemistry and the physiological and molecular actions of jasmonates. *Annu. Rev. Plant Physiol. Plant Mol. Biol.* **44**, 569 (1993).
99. Y. Koda, Possible involvement of jasmonates in various morphogenic events. *Physiol. Plant.* **100**, 639 (1997).
100. S. Gidda, O. Miersch, and L. Varin, The modulation of AtSt2a expression leads to altered flowering time in transgenic *Arabidopisis thaliana*. *Proc. Natl. Acad. Sci. USA* submitted (2001).
101. O. Miersch, H.-D. Knöfel, J. Schmidt, R. Kramell, and B. Parthier, A jasmonic acid conjugate, N[(-)-jasmonoyl]-tyramine, from *Petunia* pollen. *Phytochemistry* **47**, 327 (1998).
102. R. Kramell, O. Miersch, B. Hause, B. Ortel, B. Parthier, and C. Wasternack, Amino acid conjugates of jasmonic acid induce jasmonate-responsive gene expression in barley (*Hordeum vulgare* L.). *FEBS Lett.* **414**, 197 (1997).
103. C. Wasternack, B. Ortel, O. Miersch, R. Kramell, M. Beale, F. Greulich, I. Feussner, B. Hause, W. Krumm, W. Boland, and B. Parthier, Diversity in octadecanoid-induced gene expression of tomato. *J. Plant Physiol.* **152**, 345 (1998).
104. O. Miersch, R. Kramell, B. Parthier, and C. Wasternack, Structure-activity relations of substituted, deleted or stereospecifically altered jasmonic acid in gene expression of barley leaves. *Phytochemistry* **50**, 353 (1999).
105. M. McConn and J. Browse, The critical requirement for linolenic acid is pollen development, not photosynthesis, in an *Arabidopsis* mutant. *Plant Cell* **8**, 403 (1996).
106. M. McConn, R. A. Creelman, E. Bell, J. E. Mullet, and J. Browse, Jasmonate is essential for insect defense in *Arabidopsis*. *Proc. Natl. Acad. Sci. USA* **94**, 5473 (1997).
107. A. Stintzi, J. Weber, P. Reymond, J. A. Browse, and E. E. Farmer, Plant defense in the absence of jasmonic acid: The role of cyclopentenones. *Proc. Natl. Acad. Sci. USA* **98**, 12837 (2001).

108. T. A. Richmond and A. B. Bleecker, A defect in β-oxidation causes abnormal inflorescence development in *Arabidopsis*. *Plant Cell* **11**, 1911 (2000).
109. G. A. Howe, J. Lightner, J. Browse, and C. A. Ryan, An octadecanoid pathway mutant (JL5) of tomato is compromised in signaling for defense against insect attack. *Plant Cell* **8**, 2067 (1996).
110. D.-X. Xie, B. F. Feys, S. James, M. Nieto-Rostro, and J. G. Turner, An *Arabidopsis* gene required for jasmonate-regulated defense and fertility. *Science* **280**, 1091 (1998).
111. C. Ellis and J. G. Turner, The Arabidopsis mutant *cev*1 has constitutively active jasmonate and ethylene signal pathways and enhanced resistance to pathogens. *Plant Cell* **13**, 1025 (2001).
112. L. Xu, K. Liu, Z. Wang, W. Peng, R. Huang, D. Huang, and D. Xie, An *Arabidopsis* mutant *cex*1 exhibits constant accumulation of jasmonate-regulated At*VSP*, *Thi2.1* and *PDF2*. *FEBS Lett.* **494**, 161 (2001).
113. B. Hilpert, H. Bohlmann, R. op den Camp, D. Przybyla, O. Miersch, A. Buchala, and K. Apel, Isolation and characterization of signal transduction mutants of *Arabidopsis thaliana* that constitutively activate the octadecanoid pathway and form necrotic microlesions. *Plant J.* **26**, 435 (2001).
114. P. Hedden and A. L. Phillips, Manipulation of hormone biosynthetic genes in transgenic plants. *Curr. Opin. Biotechnol.* **11**, 130 (2000).
115. M. Martin, J. Léon, C. Dammann, J. P. Albar, G. Griffiths, and J. J. Sánchez-Serrano, Antisense-mediated depletion of potato leaf ω-3-fatty acid desaturase lowers linolenic acid content and reduces gene activation in response to wounding. *Eur. J. Biochem.* **262**, 283 (1999).
116. E. Bell, R. A. Creelman, and J. E. Mullet, A chloroplast lipoxygenase is required for wound-induced jasmonic acid accumulation in *Arabidopsis*. *Proc. Natl. Acad. Sci. USA* **92**, 8675 (1995).
117. C. Reinbothe, B. Parthier, and S. Reinbothe, Temporal pattern of jasmonate-induced alterations in gene expression of barley leaves. *Planta* **201**, 281 (1997).
118. S. Reinbothe, C. Reinbothe, C. Heintzen, C. Seidenbecher, and B. Parthier, A methyl jasmonate-induced shift in the length of the 5'untranslated region impairs translation of the plastid *rbcL* transcript in barley. *EMBO J.* **12**, 1505 (1993).
119. I. Wierstra and K. Kloppstech, Differential effects of methyl jasmonate on the expression of the early light-inducible proteins and other light-regulated genes in barley. *Plant Physiol.* **124**, 833 (2000).
120. E. Görschen, M. Dunaeva, I. Reeh, and C. Wasternack, Overexpression of the jasmonate inducible 23 kDa protein (JIP 23) from barley in transgenic tobacco leads to the repression of leaf proteins. *FEBS Lett.* **419**, 58 (1997).
121. R. A. Creelman and J. E. Mullet, Biosynthesis and action of jasmonates in plants. *Annu. Rev. Plant Physiol. Plant Mol. Biol.* **48**, 355 (1997).
122. B. Hause, U. Demus, C. Teichmann, B. Parthier, and C. Wasternack, Developmental and tissue-specific expression of JIP-23, a jasmonate-inducible protein of barley. *Plant Cell Physiol.* **37**, 641 (1996).
123. R. A. Weidhase, J. Lehmann, H. Kramell, G. Sembdner, and B. Parthier, Degradation of ribulose-1,5-bisphosphate carboxylase and chlorophyll in senescing barley leaf segments triggered by jasmonic acid methyl ester, and counteraction by cytokinin. *Physiol. Plant.* **69**, 161 (1987).
124. C. Wasternack, *in* "Plant Senescence and Apoptosis" (L. D. Noodén, ed.), in press Academic Press, New York, 2002.
125. C. A. Ryan, The search for the proteinase inhibitor inducing factor, PIIF. *Plant Mol. Biol.* **19**, 123 (1992).
126. J. Glazebrook, Genes controlling expression of defense responses in *Arabidopsis*—2001 status. *Curr. Opin. Plant Biol.* **4**, 301 (2001).

127. H. Bohlmann, The role of thionins in plant protection. *Crit. Rev. Plant Sci.* **13**, 1 (1994).
128. C. M. J. Pieterse and L. C. van Loon, Salicylic acid independent plant defence pathways. *Trends Plant Sci.* **4**, 52 (1999).
129. W. S. Chao, Y.-Q. Gu, V. Pautot, E. A. Bray, and L. L. Walling, Leucine aminopeptidase RNAs, proteins, and activities increase in response to water deficit, salinity, and the wound signals systemin, methyl jasmonate, and abscisic acid. *Plant Physiol.* **120**, 979 (1999).
130. P. E. Staswick, Storage proteins of vegetative plant tissue. *Annu. Rev. Plant Physiol. Plant Mol. Biol.* **45**, 303 (1994).
131. R. A. Weidhase, R. Kramell, J. Lehmann, H. W. Liebisch, W. Lerbs, and B. Parthier, Methyl jasmonate-induced changes in the polypeptide pattern of senescing barley leaf segments. *Plant Sci.* **51**, 177 (1987).
132. I. Andresen, W. Becker, K. Schlüter, J. Burges, B. Parthier, and K. Apel, The identification of leaf thionin as one of the main jasmonate-induced proteins of barley (*Hordeum vulgare*). *Plant Mol. Biol.* **19**, 193 (1992).
133. H. Bohlmann, S. Clausen, S. Behnke, H. Giese, C. Hiller, G. Schrader, V. Barkholt, and K. Apel, Leaf-specific thionins of barley—a novel class of cell wall proteins toxic to plant-pathogenic fungi and possibly involved in the defence mechanism of plants. *EMBO J.* **7**, 1559 (1988).
134. B. Hause, S. C. Hertel, D. Klaus, and C. Wasternack, Cultivar-specific expression of the jasmonate-induced protein of 23 kDa (JIP-23) occurs in *Hordeum vulgare* L. by jasmonate but not during seed germination. *Plant Biol.* **1**, 83 (1999).
135. J. Leopold, B. Hause, J. Lehmann, A. Graner, B. Parthier, and C. Wasternack, *Plant Cell Environ.* **19**, 675 (1996).
136. W. Becker and K. Apel, Isolation and characterization of a cDNA clone encoding a novel jasmonate-induced protein of barley (*Hordeum vulgare* L). *Plant Mol. Biol.* **19**, 1065 (1992).
137. B. Chaudhry, F. Müller-Uri, V. Cameron-Mills, S. Gough, D. Simpson, K. Skriver, and J. Mundy, The barley 60 kDa jasmonate-induced protein (JIP60) is a novel ribosome inactivating protein. *Plant J.* **6**, 815 (1994).
138. S. Reinbothe, C. Reinbothe, J. Lehmann, W. Becker, K. Apel, and B. Parthier, JIP60, a methyl jasmonate-induced ribosome-inactivating protein involved in plant stress reactions. *Proc. Natl. Acad. Sci. USA* **91**, 7012 (1994).
139. E. J. M. Van Damme, Q. Hao, Y. Chen, A. Barre, F. Vandenbussche, S. Desmyter, P. Rougé, and W. J. Peumans, Ribosome-inactivating proteins: A family of plant proteins that do more than inactivate ribosomes. *Crit. Rev. Plant. Sci.*, **20**, 395 (2002).
140. M. Dunaeva, C. Goebel, C. Wasternack, B. Parthier, and E. Görschen, The jasmonate-induced 60 kDa protein of barley exhibits N-glycosidase activity *in vivo. FEBS Lett.* **452**, 263 (1999).
141. E. Görschen, M. Dunaeva, B. Hause, I. Reeh, C. Wasternack, and B. Parthier, Expression of ribosome inactivating protein JIP60 from barley in transgenic tobacco leads to abnormal phenotype and alterations on the level of translation. *Planta* **202**, 470 (1997).
142. A. R. Menhaj, S. K. Mishra, S. Bezhani, and K. Kloppstech, Posttranscriptional control in the expression of the genes coding for high-light-regulated HL#2 proteins. *Planta* **209**, 406 (1999).
143. J. Lee, B. Parthier, and M. Löbler, Jasmonate signalling can be uncoupled from abscisic acid signalling in barley: identification of jasmonate-regulated transcripts which are not induced by abscisic acid. *Planta* **199**, 625 (1996).
144. K.-H. Kogel, B. Ortel, B. Jarosch, R. Atzorn, R. Schiffer, and C. Wasternack, Resistance in barley against the powdery mildew fungus (*Erysiphe graminis* f. sp. *hordei*) is not associated with enhanced levels of endogenous jasmonates. *Eur. J. Plant Pathol.* **110**, 319 (1995).
145. J. Lee, T. Vogt, B. Hause, and M. Löbler, Methyl jasmonate induces an O-methyltransferase in barley. *Plant Cell Physiol.* **38**, 851 (1997).

146. J. Rouster, R. Leah, J. Mundy, and V. Cameron-Mills, Identification of a methyl jasmonate-responsive region in the promoter of a lipoxygenase 1 gene expressed in barley grain. *Plant J.* **11**, 513 (1997).
147. E. Lam, F. Katagiri, and N.-H. Chua, Plant nuclear factor ASF-1 binds to an essential region of the nopaline synthase promoter. *J. Biol. Chem.* **265**, 9909 (1990).
148. A. Ishikawa, T. Yoshihara, and K. Nakamura, Jasmonate-inducible expression of a potato cathepsin D inhibitor-GUS gene fusion in tobacco cells. *Plant Mol. Biol.* **26**, 403 (1994).
149. H. S. Mason, D. B. DeWald, and J. E. Mullet, Identification of a methyl jasmonate-responsive domain in the soybean vspB promoter. *Plant Cell* **5**, 241 (1993).
150. S. R. Kim, J. L. Choi, M. A. Costa, and G. An, Identification of a G-box sequence as an essential element for methyl jasmonate response of potato proteinase inhibitor II promoter. *Plant Physiol.* **99**, 627 (1992).
151. M. Keil, J. J. Sánchez-Serrano, and L. Willmitzer, Both wound-inducible and tuber-specific expression are mediated by the promoter of a single member of the potato proteinase inhibitor II gene family. *EMBO J.* **8**, 1323 (1989).
152. R. Lorberth, C. Dammann, M. Ebneth, S. Amati, and J. J. Sánchez-Serrano, Promoter elements involved in environmental and developmental control of potato proteinase inhibitor II expression. *Plant J.* **2**, 477 (1992).
153. S. Blechert, W. Brodschelm, S. Hölder, L. Kammerer, T. M. Kutchan, M. J. Mueller, Z.-Q. Xia, and M. H. Zenk, The octadecanoic pathway: Signal molecules for the regulation of secondary pathways. *Proc. Natl. Acad. Sci. USA* **92**, 4099 (1995).
154. G. Haider, T. von Schrader, M. Füsslein, S. Blechert, and T. M. Kutchan, Structure-activity relationships of synthetic analogs of jasmonic acid and coronatine on induction of benzo[c] phenanthridine alkaloid accumulation in *Eschscholzia californica* cell cultures. *Biol. Chem.* **381**, 741 (2000).
155. F. L. H. Menke, S. Parchmann, M. J. Mueller, J. W. Kijne, and J. Memelink, Involvement of the octadecanoid pathway and protein phosphorylation in fungal elicitor-induced expression of terpenoid indole alkaloid biosynthetic genes in *Catharantus roseus*. *Plant Physiol.* **119**, 1289 (1999).
156. F. L. Menke, A. Champion, J. W. Kijne, and J. Memelink, A novel jasmonate- and elicitor-responsive element in the periwinkle secondary metabolite biosynthetic gene *Str* interacts with a jasmonate- and elicitor-inducible AP2-domain transcription factor, ORCA2. *EMBO J.* **18**, 4455 (1999).
157. L. Van der Fits, H. Zhang, F. L. Menke, M. Deneka, and J. Memelink, A *Catharanthus roseus* BPF-1 homologue interacts with an elicitor-responsive region of the secondary metabolite biosynthesis gene *Str* and is induced by elicitor via a jasmonate-independent signal transduction pathway. *Plant Mol. Biol.* **44**, 675 (2000).
158. L. Van der Fits and J. Memelink, ORCA3, a jasmonate-responsive transcriptional regulator of plant primary and secondary metabolism. *Science* **289**, 295 (2000).
159. J. L. Riechmann and E. M. Meyerowitz, The AP2/EREB family of plant transcription factors. *Biol. Chem.* **379**, 633 (1998).
160. L. Van der Fits and J. Memelink, The jasmonate-inducible AP2/ERF-domain transcription factor ORCA3 activates gene expression via interaction with a jasmonate-responsive promoter element. *Plant J.* **25**, 43 (2001).
161. D. Scheel and C. Wasternack, *in* "Plant Signal Transduction" (D. Scheel and C. Wasternack, eds.), in press. Oxford University Press, Oxford, 2002.
162. I. T. Baldwin, Jasmonate-induced responses are costly but benefit plants under attack in native populations. *Proc. Natl. Acad. Sci. USA* **95**, 8113 (1998).
163. K. L. Korth and R. A. Dixon, Evidence for chewing insect-specific molecular events distinct from a general wound response in leaves. *Plant Physiol.* **115**, 1299 (1997).

164. A. Schaller and C. Oecking, Modulation of the plasma membrane H^+-ATPase activity differentially activates wound and pathogen defense responses in tomato plants. *Plant Cell* **11**, 263 (1999).
165. G. Pearce, D. Strydom, S. Johnson, and C. A. Ryan, A polypeptide from tomato leaves induces wound-inducible proteinase inhibitor proteins. *Science* **253**, 995 (1991).
166. M. Orozco-Cardenas and C. A. Ryan, Hydrogen peroxide is generated systemically in plant leaves by wounding and systemin via the octadecanoid pathway. *Proc. Natl. Acad. Sci. USA* **96**, 6553 (1999).
167. M. Orozco-Cardenas, J. Narváez-Vásquez, and C. A. Ryan, Hydrogen peroxide acts as a second messenger for the induction of defense genes in tomato plants in response to wounding, systemin, and methyl jasmonate. *Plant Cell* **13**, 179 (2001).
168. H. Peña-Cortés, J. Fisahn, and L. Willmitzer, Signals involved in wound-induced proteinase inhibitor II gene expression in tomato and potato plants. *Proc. Natl. Acad. Sci. USA* **92**, 4106 (1995).
169. P. J. O'Donnell, C. Calvert, R. Atzorn, C. Wasternack, H. M. O. Leyser, and D. J. Bowles, Ethylene as a signal mediating the wound response of tomato plants. *Science* **274**, 1914 (1996).
170. E. E. Farmer and C. A. Ryan, Octadecanoid precursors of jasmonic acid activate the synthesis of wound-inducible proteinase inhibitors. *Plant Cell* **4**, 129 (1992).
171. M. Malone and J. J. Alarcon, Only xylem-borne factors can account for systemic wound signalling in the tomato. *Planta* **196**, 740 (1995).
172. K. Wildon, J. F. Thain, P. E. H. Minchin, I. R. Gubb, A. J. Reilly, Y. D. Skipper, H. M. Doherty, P. J. O'Donnell, and D. J. Bowles, Electrical signaling and systemic proteinase inhibitor induction in the wounded plant. *Nature* **360**, 62 (1992).
173. B. Stankovic and E. Davies, Both action potential and variation potential induce proteinase inhibitor gene expression in tomato. *FEBS Lett.* **390**, 275 (1996).
174. B. McGurl, G. Pearce, M. Orozco-Cardenas, and C. A. Ryan, Structure, expression, and antisense inhibition of the systemin precursor gene. *Science* **255**, 1570 (1992).
175. J. E. Dombrowski, G. Pearce, and C. A. Ryan, Proteinase inhibitor-inducing activity of the prohormone prosystemin resides exclusively in the C-terminal systemin domain. *Proc. Natl. Acad. Sci. USA* **96**, 12947 (1999).
176. M. Vetsch, I. Janzik, and A. Schaller, Characterization of prosystemin expressed in the baculovirus/insect cell system reveals biological activity of the systemin precursor. *Planta* **200**, 91 (2000).
177. T. Jacinto, B. McGurl, V. Franceschi, J. Delano-Freier, and C. A. Ryan, Tomato prosystemin promoter confers wound-inducible, vascular bundle-specific expression of the β-glucuronidase gene in transgenic tomato plants. *Planta* **203**, 406 (1997).
178. B. McGurl, M. Orozco-Cardenas, G. Pearce, and C. A. Ryan, Over-expression of the prosystemin gene in transgenic tomato plants generates a systemic signal that constitutively induces proteinase inhibitor synthesis. *Proc. Natl. Acad. Sci. USA* **91**, 9799 (1994).
179. C. A. Ryan and G. Pearce, Polypeptide hormones. *Plant Physiol.* **125**, 65 (2001).
180. C. P. Constabel, L. Yip, and C. A. Ryan, Prosystemin from potato, black nightshade, and bell pepper: primary structure and biological activities of the predicted systemins. *Plant Mol. Biol.* **26**, 55 (1998).
181. G. Pearce, D. S. Moura, J. Stratmann, and C. A. Ryan, Production of multiple plant hormones from a single polyprotein precursor. *Nature* **411**, 817 (2001).
182. J. Narváez-Vásquez, G. Pearce, M. Orozco-Cardenas, V. R. Franceschi, and C. A. Ryan, Autoradiographic and biochemical evidence for the systemic translocation of systemin in tomato plants. *Planta* **195**, 593 (1995).

183. T. Meindl, T. Boller, and G. Felix, The plant wound hormone systemin binds with the N-terminal part to its receptor but needs the C-terminal part of activate it. *Plant Cell* **10**, 1561 (1998).
184. G. Felix and T. Boller, Systemin induces rapid ion fluxes and ethylene biosynthesis in *Lycopersicon peruvianum* cells. *Plant J.* **7**, 381 (1995).
185. J. M. Scheer and C. A. Ryan, A 160-kD systemin receptor on the surface of *Lycopersicon peruvianum* suspension-cultured cells. *Plant Cell* **11**, 1525 (1999).
186. J. W. Stratmann, J. Scheer, and C. A. Ryan, Suramin inhibits initiation of defense signaling by systemin, chitosan, and a β-glucan elicitor in suspension-cultured *Lycopersicon peruvianum* cells. *Proc. Natl. Acad. Sci. USA* **94**, 11085 (2000).
187. M. J. Stennis, S. Chandra, C. A. Ryan, and P. S. Low, Systemin potentiates the oxidative burst in cultured tomato cells. *Plant Physiol.* **117**, 1031 (1998).
188. C. Moyen and E. Johannes, Systemin transiently depolarizes the tomato mesophyll cell membrane and antagonizes fusicoccin-induced extracellular acidification of mesophyll tissue. *Plant Cell Environ.* **19**, 464 (1996).
189. C. Moyen, K. E. Hammond-Kosack, J. J. Jones, M. R. Knight, and E. Johannes, Systemin triggers an increase of cytoplasmic calcium in tomato mesophyll cells: Ca^{2+} mobilization from intra- and extracellular compartments. *Plant Cell Environ.* **21**, 110 (1998).
190. J. W. Stratmann and C. A. Ryan, Myelin basic protein kinase activity in tomato leaves is induced systemically by wounding and increases in response to systemin and oligosaccharide elicitors. *Proc. Natl. Acad. Sci. USA* **94**, 11085 (1997).
191. D. R. Bergey, M. Orozco-Cardenas, D. S. Moura, and C. A. Ryan, A wound- and systemin-inducible polygalacturonase in tomato leaves. *Proc. Natl. Acad. Sci. USA* **96**, 1756 (1999).
192. D. R. Bergey and C. A. Ryan, Wound- and systemin-inducible calmodulin gene expression in tomato leaves. *Plant Mol. Biol.* **40**, 815 (1999).
193. A. Conconi, M. Miquel, J. A. Browse, and C. A. Ryan, Intracellular levels of free linolenic and linoleic acids increase in tomato leaves in response to wounding. *Plant Physiol.* **111**, 797 (1996).
194. M. Orozco-Cardenas, B. McGurl, and C. A. Ryan, Expression of an antisense prosystemin gene in tomato plants reduces resistance toward *Manduca sexta* larvae. *Proc. Natl. Acad. Sci. USA* **90**, 8273 (1993).
195. G. A. Howe and C. A. Ryan, Suppressors of systemin signaling identify genes in the tomato wound response pathway. *Genetics* **153**, 1411 (1999).
196. Y. Xu, P.-F. L. Chang, D. Liu, M. L. Narasimhan, K. G. Raghothama, P. M. Hasegawa, and R. A. Bressan, Plant defense genes are synergistically induced by ethylene and methyl jasmonate. *Plant Cell* **6**, 1077 (1994).
197. R. A. Winz and I. T. Baldwin, Molecular interactions between the specialist herbivore *Manduca sexta* (Lepidoptera, Sphingidae) and its natural host *Nicotiana attenuata*. IV. Insect-induced ethylene reduces jasmonate-induced nicotine accumulation by regulating putrescine *N*-methyltransferase transcripts. *Plant Physiol.* **125**, 2189 (2001).
198. S. Hiraga, H. Ito, K. Sasaki, H. Yamakawa, I. Mitsuhara, H. Toshima, H. Matsui, M. Honma, and Y. Ohashi, Wound-induced expression of a tobacco peroxidase is not enhanced by ethephon and suppressed by methyl jasmonate and coronatine. *Plant Cell Physiol.* **41**, 165 (2000).
199. E. Rojo, J. León, and J. J. Sánchez-Serrano, Cross-talk between wound signalling pathways determines local versus systemic gene expression in *Arabidopsis thaliana*. *Plant J.* **20**, 135 (1999).
200. I. A. Penninckx, K. Eggermont, F. R. Terras, B. P. Thomma, G. W. De Samblanx, A. Buchala, J.-P. Métraux, J. M. Manners, and W. F. Broekhaert, Pathogen-induced systemic activation of a plant defensin gene in *Arabidopsis* follows a salicylic acid-independent pathway. *Plant Cell* **8**, 2309 (1996).

201. I. A. M. A. Penninckx, B. P. H. J. Thomma, A. Buchala, J.-P. Métraux, and W. F. Broekaert, Concomitant activation of jasmonate and ethylene response pathways is required for induction of a plant defensin gene in *Arabidopsis*. *Plant Cell* **10**, 2103 (1998).
202. E. Carrera and S. Prat, Expression of the *Arabidopsis* abi1-1 mutant allele inhibits proteinase inhibitor wound-induction in tomato. *Plant J.* **15**, 765 (1998).
203. G. F. Birkenmeier and C. A. Ryan, Wound signaling in tomato plants. *Plant Physiol.* **117**, 687 (1998).
204. O. Herde, R. Atzorn, J. Fisahn, C. Wasternack, L. Willmitzer, and H. Peña-Cortés, Localized wounding by heat initiates the accumulation of proteinase inhibitor II in abscisic acid-deficient plants by triggering jasmonic acid biosynthesis. *Plant Physiol.* **112**, 853 (1996).
205. S. H. Doares, T. Syrovets, E. W. Weiler, and C. A. Ryan, Oligogalacturonides and chitosan activate plant defensive genes through the octadecanoid pathway. *Proc. Natl. Acad. Sci. USA* **92**, 4095 (1995).
206. T. Genoud and J.-P. Métraux, Crosstalk in plant cell signaling: structure and function of the genetic network. *Trends Plant Sci.* **4**, 503 (1999).
207. T. Genoud, M. B. T. Santa Cruz, and J.-P. Métraux, Numeric simulation of plant signaling networks. *Plant Physiol.* **126**, 1430 (2001).
208. R. Karban and I. T. Baldwin, "Induced Responses to Herbivory." University Chicago Press, Chicago, 1997.
209. H. Koiwa, R. A. Brassan, and P. M. Hasegawa, Regulation of proteinase inhibitors and plant defense. *Trends Plant Sci.* **2**, 379 (1997).
210. I. T. Baldwin and C. A. Preston, The eco-physiological complexity of plant responses to insect herbivores. *Planta* **208**, 137 (1999).
211. I. T. Baldwin, Z.-P. Zhang, N. Diab, T. E. Ohnmeiss, E. S. McCloud, G. Y. Lynds, and E. A. Schmelz, Quantification, correlations and manipulations of wound-induced changes in jasmonic acid and nicotine in *Nicotiana sylvestris*. *Planta* **201**, 397 (1997).
212. J. Kahl, D. H. Siemens, R. J. Aerts, R. Gäbler, and F. Kühnemann, Herbivore-induced ethylene supresses a direct defense but not a putative indirect defense against an adapted herbivore. *Planta* **1210**, 336 (2000).
213. W. T. G. van de Ven, C. S. LeVesque, T. M. Perring, and L. L. Walling, Local and systemic changes in squash gene expression in response to silverleaf whitefly feeding. *Plant Cell* **12**, 1409 (2000).
214. T. C. J. Turlings, J. H. Oughrin, P. J. McCall, U. S. R. Röse, W. J. Lewis, and J. H. Tumlinson, How caterpillar-damaged plants protect themselves by attracting parasitic wasps. *Proc. Natl. Acad. Sci. USA* **92**, 4169 (1995).
215. G. Vancanneyt, C. Sanz, T. Farmaki, M. Paneque, F. Ortego, P. Castanera, and J. J. Sanchez-Serrano, Hydroperoxide lyase depletion in transgenic potato plants leads to an increase in aphid performance. *Proc. Natl. Acad. Sci. USA* **98**, 139 (2001).
216. T. Koch, T. Krumm, V. Jung, J. Engelberth, and W. Boland, Differential induction of plant volatile biosynthesis in the lima bean by early and late intermediates of the octadecanoid-signaling pathway. *Plant Physiol.* **121**, 153 (1999).
217. R. Halitschke, U. Schittko, G. Pohnert, W. Boland, and I. T. Baldwin, Molecular interactions between the specialist herbivore *Manduca sexta* (Lepidoptera, Sphingidae) and its natural host *Nicotiana attenuata*. III. Fatty acid-amino acid conjugates in herbivore oral secretions are necessary and sufficient for herbivore-specific plant responses. *Plant Physiol.* **125**, 711 (2001).
218. H. T. Alborn, T. C. J. Turlings, T. H. Jones, G. Stenhagen, J. H. Loughrin, and J. H. Tumlinson, An elicitor of plant volatiles from beet armyworm oral secretion. *Science* **276**, 945 (1997).
219. D. Spiteller, K. Dettner, and W. Boland, Gut bacterial may be involved in interactions between plants, herbivores and their predators: microbial biosynthesis of *N*-acylglutamine surfactants as elicitors of plant volatiles. *Biol. Chem.* **381**, 755 (2000).

220. C. M. De Morales, W. J. Lewis, P. W. Paré, H. T. Alborn, and J. H. Tumlinson, Herbivore-infested plants selectively attract parasitoids. *Nature* **393**, 570 (1998).
221. I. T. Baldwin, R. Halitschke, A. Kessler, and U. Schittko, Merging molecular and ecological approaches in plant-insect interactions. *Curr. Opin. Plant Biol.* **4**, 351 (2001).
222. A. Kessler and I. T. Baldwin, Defensive function of herbivore-induced plant volatile emissions in nature. *Science* **291**, 2141 (2001).
223. G.-I. Arimura, R. Ozawa, T. Shimoda, T. Nishioka, W. Boland, and J. Takabayashi, Herbivory-induced volatiles elicit defence genes in lima bean leaves. *Nature* **406**, 512 (2000).
224. E. E. Farmer and C. A. Ryan, Interplant communication: airborne methyl jasmonate induces synthesis of proteinase inhibitors in plant leaves. *Proc. Natl. Acad. Sci. USA* **87**, 7713 (1990).
225. R. Karban, I. T. Baldwin, K. J. Baxter, and G. Laue, Communication between plants: induced resistance in wild tobacco plants following clipping of neighbouring sagebrush. *Oecologia* **125**, 66 (2000).
226. E. E. Farmer, Surface-to-air signals. *Nature* **411**, 854 (2001).
227. S. E. Smith and D. J. Read, "Mycorrhizal Symbiosis." Academic Press, San Diego, California, (1997).
228. M. Harrison, Molecular and cellular aspects of the arbuscular mycorrhizal symbiosis. *Annu. Rev. Physiol. Plant Mol. Biol.* **50**, 363 (1999).
229. M. Regvar, N. Gogala, and P. Zalar, Effects of jasmonic acid on myrorrhizal *Allium sativum*. *New Phytol.* **134**, 703 (1996).
230. J. Ton, C. M. Pieterse, and L. C. Van Loon, Identification of a locus in *Arabidopsis* controlling both the expression of rhizobacteria-mediated induced systemic resistance (ISR) and basal resistance against *Pseudomonas syringae* pv. tomato. *Mol. Plant Microbe Interact.* **12**, 911 (1999).
231. S. C. van Wees, E. A. de Swart, J. A. van Pelt, L. C. van Loon, and C. M. Pieterse, Enhancement of induced disease resistance by simultaneous activation of salicylate- and jasmonate-dependent defense pathway in *Arabidopsis thaliana*. *Proc. Natl. Acad. Sci. USA* **97**, 8711 (2000).
232. J. M. Park, C.-J. Park, S.-B. Lee, B.-K. Ham, R. Shin, and K.-H. Paek, Overexpression of the tobacco *Tsi1* gene encoding an EREB/AP2-type transcription factor enhances resistance against pathogen attack and osmotic stress in tobacco. *Plant Cell* **13**, 1035 (2001).
233. P. Epple, K. Apel, and H. Bohlmann, An *Arabidopsis thaliana* thionin gene is inducible via a signal transduction pathway different from that for pathogenesis-related proteins. *Plant Physiol.* **109**, 813 (1995).
234. H. Bohlmann, A. Vignutelli, B. Hilpert, O. Miersch, C. Wasternack, and K. Apel, Wounding and chemicals induce expression of the *Arabidopsis thaliana* gene *Thi2.1*, encoding a fungal defense thionin, via the octadecanoid pathway. *FEBS Lett.* **437**, 281 (1998).
235. G. Petersen, P. Brodersen, H. Naested, E. Andreasson, U. Lindhart, B. Johansen, H. B. Nielsen, M. Lacy, M. J. Austin, J. E. Parker, S. B. Sharma, D. F. Klessig, R. Martienssen, O. Mattsson, A. B. Jensen, and J. Mundy, *Arabidopsis* MAP kinase 4 negatively regulates systemic acquired resistance. *Cell* **103**, 1111 (2000).
236. B. J. Feys, L. J. Moisan, M.-A. Newman, and J. E. Parker, Direct interaction between the *Arabidopsis* disease resistance signaling proteins, EDS1 and PAD4. *EMBO J.* **20**, 5400 (2001).
237. X. Dong, SA, JA, ethylene, and disease resistance in plants. *Curr. Opin. Plant Biol.* **1**, 316 (1998).
238. J. Glazebrook, Genes controlling expression of defense responses in *Arabidopsis*. *Curr. Opin. Plant Biol.* **2**, 280 (1999).
239. P. Reymond and E. E. Farmer, Jasmonate and salicylate as global signals for defense gene expression. *Curr. Opin. Plant Biol.* **1**, 404 (1998).

240. R. A. Creelman and J. E. Mullet, Jasmonic acid distribution and action in plants: Regulation during development and response to biotic and abiotic stress. *Proc. Natl. Acad. Sci. USA* **92**, 4114 (1995).
241. T. Roitsch, R. Ehness, M. Goetz, B. Hause, M. Hofmann, and A. K. Sinha, Regulation and function of extracellular invertase from higher plants in relation to assimilate partitioning, stress responses and sugar signalling. *Austr. J. Plant Physiol.* **27**, 815 (2000).
242. H. Yamane, H. Takagi, H. Abe, T. Yokota, and N. Takahashi, Identification of jasmonic acid in three species of higher plants and its biological activities. *Plant Cell Physiol.* **22**, 689 (1981).
243. P. E. Staswick, W. Su, and S. H. Howell, Methyl jasmonate inhibition of root growth and induction of a leaf protein are decreased in an *Arabidopsis thaliana* mutant. *Proc. Natl. Acad. Sci. USA* **89**, 6837 (1992).
244. C. E. Benedetti, D. Xie, and J. G. Turner, *COI1*-dependent expression of an *Arabidopsis* vegetative storage protein in flowers and siliques and in response to coronatine or methyl jasmonate. *Plant Physiol.* **109**, 567 (1995).
245. S. Berger, E. Bell, and J. E. Mullet, Two methyl jasmonate-insensitive mutants show altered expression of *AtVsp* in response to methyl jasmonate and wounding. *Plant Physiol.* **111**, 525 (1996).
246. H. Yamane, H. Abe, and N. Takahashi, Jasmonic acid and methyl jasmonate in pollen and anthers of three *Camelia* species. *Physiol. Plant.* **52**, 305 (1982).
247. S.-A. Kim, H. J. Kwak, M.-C. Park, and S.-R. Kim, Induction of reproductive organ-preferential histone genes by wounding or methyl jasmonate. *Mol. Cell* **8**, 669 (1998).
248. H. Peña-Cortés, L. Willmitzer, and J. J. Sánchez-Serrano, Abscisic acid mediates wound induction but not developmental-specific expression of the proteinase inhibitor II gene family. *Plant Cell* **3**, 963 (1991).
249. T. Lotan, N. Ori, and R. Fluhr, Pathogenesis-related proteins are developmentally regulated in tobacco flowers. *Plant Cell* **1**, 881 (1989).
250. A. D. Neale, J. A. Wahleithner, M. Lund, H. T. Bonnett, A. Kelly, D. R. Meeks-Wagner, W. J. Peacock, and E. S. Dennis, Chitinase, β-1,3-glucanase, osmotin, and extensin are expressed in tobacco explants during flower formation. *Plant Cell* **2**, 673 (1990).
251. S. Lantin, M. O'Brien, and D. P. Matton, Pollination, wounding and jasmonate treatments induce the expression of a developmentally regulated pistil dioxygenase at a distance, in the ovary, in the wild potato *Solanum chacoense*. *Plant Mol. Biol.* **41**, 371 (1999).
252. A. N. Capella, M. Menossi, P. Arruda, and C. E. Benedetti, COI1 affects myrosinase activity and controls the expression of two flower-specific myrosinase-binding protein homologues in *Arabidopsis*. *Planta* **213**, 691 (2001).
253. A. Samach, L. Broday, D. Hareven, and E. Lifschitz, Expression of an amino acid biosynthesis gene in tomato flowers: developmental upregulation and MeJA response are parenchyma-specific and mutually compatible. *Plant J.* **8**, 391 (1995).
254. J. F. Feys, C. E. Benedetti, C. N. Penfold, and J. G. Turner, *Arabidopsis* mutants selected for resistance to the phytotoxin coronatine are male sterile, insensitive to methyl jasmonate, and resistant to a bacterial pathogen. *Plant Cell* **6**, 751 (1994).
255. M. Rodríguez-Concepción and J. P. Beltrán, Repression of the pea lipoxygenase gene *loxg* is associated with carpel development. *Plant Mol. Biol.* **27**, 887 (1995).
256. M. Fukuchi-Mizutani, K. Ishiguro, T. Nakayama, Y. Ustunomiya, Y. Tanaka, T. Kusumi, and T. Ueda, Molecular and functional characterization of a rose lipoxygenase cDNA related to flower senescence. *Plant Sci.* **160**, 129 (2000).
257. I. Kubigsteltig, D. Laudert, and E. W. Weiler, Structure and regulation of the *Arabidopsis thaliana* allene oxide synthase gene. *Planta* **208**, 463 (1999).

258. C. L. Costa, P. Arruda, and C. E. Benedetti, An *Arabidopsis* gene induced by wounding functionally homologous to flavoprotein oxidoreductases. *Plant Mol. Biol.* **44**, 67 (2000).
259. M. Yokoyama, S. Yamaguchi, S. Inomata, K. Komatsu, and S. Yoshida, Stress-induced factor involved in flower formation of Lemna is an α-ketol derivative of linolenic acid. *Plant Cell Physiol.* **41**, 110 (2000).
260. E. E. Ewing and P. C. Struik, Tuber formation in potato: Induction, initiation and growth. *Hortic. Rev.* **14**, 89 (1992).
261. Y. Koda and Y. Okazawa, Characteristic changes in the levels of endogenous plant hormones in relation to the onset of potato tuberization. *Jpn. J. Crop Sci.* **52**, 592 (1983).
262. U. Sonnewald, Control of potato tuber sprouting. *Trends Plant Sci.* **6**, 333 (2001).
263. Y. Okazawa, Studies on the relation between the tuber formation of potato and its natural gibberellin content. *Proc. Crop. Sci. Soc. Jpn.* **29**, 121 (1960).
264. G. Abdala, G. Castro, M. Guinazu, R. Tizio, and O. Miersch, Occurrence of jasmonic acid in organs of *Solanum tuberosum* L. c.v. Spunta and its effect on tuberization. *Plant Growth Regul.* **19**, 139 (1996).
265. K. Fujino, Y. Koda, and Y. Kikuta, Reorientation of cortical microtubules in the subapical region during tuberization in single-node stem segments of potato in culture. *Plant Cell Physiol.* **36**, 891 (1995).
266. Y. Koda, Changes in the levels of butanol- and water-soluble cytokinins during the life cycle of potato tubers. *Plant Cell Physiol.* **23**, 843 (1982).
267. L. Gregory, Some factors for tuberization in the potato plant. *Am. J. Bot.* **43**, 281 (1956).
268. T. Yoshihara, E.-L. A. Omer, H. Koshino, S. Sakamura, Y. Kikuta, and Y. Koda, Structure of a tuber-inducing stimulus from potato leaves (*Solanum tuberosum* L.). *Agric. Biol. Chem.* **53**, 2835 (1989).
269. Y. Koda, The role of jasmonic acid and related compounds in the regulation of plant development. *Int. Rev. Cytol.* **135**, 155 (1992).
270. T. Yoshihara, M. Amanuma, T. Tsutsumi, Y. Okumura, H. Matsuura, and A. Ichihara, Metabolism and transport of [2-^{14}C]($\pm$) jasmonic acid in the potato plant. *Plant Cell Physiol.* **37**, 586 (1996).
271. H. Helder, O. Miersch, D. Vreugdenhil, and G. Sembdner, Occurrence of hydroxylated jasmonic acids in leaflets of *Solanum demissum* plants grown under long- and short-day conditions. *Physiol. Plant.* **88**, 647 (1993).
272. Y. Koda, Y. Kikuta, H. Tazaki, Y. Tsujino, S. Sakamura, and T. Yoshihara, Potato tuber-inducing activities of jasmonic acid and related compounds. *Phytochemistry* **30**, 1435 (1991).
273. K. Takahashi, K. Fujino, Y. Kikuta, and Y. Koda, Expansion of potato cells in response to jasmonic acid. *Plant Sci.* **100**, 3 (1994).
274. M. V. Kolomiets, D. J. Hannapel, H. Chen, M. Tymeson, and R. J. Gladon, Lipoxygenase is involved in the control of potato tuber development. *Plant Cell* **13**, 613 (2001).
275. M. Hamberg, New cyclopentenone fatty acids formed from linoleic and linolenic acids in potato. *Lipids* **35**, 353 (2000).
276. B. Klüsener, G. Boheim, H. Liss, J. Engelberth, and E. W. Weiler, Gadolinium-sensitive, voltage-dependent calcium release channels in the endoplasmic reticulum of a higher plant mechanoreceptor organ. *EMBO J.* **14**, 2708 (1995).
277. R. Ehret, J. Schab, and E. W. Weiler, Lipoxygenases in *Bryonia dioica* Jacq. *Planta* **144**, 175 (1994).
278. E. Falkenstein, B. Groth, A. Mithofer, and E. W. Weiler, Methyljasmonate and α-linolenic acid are potent inducers of tendril coiling. *Planta* **185**, 316 (1991).
279. E. W. Weiler, T. M. Kutchan, T. Gorba, W. Brodschelm, U. Niesel, and F. Bublitz, The *Pseudomonas* phytotoxin coronatine mimics octadecanoid signalling molecules of higher plants. *FEBS Lett.* **345**, 9 (1994).

280. S. Blechert, C. Bockelmann, M. Füsslein, T. v. Schrader, B. Stelmach, U. Niesel, and E. W. Weiler, Structure activity analyses reveal the existence of two separate groups of active octadecanoids in elicitation of the tendril-coiling response of *Bryonia dioica* Jacq. *Planta* **207**, 470 (1999).
281. E. W. Weiler, Grundlagen der pflanzlichen Mechanosensorik. *Naturw. Rundsch.* **50**, 337 (1997).
282. L. D. Noodén, in "Aging in Plants" (L. N. Noodén and A. C. Leopold, eds.), p. 1. Academic Press, San Diego, California, 1988.
283. J. L. Dangl, R. A. Dietrich, and H. Thomas, in "Biochemistry and Molecular Biology of Plants" (B. Buchanan, W. Gruissem, and R. Jones, eds.), p. 1044. American Society of Plant Physiologists, Rockville, Maryland, 2000.
284. L. D. Noodén and J. J. Guiamét, in "Handbook of the Biology of Aging" (E. L. Schneider and J. W. Rowe, eds.), p. 94. Academic Press, Orlando, Florida, 1996.
285. L. D. Noodén, J. J. Guiamét, and I. John, Senescence mechanisms. *Physiol. Plant.* **101**, 746 (1997).
286. C. M. Smart, Gene expression during leaf senescence. *New Phytol.* **126**, 419 (1994).
287. B. F. Quirino, Y.-S. Noh, E. Himelblau, and R. M. Amasino, Molecular aspects of leaf senescence. *Trends Plant Sci.* **5**, 278 (2000).
288. S. Berger, H. Weichert, A. Porzel, C. Wasternack, H. Kühn, and I. Feusner, Enzymatic and non-enzymatic lipid peroxidation in leaf development. *Biochim. Biophys. Acta* **1533**, 266 (2001).
289. T. Tsuchiya, H. Ohta, K. Okawa, A. Iwamastu, H. Shimada, T. Masuda, and K. Takamiya, Cloning of chlorophyllase, the key enzyme in chlorophyll degradation: finding of a lipase motif and the induction by methyl jasmonate. *Proc. Natl. Acad. Sci. USA* **96**, 15362 (1999).
290. P. Matile, S. Hörtensteiner, and H. Thomas, Chlorophyll degradation. *Annu. Rev. Plant Physiol. Plant Mol. Biol.* **50**, 67 (1999).
291. J. J. Guiamét, E. Pichersky, and L. D. Noodén, Mass exodus from senescing soybean chloroplasts. *Plant Cell Physiol.* **40**, 986 (1999).
292. S. Gan and R. M. Amasino, Inhibition of leaf senescence by autoregulated production of cytokinin. *Science* **270**, 1986 (1995).
293. V. Buchanan-Wollaston and C. Ainsworth, Leaf senescence in *Brassica napus*: cloning of senescence related genes by subtractive hybridization. *Plant Mol. Biol.* **3**, 821 (1997).
294. J.-H. Park, S. A. Oh, Y. H. Kim, H. R. Woo, and H. G. Nam, Differential expression of senescence-associated mRNAs during leaf senescence induced by different senescence-inducing factors in *Arabidopsis*. *Plant Mol. Biol.* **37**, 445 (1998).
295. L. M. Weaver, A comparison of the expression patterns of several senescence-associated genes in response to stress and hormone treatment. *Plant Mol. Biol.* **37**, 455 (1998).
296. Y. He, W. Tang, J. D. Swain, G. A.L. T. P. Jack, and S. Gan, Networking senescence-regulating pathways by using *Arabidopsis* enhancer trap lines. *Plant Physiol.* **126**, 707 (2001).
297. H. R. Woo, K. M. Chung, J.-H. Park, S. A. Oh, T. Ahn, S. H. Hong, S. K. Jang, and H. G. Nam, ORE9, an F-box protein that regulates leaf senescence in *Arabidopsis*. *Plant Cell* **13**, 1779 (2001).
298. J. C. del Pozo and M. Estelle, F-box proteins and protein degradation: An emerging theme in cellular regulation. *Plant Mol. Biol.* **44**, 123 (2000).
299. W. Xiao and J.-C. Jang, F-box proteins in *Arabidopsis*. *Trends Plant Sci.* **5**, 454 (2000).

The Ubiquitous Nature of RNA Chaperone Proteins

GAËL CRISTOFARI AND
JEAN-LUC DARLIX

LaboRetro, Unité de Virologie Humaine
INSERM #412
Ecole Normale Supérieure de Lyon
69364 Lyon Cedex, France

I. Introduction... 224
II. RNA Structure and the Folding Problem 226
   A. Secondary and Tertiary Structures............................. 226
   B. RNA Misfolding: Kinetic and Thermodynamic Problems............ 228
   C. Role of Metal Ions... 228
III. RNA-Binding Proteins 229
   A. RNA Binding Rules... 229
   B. RNA-Binding Proteins and the RNA Folding Problem 231
   C. RNA Binding Motifs Found in RNA Chaperone Proteins 233
IV. Investigating the Biochemical Properties of Nucleic Acid Chaperone
   Proteins 234
   A. Preliminary Experiments 234
   B. Simple Assays for Identifying Proteins with Nucleic Acid Chaperone
      Properties 235
   C. Advanced Assays for Examining Properties of Nucleic Acid
      Chaperone Proteins 239
   D. Annealing versus Unwinding: Conditions for Investigating Chaperone
      Properties 245
V. How Do RNA Chaperone Proteins Work? 246
   A. RNA Occupancy Level and Window of Activity of RNA Chaperones ... 246
   B. RNA Chaperones in Action.................................... 246
   C. Structural and Functional Diversity of RNA Chaperone Proteins 250
VI. Conclusions and Future Prospects............................... 258
   References.. 263

RNA chaperones are ubiquitous and abundant proteins found in all living organisms and viruses, where they interact with various classes of RNA. These highly diverse families of nucleic acid-binding proteins possess activities enabling rapid and faithful RNA–RNA annealing, strand transfer, and exchange and RNA ribozyme-mediated cleavage under physiological conditions. RNA chaperones appear to be critical to functions as important as maintenance of chromosome ends, DNA transcription, preRNA export, splicing and modifications, and mRNA translation and degradation. Here we review some of the properties of RNA chaperones in RNA–RNA interactions that take place during cellular processes and retrovirus replication. Examples of cellular and viral proteins are dicussed vis à vis the relationships between RNA chaperone activities *in vitro* and functions. In this new "genomic era" we discuss the possible use of small RNA chaperones to improve the synthesis of cDNA libraries for use in large screening reactions using DNA chips. © 2002, Elsevier Science (USA).

I. Introduction

A crystal is the classic archetype of repeated order, symmetry, a structure of permanence, dense and compacted. Smoke, on the other hand, is in a state of constant flux, unpredictable in its movements, endlessly changing shape and direction. Crystal and smoke, like order and entropy, the regular and the chaotic— Henri Atlan, *Entre le cristal et la fumée*, Seuil, Paris, 1979.

RNA and chaperones have their smoky moments when on their own, and their crystal-like periods when they come together.

RNA is a central player in life, and probably is always associated with proteins in modern living organisms. The biological functions of RNA are very diverse, from cell DNA maintenance to genetic expression in the form of proteins and enzymes. In eukaryotes, RNA is critical for the maintenance of DNA ends by telomerases (*1–3*) and the reshuffling of the DNA genome through replication of a large family of mobile genetic elements called retroelements (*4, 5*).

As a protein-coding molecule, RNA is a vector between the genetic DNA material of the cell and the decoding ribosome machinery. In eukaryotes, the coding RNA is transported in the form of heterogeneous nuclear ribonucleoparticles (hnRNPs) from the nucleus, where they are synthesized, to the cytoplasm, where they are translated by ribosomes, which are themselves highly specialized RNPs. During transportation within the cell, either coding or noncoding RNAs are posttranscriptionally modified. To this end they interact with small, noncoding RNAs known as small nuclear and cytoplasmic RNAs in the form of small RNPs. These events are known as splicing, editing, and nucleotide modification and appear to be essential for RNA decoding by ribosomes and additional regulation (*6–8*).

RNA can also have a catalytic activity that is critical in group I and II intron elimination (*9*) and perhaps in the spliceosome (*10*) and in tRNA (*11, 12*) and viral RNA maturation (*13*). Recently the peptidyl transferase activity of the large ribosomal subunit was directly attributed to the 23S rRNA (*14, 15*), and therefore the ribosome itself can be considered as a ribozyme.

In the virus world, RNA is the genetic material of many viruses, for example, that of the widespread family of retroviruses and retroelements found as multiple copies in the genome of vertebrates (*16, 17*).

Cellular RNAs can be classified into major types according to their function, such as transfer RNA (tRNA), small nuclear, nucleolar, and cytoplasmic RNA (snRNA, snoRNA, and scRNA respectively), ribosomal RNA (rRNA), and pre- and messenger RNA (mRNA) for protein-coding RNAs as well as the recently discovered large family of microRNAs (miRNAs), which includes small temporal RNAs (stRNAs) and small interfering RNAs (siRNAs) 21–24 nucleotides (nt) in length (*18–20*), the role of which is to regulate the expression of mRNAs. Virus RNAs form distinct classes of RNA, and can be single- or double-stranded and either coding or noncoding. Viroid RNAs are very unusual in that they are single-stranded and circular and contain a large number of highly modified residues (*13, 21*).

As already pointed out, RNAs have many essential functions in the cell and this is due, at least in part, to their ability to form functionally active secondary and tertiary structures. On the other hand, double-stranded DNA is much less reactive and is the reliable carrier of the genetic information.

Single-stranded RNAs are flexible macromolecules and can take a wide variety of alternative conformations. However, only a given conformation is functionally relevant. This highlights the RNA folding problem, similar in nature to that of protein folding, albeit much simpler, since there are only four residues. Within the RNA, interactions are governed by hydrogen bonding and base stacking. With regard to RNA folding, five major types of secondary structures govern the process: helices with canonical and noncanonical base pairings, loops, bulges, junctions, and pseudoknots (see Fig. 1).

RNA folding, especially that of large RNAs, must be assisted in order for the RNA to reach the biologically relevant conformation and not be trapped in one of the many potential and incorrect structures. There are many RNA-binding proteins that recognize RNA with a broad specificity, also termed nonspecific nucleic acid-binding proteins (NABP) (*22*). These proteins have a biological role as RNA chaperones and are thought to direct correct folding of RNA by preventing and resolving misfoldings (see below) (*23*).

Two major classes of RNA chaperones have been identified: (1) The cofactor or ligand proteins recognize and capture a correctly folded RNA conformation that is unstable. Alternatively, the proteins bind to the unfolded RNA and nucleate the tertiary structure. In all cases, this leads to the recruitment of the RNA in the proper conformation, and consequently the protein behaves as a ligand (*24*). (2) The chaperone class is formed of proteins that recognize RNA with broad specificity. Upon binding, these proteins destabilize or disrupt internal bonds and incorrect folding, giving rise to RNP complexes where the RNA is in the correct conformation. Sometimes this process involves hybridization to another RNA with complementary sequences by virtue of the annealing activity of the protein. Examples of chaperones of this class are the mammalian hnRNPs, ribosomal protein S12, and retroviral nucleocapsid proteins (see Table III). When protein–protein interactions play an important role in RNA:RNA interaction, this is called matchmaker activity (25).

The purpose of this review is to focus on RNA chaperone proteins, also called RNA chaperones, *stricto sensu*.

II. RNA Structure and the Folding Problem

For the purpose of this review, only basic aspects of RNA structures and folding will be discussed. The reader can consult recent reviews on the subject of how RNA can gain a proper conformation, which is thought to proceed via several pathways and by iterative processes, the details of which are poorly understood (26–30).

A. Secondary and Tertiary Structures

Free nucleic acids are highly flexible macromolecules in solution, as can be seen by electron microscopy. Single-stranded DNA and RNA molecules adopt many different conformations, whereas double-stranded DNA is usually found in the classical B-helix conformation.

Folding of single-stranded RNA molecules appears to be governed by five major types of secondary structures: helices with canonical and noncanonical base pairings, loops, bulges, junctions, and pseudoknots. Most of the time, all these types of secondary structures are found in functional RNA domains. This is exemplified in Fig. 1A, showing the overall secondary structure of the 5′ leader of human immunodeficiency virus 1 (HIV-1) genomic RNA. This plays key roles in the early and the late steps of virus replication, such as proviral DNA synthesis, recombination, and integration, genomic RNA dimerization, and packaging (31). Figure 1B shows another secondary structure, found in the retroviral genome that directs Pol translation. In fact, the mouse mammary tumor virus (MMTV) pseudoknot is part of the −1 frameshift signal required to express the viral enzymes.

Tertiary structures are derived from the packing of previously formed secondary structures. They often involve metal ions (see below) that neutralize phosphate negative charges, Watson–Crick or non-Watson–Crick base pairing between distant single-stranded regions, and non-Watson–Crick base pairing between distant single-stranded and double-stranded regions giving rise to triple-helix formation (32, 33). GNRA tetraloop sequences capping RNA stems are very stable secondary structures and are common interacting modules involved in the tertiary structure folding (34–39).

A

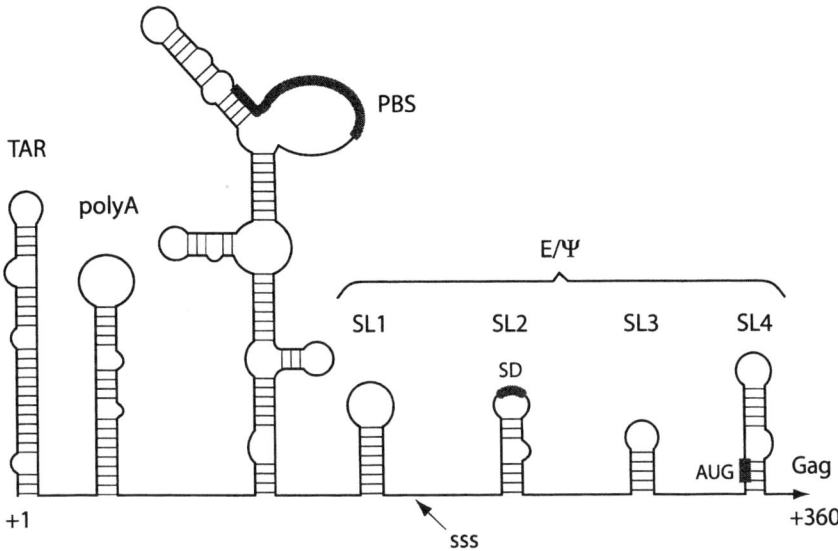

B

FIG. 1. Major RNA secondary structures found in retroviruses. (A) Secondary RNA structure model for HIV-1 5′-leader RNA. Stem–loops (SL), bulge (PBS), single-stranded sequences (sss), and junctions are present in the overall structure. This viral domain directs key functions in the course of HIV-1 replication: TAR and poly(A) SL for transcription initiation and poly(A) addition, respectively (224, 225); PBS SL for replication primer tRNA Lys3 annealing, initiation of reverse transcription (88, 226, 227), and plus-strand DNA transfer (228, 229); TAR for minus-strand DNA transfer (83, 84, 98, 230); SL1, also called DIS (for dimerization initiation sequence), for the initial step of genomic RNA dimerization (231, 232); SL2 and SL3 for genomic RNA dimerization and packaging (81, 226, 233, 234, 235); SL4 for genomic RNA packaging and translation (233–241); SL2, SL3, and SL4 for Gag translation by means of an internal ribosome entry site (IRES)-mediated process as in simian immunodeficiency virus (242, 243). (B) Secondary structure of MMTV pseudoknot. Translation of the Gag–Pol polyprotein occurs by means of a −1 frameshift, which requires two *cis*-acting signals: a heptanucleotide shifting sequence and a pseudoknot structure positioned downstream. The two helices of the pseudoknot are not coaxial, because of a bulged adenylate residue, which directs the bending of the pseudoknot at the stem–stem junction. This residue is absolutely necessary to frameshift efficiency (244, 245).

B. RNA Misfolding: Kinetic and Thermodynamic Problems

In physiological conditions and on the time scale compatible with life, RNA has difficulty reaching its single or few biologically relevant conformation(s). There are two major causes for misfolding: either too much energy or too little. The first situation results in the quick formation of stable, local secondary structures. If such secondary structures do not belong to the final and active RNA structure, then the RNA gets trapped in local minima in the conformational free energy surface. Indeed the formation of a local secondary structure is kinetically favored, whereas the opening of a formed secondary structure is strongly unfavored from a thermodynamic point of view. Therefore only a small fraction of the RNA molecule population can spontaneously reach the secondary structure with the minimal free energy, which is thought to be the biologically active one (see example given in Ref. *40*).

The second problem encountered during RNA folding is the difficulty in specifying a unique tertiary conformation that is thermodynamically favored. RNA tertiary structures appear to be governed by interactions between different loops, or loops and bulges, and this can take place by means of long-range interactions. Furthermore, the packing of helices is unfavored due to the negative charge repulsion of the phosphate backbone.

C. Role of Metal Ions

Metal ions are known to play important roles in RNA structures. There are two types of specifically bound metal ions. Either they bind to preformed sites in the secondary structure and are not involved in important structural rearrangements, or they are part of the tertiary structure, which means that without the metal ion the correct tertiary structure would not exist. Well-known examples include tRNAs and group I intron ribozymes, which fold around an ion (*41*). These Mg^{2+} bridges allow the close packing of phosphate backbones (*42*).

In the splicing machinery, the metal ions appear to play an even more critical role, since they are thought to be directly involved in the catalysis of the reaction (*43*). In the presence of metal ions, RNA folds to form a scaffold, which is then used by the metal ion to cleave the phosphodiester bond (*44*). More general and less specific effects of positively charged ions on nucleic acids are the so-called polyelectrolyte effects, which can drastically alter RNA folding conditions (*40, 45*). Therefore different salt conditions give rise to different conformations with minimal energy. For example, the *Tetrahymena* group I ribozyme (TR) reaches a native conformation step by step and this requires a precise Mg^{2+} concentration, close to that considered to exist in the cell, at which a native structure is stabilized, but kinetic traps are largely avoided. Indeed, an excess of Mg^{2+} slows down folding by stabilizing folding intermediates (*46, 47*). Thus the rule is neither too much nor too little. These findings on TR further show how complex is the interplay between Mg^{2+} concentration, secondary and tertiary structures that govern folding, and the ultimate conformations.

A K^+ ion has been also localized in the catalytic core of this ribozyme. It seems to stabilize the tertiary structure of the core and thus increases its activity. This shows that monovalent metal ions may also play an important role in RNA folding (48).

In summary, the RNA folding follows a hierarchical pathway: (1) Local secondary structures are formed quickly (i.e., hairpins, loops, bulges, pseudoknots). (2) These are organized in tertiary structures within small RNA modules (tRNA-like, small ribozymes). (3) Long-range interactions, which are slowly formed by means of iterative processes, occur (i.e., intramolecular/*cis* interactions). (4) Intermolecular/*trans* interactions occur, which can be either simple and rapid, as when a limited number of nucleotides participates, or complex and slow, as when more than one site and/or a large number of nucleotides are involved.

As already pointed out, RNA folding, in order to achieve a biologically relevant structure harboring specific *cis*-acting motifs able to conduct intermolecular interactions, is critical in many basic cellular processes, such as transcription, posttranscriptional RNA modification, splicing, and translation, as well as virus replication and variability. Thus, solution of the RNA folding problem means not only reaching a good conformation, but also allowing dynamic intermolecular RNA:RNA interactions.

III. RNA-Binding Proteins

Until now we have only considered the folding of free RNA in solution. However, in living cells, RNAs are never found alone, but are rather channeled from one RNP to another one. Thus RNA-binding proteins will profoundly influence the folding process and RNA accessibility to other proteins, especially enzymes and RNAs.

After describing the general rules guiding RNA–protein interactions, we will focus our attention in structural changes observed after binding and how this can help solve the RNA folding problem. Finally, a short description of known RNA-binding domains present in RNA chaperones will be given.

A. RNA Binding Rules

1. AMINO ACID/RNA CONTACTS AND BINDING SURFACE

Nucleic acid-binding proteins (NABPs) usually contain a number of basic and aromatic residues, some of which are organized in RNA-binding motifs known as RRM, dsRBD, KH, RGG, CSD, or zinc fingers (see Table II for definitions). Aromatic and aliphatic residues intercalate between the RNA bases allowing base-stacking interactions or van der Waals interactions (base sandwiching). Basic residues are involved in charge interactions with the negatively charged phosphoribose backbone. Furthermore, both charged or polar amino acid side chains interact through hydrogen bonds with the RNA bases and participate in sequence-specific recognition of the RNA or with ribose 2'-OH and participate in the specific binding to RNA versus DNA.

These RNA-binding residues define in most cases a β-sheet surface that contacts single-stranded RNA or that sits in the major or minor groove of double-stranded RNA helices (49). However, KH domains display a mixed α/β surface (50) and zinc fingers display a unique surface (51).

2. COOPERATIVITY

Often multiple copies of these motifs are present on one NABP and are necessary for function. In addition, more than one protein molecule binds to one molecule of RNA and it does so by binding in a cooperative manner to RNA leading to RNP formation. Some of these multimeric or multidomain complexes have been extensively studied and elegant structures proposed. For example, *Escherichia coli* transcription termination factor Rho is organized as hexamers where poly(rC) stretches bind on the inside (52). In contrast, *Bacillus subtilis* Trp RNA-binding attenuation protein (TRAP) forms 11-mers where GAG and UAG triplets bind on the outside (53). The poly(A)-binding protein (PABP), the *Drosophila melanogaster* sex-lethal protein, and the UP1 fragment of the hnRNP A1 bind nucleic acids through the cooperation between several RRMs (54–56). Interestingly, one RRM alone binds weakly RNA or does not bind RNA at all, reinforcing the notion that these domains need to cooperate in order to bind to RNA with high affinity and specificity (57–61).

3. RECIPROCAL INDUCED FIT

Protein binding to nucleic acid not only induces a structural rearrangement of the RNA component, but also directs the folding of protein regions that were not folded before. Thus RNA molecules can provide structural scaffolds for protein folding (62).

There is a growing number of examples that illustrate this notion. In RRM-containing proteins, the peptidic linker between two RRMs appears to be folded when bound to the RNA, although it is unfolded when the protein is free, and many residues of the linker directly contact the RNA (54–56). In the case of the nucleocapsid protein of HIV-1, the central zinc fingers are the only parts of the free protein to be folded, whereas the N- and C-terminal parts are unfolded. Again RNA induces the folding of these regions (51, 63–65). Another interesting example comes from studies on the viral protein Tat, which is a viral transcriptional transactivator. Tat contains an arginine-rich RNA-binding motif that folds differently depending on the target Tar RNA sequence. On the basis of this property, it has been called a "molecular chameleon" (66).

4. PLATFORMS FOR FURTHER INTERACTIONS

Outside the RNA-binding domain, RNA-binding proteins often contain domains that allow interactions with other proteins. Thus, they act as landing platforms to recruit other factors within RNPs. Examples are given in Table III. The RNA is not fully embedded in the protein, but instead contacts the protein surface. This means that the other side of the RNA can remain accessible to other RNA-binding factors, including both nucleic acids and proteins. This issue will be further discussed in the next sections.

B. RNA-Binding Proteins and the RNA Folding Problem

As already pointed out, proper RNA folding needs assistance to overcome the problem of kinetically trapped intermediates, a critical problem for very large RNAs like messenger and ribosomal RNAs. NABPs appear to be major actors in directing the proper folding of RNA. A great number of notions and terms have been introduced to describe the role of RNA-binding proteins in RNA folding. This profusion of terms is somewhat confusing. An attempt is made to precisely define them in Table I.

There are two main ways for NABPs to influence or assist the folding of a nucleic acid, depending on their binding specificity. Some RNA-binding proteins are highly specific for an RNA motif or RNA structure. Other are less sequence-specific and can bind RNA with high affinity, whatever its sequence, by a highly cooperative mode of interaction. However, these latter proteins may also have high-affinity binding for sequence-specific sites, which can then be used as nucleation sites for RNP formation.

1. SPECIFIC RNA-BINDING PROTEINS

Specific RNA-binding proteins can assist the folding of RNA by at least three different mechanisms.

First, they can bind to a specific sequence in the RNA. This sequence is then masked with respect to other RNA regions, avoiding the formation of misleading secondary structures and allowing the formation of active secondary and tertiary structures. The *trp* antitermination and transcriptional attenuation are well-known mechanisms that are based on the selective formation of either of two alternative stem–loops in a nascent transcript, one of which causes transcription termination. Selection of one structure over the other is due to ribosome pausing when tryptophan is present in limiting quantities in bacteria (67). In this case, mRNA folding is assisted by ribosomes.

A group I intron present in the *Saccharomyces cerevisiae* mitochondrial *COB* gene is self-spliced *in vitro* under high Mg^{2+} concentration. However, the Cbp2 protein is required for efficient splicing both *in vivo* and *in vitro* under physiological salt conditions. Cbp2 seems to act by two different, but complementary ways. It is able to bind the intron and stabilize tertiary interactions between

TABLE I
DEFINITIONS OF TERMS RELATED TO THE FOLDING OF RNAs

Term	Definition
Chaperoning	A process by which a molecule helps another molecule to reach its native and functional conformation (the term does not presume which kind of molecules are involved and how they perform their functions)
Protein-assisted folding	A process by which a protein helps an RNA to reach its native and functional conformation
Protein ligand or cofactor	A protein that binds to RNA in a stoichiometric proportion and helps RNA to fold by stabilizing poorly stable structures; the presence of the protein is needed to maintain the native RNA structure
Matchmaker	A protein that helps RNA folding through (often homotypic) protein–protein interactions
Annealing activity	An activity that allows hybridization of two complementary nucleic acid strands
Helix-destabilizing or -melting activity	An activity that facilitates the destabilizing of two complementary nucleic acid strands
Strand-exchange activity	This is the result of the combined action of both annealing and melting activities; it causes destabilization of a nucleic acid duplex and formation of a more stable duplex with another nucleic acid strand where complementarities are more extended, if not complete
Iterative annealing	A process by which RNA reaches its native conformation through repeated cycles of annealing and melting (i.e., strand exchange); this avoids folding traps and allows the RNA to reach the free energy minimum of the RNA folding landscape
RNA chaperone	A protein that binds to RNA in a more-than-stoichiometric proportion and helps RNA to fold by iterative annealing and/or matchmaker activities; once the RNA is well-folded, the RNA chaperone is no longer needed and can be removed without altering the native structure of the RNA

the substrate domain and the core by trapping the native tertiary structure (68–70). It acts not only as a protein ligand for the structured RNA, but also as a cotranscriptional scaffold for RNA folding (71). Indeed it restricts the folding pathway of the transcribed RNA by masking or securing RNA substructures that are ultimately required for the final tertiary structure. Only a narrow range of secondary structures is thus explored, which avoids misfolding.

The tyrosyl-tRNA synthetase of *Neurospora crassa*, CYT-18, acts as a splicing cofactor for the mitochondrial large rRNA intron. CYT-18 is also capable of binding group I introns from other species, like T4 phage *thymidylate synthase* (*td*) intron. Expression of CYT-18 in this latter system suppresses mutations that destabilize the *td* intron structure and inhibit splicing. This genetic evidence, added to biochemical observations, led to the proposal that CYT-18 promotes splicing by stabilizing the catalytically active structure of the RNA, even if this structure is not necessarily the most stable structure by itself (72–77).

In prokaryotes, mRNAs are cotranscriptionally translated and in eukaryotes, they are cotranscriptionally spliced. Thus RNA is channeled from one biological process to another, from one RNP to another, and this influences the folding process of large RNAs: mRNAs undergo transconformation processes from the site of synthesis to that of decoding by ribosomes.

2. RNA-BINDING PROTEINS WITH BROAD SPECIFICITY

An increasing number of proteins have been described that bind RNA and/or DNA with poor sequence specificity, at least *in vitro,* and help nucleic acid folding. These proteins, known as RNA chaperones, promote the rapid annealing of complementary nucleic acid strands, even when they are already folded through intramolecular interactions. They are also able to favor the most stable interaction by catalyzing strand exchange between a preformed duplex and a single-stranded nucleic acid. This process is thought to allow the proper folding of RNAs by destabilizing helices until the most stable structure—with the global free energy minimum—is reached (78). This model can be compared to the "iterative annealing" model of protein folding. However, this model does not account for the increased rate of annealing in the case of small and unfolded nucleic acids [like oligo(ribo)nucleotides]. We need to consider the possible involvement of protein–protein interactions that can strongly augment the local concentration of nucleic acids and therefore accelerate the pairing process, through their so-called "matchmaker" activity. This is supported by the observation that RNA chaperones are usually able to contract homotypic interactions (cf. Table III).

In all cases, RNA chaperones can be differentiated from other types of proteins involved in RNA folding, since the protein is no longer needed once the RNA has been folded and can be removed without altering the RNA conformation. Moreover, RNA chaperones do not need ATP or GTP hydrolysis to achieve their function.

It should be noticed that a number of proteins have been called RNA chaperones on the basis of their ability to help the maturation process of a particular transcript (like La protein and RNA–pol III transcripts or the RAC complex associated with unprocessed rRNAs) and on the assumption that the folding and maturation processes are interrelated. Such uses of the term "RNA chaperone" do not correspond to the definition given in Table I.

C. RNA Binding Motifs Found in RNA Chaperone Proteins

Table II summarizes common RNA binding motifs found in RNA chaperones; the reader is encouraged to consult more specialized reviews and minireviews (49, 79, 80). It should be noticed that RNA chaperone proteins do not share

TABLE II
RNA BINDING MOTIFS FOUND IN RNA CHAPERONE[a]

RNA-binding domain	Alternative name	InterPro entry	Average size (amino acids)	Predictive pattern	Example of protein	Ref. for structure
RRM	RNP motif	IPR000504	90	Yes	hnRNP AI	54–56
RGG box	—	NA	25	Yes	hnRNP U	NA
KH motif	—	IPR004087	70	Yes	FMRI	50
		IPR004044	—	—	—	—
		IPR004088	—	—	—	—
dsRBD		IPR001159	70	Yes	Xlrbpa	175–178
Zinc finger	Zinc knuckle	IPR001878	18	Yes	NCp7	51, 63–65, 179–183
CSD	—	IPR002059	70	Yes	CspA	184–187

[a] Abbreviations: RRM, RNA recognition motif; RNP, ribonucleoprotein; RGG, arginine–glycine–glycine; KH, hnRNP K homology; dsRBD, double-stranded RNA-binding domain; CSD, cold shock domain. InterPro is an integrative database of protein families and domains (174) and is available at the European Bioinformatics Institute web site (http://www.ebi.ac.uk/interpro/).

a common structural signature. They belong to different classes of RNA-binding proteins. In many cases it is not possible to predict a known RNA-binding domain (RBD) from the sequence. As already underlined, many of these proteins are not even folded in the absence of RNA (62). Thus we describe here the only RBDs known to occur in RNA chaperones and which are clearly identified.

IV. Investigating the Biochemical Properties of Nucleic Acid Chaperone Proteins

A. Preliminary Experiments

Before carrying out an extensive investigation on the putative nucleic acid chaperone properties of a protein, very simple assays should be carried out to clearly show that a given protein binds nucleic acids. Single- and double-stranded nucleic acids 50–100 nt in length are easily generated *in vitro* by means of chemical or enzymatic synthesis. The purified nucleic acids are ^{32}P-labeled at the 5' or 3' end or internally with ^{32}P-NMP (nucleoside 5'-monophosphate). Proteins are separated by polyacrylamide gel electrophoresis (PAGE) in sodium dodecyl sulfate (SDS) (PAGE–SDS), and subsequently transferred onto a synthetic membrane and renatured. They are probed with the labeled nucleic acid (North–Western probing) under conditions of low or high stringency. When nuclease-free extracts are available, gel retardation assays can be carried out. To this end, protein(s) is mixed with the labeled nucleic acid at increasing protein-to-nucleotide molar ratios (usually from 1:100 to 1:1) and nucleoprotein complexes are analyzed by 3% to 5% polyacrylamide gel electrophoresis under nondenaturing conditions. Binding of the protein to the nucleic acid will result in the formation of nucleoprotein complexes, the gel migration of which will be clearly retarded as compared with the free nucleic acid. Using this technical approach, one can also assess a possible cooperative binding of the protein to the nucleic acid.

Another approach is the well-known ultraviolet light (UV at 258 nm) cross-linking protocol. ^{32}P-Labeled nucleoprotein complexes are formed and subsequently UV-irradiated for a brief period of time (on the order of seconds). Complexes are heat-denatured and analyzed by PAGE–SDS. Direct autoradiography of the wet gel is generally carried out. ^{32}P-Labeled complexes can be recovered and the protein-binding site along the nucleic acid characterized. Examples are given in Refs. 81 and 82. For the following assays, references are cited in Table III.

B. Simple Assays for Identifying Proteins with Nucleic Acid Chaperone Properties

A number of simple assays have been developed for examining the nucleic acid chaperone properties of RNA-binding proteins. Proteins can be partially purified from cells or virus preparations. Of course the preliminary results on the nucleic acid annealing properties of a given protein should be confirmed using the protein in a highly purified recombinant form.

As will be discussed later in this review, it is important that studies investigating the chaperone properties of nucleic acid-binding proteins be carried out over a wide range of protein-to-nucleotide molar ratios, typically 1–100 to 1–5.

1. THE ANNEALING ASSAY

The first and most widely used assay is the annealing assay. Typically two nucleic acids with complementary sequences of about 60 nt in length are used. The simplest assay uses two DNA oligonucleotides (ODNs), one of which is ^{32}P-labeled, and carries out the annealing assays for a short period of time (5 min) at 25–30°C to eventually minimize spontaneous hybridization of the oligonucleotides in absence of any added protein. Subsequently, SDS, proteinase K, and carrier tRNA are added and assays incubated for another 5 min at 20°C. Single-stranded and double-stranded DNA are analyzed by native 6% PAGE. Results are revealed by autoradiography of the wet or dried gel. A very similar assay can be set up using two *in vitro*-generated RNAs (60–100 nt in length), one of which has been internally ^{32}P-labeled during *in vitro* synthesis by a bacteriophage RNA polymerase.

A more sophisticated assay involves two *in vitro*-generated RNAs where the ^{32}P-labeled RNA probe is about 100 nt in length and flanked by

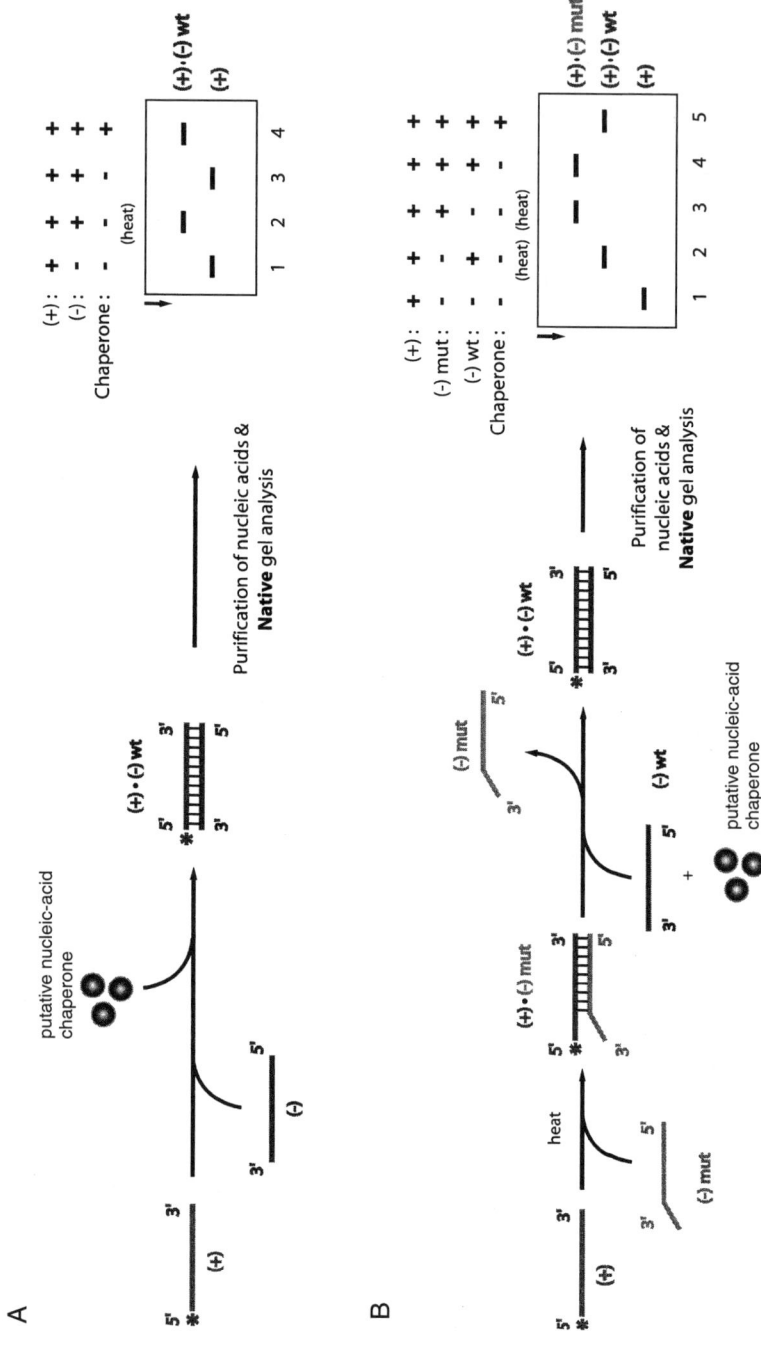

noncomplementary sequences. Upon RNA:RNA duplex formation, nonhybridized RNAs and single-stranded overhangs of the ^{32}P probe are degraded by T1 RNase. The RNA probe is subsequently purified by SDS/proteinase K digestion and phenol extraction and finally analyzed by 10% PAGE in denaturing conditions (25).

Among variations on the same theme, formation of ^{32}P-labeled DNA:RNA hybrids has also been used to monitor the annealing activity of cellular and viral proteins (for examples see Refs. 83–89).

Figure 2A illustrates principles of the annealing assay where two complementary oligonucleotides are used. Oligonucleotides are often DNA generated by chemical synthesis.

To carry out very simple assays, two DNA oligonucleotides with complementary sequences are used, one of which has been 5'-^{32}P-labeled. For example, ODNs derived from HIV-1 TAR (+) and TAR (−) sequences 56 nt in length have been used in the picomolar range in several reports (86, 89, 90). After 5 min at 30°C, reactions are stopped by SDS and carrier tRNA and formation of short, double-stranded DNA analyzed by 6–8% PAGE under nondenaturing conditions (50 mM Tris–borate, pH 8.3, 0.1 mM ethylenediaminetetra-acetic acid, EDTA).

2. THE STRAND EXCHANGE ASSAY

This assay utilizes three ODNs: a plus-strand one (+), which is ^{32}P-labeled, a wild-type minus strand [(−) wt], and a mutated minus strand [(−) mut]. ODNs (+) and (−) wt are complementary, whereas (−) mut is only partially

---

FIG. 2. Investigating the nucleic acid annealing and strand exchange activities of a RNA chaperone. (A) Nucleic acid annealing. Two oligonucleotides (ODN) 50–80 nt in length with complementary sequences, one of which is 5'-^{32}P-labeled, are incubated in a small volume (10 µl) under physiological salt conditions and in the picomolar range. Increasing protein-to-nucleotide molar ratios are used in the assays. After 3 min at 30°C reactions are stopped by addition of SDS, EDTA, and carrier tRNA (1 µg). Nucleic acids are analyzed by native polyacrylamide gel electrophoresis (PAGE), using 50 mM Tris–borate (TB). Autoradiography is usually performed on the wet gel at 4°C for 3–4 h. Lane 1, 5'-^{32}P-labeled ODN. Lane 2, positive control with incubation at 65°C for 10–30 min. Lane 3, negative control without incubation. Lane 4, the assay for annealing activity. (B) Nucleic acid strand exchange. Incubation conditions are very similar to A, except that the starting nucleic acid material is a double-stranded ODN carrying nonpaired residues (usually 3' to 5'). ODN completely complementary to the 5'-^{32}P-labeled ODN is added together with the chaperone protein (step 2). Reaction conditions are as in A (5 min at 30°C) and analyses are performed by native PAGE in TB. Lane 1, ^{32}P-labeled ODN. Lane 2, positive control for double-stranded ODN generated by heat annealing. Lane 3, positive control for mutated double-stranded ODN generated by heat annealing. Lane 4, wt ODN added to the preformed, mutated double-stranded ODN. Lane 5, chaperone added together with wt ODN to the preformed, mutated double-stranded ODN. As shown, only addition of the chaperone allows strand exchange to take place under physiological conditions.

A Without RNA chaperone

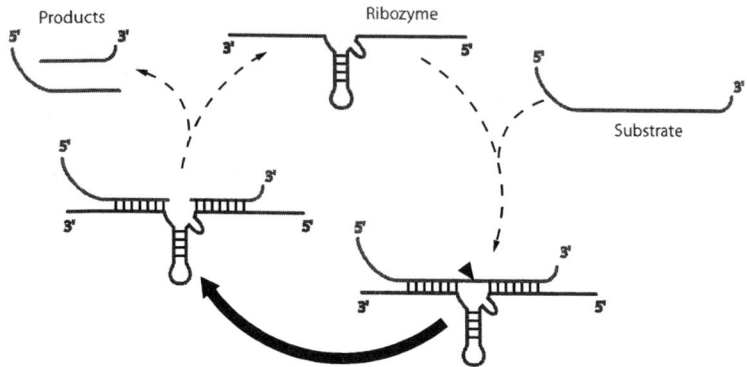

B With RNA chaperone

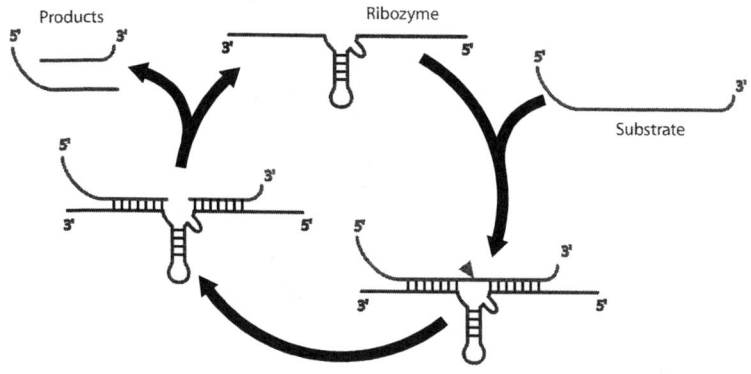

C

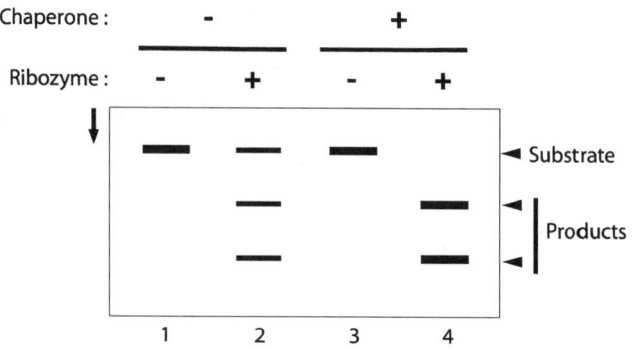

complementary to (+), as shown in Fig. 2B. ODNs (+) and (−) mut are first heat-annealed, giving rise to a double-stranded molecule containing mismatches. Next ODN (−) wt and the protein are added to preformed (+):(−) mut molecules and incubated for 5 min under the same conditions as for the annealing assay. Then nucleic acids are deproteinized and analyzed by PAGE under native conditions. The strand exchange activity of the protein will be seen in the formation of double-stranded (+):(−) wt molecules at the expense of (+):(−) mut (see Fig. 1B). In other words, the chaperone protein is able to substitute ODN (−) wt for (−) mut in the double-stranded molecules, thus favoring formation of the most stable double-stranded molecules (86, 91–93).

C. Advanced Assays for Examining Properties of Nucleic Acid Chaperone Proteins

A number of more advanced assays have also been set up to examine some of the properties of nucleic acid chaperone proteins. These include enhancement of ribozyme cleavage and inhibition of reverse transcription by self-priming.

1. ENHANCEMENT OF HAMMERHEAD RIBOZYME CLEAVAGE

Nucleic acid chaperone proteins have been shown to enhance hammerhead ribozyme-directed cleavage of an RNA substrate (94–97). This assay requires a hammerhead ribozyme and a ^{32}P-labeled RNA substrate, which are generated

---

FIG. 3. Enhancement of ribozyme cleavage by a chaperone protein. The importance of hammerhead ribozymes is based on the fact that they are able to completely inactivate their target RNA and also to turn over (95). (A) A *trans*-acting hammerhead ribozyme is made of two parts: (1) a catalytic core formed of conserved nucleotides and a stem–loop structure, and (2) two variable sequences, located on either side of the catalytic core, forming the specifier sequence 15–16 nt in length. This specifier element allows the ribozyme to hybridize to a target sequence present in the target RNA substrate. The ribozyme/template hybrid structure allows restoration of the complete hammerhead ribozyme structure (thin dashed arrow). Due to formation of a competent structure, cleavage can take place (thick black arrow) and products generated can be released from the ribozyme (thin dashed arrow), allowing turnover to take place. The limiting steps of ribozyme cleavage appear to be template/ribozyme hybridization and release of the cleaved products. (B) Addition of a chaperone protein accelerates hybridization of the template to the ribozyme and thus formation of a competent structure. Upon cleavage, the chaperone accelerates the release process, since there is preference for hybridization of the uncleaved substrate for the complementary sequence of the ribozyme (see discussion of strand exchange activity). (C) *In vitro* assays. Ribozyme and ^{32}P-labeled substrate RNA are generated by *in vitro* transcription, purified by gel electrophoresis, and ribozyme-renatured in presence of MgCl$_2$. Substrate-to-ribozyme molar ratios are from 1:3 to 1:10. Substrate is in excess in order to examine the turnover rate. Incubation conditions are similar to the annealing reaction conditions and incubation time is 20 min at 30°C. Reactions are stopped by SDS, and products are purified by phenol extraction and are analyzed by denaturing PAGE in 7 M urea and in TB. Lane 1, ^{32}P-labeled substrate. Lane 2, ^{32}P-labeled substrate and ribozyme. Lane 3, substrate with chaperone. Lane 4, ^{32}P-labeled substrate, ribozyme, and chaperone. It is noteworthy that cleavage is complete. In fact, the chaperone accelerates the turnover rate by probably more than 100-fold.

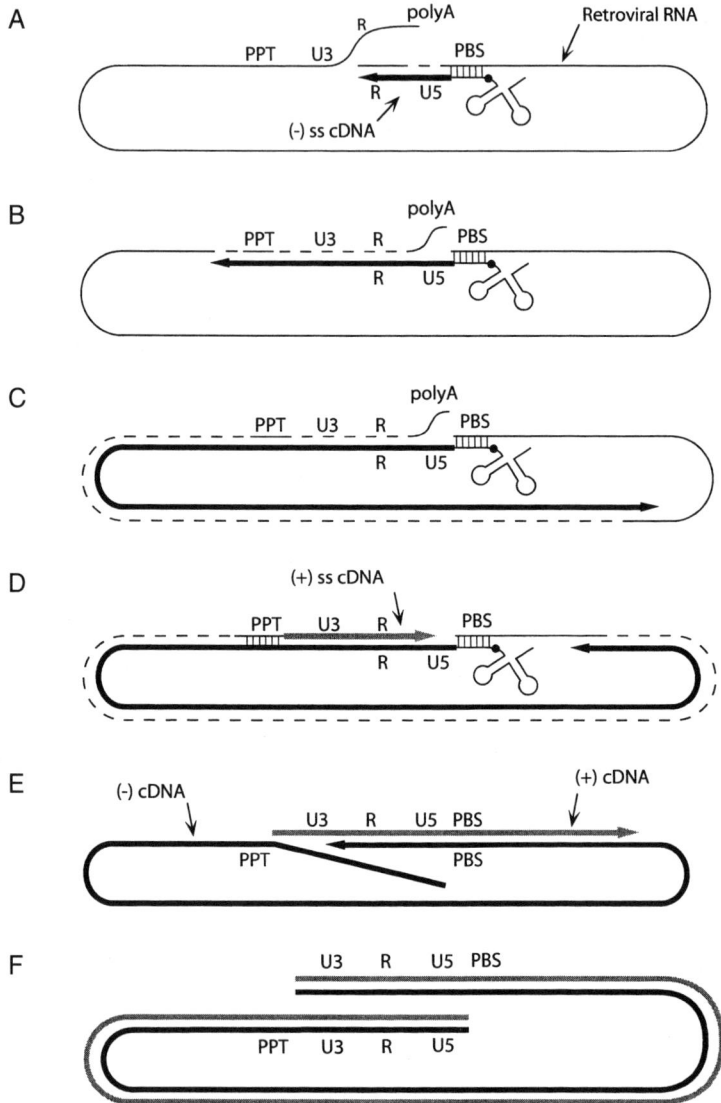

FIG. 4. Scheme of the reverse transcription process. The retroviral genomic RNA is dimeric in the virus (previously described as 60S RNA). For the sake of simplicity, only one viral RNA is represented. The main elements acting in *cis* during the reverse transcription process are the primer tRNA-binding site (PBS) of 18 nt, the repeated R sequences (21 nt in avian sarcoma and leukemia viruses, ASLV; 65 nt in murine leukemia viruses, MLV; 96 nt in HIV-1) located at both ends of the viral RNA, and the polypurine track (PPT) close to the 3′ end. U5 and U3 are the 5′- and 3′-untranslated regions, respectively. In addition to the 3′ PPT, a central PPT (cPPT) also exists in lentiviruses and yeast retrotransposons (246–250). Conversion of the single-stranded viral genome into double-stranded proviral DNA is carried out by the viral DNA polymerase reverse transcriptase

by *in vitro* transcription and gel-purified. As shown in Fig. 3A, cleavage of the labeled RNA substrate by the ribozyme appears to first necessitate hybridization of the ribozyme to the substrate. After substrate cleavage, the RNA products must be released to allow a cyclic reuse of the ribozyme. The ribozyme and the labeled RNA substrate are mixed and incubated at 25–35°C for 10–60 min. At the end of the reaction period, RNAs are deproteinized and analyzed by PAGE in denaturing conditions to visualize the ^{32}P-labeled RNA products. In the absence of a nucleic acid chaperone, hybridization of the substrate to the ribozyme and release of the RNA products appear to be slow.

Addition of a nucleic acid chaperone accelerates hybridization of the substrate to the ribozyme and dissociation of the RNA products (Fig. 3B), thus causing a rapid ribozyme turnover. As indicated in Fig. 3C, the products of RNA substrate cleavage tend to rapidly accumulate in the presence of an RNA chaperone, whereas they do not in its absence.

2. INHIBITION OF SELF-PRIMED CDNA SYNTHESIS

A succession of specific events is required for the conversion of the retroviral genomic RNA into a complete double-stranded proviral DNA with duplication of the long terminal repeats (LTRs). This is depicted in Fig. 4.

Reverse transcriptase (RT) first elongates primer tRNA annealed to the viral primer-binding site (PBS) and synthesizes minus-strand strong-stop cDNA

(RT), which is heterodimeric in ASLV and HIV and homodimeric in MLV. In addition to its RNA- and DNA-dependent polymerase activities, RT possesses an RNase H activity required for the minus- and plus-strand DNA transfers and thus synthesis of the complete proviral DNA. The other viral protein required for proviral DNA synthesis is nucleocapsid protein, NC (NCp12 of 87 residues in ASLV; NCp10 of 56 residues in MLV; and NCp7 of 71 or 55 residues in HIV-1). This extensively studied small basic viral protein is required for cellular primer tRNA annealing to the PBS (see Ref. 147 for a recent review). The main steps of proviral DNA synthesis are: (A) Reverse transcriptase (RT) first elongates primer tRNA annealed to the viral PBS and synthesizes minus-strand strong-stop cDNA (minus ss-cDNA) (*114*). (B) Minus ss-cDNA is transferred to the 3' end of the genomic RNA, a process that requires the R sequences, RT-associated RNase H activity, and the nucleic acid chaperone activity of NC (*83, 84*). (C) RT completes synthesis of minus-strand cDNA. (D) RT initiates plus-strand DNA synthesis by elongation of the RT RNase H-resistant PPT sequence located at the 5' end of the U3 region and synthesizes plus-strand strong-stop DNA (ss-DNA+). (E) ss-cDNA+ is transferred to the 3' end of minus-strand cDNA, which requires PBS, primer tRNA, RT-associated RNase H activity, and the nucleic acid chaperone activity of NC (*228, 229*). (F) RT completes synthesis of plus- and minus-strand DNA, which results in 3' LTR duplication and thus generates the complete proviral DNA with one LTR at each end. Nonspecific self-replicating events of the viral RNA and the newly formed cDNA can take place during proviral DNA synthesis, as observed *in vitro* using solely RNA or DNA template and RT. However, self-initiation of reverse transcription is strongly inhibited by NC in the MLV and HIV-1 systems *in vitro* (*88, 98–100, 165, 166*). It should also be pointed out that two viral RNA copies form the virus genome, allowing a high level of genetic recombination during proviral DNA synthesis, thus fueling viral diversity, as has been shown for ASLV, MLV, and HIV-1 (reviewed in Ref. *147*).

(minus ss-cDNA). However, nonspecific replication events can take place during conversion of the single-stranded genomic RNA into minus-strand viral DNA, and subsequently during copying of minus-strand DNA to generate the double-stranded proviral DNA. These events include self-initiation of reverse transcription and formation of artefactual double-stranded DNA products by RT going in the opposite direction at any point during minus-strand DNA synthesis (see Fig. 4, steps A–C). Such nonspecific initiation events appear to be promoted by template interactions, folding back of the template, and nicks in the genome or in the newly made minus-strand DNA. Nucleocapsid (NC) proteins have been found to strongly diminish, but not completely abolish, such nonspecific RT-dependent replication events in the murine leukemia virus (MLV) and the HIV-1 systems *in vitro* (*88, 98–100*). The partial nucleic acid helix-destabilizing activity of HIV-1 NCp7 is probably a mechanism through which nonspecific replication events are avoided in the course of proviral DNA synthesis (*101–103*).

A simple reverse transcription assay can easily be set up using an *in vitro*-generated RNA and a primer oligonucleotide ODN (Fig. 5A). RT of Moloney murine leukemia virus (MoMLV; commercial source) and ^{32}P-dNTPs are added to the template RNA with or without primer and reaction allowed to proceed for 30 min at 37°C. cDNA products are purified and analyzed by gel electrophoresis under denaturing conditions (Fig. 5, bottom). In the absence of a chaperone and of a specific primer, self-primed cDNAs are by far the major products (lane 1), but reaction is completely inhibited by addition of a chaperone as, for example, NCp7 of HIV-1 (lane 2) (*88, 98, 99*). Addition of specific primer results in the synthesis of a small amount of ODN-primed cDNA product together with large amounts of self-primed products (lane 3). Addition of both a specific primer and a chaperone will lead to the nearly exclusive synthesis of the ODN-primed cDNA (Fig. 5B and bottom, lane 4). This result can be explained, at least in part, by properties of the chaperone that promote primer annealing to the RNA and destabilization of RNA secondary structures (Fig. 5B, top) and, moreover, it implies that RNA remains accessible to the RT even in the presence of the RNA chaperone.

3. HELIX-DESTABILIZING ACTIVITY OF CHAPERONES

Nucleic acid chaperone proteins cause a rapid hybridization of complementary sequences and this appears to be due, at least in part, to their ability to destabilize secondary and tertiary structures within each strand (*102, 103*; Y. Mely, personal communication). A straightforward assay can be used to examine the helix-destabilizing activity of a chaperone protein. A labeled single-stranded RNA is generated *in vitro* and gel-purified. Conditions are set up so that the labeled RNA is partially cleaved at G residues by T1 RNase in physiological ion concentrations *in vitro* (Fig. 6, bottom, lanes 1 and 2). The putative chaperone is added to the labeled RNA at increasing protein-to-RNA molar ratios. At high

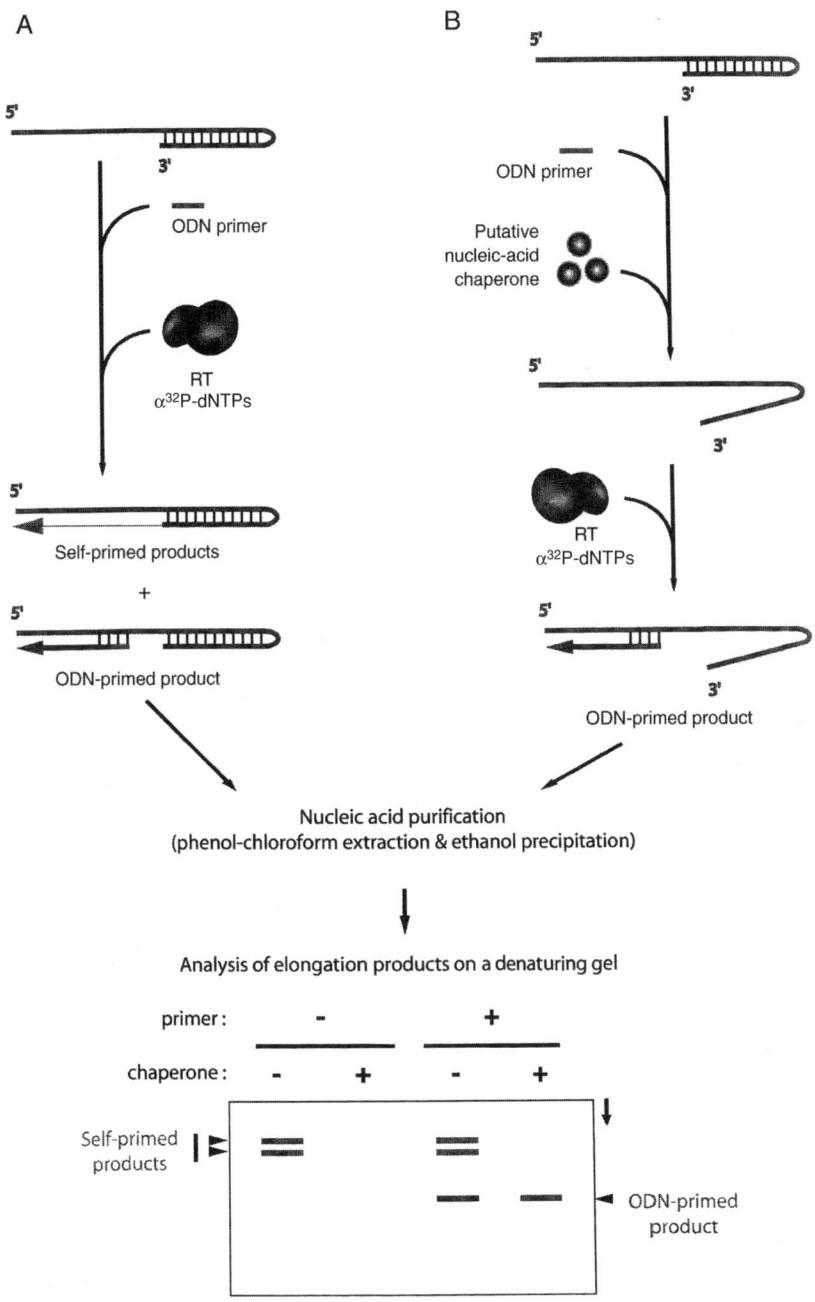

FIG. 5. Inhibition of self-primed reverse transcription by RNA chaperones. A template RNA is easily generated by *in vitro* transcription and subsequently purified. ODNs 18–30 nt in length and RT (MLV or HIV) are from commercial sources.

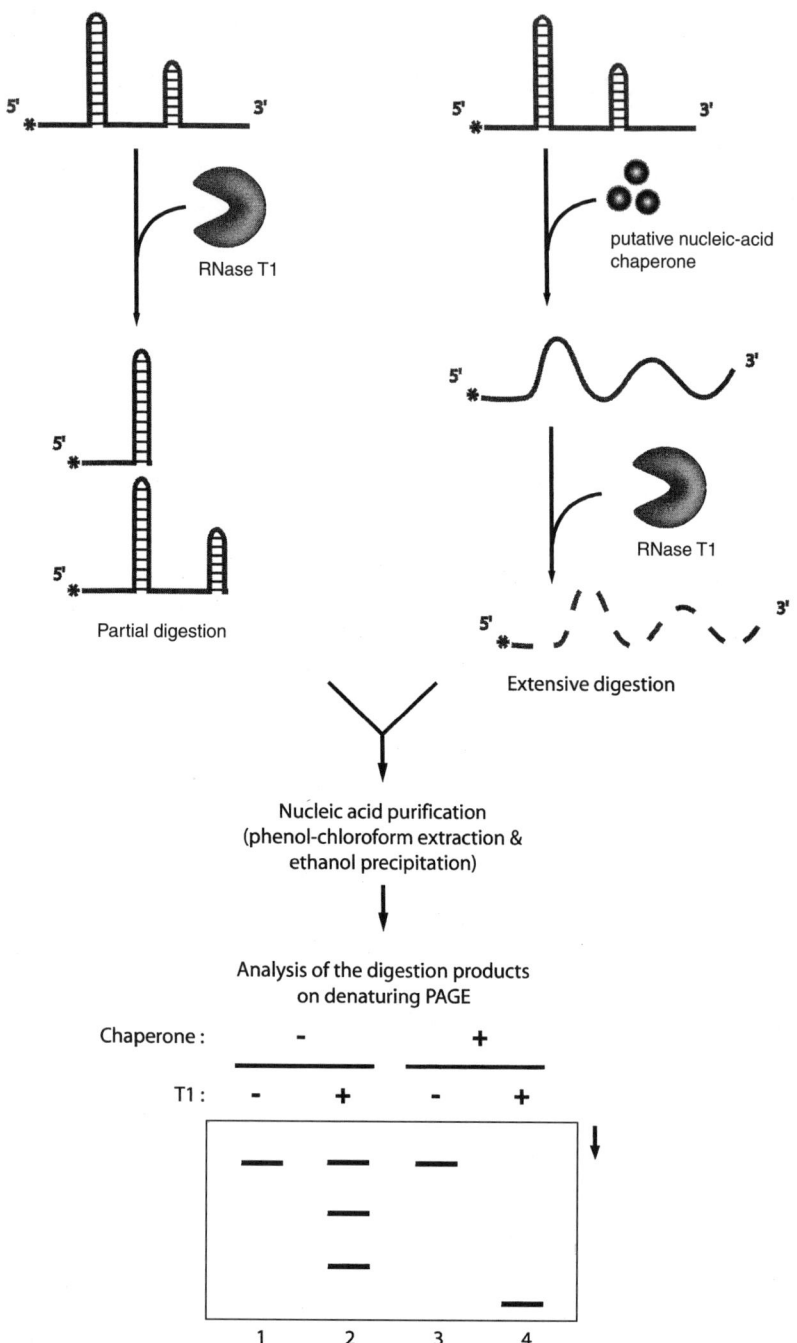

chaperone concentration, its helix-destabilizing activity will unwind the RNA secondary structures, allowing the small nuclease T1 RNase to cleave at probably most, if not all, G residues (shown in lanes 3 and 4; see Table III for concrete examples).

D. Annealing versus Unwinding: Conditions for Investigating Chaperone Properties

Nucleic acid chaperone proteins can direct annealing of complementary nucleic acids and at the same time unwinding of hybridized sequences. However, this turns out to be strictly dependent upon the protein-to-RNA ratio (see next section and Fig. 7). At a low protein-to-RNA ratio the chaperone enhances hybridization of nucleic acids with complementary sequences, whereas at a high protein-to-RNA ratio the reaction does not take place due to the extensive coating of the RNAs. Otherwise the protein might simply destabilize the specific domain or unwind the hybridized RNA due to a small number of interacting nucleotides. Interestingly, windows of activity for annealing versus unwinding and vice versa appear to be very narrow. For example, when protein-to-RNA ratios are increased by a factor of 2–3, annealing activity switches from background levels to an optimum for hnRNP C1 and U and for HIV-1 NCp7, but a further increase of 2- to 3-fold of the ratio has a strong inhibitory effect (e.g., see Refs. 25, 104). This can be explained by the unwinding activity of the chaperone, which is exerted only at a high protein-to-RNA ratio. For this reason, experiments on RNA chaperone proteins have sometimes given rise to apparent contradictory results (101–103, 105).

Therefore, in all cases of where chaperones are examined one should vary the protein-to-RNA molar ratio due to expected narrow windows of activity. In subsequent steps, one should use appropriate nucleic acid sequences in order to examine the chaperone properties in a context as relevant as possible to function. For example, properties of HIV-1 NC have been examined using 5′ and 3′ RNA accurately representing the ends of the genome, known to contain sequences required for replication and high-affinity binding sites for NC (81–83, 85, 98, 106, 107).

FIG. 6. A simple assay for examining the helix-destabilizing activity of chaperone proteins. ^{32}P-labeled RNA is generated *in vitro* and gel-purified. T1 RNase is from a commercial source. Conditions have been set to generate partial cleavage by T1 RNase (10 min at 30°C). The chaperone–RNA complex is formed at protein-to-nucleotide ratios of 1:5 to 1:2 and T1 RNase is subsequently added. Reactions are stopped by addition of 0.5% SDS and cleavage products are phenol-extracted. Analysis of the heat-denatured T1 products is carried out by PAGE in 7 M urea and 50 mM Tris–borate–EDTA (TBE).

V. How Do RNA Chaperone Proteins Work?

A. RNA Occupancy Level and Window of Activity of RNA Chaperones

Several windows of activity for RNA chaperone proteins can be considered according to the following scheme: (1) Under conditions where the protein is in limiting concentration, proteins bind to a fraction of the RNA leading to the formation of simple RNP complexes ("binding mode" in Fig. 7). (2) As more protein is available, all RNA molecules interact with chaperone molecules, resulting in the formation of high molecular weight RNPs. Chaperone molecules coat most, but not all, RNA sequences, causing a high degree of RNA occupancy. This is the "chaperone mode," which allows reconformation of RNA structures, strong interactions with other RNA molecules, and their hybridization if complementary sequences are available. Such RNP complexes may also recruit proteic factors and enzymes and make contacts with other RNPs. (3) When the chaperone protein is present at saturating levels, all RNA molecules are completely coated and this results in the unwinding of inter- and intramolecular RNA secondary structures. It may be possible in some cases that interactions with other RNAs or proteins become difficult if not impossible, because of the completeness of RNA occupancy. However, at these very high concentrations, RNA becomes even more sensitive to RNase than free RNA and thus probably remains accessible to small proteins.

B. RNA Chaperones in Action

Upon binding of the chaperone to the RNA molecule a stable ribonucleoprotein complex is formed. Multiple molecular connections take place between the RNA and the chaperone and involve hydrogen-bonding, hydrophilic, and hydrophobic interactions. It is as if many residues spread throughout the chaperone could sense nucleotides of the RNA sequence. These multiple interactions induce reciprocal conformations of both the protein and the RNA ("best induced fit"; see below). However, our present knowledge is restricted to the propensity of the RNA chaperone protein to form high molecular weight complexes in presence of nucleic acids (referred to as "aggregates" in biophysics) (82, 85, 89, 90, 104, 108–111). Thus all the structural data on RNA chaperone proteins obtained so far are limited to a single protein molecule interacting with a very short RNA or DNA molecule. It remains that such molecular model structures indicate the fact that residues located in distant regions of the RNA chaperone contact a number of nucleotides present in the model nucleic acid. Information provided by these structural models is in agreement with biochemical and genetic data showing that several domains of the chaperone protein, whether well-structured or not, cooperate and are critical for chaperone activity *in vitro* and for known function(s) *in vivo*.

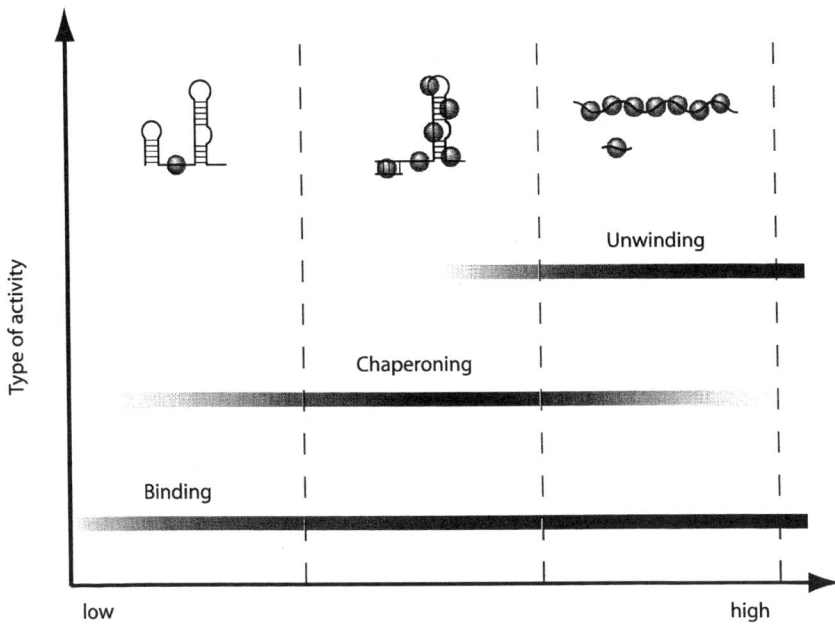

FIG. 7. Relationship between RNA occupancy by an RNA chaperone protein and its activity. This scheme illustrates the different types of activity associated with RNA chaperone proteins according to the degree of RNA occupancy. In other words, the three modes depicted in the scheme can be classified with respect to the chaperone protein as noncrowded, crowded, and overcrowded. Among the many cellular and viral RNA chaperones known, nucleocapsid proteins (NCp) of retroviruses and certain retrotransposons typically exhibit these three types of activity. These small basic proteins have molecular weight between 7 and 12 kDa with one or two "CCHC" zinc finger(s) (also known as "zinc knuckles") and occupy 6–8 nt on RNA molecules. At protein-to-nucleotide molar ratios of 1:100 to 1:40, all NC molecules bind to the RNA, but do not exhibit any chaperone activity. At molar ratios of 1:30 to 1:10, all NC molecules are bound and form high molecular weight RNP complexes. Optimal chaperone activities are reached at ratios between 1:12 and 1:8. Under these conditions, stem–loop structures are destabilized according to biophysical data. This can lead to important transconformation processes for an RNA molecule, like the 5′-leader of HIV-1 RNA, which goes from extended closed hairpin structures to an open, cruciform-like conformation (see Fig. 1). This also leads to intermolecular interactions between two identical RNA molecules (dimerization) as well as with a cellular tRNA (primer tRNA annealing to the initiation site of reverse transcription called PBS). At a ratio above 1:5, the chaperone activity is progressively turned off and the unwinding activity becomes apparent. Even very stable stem–loop structures, like HIV-1 TAR, 56 nt in length, appear to be single-stranded, and small RNA molecules of 20 nt hybridized to a large RNA are released.

These structural models also indicate that the two components (the protein and the RNA) interact through one of their sides while the other side remains free and thus should be available for other nucleic acid and protein interactions Fig. 9).

Based on these simplified model structures, how can we envision more complex nucleoprotein structures that function in nucleic acid annealing, strand transfer, and exchange?

This can be illustrated by the hybridization reaction that takes place at the beginning of the reverse transcription process in all retroviruses and retrotransposons and that corresponds to minus-strand DNA transfer (cf. Fig. 8). In HIV-1, the *trans*-activating region (Tar) of 57 nt is present at both ends of

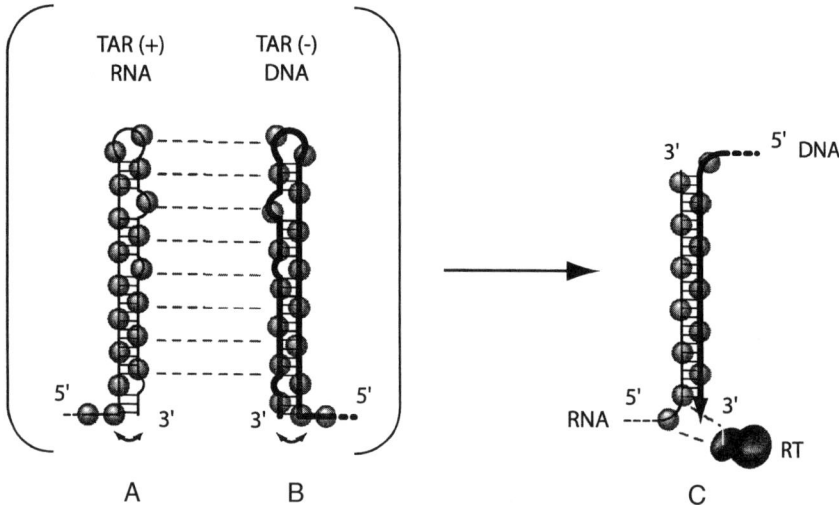

FIG. 8. Nucleic acid annealing during HIV-1 minus-strand DNA synthesis is chaperoned by viral NC protein. Tar (+) sequences are present at both ends of HIV and SIV RNA genomes. In HIV-1, Tar is 57 nt in length and forms a stable hairpin loop structure as shown in A. At the early stage of the reverse transcription process, 5′ Tar (+) is converted into DNA Tar (−) by reverse transcriptase and is located at the 3′ end of the nascent cDNA transcript (also called minus-strand strong-stop cDNA; see B and Fig. 4). NC protein molecules are bound to both DNA Tar (−) and RNA Tar (+) every 12 nt, on the average, and are thought to destabilize the double helix of Tar (+) and Tar (−) (arrows). Due to NC-to-NC interactions (dashed lines), complementary Tar sequences are in close proximity, giving rise to a "molecular crowding" effect (R. Kornberg). The hybridization reaction is probably initiated through loop–loop and end–end interactions between the complementary sequences. Indeed, deletion of the loops and/or the ends causes strong inhibition of (−)-strand DNA transfer *in vitro*. Once Tar (−) is annealed to the 3′ end of the genomic RNA, RT can resume viral cDNA synthesis (C). It should be also pointed out that Tar (+) and Tar (−) must be in the same nucleoprotein complex (brackets), since hybridization works poorly when the two nucleic acids belong to two different preformed nucleoprotein complexes.

the genomic RNA. During minus-strand DNA transfer, the antisense Tar (−) sequence present at the 3′ end of the nascent minus-strand viral DNA must be hybridized to the sense Tar (+) RNA sequence at the genome 3′ end for RT to resume proviral DNA synthesis. NC has been shown to chaperone this hybridization reaction, accelerating the process by about 3000-fold in physiological temperature and salt conditions (82–84, 86, 98).

Two factors must be taken into account when trying to draw simple schemes depicting the ongoing hybridization reaction. First, more than one NC molecule binds to Tar and since optimal chaperone activity occurs at a molar ratio of about 1 NC per 15 nucleotides, 4 NC molecules, on the average, should be bound to Tar. Second, NC molecules appear to be engaged in NC-to-NC interactions upon binding to RNA or DNA. Therefore, one can imagine that in such nucleoprotein complexes NC molecules destabilize the ends of nucleic acid molecules. At the same time, a high number of NC-to-NC connections causes the rapid hybridization of Tar (−) to Tar (+) sequences by a "molecular crowding" effect (the matchmaker activity). Finally, in agreement with the structural models, NC appears to interact with only one side of nucleic acids. Such a structural conformation should allow the viral DNA polymerase to interact with both the 3′ end of viral cDNA and genomic RNA. In addition, interactions between the small viral NC (7 kDa) and heterodimeric reverse transcriptase (66 and 51 kDa, respectively) have been reported and found to help viral DNA synthesis by RT (112, 113).

In an attempt to understand the global conformation of nucleoprotein complexes with chaperone activity, one can draw a picture in which chaperone molecules are bound onto one side of the nucleic acid. Although protein molecules are probably not exactly next to each other, the notion is that they are engaged in homotypic interactions. This should create a platform able to recruit other proteic factors either in a free form or bound to a nucleic acid. At the same time, the nucleic acid on the other side of the nucleoprotein complex would be in a position to contact other nucleic acid molecules and hybridize either in a transient manner as for ribozymes (94–96) or more stably like a cellular tRNA annealed to a single site located at the 5′ end of the genomic RNA of animal retroviruses (114).

It is also likely that both sides of the complex are required to recruit enzymes and enhance their activities, similar to the action of retroviral reverse transcriptase and integrase enzymes by NC–retroviral nucleic acid complexes (82, 99, 112, 113, 115, 116).

It is noteworthy that this proposed picture of nucleoprotein complexes is highly dynamic and also amenable to controls. For example, when chaperone protein molecules overload RNA, the chaperone properties of the complex are lost and RNA functions appear to progressively be "frozen" (see Fig. 7). For example, excess binding of p50 to a mRNA has the interesting consequence of switching the translation level of that RNA from high to low (117–120).

The mRNA becomes, at least in part, hidden within the complex and its ability to recruit translation initiation factors like eIF4E is lost (121). Similarly, the presence of excess NC molecules on HIV-1 genomic RNA causes a drastic reduction in reverse transcriptase activity and in the amount of newly made cDNA *in vitro* (99, 104).

Thus, the emerging picture is that the degree of RNA occupancy by chaperone proteins probably finely regulates the biological properties and functions of the RNA. The structural switch of a nucleoprotein complex from a crowded state to an overcrowded one will result in the loss of RNA function.

Another issue of importance is the unwinding of RNA secondary structures when the degree of occupancy is high (Fig. 7). Under these conditions, RNA is completely coated by chaperone molecules, and other nucleoprotein complexes and large protein molecules can no longer interact with the compacted RNP structure. These losses of functions and RNA secondary structures will render the RNA more susceptible to degradation by small RNases.

In an attempt to draw (probably oversimplified) conclusions on how chaperone proteins work, one should keep in mind that nucleoprotein complexes can be everything but "beads on a string." The starting process is the best induced fit, that is, when the RNA and the protein come together and induce the most appropriate configuration of the respective components. Next, we need to go from one to dozens and hundreds of chaperone molecules bound onto a single RNA, and sometimes on more than one RNA. Considering a single nucleoprotein complex, chaperone molecules accumulate along the RNA, most probably in an ordered vectored fashion, thus forming oriented oligomeric proteic structures held together by additional protein interactions. Last, due to strong interactions between nucleoprotein complexes, molecular crowding emerges, generating an environment suitable for proper and rapid annealing, strand exchange, and ribozyme enhancement reactions under physiological conditions. The narrow window of activity might be the mechanism by which RNA chaperones, in the form of oligomeric structures, have evolved in order to provide a fine tuning of their activities and functions (see Section VI).

C. Structural and Functional Diversity of RNA Chaperone Proteins

Examples of proteins with RNA chaperone properties are listed in Table III, which gives a protein identity card: origin and SwissProt reference number, experimental evidence of nucleic acid chaperone activities, RNA binding motif, interaction with other proteins, isoforms and posttranslational modifications, subcellular localization, and possible functions.

An emphasis is made here on cellular proteins that are abundant and thought to achieve more than one key function in the cell. Accordingly these proteins are localized in different cellular compartments.

TABLE III
RNA CHAPERONE PROTEINS

Name	Accession number[a]	Experimental evidence for nucleic acid chaperone activity[b]	Known RNA-binding motif(s)[c]	Protein interactions[d]	Isoforms and post translational modifications	Functions	Subcellular localization	Comments	Ref.[e]
hnRNP A1 (H. sapiens)	P09651	Annealing, strand exchange, ribozyme, folding trap resolution	RRM (2), RGG	Homotypic	Proteolytic (UP1)	Pre-mRNA processing and mRNA export, part of the ribonucleosome	Nucleus, nuclear–cytoplasmic shuttling	—	25, 95, 96, 122, 127, 188–195
hnRNP C1/C2 (H. sapiens)	P07910	Annealing	RRM	Homotypic	Alternative splicing	Part of the ribonucleosome	Nucleus	—	25
hnRNP U (H. sapiens)	Q00839	Annealing	RGG	Homotypic, hnRNP K, glucocorticoid receptor	Alternative splicing, phosphorylation	Part of the ribonucleosome, bind to scaffold-attached region (SAR) DNA, regulation of transcription	Nucleus	—	25
SF2 (H. sapiens)	Q07955	Annealing	RRM (2)	Homotypic, and UT U2AF (splicing factors)	Alternative splicing, phosphorylation	Splicing accuracy	Nucleus	—	196

(continued)

TABLE III (Continued)

Name	Accession number[a]	Experimental evidence for nucleic acid chaperone activity[b]	Known RNA-binding motif(s)[c]	Protein interactions[d]	Isoforms and post translational modifications	Functions	Subcellular localization	Comments	Ref.[e]
U2AF65 (H. sapiens)	P26368	Annealing	RRM (3)	U2AF35, SF2, UAP56	—	Splicing (intronic polypyrimidine tract recognition and prespliceosome formation)	Nucleus, nuclear–cytoplasmic shuttling	—	197
PSF (H. sapiens)	P23246	Annealing, D-loop formation	RRM (2)	PTB	Two isoforms produced by alternative splicing	Early splicing factor	Nucleus	—	198–201
FUS (H. sapiens)	P35637	Annealing, D-loop formation	RRM	—	—	Unknown	Nucleus	Protooncogene	198–200, 202
Yralp (S. cerevisiae)	Q12159	Annealing	RRM	Mex67p (mRNA export receptor)	—	RNA and export factor binding proteins (REF family), links pre-messenger RNA splicing to nuclear export	Nucleus	Yeast homolog of mammalian Aly protein	203
gBP21 (T. brucei)	P90629	Annealing	—	RNP editing complex	—	RNA editing (annealing of gRNAs to cognate pre-mRNAs)	Mitochondria	—	204

Nucleolin (H. sapiens)	P19338	Annealing	RRM (4), RGG	Ribosomal proteins	Phosphorylation	pre-rRNA transcription and early steps of ribosome assembly	Nucleolus, nucleus	Nuclear–cytoplasmic shuttling	205, 206
p50 (H. sapiens)	Q28618	Rnase T1, annealing, strand exchange, ribozyme	CSD	Homotypic	—	mRNP major core protein, regulation of translation	Cytoplasm	—	93, 137
FMRP (H. sapiens)	Q06787	Annealing	KH (2), RGG	FXR1, FXR2	Many isoforms produced by alternative splicing	Translation repressor, mRNA transport from nucleus to cytoplasm	Cytoplasm, nucleus	Absence of protein FMRP leads to fragile X syndrome	139 and E. W. Khandjian, personal communication
eIF4B (H. sapiens)	P23588	Annealing, strand exchange	RRM	Homotypic, eIF3, eIF4A, eIF4F, PABP	—	Translation initiation, enhances the RNA helicase activity of both EIF4A and EIF4F	Cytoplasm	—	207
Ribosomal protein S12 (E. coli)	P02367	Folding trap resolution, self-splicing, ribozyme	—	—	—	Translation initiation, splicing of T4 phage introns	Small ribosomal subunit (prokaryote)	Homologous to eukaryotic s23p	208–210

(*continued*)

TABLE III (*Continued*)

Name	Accession number[a]	Experimental evidence for nucleic acid chaperone activity[b]	Known RNA-binding motif(s)[c]	Protein interactions[d]	Isoforms and post translational modifications	Functions	Subcellular localization	Comments	Ref.[e]
Xlrbpa (*X. laevis*)	Q91836	Annealing	dsRBD (3)	—	—	Associated with the majority of cellular RNAs, ribosomal RNAs, and hnRNAs either alone or as part of an hnRNP complex	Nucleus, cytoplasm	Human homolog of TAR RNA-binding protein (TRBP)	*211*
Retroviral NCp	Q9PY17 (HIV-1)	Annealing, strand exchange, ribozyme, specific priming of reverse transcription, folding trap resolution	"CCHC" ZF (1 or 2)	Homotypic, RT	Derived from the Gag polyprotein	Viral genome packaging, reverse transcription, virus assembly	Cytoplasm, virions	—	*82, 91, 94–96, 112, 114, 147, 149, 150, 208, 212*
Retrotransposon Ty3 NCp9	gb: M34549	Annealing, specific priming of reverse transcription	"CCHC" ZF	—	Derived from the Gag polyprotein	Retrotransposition	Cytoplasm	—	*108, 163*
I element ORFIp (*Drosophila*)	gb: AAA70221	Annealing	"CCHC" ZF	Homotypic	—	Retroposition	Cytoplasm	—	*109*

254

ORFIp L1 (*M. musculus*)	gb: AAA67726	Annealing, strand exchange	—	Homotypic	—	Retroposition	Cytoplasm	—	92, 213
TYA1 (*S. cerevisiae*)	gb: AAA66937	Annealing, specific priming of reverse transcription	—	Homotypic, Ty1B Ty1 enzymes)	Proteolytic maturation	Virus-like particle formation, RNA genome packaging, reverse transcription	Cytoplasm	—	90
Delta hepatitis, antigen HDAg	P25989	Ribozyme	—	Homotypic, nucleolin, RNA—pol II	Two isoforms, phosphorylation	Regulation of transcription (elongation), HDV RNA replication, nuclear import of HDV RNA	Nucleus	—	214, 215
DnaX (*E. coli*)	P06710	Annealing	—	Homotypic, DNA pol III holoenzyme	—	Part of DNA pol III holoenzyme, stabilizes the primer-template interaction during DNA replication	Prokaryote	Tau subunit (DNA polymerase III holoenzyme)	87
StpA	P30017	Annealing, strand exchange, self-splicing, folding trap resolution	—	—	—	Histone-like nucleoid-structuring (H-NS) protein, regulation of transcription	Prokaryote	—	208, 212, 216–218

(*continued*)

TABLE III (Continued)

Name	Accession number[a]	Experimental evidence for nucleic acid chaperone activity[b]	Known RNA-binding motif(s)[c]	Protein interactions[d]	Isoforms and post translational modifications	Functions	Subcellular localization	Comments	Ref.[e]
CspA	P15277	Rnase T1 susceptibility	CSD	—	—	Regulation of transcription, transcription antiterminator	Prokaryote	—	219–221
La antigen (*H. sapiens*)	P05455	No direct evidence, most probably capture of an RNA conformation	RRM	—	Phosphorylation	Pol III transcript maturation, associated with many RNPs	Nucleus, cytoplasm	—	222
Prion protein (*H. sapiens*)	P04156	Annealing, strand exchange, ribozyme, specific priming of reverse transcription	—	—	GPI-anchored	Copper metabolism?, signal transduction?, nucleic acid metabolism?	Membrane, cytoplasm, nucleus	Involved in the propagation of several prion diseases	89, 104 and C. Gabus and J. L. Darlix, unpublished results
DdRBP1 (*D. discoideum*)	Q94467	Annealing	RRM	—	—	Unknown	Nucleus	—	223

[a] In SwissProt database unless gb is indicated, for Genbank.
[b] Annealing, strand exchange, and specific priming of reverse transcription refer to assays described in the text. Ribozyme: The ability of the protein to enhance ribozyme catalysis is also described in the text. RNase T1: The ability of the protein to enhance RNase T1 digestion due to its unwinding activity. Folding trap resolution: An *in vitro* system for assessing RNA chaperone protein activity based on the resolution of a *td* intron folding trap resolution in *E. coli* (208).
[c] See Table II for abbreviations. ZF, Zinc finger.
[d] PTB, Polypyrimidine-track-binding protein; eIF, translation initiation factor; PABP, poly (A)-binding protein.
[e] References have been selected according to the RNA chaperone activity.

hnRNP A1 is an abundant and ubiquitous cellular protein and, like many other hnRNP proteins, it exhibits strong chaperone activities (25). HnRNP A1 shuttles from the cytoplasm to the nucleus, and is tightly associated with pre-mRNAs as part of hnRNP structures. It is involved in the splicing (122–129) and nuclear export (130, 131) of cellular RNAs as well as in telomere maintenance (54, 132–134), and such functions most probably require the chaperone activities of the protein.

Nucleolin is the most abundant and ubiquitous protein of the nucleolus and an important driving force of pre-rRNA synthesis and processing and in ribosome assembly (135). Nucleolin shuttles from the nucleolus to the cytoplasm and is also found at the plasma membrane, where it has been reported to interact with glucocorticoid receptor (135). During budding from mouse cells murine leukemia viruses can incorporate nucleolin by virtue of Gag–nucleolin interactions (136). It has been argued that the function of nucleolin in MLV virions could be to attenuate virus replication (136).

P50 is the most abundant and ubiquitous protein of ribonucleoprotein particles and regulates the level of translation of mRNAs in general (93, 117–119, 121, 137). In agreement with its functions, it associates with actin fibers and polysomes in the cytoplasm (119, 138).

FMRP, the fragile X mental retardation protein, is also an important protein associated with RNPs and ribosomes (139). Lack or even a low level of protein synthesis results in mental retardation, a dominant genetic trait (140, 141). FMRP regulates protein synthesis, possibly through the recognition of G-quartets (142). Interestingly, many isoforms have been synthesized, but the function of each of them is unknown (143).

PrP, the prion protein, is a ubiquitous protein, rather abundant in the brain and spinal cord as well as in cells of the immune system. Transconformation of PrP structure, from a majority of α helices in the C-terminal domain to β sheets, has been proposed to be the major cause of spongiform neuronal degeneration in sheep, cattle, and humans (bovine spongiform encephalopathy, BSE; scrapie; and Creutz feldt–Jakob disease, CJD) (144). However, this awaits a direct demonstration. In addition, the cellular function of PrP remains a matter of speculation, since PrP ($-/-$) laboratory mice develop and live normally (145). Recently, PrP has been found to bind DNA and RNA, to form large oligomeric structures upon binding, and to exhibit extensive RNA chaperone properties in vitro (89, 104). These findings prompted the authors to propose that PrP might be involved in RNA metabolism and that nucleic acid binding may be implicated in the aggregation of PrP and thus in the scrapie process. However, this awaits direct demonstration. It is noteworthy that a first step has been achieved by showing that DNA binding induces a conformational change from α helix to β sheet in vitro (146).

Retroviral NC proteins are canonical examples of viral proteins that exhibit strong chaperone activities in vitro (147). NC protein is the sole nucleic acid-binding protein encoded by oncoretroviruses, lentiviruses, and related retroelements like endogenous and retrotransposons that are present at high copy numbers in the genomes of yeast to humans (5, 16, 17). NC protein directs recruitment and dimerization of the retroviral genome during virion assembly. The viral RNA packaging process seems to be mediated by NC–RNA stem–loop interactions involving RNA Ψ sequences. To explain genomic RNA recruitment at the molecular level, it is has been stressed that NC–zinc finger structures tightly bind to G residues that are at the tip of the Ψ RNA stem–loop structures (51). In the virus core, 2000 molecules or so of NC protein extensively coat the dimeric RNA genome, forming a chromatin-like structure (148). Upon cell infection, proviral DNA synthesis by RT takes place and NC protein has been shown to chaperone most, if not all, of the reverse transcription reactions, at least in HIV and oncoretroviruses like MLV and avian sarcoma and leukemia viruses (ASLV) (see Figs. 4 and 8) (147, 149, 150). NC protein is also thought to chaperone integration of HIV-1 proviral DNA (115, 116).

A special mention should be made regarding the tumor suppressor protein p53, the functions of which have been and still are of great interest and the subject of constant efforts. This key cellular protein is multifunctional and plays a critical role in modulating cellular responses upon DNA damage. Indeed, after DNA damage, normal mammalian cells exhibit G1 cell cycle arrest and inhibition of DNA replication. This mechanism, which requires the wild-type form of p53, allows normal cells to undertake DNA repair, thus avoiding fixation of mutations. P53 binds to DNA and RNA and more tightly to ssRNA than to ssDNA and possesses strong RNA and DNA chaperone activities *in vitro* (151). The DNA chaperone properties of p53 are most probably implicated in the mechanism by which DNA damage is sensed and repaired by means of recombination repair. Well-known oncogenic viral proteins such as the simian virus 40 large tumor antigen (SV40 T) and protein E6 of human papilloma virus 16 (HPV-16 E6) (152) interact with p53 and a high level of their expression releases the G1 cell cycle arrest caused by p53 upon DNA damage. The notion is that such oncogenic viral proteins, through their interaction with p53, impair the chaperone activities of p53, thus overcoming the inhibition of cellular DNA synthesis exerted by this cellular protein.

VI. Conclusions and Future Prospects

RNA chaperones are widespread and abundant cellular and viral proteins and exhibit a wide variety of biological activities, such as nucleic acid binding, annealing, strand exchange, ribozyme enhancement, and helix destabilization (see discussion of the narrow window of activity). Explaining these activities at the molecular level needs an extensive understanding of the structural parameters governing formation of RNA–chaperone complexes. With respect to RNA binding, a number of structures are known, but are limited to the model of a single chaperone molecule bound to a very small nucleic acid. As discussed before, many contacts are established between the protein and the nucleic acid and involve hydrophilic and hydrophobic interactions and intercalation. Interestingly, each component of the complex seems clearly to be more structured than when on its own, and thus there is a loss of free energy (see discussion of NC, hnRNPs). In addition, protein and RNA tend to segregate on one side, or hemisphere, each of the ribonucleoprotein complex. The following example reports the interaction between HIV-1 nucleocapsid protein and a stem–loop structure called SL3, part of the HIV-1 genomic RNA packaging Ψ signal, and the DLS dimerization sequence (see Fig. 1A). The viral RNA–chaperone structure clearly illustrates that (1) NC protein lies on the RNA surface with sections like the N-terminal domain going into the helix while the C-terminus contacts the upper loop (Fig. 9), (2) although a series of contacts takes place, the two components of this simple ribonucleoprotein complex appear to occupy separated domains in space, and (3) NC domains such as the basic C-terminus and sequence between the two fingers, and the 5'- and 3'-terminal RNA sequences, appear to be available for more interactions.

Yet, these structures at the atomic level do not explain how chaperones strongly accelerate hybridization of two complementary RNAs. Since it is known that a single chaperone molecule is insufficient to accelerate the annealing reaction, we need more sophisticated structures where several chaperone molecules and a longer RNA interact with each other. When available, these structures at the atomic level should contribute to answering a number of key questions: (1) Do all chaperone molecules interact the same way with the RNA molecule when they are next to each other? Since cooperative binding of the chaperone to RNA has been observed *in vitro*, is there a leading chaperone molecule in the ribonucleoprotein complex? Is this dictated by nucleic acid sequences or structural motifs? (2) Chaperone molecules are thought to set up homotypic interactions when bound to the RNA (see Table III and references therein); thus, how real and tight are these chaperone–chaperone interactions? If tight homotypic interactions exist, they will generate oligomers, but are there one or more types of oligomers? How do they influence the architecture of the ribonucleoprotein particle? (3) Based on single-molecule structure data, the protein and the RNA appear to be on either side of the complex; therefore, are there mini- or large platforms or both on each side of large RNP complexes to initiate contacts with other nucleic acids and proteins? Existence of such interacting platforms could be linked to the multifunctional nature of chaperones. To add to this list, it would be of great interest to decipher the atomic structure of RNP complexes interacting with an enzyme, particularly that of the nucleocapsid of pathogenic viruses like HIV, hepatitis C virus (HCV), and hanta virus, where

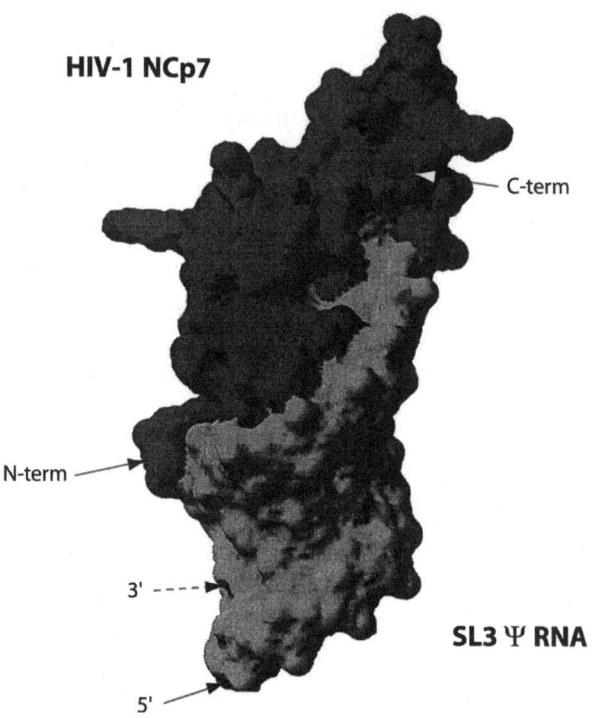

FIG. 9. Three-dimensional structure of HIV-1 NCp7–SL3-Ψ-RNA complex. The 5′-to-3′ direction of the SL3-like RNA (5′-GGACUAGCGGAGGCUAGUCC-3′) is indicated. The RNA stem corresponds to sequences 5′-GGACUAGC...GCUAGUCC-3′ and loop is 5′-GGAG-3′. The N- and C-termini of NCp7 (1–55) are indicated from bottom to top. Tight binding of NCp7 (K_d of about 100 nM) is mediated by specific interactions between the two amino- and carboxy-terminal zinc fingers of NCp7 and the two G residues flanking the A residue in the GGAG RNA tetraloop. Interestingly, this tetraloop sequence is highly conserved in all HIV-1 strains and such NCp7–RNA tetraloop interactions are thought to direct, at least in part, efficient genomic RNA recognition by NC during virion formation.

the replicating viral enzyme interacts with the virus RNP. This would facilitate design of new drugs aimed at irreversibly inhibiting viral replication.

Leaving aside this complicated issue of ribonucleoprotein particle structure, we can approach the functions of RNP from a mechanistic and dynamic point of view, from apparently simple RNPs like retroviral/retrotransposon nucleocapsids to sophisticated and highly dynamic structures like ribosomes and spliceosomes. In retroviral NC–RNA complexes, it has recently been proposed that bound NC molecules move along the RNA, but rarely leave the nucleic acid. Is this carried out according to a rolling or an inchworm model? Functional NC–RT interactions during reverse transcription of the viral RNA template favor the notion that NC molecules move along the RNA probably in an inchworm manner facilitating RT progression (*85, 112, 113*). After reverse transcription and RNase H-mediated degradation of the RNA template, NC molecules are bound to the newly made minus-strand DNA, preventing nonspecific DNA synthesis by RT.

In living cells, RNPs are probably far more functionally sophisticated than a viral nucleocapsid structure. Biological processes need highly dynamic and transient RNA–DNA, RNA–RNA, and RNA–protein interactions that take place within nucleoprotein structures. In other words, small and large RNA molecules are channeled from RNPs to RNPs. Therefore all these RNAs must accommodate their functions through efficient, ordered, specific, and most probably transient interactions with nucleic acid-binding proteins, a set of which is made up of chaperones. How is this done? Among NABPs, the DExD/H-box RNA helicase family has been found to alter in an energy-dependent fashion RNA–DNA, RNA–RNA, and RNA–protein interactions. The DExD/H RNA helicase family is a large and widely dispersed group of motor-like proteins that catalyze transconformational rearrangements of RNA interactions. How do they work to sequentially drive the reactions needed to rearrange RNA interactions? Two models have been proposed, the active rolling model and the inchworm model. In the latter model, which, to date, appears to best explain RNA helicase activity, the helicase moves step by step down the RNA strand causing the unwinding of double-stranded RNA sections or the transient dissociation of the chaperone from the RNA substrate. This is why helicases have also been named RNPases. Helicase/RNPase transit along the RNA will render RNA sections free and thus the RNA will be able to set up new and transient interactions with other RNPs, enzymes, and RNAs (*153, 154*).

Interestingly, a recent report shows that two closely related RNA helicases, p68 and p72 proteins, exhibit an RNA annealing activity in addition to their helicase activity (*155*). Such dynamic interactions between RNA, chaperones, and helicases are key to basic biological processes, which include DNA transcription, pre-mRNA splicing and editing, ribosome biogenesis, RNA export from the nucleus, and RNA translation and degradation. However, although helicases are widely distributed in living cells, with some of them well characterized by means of biochemical analyses *in vitro* and genetics, a clear demonstration of the motor role of the enzymes *in vivo* is missing. Also, the future will tell us if chaperones and helicases are involved in the modulation of RNA functions by small interfering RNAs (siRNAs) and small temporal RNAs (stRNAs) (or microRNAs) possibly at the level of translation and degradation (*156–162*). Based on recent findings (*18–20*), it can be hypothesized that microRNAs will contact a unique type of mRNA by virtue of single-site hybridization mediated by a chaperone, thus interfering with the functions of that particular RNA. Similarly, chaperones could hybridize microRNAs to series of mRNAs by means of multiple-site hybridization, similar to primer tRNA in yeast mobile genetic elements (163, 164). In the virus world, chaperone and helicase activities are essential to genome replication and virus propagation, as exemplified by NC proteins of exogenous and endogenous retroviruses.

With regard to biotechnological applications, there are two domains where RNA chaperone proteins could be useful. In this new "genomic era" very strong efforts are being made by public laboratories and private companies in order to pave the way for an understanding of how the myriad of genes are expressed and how networks of gene expression are set up, regulated, and modulated, for example, in cancer cells versus healthy cells, in inflammatory organs versus fully functional organs, in the course of disease progression, following therapeutic treatments, and in aging. In work with DNA chips, one relies on hybridization reactions carried out on the micro-range level and probably soon at the nano-range level. However, starting from the beginning of such investigations, reverse transcription of small samples of RNA populations has been done *in vitro,* but appears to be loosely controlled according to available protocols. RNA chaperones of retroviral origin have been shown to facilitate and augment the specificity of cDNA synthesis by RT *in vitro* (88, 90, 98–100, 108, 165, 166). Thus, the use of chemically synthesized peptides (90, 108, 163, 167) derived from retroviral NC proteins should improve both the yield and the specificity of RT-generated cDNA libraries intended to be used in large-scale screening reactions using DNA chips.

Another domain of biotechnological interest is the use of viral or synthetic DNA vectors in gene transfer of cells *ex vivo* and gene therapy *in vivo.* For example, retroviral vectors are currently being used in several hundred clinical protocols with the aim of curing cancers (168, 169) and correcting genetic diseases (170). Both retroviral and lentiviral vector helper functions are limited in their capacity to package a recombinant RNA vector and deliver it into target cells. The viral NC protein is the driving chaperone force during recombinant RNA packaging and vector production; subsequently it assists RT and integrase during recombinant viral DNA synthesis and integration in the target cells. Using *in vitro* genetic approaches to retroviral NC, especially involving zinc fingers, which have been shown to dictate the selectivity of genomic RNA packaging (171–173), we should reasonably be able to improve NC performance and in the future that of recombinant retroviral and lentiviral vectors for gene transfer into living cells and organisms.

Acknowledgments

We thank Bernard Roques (Paris) and Michael Summers (University of Maryland, Baltimore County, Baltimore, MD) for discussion and collaboration and for providing unpublished results, respectively. Damien Ficheux (CNRS Lyon, France) and Witold Surewicz (Case Western Reserve University, Cleveland, OH) are gratefully acknowledged for their collaboration. Work in the laboratory is supported by INSERM (France), ANRS (France), Sidaction (France), ARC (France), and BioMed (Europe).

REFERENCES

1. D. Shippen-Lentz and E. H. Blackburn, *Science* **247**, 546 (1990).
2. C. W. Greider and E. H. Blackburn, *Nature* **337**, 331 (1989).
3. C. W. Greider and E. H. Blackburn, *Cell* **51**, 887 (1987).
4. J. D. Boeke and J. P. Stoye, in "Retroviruses" (J. M. Coffin, S. H. Hughes, and H. E. Varmus, eds.), p. 343. Cold Spring Harbor Laboratory Press, Cold Spring Harbor, New York, 1997.
5. E. T. Prak and H. H. Kazazian, Jr., *Nat. Rev. Genet.* **1**, 134 (2000).
6. G. L. Eliceiri, *Cell. Mol. Life Sci.* **56**, 22 (1999).
7. T. Kiss, *EMBO J.* **20**, 3617 (2001).
8. C. L. Will and R. Luhrmann, *Curr. Opin. Cell Biol.* **13**, 290 (2001).
9. T. R. Cech, *Annu. Rev. Biochem.* **59**, 543 (1990).
10. S. Valadkhan and J. L. Manley, *Nature* **413**, 701 (2001).
11. J. C. Kurz and C. A. Fierke, *Curr. Opin. Chem. Biol.* **4**, 553 (2000).
12. D. N. Frank and N. R. Pace, *Annu. Rev. Biochem.* **67**, 153 (1998).
13. M. M. Lai, *Annu. Rev. Biochem.* **64**, 259 (1995).
14. N. Ban, P. Nissen, J. Hansen, P. B. Moore, and T. A. Steitz, *Science* **289**, 905 (2000).
15. P. Nissen, J. Hansen, N. Ban, P. B. Moore, and T. A. Steitz, *Science* **289**, 920 (2000).
16. E. S. Lander *et al.*, *Nature* **409**, 860 (2001).
17. J. C. Venter *et al.*, *Science* **291**, 1304 (2001).
18. R. C. Lee and V. Ambros, *Science* **294**, 862 (2001).
19. N. C. Lau, E. P. Lim le, E. G. Weinstein, and V. P. Bartel da, *Science* **294**, 858 (2001).
20. M. Lagos-Quintana, R. Rauhut, W. Lendeckel, and T. Tuschl, *Science* **294**, 853 (2001).
21. R. H. Symons, *Nucleic Acids Res.* **25**, 2683 (1997).
22. G. Dreyfuss, M. J. Matunis, S. Pinol-Roma, and C. G. Burd, *Annu. Rev. Biochem.* **62**, 289 (1993).
23. D. Herschlag, *J. Biol. Chem.* **270**, 20871 (1995).
24. K. M. Weeks, *Curr. Opin. Struct. Biol.* **7**, 336 (1997).
25. D. S. Portman and G. Dreyfuss, *EMBO J.* **13**, 213 (1994).
26. I. Tinoco, Jr. and C. Bustamante, *J. Mol. Biol.* **293**, 271 (1999).
27. U. Z. Littauer, *Biophys. Chem.* **86**, 259 (2000).
28. J. R. Williamson, *Nat. Struct. Biol.* **7**, 834 (2000).
29. S. A. Woodson, *Cell. Mol. Life Sci.* **57**, 796 (2000).
30. D. K. Treiber and J. R. Williamson, *Curr. Opin. Struct. Biol.* **11**, 309 (2001).
31. B. Berkhout, *Prog. Nucleic Acid Res. Mol. Biol.* **54**, 1 (1996).
32. B. L. Golden, A. R. Gooding, E. R. Podell, and T. R. Cech, *Science* **282**, 259 (1998).
33. A. A. Szewczak, L. Ortoleva-Donnelly, S. P. Ryder, E. Moncoeur, and S. A. Strobel, *Nat. Struct. Biol.* **5**, 1037 (1998).
34. C. R. Woese, S. Winker, and R. R. Gutell, *Proc. Natl. Acad. Sci. USA* **87**, 8467 (1990).
35. M. Costa and F. Michel, *EMBO J.* **14**, 1276 (1995).
36. L. Jaeger, F. Michel, and E. Westhof, *J. Mol. Biol.* **236**, 1271 (1994).
37. M. A. Tanner and T. R. Cech, *RNA* **1**, 349 (1995).
38. M. E. Robertson, R. A. Seamons, and G. J. Belsham, *RNA* **5**, 1167 (1999).

39. S. Lopez de Quinto and E. Martinez-Salas, *J. Virol.* **71**, 4171 (1997).
40. J. Pan and S. A. Woodson, *J. Mol. Biol.* **280**, 597 (1998).
41. M. Wu and I. Tinoco, Jr., *Proc. Natl. Acad. Sci. USA* **95**, 11555 (1998).
42. J. H. Cate *et al.*, *Science* **273**, 1678 (1996).
43. S. L. Yean, G. Wuenschell, J. Termini, and R. J. Lin, *Nature* **408**, 881 (2000).
44. T. A. Steitz and J. A. Steitz, *Proc. Natl. Acad. Sci. USA* **90**, 6498 (1993).
45. T. Pan and T. R. Sosnick, *Nat. Struct. Biol.* **4**, 931 (1997).
46. M. S. Rook, D. K. Treiber, and J. R. Williamson, *Proc. Natl. Acad. Sci. USA* **96**, 12471 (1999).
47. J. Pan, D. Thirumalai, and S. A. Woodson, *Proc. Natl. Acad. Sci. USA* **96**, 6149 (1999).
48. S. Basu *et al.*, *Nat. Struct. Biol.* **5**, 986 (1998).
49. A. A. Antson, *Curr. Opin. Struct. Biol.* **10**, 87 (2000).
50. H. A. Lewis *et al.*, *Cell* **100**, 323 (2000).
51. R. N. De Guzman *et al.*, *Science* **279**, 384 (1998).
52. C. E. Bogden, D. Fass, N. Bergman, M. D. Nichols, and J. M. Berger, *Mol. Cell* **3**, 487 (1999).
53. A. A. Antson *et al.*, *Nature* **401**, 235 (1999).
54. J. Ding *et al.*, *Genes Dev.* **13**, 1102 (1999).
55. R. C. Deo, J. B. Bonanno, N. Sonenberg, and S. K. Burley, *Cell* **98**, 835 (1999).
56. N. Handa *et al.*, *Nature* **398**, 579 (1999).
57. W. Nietfeld, H. Mentzel, and T. Pieler, *EMBO J.* **9**, 3699 (1990).
58. C. G. Burd, E. L. Matunis, and G. Dreyfuss, *Mol. Cell. Biol.* **11**, 3419 (1991).
59. Y. Shamoo *et al.*, *Biochemistry* **33**, 8272 (1994).
60. R. Kanaar, A. L. Lee, D. Z. Rudner, D. E. Wemmer, and D. C. Rio, *EMBO J.* **14**, 4530 (1995).
61. U. Kuhn and T. Pieler, *J. Mol. Biol.* **256**, 20 (1996).
62. P. E. Wright and H. J. Dyson, *J. Mol. Biol.* **293**, 321 (1999).
63. N. Morellet *et al.*, *EMBO J.* **11**, 3059 (1992).
64. N. Morellet *et al.*, *J. Mol. Biol.* **283**, 419 (1998).
65. G. K. Amarasinghe *et al.*, *J. Mol. Biol.* **301**, 491 (2000).
66. C. A. Smith, V. V. Calabro, and A. D. Frankel, *Mol. Cell* **6**, 1067 (2000).
67. C. Yanofsky, *J. Bacteriol.* **182**, 1 (2000).
68. L. C. Shaw and A. S. Lewin, *J. Biol. Chem.* **270**, 21552 (1995).
69. K. M. Weeks and T. R. Cech, *Cell* **82**, 221 (1995).
70. K. M. Weeks and T. R. Cech, *Science* **271**, 345 (1996).
71. A. S. Lewin, J. Thomas, Jr., and H. K. Tirupati, *Mol. Cell. Biol.* **15**, 6971 (1995).
72. J. D. Kittle, Jr., G. Mohr, J. A. Gianelos, H. Wang, and A. M. Lambowitz, *Genes Dev.* **5**, 1009 (1991).
73. G. Mohr, A. Zhang, J. A. Gianelos, M. Belfort, and A. M. Lambowitz, *Cell* **69**, 483 (1992).
74. G. Mohr, M. G. Caprara, Q. Guo, and A. M. Lambowitz, *Nature* **370**, 147 (1994).
75. M. G. Caprara, G. Mohr, and A. M. Lambowitz, *J. Mol. Biol.* **257**, 512 (1996).
76. X. Chen, R. R. Gutell, and A. M. Lambowitz, *J. Mol. Biol.* **301**, 265 (2000).
77. A. E. Webb, M. A. Rose, E. Westhof, and K. M. Weeks, *J. Mol. Biol.* **309**, 1087 (2001).
78. M. C. Williams *et al.*, *Proc. Natl. Acad. Sci. USA* **98**, 6121 (2001).
79. C. G. Burd and G. Dreyfuss, *Science* **265**, 615 (1994).
80. G. Varani and K. Nagai, *Annu. Rev. Biophys. Biomol. Struct.* **27**, 407 (1998).
81. J. L. Darlix, C. Gabus, M. T. Nugeyre, F. Clavel, and F. Barre-Sinoussi, *J. Mol. Biol.* **216**, 689 (1990).
82. V. Tanchou, C. Gabus, V. Rogemond, and J. L. Darlix, *J. Mol. Biol.* **252**, 563 (1995).
83. J. L. Darlix, A. Vincent, C. Gabus, H. de Rocquigny, and B. Roques, *C. R. Acad. Sci. III* **316**, 763 (1993).
84. B. Allain, M. Lapadat-Tapolsky, C. Berlioz, and J. L. Darlix, *EMBO J.* **13**, 973 (1994).
85. M. Lapadat-Tapolsky *et al.*, *Nucleic Acids Res.* **21**, 831 (1993).

86. M. Lapadat-Tapolsky, C. Pernelle, C. Borie, and J. L. Darlix, *Nucleic Acids Res.* **23**, 2434 (1995).
87. S. Kim and K. J. Marians, *Nucleic Acids Res.* **23**, 1374 (1995).
88. X. Li *et al.*, *J. Virol.* **70**, 4996 (1996).
89. C. Gabus *et al.*, *J. Mol. Biol.* **307**, 1011 (2001).
90. G. Cristofari, D. Ficheux, and J. L. Darlix, *J. Biol. Chem.* **275**, 19210 (2000).
91. Z. Tsuchihashi and P. O. Brown, *J. Virol.* **68**, 5863 (1994).
92. S. L. Martin and F. D. Bushman, *Mol. Cell. Biol.* **21**, 467 (2001).
93. M. A. Skabkin, V. M. Evdokimova, A. A. Thomas, and L. P. Ovchinnikov, *J. Biol. Chem.* **3**, 3 (2001).
94. Z. Tsuchihashi, M. Khosla, and D. Herschlag, *Science* **262**, 99 (1993).
95. E. L. Bertrand and J. J. Rossi, *EMBO J.* **13**, 2904 (1994).
96. D. Herschlag, M. Khosla, Z. Tsuchihashi, and R. L. Karpel, *EMBO J.* **13**, 2913 (1994).
97. W. Nedbal and G. Sczakiel, *Biochem. Soc. Trans.* **24**, 615 (1996).
98. J. Guo, L. E. Henderson, J. Bess, B. Kane, and J. G. Levin, *J. Virol.* **71**, 5178 (1997).
99. M. Lapadat-Tapolsky, C. Gabus, M. Rau, and J. L. Darlix, *J. Mol. Biol.* **268**, 250 (1997).
100. J. B. Rascle, D. Ficheux, and J. L. Darlix, *J. Mol. Biol.* **280**, 215 (1998).
101. R. Khan and D. P. Giedroc, *J. Biol. Chem.* **267**, 6689 (1992).
102. B. Chan, K. Weidemaier, W. T. Yip, P. F. Barbara, and K. Musier-Forsyth, *Proc. Natl. Acad. Sci. USA* **96**, 459 (1999).
103. M. R. Hargittai, A. T. Mangla, R. J. Gorelick, and K. Musier-Forsyth, *J. Mol. Biol.* **312**, 985 (2001).
104. C. Gabus *et al.*, *J. Biol. Chem.* **276**, 19301 (2001).
105. C. J. Gregoire, D. Gautheret, and E. P. Loret, *J. Biol. Chem.* **272**, 25143 (1997).
106. B. Berkhout and J. L. van Wamel, *RNA* **6**, 282 (2000).
107. H. Huthoff and B. Berkhout, *RNA* **7**, 143 (2001).
108. G. Cristofari *et al.*, *J. Biol. Chem.* **274**, 36643 (1999).
109. A. Dawson, E. Hartswood, T. Paterson, and D. J. Finnegan, *EMBO J.* **16**, 4448 (1997).
110. E. Le Cam *et al.*, *Biopolymers* **45**, 217 (1998).
111. S. P. Stoylov *et al.*, *Biopolymers* **41**, 301 (1997).
112. D. Lener, V. Tanchou, B. P. Roques, S. F. Le Grice, and J. L. Darlix, *J. Biol. Chem.* **273**, 33781 (1998).
113. S. Druillennec, A. Caneparo, H. de Rocquigny, and B. P. Roques, *J. Biol. Chem.* **274**, 11283 (1999).
114. A. C. Prats *et al.*, *EMBO J.* **7**, 1777 (1988).
115. S. Carteau *et al.*, *J. Virol.* **71**, 6225 (1997).
116. S. Carteau, R. J. Gorelick, and F. D. Bushman, *J. Virol.* **73**, 6670 (1999).
117. V. M. Evdokimova *et al.*, *J. Biol. Chem.* **273**, 3574 (1998).
118. E. K. Davydova, V. M. Evdokimova, L. P. Ovchinnikov, and J. W. Hershey, *Nucleic Acids Res.* **25**, 2911 (1997).
119. W. B. Minich, I. P. Maidebura, and L. P. Ovchinnikov, *Eur. J. Biochem.* **212**, 633 (1993).
120. W. B. Minich, E. V. Volyanik, N. L. Korneyeva, Y. V. Berezin, and L. P. Ovchinnikov, *Mol. Biol. Rep.* **14**, 65 (1990).
121. V. Evdokimova *et al.*, *EMBO J.* **20**, 5491 (2001).
122. M. Buvoli, F. Cobianchi, and S. Riva, *Nucleic Acids Res.* **20**, 5017 (1992).
123. A. Mayeda and A. R. Krainer, *Cell* **68**, 365 (1992).
124. A. Mayeda, D. M. Helfman, and A. R. Krainer, *Mol. Cell. Biol.* **13**, 2993 (1993).
125. C. G. Burd and G. Dreyfuss, *EMBO J.* **13**, 1197 (1994).
126. J. F. Caceres, S. Stamm, D. M. Helfman, and A. R. Krainer, *Science* **265**, 1706 (1994).
127. A. Mayeda, S. H. Munroe, J. F. Caceres, and A. R. Krainer, *EMBO J.* **13**, 5483 (1994).

128. X. Yang et al., *Proc. Natl. Acad. Sci. USA* **91**, 6924 (1994).
129. M. Blanchette and B. Chabot, *EMBO J.* **18**, 1939 (1999).
130. E. Izaurralde et al., *J. Cell. Biol.* **137**, 27 (1997).
131. W. M. Michael, M. Choi, and G. Dreyfuss, *Cell* **83**, 415 (1995).
132. H. LaBranche et al., *Nat. Genet.* **19**, 199 (1998).
133. F. Dallaire, S. Dupuis, S. Fiset, and B. Chabot, *J. Biol. Chem.* **275**, 14509 (2000).
134. S. Fiset and B. Chabot, *Nucleic Acids Res.* **29**, 2268 (2001).
135. H. Ginisty, H. Sicard, B. Roger, and P. Bouvet, *J. Cell Sci.* **112**, 761 (1999).
136. E. Bacharach, J. Gonsky, K. Alin, M. Orlova, and S. P. Goff, *J. Virol.* **74**, 11027 (2000).
137. V. M. Evdokimova et al., *J. Biol. Chem.* **270**, 3186 (1995).
138. P. V. Ruzanov, V. M. Evdokimova, N. L. Korneeva, J. W. Hershey, and L. P. Ovchinnikov, *J. Cell Sci.* **112**, 3487 (1999).
139. E. W. Khandjian, F. Corbin, S. Woerly, and F. Rousseau, *Nat. Genet.* **12**, 91 (1996).
140. Y. Feng et al., *Mol. Cell* **1**, 109 (1997).
141. C. T. Ashley, Jr., K. D. Wilkinson, D. Reines, and S. T. Warren, *Science* **262**, 563 (1993).
142. C. Schaeffer et al., *EMBO J.* **20**, 4803 (2001).
143. E. W. Khandjian et al., *Hum. Mol. Genet.* **4**, 783 (1995).
144. J. Collinge, *Annu. Rev. Neurosci.* **24**, 519 (2001).
145. H. Bueler et al., *Nature* **356**, 577 (1992).
146. Y. Cordeiro et al., *J. Biol. Chem.* **16**, 16 (2001).
147. J. L. Darlix et al., *Adv. Pharmacol.* **48**, 345 (2000).
148. M. Chen, C. F. Garon, and T. S. Papas, *Proc. Natl. Acad. Sci. USA* **77**, 1296 (1980).
149. J. L. Darlix, M. Lapadat-Tapolsky, H. de Rocquigny, and B. P. Roques, *J. Mol. Biol.* **254**, 523 (1995).
150. A. Rein, L. E. Henderson, and J. G. Levin, *Trends Biochem. Sci.* **23**, 297 (1998).
151. P. Oberosler, P. Hloch, U. Ramsperger, and H. Stahl, *EMBO J.* **12**, 2389 (1993).
152. T. D. Kessis et al., *Proc. Natl. Acad. Sci. USA* **90**, 3988 (1993).
153. N. K. Tanner and P. Linder, *Mol. Cell* **8**, 251 (2001).
154. P. Linder, N. K. Tanner, and J. Banroques, *Trends Biochem. Sci.* **26**, 339 (2001).
155. O. G. Rossler, A. Straka, and H. Stahl, *Nucleic Acids Res.* **29**, 2088 (2001).
156. T. Tuschl, P. D. Zamore, R. Lehmann, D. P. Bartel, and P. A. Sharp, *Genes Dev.* **13**, 3191 (1999).
157. S. M. Hammond, E. Bernstein, D. Beach, and G. J. Hannon, *Nature* **404**, 293 (2000).
158. P. D. Zamore, T. Tuschl, P. A. Sharp, and D. P. Bartel, *Cell* **101**, 25 (2000).
159. S. M. Elbashir et al., *Nature* **411**, 494 (2001).
160. S. M. Elbashir, W. Lendeckel, and T. Tuschl, *Genes Dev.* **15**, 188 (2001).
161. A. E. Pasquinelli et al., *Nature* **408**, 86 (2000).
162. B. J. Reinhart et al., *Nature* **403**, 901 (2000).
163. C. Gabus et al., *EMBO J.* **17**, 4873 (1998).
164. S. Friant, T. Heyman, A. S. Bystorm, M. Wilhelm, and F. X. Wilhelm, *Mol. Cell. Biol.* **18**, 799 (1998).
165. M. D. Driscoll, M. P. Golinelli, and S. H. Hughes, *J. Virol.* **75**, 672 (2001).
166. M. P. Golinelli and S. H. Hughes, *Virology* **285**, 278 (2001).
167. H. de Rocquigny et al., *Biochem. Biophys. Res. Commun.* **180**, 1010 (1991).
168. D. Klatzmann et al., *Hum. Gene Ther.* **9**, 2595 (1998).
169. D. Klatzmann et al., *Hum. Gene Ther.* **9**, 2585 (1998).
170. M. Cavazzana-Calvo et al., *Science* **288**, 669 (2000).
171. R. J. Gorelick, D. J. Chabot, A. Rein, L. E. Henderson, and L. O. Arthur, *J. Virol.* **67**, 4027 (1993).
172. V. Tanchou et al., *J. Virol.* **72**, 4442 (1998).

173. L. Berthoux, C. Pechoux, M. Ottmann, G. Morel, and J. L. Darlix, *J. Virol.* **71**, 6973 (1997).
174. R. Apweiler *et al.*, *Nucleic Acids Res.* **29**, 37 (2001).
175. A. Ramos *et al.*, *EMBO J.* **19**, 997 (2000).
176. S. Nanduri, B. W. Carpick, Y. Yang, B. R. Williams, and J. Qin, *EMBO J.* **17**, 5458 (1998).
177. M. Bycroft, S. Grunert, A. G. Murzin, M. Proctor, and D. St Johnston, *EMBO J.* **14**, 3563 (1995).
178. J. M. Ryter and S. C. Schultz, *EMBO J.* **17**, 7505 (1998).
179. H. Demene *et al.*, *J. Biomol. NMR* **4**, 153 (1994).
180. Y. Gao, K. Kaluarachchi, and D. P. Giedroc, *Protein Sci.* **7**, 2265 (1998).
181. Y. Kodera *et al.*, *Biochemistry* **37**, 17704 (1998).
182. W. Schuler, C. Dong, K. Wecker, and B. P. Roques, *Biochemistry* **38**, 12984 (1999).
183. D. J. Klein, P. E. Johnson, E. S. Zollars, R. N. De Guzman, and M. F. Summers, *Biochemistry* **39**, 1604 (2000).
184. H. Schindelin, W. Jiang, M. Inouye, and U. Heinemann, *Proc. Natl. Acad. Sci. USA* **91**, 5119 (1994).
185. W. Kremer *et al.*, *Eur. J. Biochem.* **268**, 2527 (2001).
186. A. Schnuchel *et al.*, *Nature* **364**, 169 (1993).
187. U. Mueller, D. Perl, F. X. Schmid, and U. Heinemann, *J. Mol. Biol.* **297**, 975 (2000).
188. A. Kumar, K. R. Williams, and W. Szer, *J. Biol. Chem.* **261**, 11266 (1986).
189. A. Kumar and S. H. Wilson, *Biochemistry* **29**, 10717 (1990).
190. B. W. Pontius and P. Berg, *Proc. Natl. Acad. Sci. USA* **87**, 8403 (1990).
191. S. H. Munroe and X. F. Dong, *Proc. Natl. Acad. Sci. USA* **89**, 895 (1992).
192. B. W. Pontius and P. Berg, *J. Biol. Chem.* **267**, 13815 (1992).
193. J. R. Casas-Finet *et al.*, *J. Mol. Biol.* **229**, 873 (1993).
194. F. Cobianchi, C. Calvio, M. Stoppini, M. Buvoli, and S. Riva, *Nucleic Acids Res.* **21**, 949 (1993).
195. H. Idriss *et al.*, *Biochemistry* **33**, 11382 (1994).
196. A. R. Krainer, G. C. Conway, and D. Kozak, *Genes Dev.* **4**, 1158 (1990).
197. C. G. Lee, P. D. Zamore, M. R. Green, and J. Hurwitz, *J. Biol. Chem.* **268**, 13472 (1993).
198. A. T. Akhmedov, P. Bertrand, E. Corteggiani, and B. S. Lopez, *Proc. Natl. Acad. Sci. USA* **92**, 1729 (1995).
199. E. A. Namsaraev, V. A. Lanzov, and A. T. Akhmedov, *Biochim. Biophys. Acta* **1305**, 172 (1996).
200. P. Bertrand, A. T. Akhmedov, F. Delacote, A. Durrbach, and B. S. Lopez, *Oncogene* **18**, 4515 (1999).
201. A. T. Akhmedov and B. S. Lopez, *Nucleic Acids Res.* **28**, 3022 (2000).
202. H. Bacchtold *et al.*, *J. Biol. Chem.* **274**, 34337 (1999).
203. D. S. Portman, J. P. O'Connor, and G. Dreyfuss, *RNA* **3**, 527 (1997).
204. U. F. Muller, L. Lambert, and H. U. Goringer, *EMBO J.* **20**, 1394 (2001).
205. L. A. Hanakahi, Z. Bu, and N. Maizels, *Biochemistry* **39**, 15493 (2000).
206. F. H. Allain, P. Bouvet, T. Dieckmann, and J. Feigon, *EMBO J.* **19**, 6870 (2000).
207. M. Altmann, B. Wittmer, N. Methot, N. Sonenberg, and H. Trachsel, *EMBO J.* **14**, 3820 (1995).
208. E. Clodi, K. Semrad, and R. Schroeder, *EMBO J.* **18**, 3776 (1999).
209. T. Coetzee, D. Herschlag, and M. Belfort, *Genes Dev.* **8**, 1575 (1994).
210. K. Semrad and R. Schroeder, *Genes Dev.* **12**, 1327 (1998).
211. E. Hitti, A. Neunteufl, and M. F. Jantsch, *Nucleic Acids Res.* **26**, 4382 (1998).
212. M. Negroni and H. Buc, *Proc. Natl. Acad. Sci. USA* **97**, 6385 (2000).
213. S. L. Martin, J. Li, and J. A. Weisz, *J. Mol. Biol.* **304**, 11 (2000).
214. Z. S. Huang and H. N. Wu, *J. Biol. Chem.* **273**, 26455 (1998).
215. K. S. Jeng, P. Y. Su, and M. M. Lai, *J. Virol.* **70**, 4205 (1996).
216. M. E. Cusick and M. Belfort, *Mol. Microbiol.* **28**, 847 (1998).
217. M. Negroni and H. Buc, *J. Mol. Biol.* **286**, 15 (1999).

218. A. Zhang, V. Derbyshire, J. L. Salvo, and M. Belfort, *RNA* **1,** 783 (1995).
219. W. Bae, P. G. Jones, and M. Inouye, *J. Bacteriol.* **179,** 7081 (1997).
220. W. Bae, B. Xia, M. Inouye, and K. Severinov, *Proc. Natl. Acad. Sci. USA* **97,** 7784 (2000).
221. W. Jiang, Y. Hou, and M. Inouye, *J. Biol. Chem.* **272,** 196 (1997).
222. B. K. Pannone, D. Xue, and S. L. Wolin, *EMBO J.* **17,** 7442 (1998).
223. P. Oberosler and W. Nellen, *Biol. Chem.* **378,** 1353 (1997).
224. B. Berkhout and K. T. Jeang, *J. Virol.* **63,** 5501 (1989).
225. B. Berkhout, R. H. Silverman, and K. T. Jeang, *Cell* **59,** 273 (1989).
226. C. Barat et al., *EMBO J.* **8,** 3279 (1989).
227. N. Beerens, F. Groot, and B. Berkhout, *J. Biol. Chem.* **276,** 31247 (2001).
228. S. Auxilien, G. Keith, S. F. Le Grice, and J. L. Darlix, *J. Biol. Chem.* **274,** 4412 (1999).
229. T. Wu, J. Guo, J. Bess, L. E. Henderson, and J. G. Levin, *J. Virol.* **73,** 4794 (1999).
230. J. A. Peliska and S. J. Benkovic, *Science* **258,** 1112 (1992).
231. E. Skripkin, J. C. Paillart, R. Marquet, B. Ehresmann, and C. Ehresmann, *Proc. Natl. Acad. Sci. USA* **91,** 4945 (1994).
232. F. Baudin et al., *J. Mol. Biol.* **229,** 382 (1993).
233. J. F. Kaye, J. H. Richardson, and A. M. Lever, *J. Virol.* **69,** 6588 (1995).
234. J. H. Richardson, L. A. Child, and A. M. Lever, *J. Virol.* **67,** 3997 (1993).
235. A. Lever, H. Gottlinger, W. Haseltine, and J. Sodroski, *J. Virol.* **63,** 4085 (1989).
236. B. Berkhout and J. L. van Wamel, *J. Virol.* **70,** 6723 (1996).
237. S. Hoglund, A. Ohagen, J. Goncalves, A. T. Panganiban, and D. Gabuzda, *Virology* **233,** 271 (1997).
238. M. S. McBride, M. D. Schwartz, and A. T. Panganiban, *J. Virol.* **71,** 4544 (1997).
239. J. I. Sakuragi and A. T. Panganiban, *J. Virol.* **71,** 3250 (1997).
240. M. S. McBride and A. T. Panganiban, *J. Virol.* **71,** 2050 (1997).
241. M. S. McBride and A. T. Panganiban, *J. Virol.* **70,** 2963 (1996).
242. C. B. Buck et al., *J. Virol.* **75,** 181 (2001).
243. T. Ohlmann, M. Lopez-Lastra, and J. L. Darlix, *J. Biol. Chem.* **275,** 11899 (2000).
244. X. Chen et al., *EMBO J.* **14,** 842 (1995).
245. L. X. Shen and I. Tinoco, Jr., *J. Mol. Biol.* **247,** 963 (1995).
246. O. Hungnes, E. Tjotta, and B. Grinde, *Virology* **190,** 440 (1992).
247. P. Charneau, M. Alizon, and F. Clavel, *J. Virol.* **66,** 2814 (1992).
248. O. Hungnes, E. Tjotta, and B. Grinde, *Arch. Virol.* **116,** 133 (1991).
249. T. Heyman, B. Agoutin, S. Friant, F. X. Wilhelm, and M. L. Wilhelm, *J. Mol. Biol.* **253,** 291 (1995).
250. P. Pochart et al., *Nucleic Acids Res.* **21,** 3513 (1993).

Mechanisms of Basal and Kinase-Inducible Transcription Activation by CREB

PATRICK G. QUINN

Department of Cellular and Molecular
Physiology
The Pennsylvania State University,
College of Medicine
Hershey, Pennsylvania 17033

I. Introduction... 270
II. Distinct CREB Activation Domains Mediate Constitutive and
   Kinase-Inducible Transcription 272
   A. Evidence for Distinct Constitutive and Kinase-Inducible Activities..... 273
   B. Definition of the Constitutive Activation Domain (CAD) and the
      Kinase-Inducible Domain (KID)............................... 274
   C. Characterization of the CAD and Its Interactions with General
      Transcription Factors... 279
III. A Concerted Mechanism of Transcription Activation Involving
   Stimulation of Sequential Steps in Transcription Initiation by the CAD
   and P-KID.. 284
   A. Role of the CAD in the Recruitment of an RNA Polymerase II Complex 289
   B. Role of Phosphorylation of the KID in Activation of the RNA
      Polymerase II Complex.. 293
   C. A Model for a Concerted Mechanism of Transcription Activation by
      Phosphorylated CREB 296
IV. Perspectives... 298
   References.. 299

The cAMP response element (CRE)-binding protein (CREB) stimulates basal transcription of CRE-containing genes and mediates induction of transcription upon phosphorylation by protein kinases. The basal activity of CREB maps to a carboxy-terminal constitutive activation domain (CAD), whereas phosphorylation and inducibility map to a central, kinase-inducible domain (KID). The CAD interacts with and recruits the promoter recognition factor TFIID through an interaction with a specific TATA-binding-protein-associated factor (TAF), dTAF$_{II}$110/hTAF$_{II}$135. Interaction between the TAF and the CAD is mediated by a central cluster of hydrophobic amino acids, mutation of which disrupts TAF binding, polymerase recruitment, and transcription activation. Assessment of the contributions of the CAD and KID to recruitment of the polymerase complex versus enhancement of subsequent reaction steps (isomerization, promoter clearance, and reinitiation) showed that the CAD and P-KID act in a concerted mechanism to stimulate transcription. The CAD, but not the KID, mediated recruitment of a complex containing components of a transcription initiation complex, including pol II, IIB, and IID. However, the CAD was relatively ineffective in stimulating subsequent steps in the reaction mechanism. In contrast, phosphorylation of the KID in CREB effectively stimulated isomerization of the recruited polymerase complex and multiple-round transcription. A model for the activation of transcription by phosphorylated CREB is proposed, in which the polymerase is recruited by interaction of the CAD with TFIID and the recruited polymerase is activated further by phosphorylation of the KID in CREB. © 2002, Elsevier Science (USA).

I. Introduction

CREB, cAMP response element (CRE)-binding protein, mediates basal and kinase-inducible transcription of target genes in a variety of cells and tissues. For any gene to be transcribed, there must be recruitment of a polymerase complex, isomerization to expose the template strand, and promoter clearance of the polymerase to transcribe the body of the gene (1). Studies based on the binding of endogenous proteins to CREs produced conflicting results due to heterogeneity of CRE-binding proteins and the influence of different promoter contexts for the CRE. However, early studies did demonstrate that the presence of a CRE in a promoter produced an extended footprint due to a complex containing RNA polymerase II and TFIIB, indicating that CRE-binding proteins may facilitate recruitment (2, 3). In contrast, *in vitro* transcription experiments with nuclear extracts suggested that factors associated with CREs could promote any (2, 4, 5) or all (6) of the steps in transcription initiation in response to cAMP. Analysis of the CREB protein identified distinct domains that mediate basal and kinase-inducible transcription activation (7, 8). The focus of this review will be on the definition and characterization of the activities of these domains and characterization of the mechanisms employed by CREB to mediate basal and kinase-inducible gene transcription.

Activation of protein kinase A (PKA) by cAMP was first shown to stimulate run-on transcription of the prolactin gene in 1981 by Maurer (9). Over the next decade, the promoter regions of many genes were studied in detail using the newly developed chloramphenicol acetyltransferase (CAT) reporter assay. The CRE, an 8-base pair (bp)-palindrome, TGACGTCA, containing two inverted CGTCA half sites, was first identified in the somatostatin and phosphoenolpyruvate carboxykinase (PEPCK) genes (10, 11) and later found to be present in the promoters of many genes regulated by hormones and neurotransmitters that bind receptors coupled to adenylate cyclase (for review see Ref. 12). The CRE is also found in viral promoters, where it is referred to as an activating transcription factor (ATF) site (13, 14). The transcription factors binding CRE/ATF sites make up a large family of proteins (CREB/ATF factors; recently reviewed in Ref. *15*) related by virtue of a common DNA-binding domain, but distinguished by highly divergent activation domains (*16, 17*). Of these, CREB, CREM, and ATF-1 are regulated by cAMP-activated PKA. CREB and ATF-1 are ubiquitously expressed, but ATF-1 has lower basal activity and is a weaker activator than CREB (*18*). CREM is equipotent with CREB as an activator, but expression of the CREM gene is more restricted (*19, 7*). All three factors are regulated in the same way by protein kinases that phosphorylate a serine residue (S133 in CREB) within a highly conserved region of these factors.

In the PKA holoenzyme, the activity of the two catalytic subunits is restrained by their association with two regulatory subunits. Binding of cAMP to the regulatory subunits of PKA induces a conformational change, resulting in the release of free, active catalytic subunit (PKAc) (*20, 21*). Some of this PKAc is translocated to the nucleus, where it phosphorylates CREB (*22*). Evidence that PKAc is necessary and sufficient for stimulation of gene transcription was provided by experiments in which the introduction of recombinant PKAc into cells by microinjection (*23*) or transfection (*24, 25*) specifically stimulated transcription of CRE-containing genes and by experiments in which the catalytic subunit stimulated *in vitro* transcription of CRE-containing genes (*26–29*). Protein kinase A stimulates the transcriptional activity of CREB by phosphorylating Ser133 in the CREB activation domain (*30*). Subsequently, several other signaling pathways were demonstrated to activate protein kinases that phosphorylate Ser133 and enhance transcription of CRE-containing genes in a variety of cells and tissues (Fig. 1; CaMK, MAPK/p90rsk, PKC, p38, PKB/Akt) (*31, 32*).

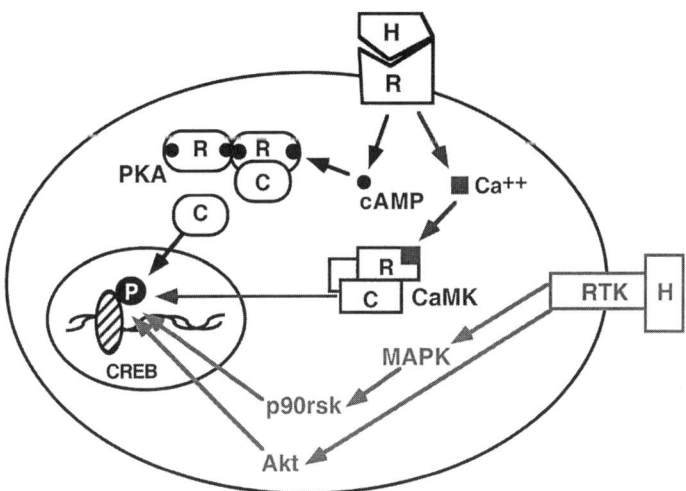

FIG. 1. Regulation of CREB activation by extracellular signals.

A key observation from early studies of gene regulation by cAMP was that mutation of the PEPCK CRE not only abolished induction, but also diminished basal activity of the reporter gene (33). Similar observations were made with other genes (34–36), suggesting that a common mechanism of regulation was used. Gonzalez et al. (37) used reverse genetics to obtain a cDNA encoding an active CREB molecule of 341 amino acids. Hoeffler et al. (38) used recognition site screening of a λgt11 human placental cDNA library to identify a CREB cDNA of 327 amino acids, resulting from translation of an alternately spliced form of the CREB mRNA. The long and short isoforms activate transcription equivalently (39, 8). Subsequently, Hai et al. (17) used a similar screen with an ATF/CRE probe to identify several ATF cDNAs that have a conserved DNA-binding domain, but widely divergent transcription regulatory domains. These findings suggested a plausible mechanism for regulation of basal and PKA-inducible activity through the CRE: regulation of the binding of different ATF factors by PKAc. We had shown that CRE binding was unaffected by PKAc (29). However, the low resolution of mobility shift gels does not preclude the possibility that PKAc stimulated the exchange of a weaker activator for a stronger one.

II. Distinct CREB Activation Domains Mediate Constitutive and Kinase-Inducible Transcription

An alternative hypothesis to explain the observation that mutation of the CRE affected both basal and inducible gene expression was that CREB exhibited basal activity that could be increased by phosphorylation, either by increasing the efficacy of a single domain or through distinct domains. We employed two strategies to test the possibility that the CREB activation domain possessed both constitutive and kinase-inducible activity. In one case, activity of PKA was specifically inhibited, whereas in the other, the serine in CREB (S133) that is phosphorylated by PKA was changed to alanine (S133A). Subsequently, we employed extensive mapping of the CREB activation domain to determine whether single or independent domains mediated basal and inducible gene expression.

The experimental model we used here and in subsequent studies to test activities of the CREB activation domain employed the now common strategy of using a hybrid transcription factor, CREB–Gal4 (CRG), in combination with a reporter gene in which the CREs are replaced with Gal4 binding sites. In this procedure, the activation domain of the factor to be tested, CREB in this case, is fused in-frame to a heterologous DNA-binding domain not expressed in the test cells, the yeast Gal4 regulator in this case. This allows analysis of both gain of function and loss of function for the activation domain, even in the presence of endogenous CREB, which is expressed ubiquitously. The C-terminal positioning of the Gal4 DNA-binding domain, previously demonstrated to be effective by Raycroft et al. (40), keeps the CREB activation domain in its native orientation with respect to the DNA and other factors bound to it. The JEG3 cell line was chosen for these experiments, because they are readily transfected and are highly responsive to stimulation by PKAc (35, 41, 42). Because cotransfection of a vector encoding the catalytic subunit of PKA (PKAc) provides stronger stimulation than treatment with cAMP, cotransfection with PKAc was used in all of our experiments.

A. Evidence for Distinct Constitutive and Kinase-Inducible Activities

The activity of the catalytic subunit of PKA is potently inhibited by a small, heat-stable peptide inhibitor, PKI (43). Cotransfection of a vector encoding PKI had been shown to at least partially inhibit expression of cAMP-regulated promoters in some circumstances (44–46), suggesting that partial activation of PKA could explain basal expression of some genes. To determine whether CREB had constitutive activity in the absence of activation of PKA, we cotransfected PKAc, −/+ PKI, and CRG with G4tk-CAT and measured its activity (8). PKI reversed CRG-mediated induction by PKAc, but reduced expression only to the level seen in the absence of PKAc, not to the level seen in the absence of CRG. This result suggested that constitutive activity was a fundamental property of CRG. A more specific approach to the question of constitutive activity of CREB involves site-specific mutation of Ser133 to an alanine, which had been shown to prevent phosphorylation and transcription activation by PKAc of the native CREB (37). Cotransfection of cells with PKAc and CRG-S133A produced a level of transcription activation identical to that produced by transfection with CRG-S133A alone (8). This result directly demonstrated that the CREB activation domain exhibits constitutive activity, independently of phosphorylation by PKAc.

The G4tk reporter vector used for the analyses described above contains sites for other transcription factors (C/EBP and Sp1), which may have interacted with CRG to produce the regulation observed. To determine whether the constitutive activity of the CREB activation domain was due to interactions with other factors, we analyzed CRG in the context of promoters containing only Gal4 sites and the minimal promoter spanning the region encompassing the TATA box through the start site of the E1b and PEPCK promoters (8). Analysis of CRG activation of these minimal promoters showed that CRG enhanced basal activity, indicating that constitutive activity is a fundamental property of the CREB activation domain. Kinase-inducible activity was particularly robust with these minimal promoters, which have significantly lower basal activity than the G4tk promoter. This experiment demonstrated that both constitutive and inducible activities of the CREB activation domain can operate independently of interaction with other regulators. However, interactions between CREB and other factors contribute significantly to basal and inducible activity in different promoter contexts (47–52). The basal and kinase-inducible activities of the CREB activation domain could result from a single domain, the activity of which is enhanced by phosphorylation, or from distinct domains mediating basal and kinase-inducible transcription. To distinguish between these possibilities, we undertook extensive mapping of the CREB activation domain in CRG.

B. Definition of the Constitutive Activation Domain (CAD) and the Kinase-Inducible Domain (KID)

At the outset of our work on the mapping of activation domains, it had been reported that phosphorylation of Ser133 by PKAc was required for induction by cAMP and that inclusion of an alternatively spliced exon in the amino terminus of CREB conferred additional activity (53). The shorter CREB isoform, designated ΔCREB, is more abundant in most tissues (53) and was the clone isolated in the library screen of Hoeffler et al. (38), whereas Montminy and Bilezeikjian isolated a clone for the longer isoform (54). Using primers encompassing the termini of the rat pheochromocytoma cDNA, we found that both CREB isoforms were obtained when cDNA from rat liver and brain was amplified by polymerase chain reaction (PCR) [29]. Further investigation showed that the alternatively spliced isoforms of CREB had equivalent capacities for transcription activation (39, 8). In addition, the ATF-1 protein, which is highly homologous with CREB outside of its amino-terminal region, had been shown to confer activation by cAMP (55), although its transactivation potential was less than that of CREB (18). Beyond this, the existence and significance of constitutive activation by CREB had not been addressed.

The exon/intron boundaries of the CREB protein were not known, so we based our analysis on the identification of characteristic amino acid motifs found in the activation domains of Sp1, CTF, and other transcription factors (reviewed in Ref. 56). We targeted motifs rich in serine/threonine (S/T), glutamine (Q), or proline (P) as well as a region implicated in kinase inducibility (X) and an undefined region (Y) for deletion analysis, as an initial step in the characterization of the CREB activation domain (see Fig. 2). Restriction sites for the enzyme Spe I were introduced at the boundaries of these motifs such that the intervening sequence could be deleted without changing the reading frame. We truncated the activation domain successively from the carboxy terminus (Fig. 2). Based on these analyses of loss of function, we then tested putative activation domains for gain of function (Fig. 3). In all cases, the expression of deletion mutants was comparable to that of wild type, as judged by western blotting with antibody directed against the Gal4 DNA-binding domain of the fusion proteins. Expression of the Gal4 reporter gene was normalized to a second reporter gene containing

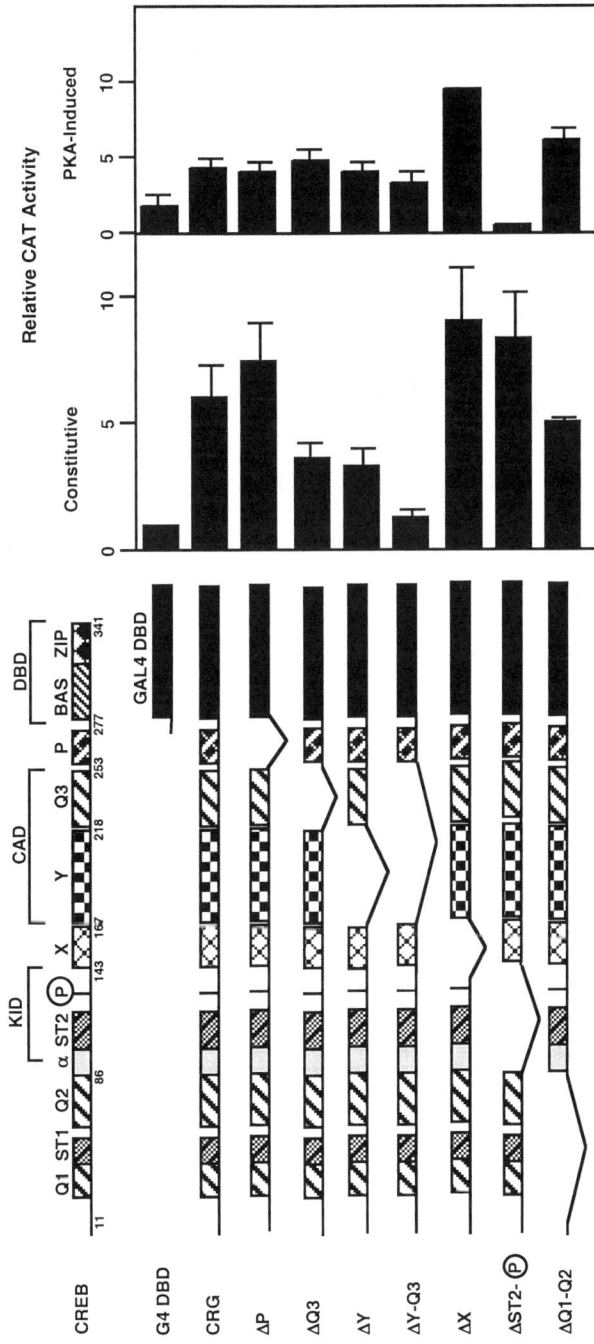

FIG. 2. Segregation of constitutive and kinase-inducible CREB activity into separate domains. The structures of wild-type and mutated forms of CRG expression plasmids are shown on the left. JEG3 cells were cotransfected with 10 μg of G4tk-CAT, 2 μg of RSV β-galactosidase, and 1 μg of the indicated CRG plasmid, in the absence or presence of 1 μg of PKAc. The basal CAT activity shown on the left is the CAT activity corrected for β-galactosidase activity. PKA-induced activity on the right represents the CAT activity obtained in the presence of PKAc divided by that obtained in its absence. The results shown are the mean ± SEM from five to eight independent experiments. Adapted from Ref. 8.

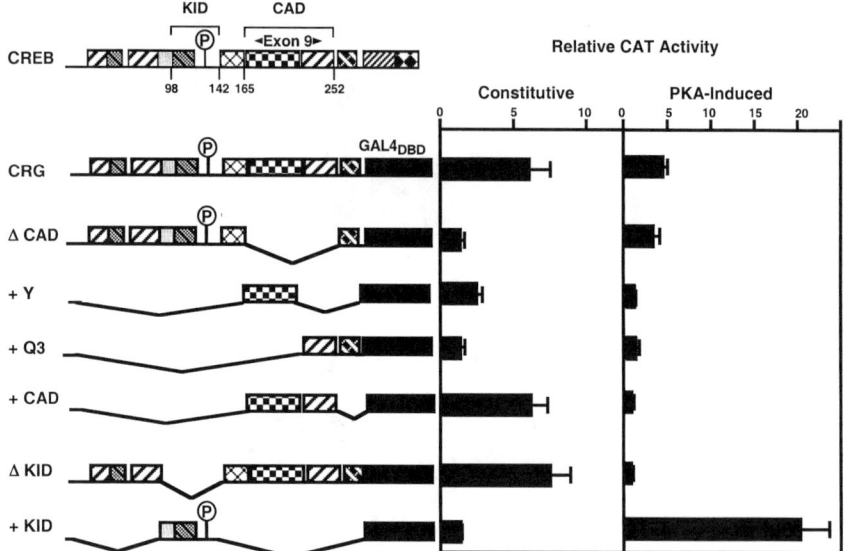

FIG. 3. Independent constitutive and kinase-inducible domains mediate induction by CREB. The structure of wild-type and mutated forms of CRG expression plasmids are shown on the left. JEG3 cells were cotransfected with 10 μg of G4tk-CAT, 2 μg of RSV β-galactosidase, and 1 μg of the indicated CRG plasmid, in the absence or presence of 1 μg of PKAc. The basal CAT activity shown on the left is the CAT activity corrected for β-galactosidase activity. PKA-induced activity on the right represents the CAT activity obtained in the presence of PKAc divided by that obtained in its absence. The results shown are the mean ± SEM from five to eight independent experiments. Adapted from Ref. 8.

a promoter from Rous sarcoma virus (RSV) that is unaffected by expression of CRG or PKAc to derive basal activity. The induction ratio was calculated from the activity obtained in cells cotransfected with PKAc relative to those without PKAc. An extensive set of deletion mutants was analyzed (8). The evidence for the definition of the constitutive and kinase-inducible domains in CREB is summarized in Figs. 2 and 3.

Deletion of the carboxy-terminal proline-rich domain had no effect on either basal or inducible activity of CRG (Fig. 2). However, deletion of either Q3 or Y reduced basal activity without affecting inducibility, and deletion of Q3 and Y together abolished basal activity. These results provided the first evidence that basal and inducible activity may be mediated by distinct domains in CREB and suggested that the carboxy-terminal portion of the activation domain provides the majority of basal activity. Deletion of the region denoted X enhanced inducible activity. In an analysis of a much larger series of deletion constructs, any construct lacking X was superinducible, indicating that X represses the activity of the activation domain (8). Deletion of the phosphorylation site and the ST2 region abolished kinase-inducible activity, but had no effect upon constitutive activity. Finally, deletion of amino-terminal sequences reduced basal activity modestly, but did not exhibit any independent activity when fused to the Gal4 DNA-binding domain. This suggests that the N-terminal amino acids contribute to a more active or stable conformation of the activation domain, but are not essential.

The previous experiments represent loss of function and allowed us to conclude only that these regions of the activation domain were necessary for constitutive and kinase-inducible activation, not that they were sufficient. The activation domain regions identified by loss of function were then analyzed for their ability to confer function upon the inert Gal4 DNA-binding domain (Fig. 3). Fusion of either Y or Q3 alone to the Gal4 DNA-binding domain provided little, if any, constitutive activity. Together, these regions provide the majority of constitutive activity seen with the complete activation domain, indicating that amino acids 165–252 of CREB make up a constitutive activation domain (CAD) that acts independently of phosphorylation of CRG by PKAc. Fusion of amino acids 98–142 of CREB to the Gal4 DNA-binding domain conferred inducible activity upon a Gal4 reporter, in spite of the fact that constitutive activity was quite low. Thus, this region of the CREB activation domain is a kinase-inducible domain (KID) capable of acting autonomously to confer induction in the absence of any measurable constitutive activity.

A prediction of these results is that the constitutive and kinase-inducible domains of CRG should be able to act in concert with heterologous transcription factors. We have tested this in the context of the complex PEPCK promoter, where CRG contributes to constitutive and kinase-inducible activation in concert with other factors, and in the context of the 5XGT minimal promoter, where CRG mediates essentially all constitutive and kinase-inducible activation (Fig. 4) (50). In G4-PEPCK, the endogenous CRE sequence is replaced with a Gal4 site, which, in the absence of CRG, is effectively a site-directed mutation of the CRE that reduces basal activity. In contrast, the minimal 5XGT promoter, which contains only Gal4 sites ligated to the TATA-containing region through the start site of the PEPCK promoter ($-40/+1$), relies entirely upon CRG for both basal and inducible activity. Cotransfection with CRG or S133A, but not KID-G4, enhanced basal activity of G4-PEPCK and stimulated basal activity of 5XGT to an even greater extent due to the complete reliance upon CRG for activity. Cotransfection of CRG and PKAc resulted in further stimulation of activity for both promoters. Cotransfection of KID-G4 and PKAc stimulated a large increase in 5XGT activity, but it starts from such a low basal level that the overall level of activation remains low. However, KID-G4 allowed PKAc to stimulate G4-PEPCK to nearly the same extent as CRG, because other factors contribute to basal activity, even in the absence of the CAD. Thus, in the context of complex promoters, other factors that promote basal activity can cooperate with the kinase-inducible domain in CREB to mediate high-level induced

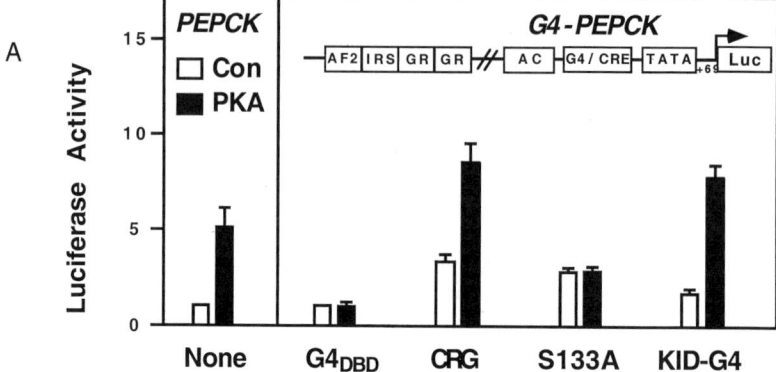

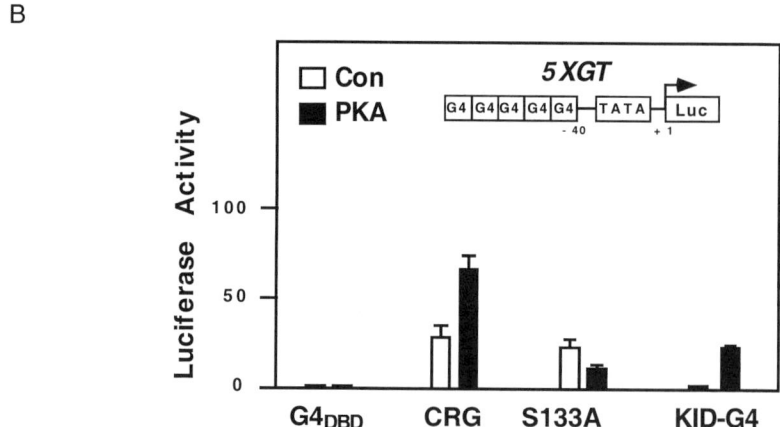

FIG. 4. Heterologous factors can substitute for the CAD in promoting basal expression. (A) H4IIe cells were cotransfected with either PEPCK-Luc or G4-PEPCK-Luc + CRG expression vectors in the absence and presence of a PKAc expression vector, as described in the text. Each precipitate was split into two dishes and half of them were treated with 10 nM insulin for the final 20 h of the experiment. The results shown represent four independent experiments. (B) H4IIe cells were cotransfected with 5XGT-Luc + CRG expression vectors and treated as above. The results shown represent five independent experiments. Adapted from Ref. 50.

transcription. A related analysis showed that fusion of the CAD and KID to the Lex A and Gal4 DNA-binding domains allowed reconstitution of basal and inducible activation of a promoter containing both LexA and Gal4 regulatory sites (7).

The results of the CREB mapping studies are summarized in Fig. 5, along with a map of the exon/intron boundaries of the CREB gene (57, 58). Notably,

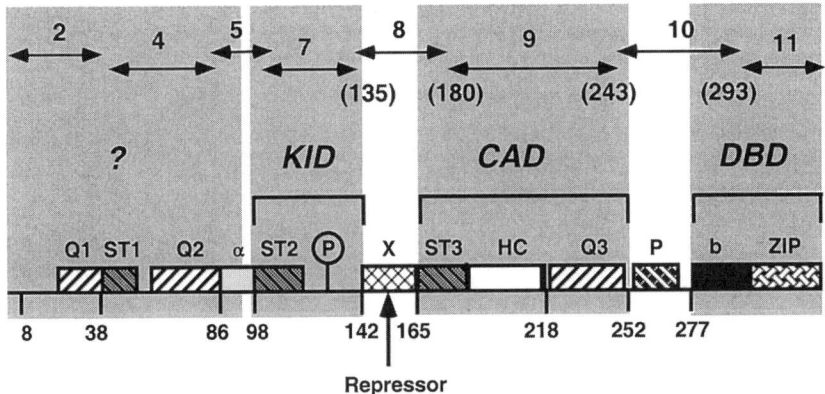

FIG. 5. Comparison of CREB exons and functional domains. The location of CREB exons is shown relative to amino acid motifs targeted for deletion in mapping the functional domains within CREB. The shaded bars show the relationship between the exon boundaries of the CREB gene (arrows) and the functional CREB domains (brackets). Adapted from Ref. 50.

none of the activation domains appears to be contained within a single exon. The DNA-binding domain (DBD) is split between two exons and the KID has been shown to rely upon an LXXLL motif beyond amino acid 135 for interaction with CBP sequences (59, 60). The CAD includes exon 9, but further analysis showed that part of exon 8 is also required for function (see below). An independent analysis of the activities and boundaries of the CREB activation domains which is in excellent agreement with the results described above was published at the same time by Brindle et al. (7). They described similar boundaries for the CAD, which they named Q2, and the KID and noted, as we did, the significance of the observation that the CAD is spliced out in CREMα, which fails to activate transcription effectively and thus behaves as a competitive inhibitor with CREMτ or CREB for CRE sites (61). Thus, the CAD is functionally important for maximal activation by CREB, in spite of the fact that it plays no direct part in regulation by protein kinases activated by cAMP and other signals generated by hormones and neurotransmitters.

C. Characterization of the CAD and Its Interactions with General Transcription Factors

The observation that the CAD overlapped and included exon 9, which is spliced out of the CREM protein with impaired activation potential (42, 7), suggested that exon 9 may encode a functional CAD. To investigate this possibility, we examined loss of function and gain of function for either exon 9 or the CAD. Deletion of either exon 9 or the CAD resulted in loss of basal activity. However, the CAD provided gain of function, but exon 9 did not. This disparity indicated that either or both of the flanking exons contributed to basal activity. Truncation of the exon 10 fragment flanking the 3′ terminus of exon 9 was without effect, but truncation of the exon 8 fragment flanking the 5′ terminus impaired induction, indicating that the functional CAD spans more than one exon, like the other functional domains in CREB. The exon 8/9 boundary within the CAD divides two amino acid motifs, a central hydrophobic cluster of amino acids (HC) and an amino-terminal region rich in serine and threonine (ST3). Deletion analysis showed that each of these motifs is indispensable for CAD function, as is the carboxy-terminal glutamine-rich region (Q3) (62). As in the previous analysis, deletion of these CAD motifs did not affect inducibility. None of the motifs can act alone or in concert with another. Rather, all three are required, indicating that the CAD contacts more than one target in the general transcription machinery or that more than one motif is required for effective binding to a single target.

The CAD in CREB is responsible for basal activation of transcription, which suggested that it may play a role in the recruitment or assembly of the polymerase complex. Many factors had been shown to interact with the promoter recognition factor, TFIID. Some transcription factors bound TATA-binding protein (TBP) directly, whereas others bound TFIID indirectly through TBP-associated factors (TAFs), which had been shown to be required for activator-dependent transcription activation (63). Gill *et al.* (64) characterized an interaction between Sp1 and dTAF$_{II}$110, which was required for activation by Sp1; by analyzing the interacting pair in a yeast two-hybrid system. Like CREB, the Sp1 activation domain contains a glutamine-rich region (65), and subsequent analysis of CREB in the two-hybrid assay showed that it also interacted with dTAF$_{II}$110 (66). We demonstrated that CREB interacted physically with TFIID, but not TBP, in a coimmunoprecipitation assay, indicating that the interaction was mediated by one or more TAFs in TFIID, and that the CAD was sufficient to mediate TFIID binding (67). Therefore, we took advantage of this interaction characterized in yeast to analyze the CREB interaction surface required for interaction with the TAF.

Transformation of yeast with CRG, CAD-G4, or dTAF$_{II}$110-AD alone failed to activate transcription of an integrated reporter under control of Gal4 sites (G4–*lacZ*). However, when yeast were cotransformed with dTAF$_{II}$110–AD and either CRG or CAD-G4, transcription of the G4–*lacZ* reporter was stimulated. Quantitation of the strength of the interactions showed that CRG interacted more strongly than CAD-G4 with dTAF$_{II}$110, which is consistent with results showing more effective recruitment of a polymerase complex (Fig. 11) and greater stimulation of basal activity by CRG than CAD-G4 of a strictly Gal4-dependent minimal promoter (68, 69). Deletion of the glutamine-rich region (Q3) or the hydrophobic cluster region (HC) abolished interaction with dTAF$_{II}$110 in the yeast two-hybrid assay, whereas deletion of ST3 reduced activity to one-third of that provided by CAD–G4. Further truncation of the core HC/Q3 region from either end resulted in progressive decreases in the remaining activity (70). Taken together, these data indicate that the entire CAD is required for effective interaction with dTAF$_{II}$110, either because the separate amino acid motifs form an integrated interaction surface or they are required for a stable conformation of the activation domain.

To determine which amino acids within the CAD are crucial for interaction with dTAF$_{II}$110, we employed a reverse two-hybrid strategy in yeast. In this assay, activation of the G4-URA3 gene in the presence of the drug 5-fluoroorotic acid (5-FOA) is toxic (Fig. 5A), as shown by killing of cells transformed with Gal4 or cotransformed with the interacting pair CAD-G4 and dTAF$_{II}$110 (70). This technique provides a useful screen for interaction-defective point mutants, because they will be the only transformants to survive in the presence of 5-FOA. Plasmids with frameshift or termination mutations were not considered further, nor were rare plasmids with more than one mutation, or plasmids which failed to express a stable protein, as judged by western blotting of transformed yeast. All of the mutations isolated at 0.015% 5-FOA had wild-type (wt) or near wt activity when analyzed in a standard two-hybrid assay for interaction with dTAF$_{II}$110 and were scattered among all three CAD regions, ST3, HC, and Q3 (Fig. 5C). In contrast, most of the mutants isolated at 0.025% 5-FOA exhibited severely impaired interaction with dTAF$_{II}$110 and all came from within HC (Fig. 6). The most severely impaired mutants resulted from changes of hydrophobic amino acids within HC to polar or charged amino acids. Analysis of codon usage and probabilities of obtaining polar or charged versus neutral amino acids from changing a single base within the codons of the hydrophobic amino acids showed that only half were even 50% probable, whereas three others ranged from 17% to 33% probability. This result suggests that more probable conservative mutations of these amino acids did not impair interaction and were not identified in the screen. These data may indicate that mutation of hydrophobic amino acids in the CAD disrupts a hydrophobic interaction surface between the CREB CAD and dTAF$_{II}$110. This interpretation is strengthened by the finding that a similar hydrophobic surface, which is required for interaction with Sp1 and CREB, was mapped in hTAF$_{II}$135, the human homolog of dTAF$_{II}$110, which shows a high degree of conservation in the amino-terminal activator-binding domain used in these studies (71). Thus, the CREB and TAF proteins appear to interact through juxtaposed hydrophobic surfaces. We have screened other TAFs (TAF55, 105, 250) and general transcription factors (GTFs IIB, IIE, IIF), but have found no evidence for interaction. In addition, subsequent analyses showed that these interaction-defective HC mutants were impaired for both recruitment of a polymerase complex and augmentation of transcription activation in concert with dTAF$_{II}$110 in mammalian cells (72). Our data, together with evidence

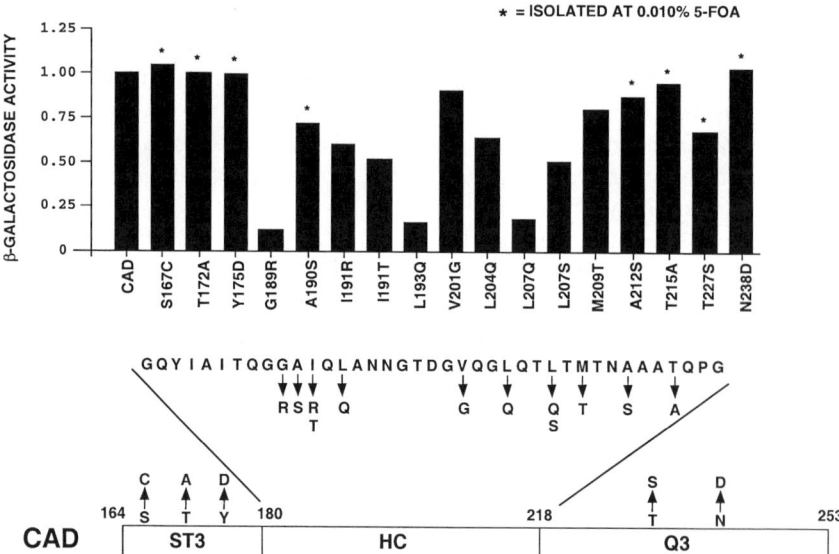

FIG. 6. Screening of a CAD point mutation library identifies interaction-defective mutants. The CREB CAD coding region was amplified, using relaxed stringency conditions and increased amplification cycle number to introduce ≤1 point mutation per CAD sequence of 300 bp amplified. The resulting amplicon pool was subcloned into the yeast expression vector in fusion with a wild-type Gal4 DBD. The library of CAD point mutations was transformed along with the TAF110 1-308-AD expression plasmid into the reverse two-hybrid yeast strain MaV203 and plated on double-selective media containing increasing concentrations of 5-fluoroorotic acid (5-FOA). Plasmid DNA was recovered from the yeast and sequenced to determine the presence and identity of the mutations shown. Adapted from Ref. 70.

that deletion of the CAD renders CREB (and the related CREMα) poor activators (61, 7, 8) due to low basal activity, indicates that the interaction between the CAD and TAF$_{II}$110 is important to transcription initiation.

Taken together, our results suggested a model for sequential polymerase recruitment involving recruitment of the promoter recognition factor TFIID by interaction of the dTAF$_{II}$110/hTAF$_{II}$135 subunit with the CAD in CREB, followed by TFIID-directed assembly of a polymerase complex on the promoter (Fig. 7A). This model is consistent with early studies of polymerase complex recruitment in which TFIID directed preferential assembly of a polymerase complex on templates it bound in a process termed template commitment (73). An alternative model (Fig. 7B), suggesting that recruitment was bivalent, with the CAD recruiting TFIID and P-KID recruiting holoenzyme, was proposed by Nakajima et al. (74, 75). In those experiments, CREB was immobilized for binding studies, followed by the addition of template, missing factors, and nucleotides

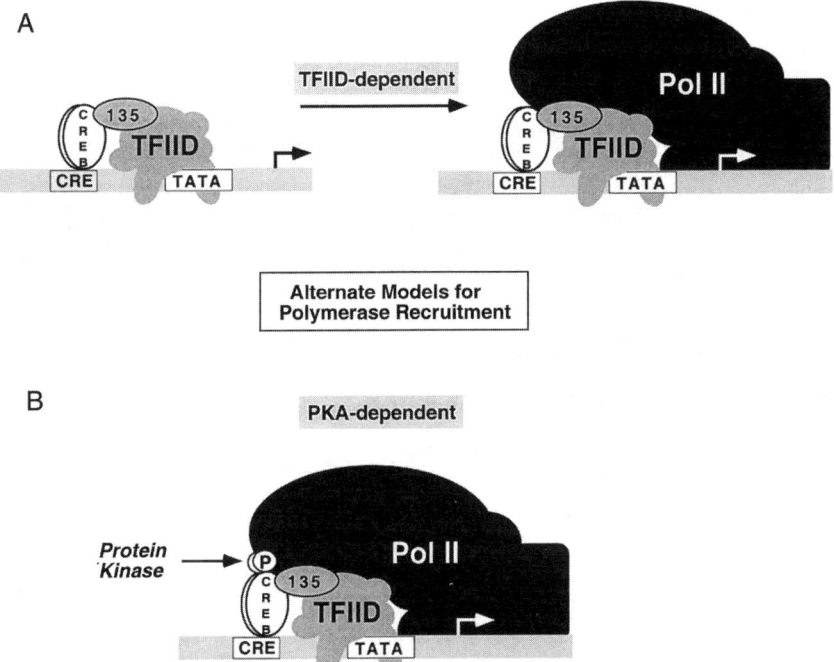

FIG. 7. Alternate models for recruitment of functional RNA polymerase II complexes to promoters. (A) A sequential, template commitment model, based on facilitated recruitment of pol II as a consequence of IID recruitment. (B) A concerted model for recruitment, based on interactions between the CAD and IID and between P-KID and CBP-pol II. Adapted from Ref. 69.

for the assay of transcriptional activity. Nakajima et al. (74, 75) found that immobilization of unphosphorylated CREB promoted association of TFIID, but not holoenzyme. In contrast, immobilization of P-KID promoted association of holoenzyme, but not TFIID. However, this is not a direct assay of recruitment, because it measures association of TFIID and polymerase with CREB rather than with the template to be transcribed. In addition, it would be necessary for P-KID to interact with the polymerase complex in order for it to affect any step in transcription initiation. Thus, reasonable but incomplete data existed to support either of two models for polymerase recruitment by CREB. Our hypothesis suggested that the CAD in CREB was sufficient to recruit a polymerase complex through its interaction with TFIID and that phosphorylation of the KID in CREB may facilitate subsequent steps in transcription initiation, involving promoter melting and disassembly of the polymerase complex to facilitate polymerase escape. To distinguish between these alternative models, we employed and developed assays to directly evaluate the contributions of the constitutive and kinase-inducible domains in CREB to recruitment and subsequent steps in transcription initiation (Fig. 8).

III. A Concerted Mechanism of Transcription Activation Involving Stimulation of Sequential Steps in Transcription Initiation by the CAD and P-KID

The general transcription factors (GTFs: IID, IIB, IIE, IIF, and IIH) required for accurate initiation of basal transcription by RNA polymerase II (pol II) have been identified, purified, characterized, and cloned (for review see Ref. 76). In some cases, their structures have been solved by X-ray crystallography in complexes with each other and DNA templates (77–82). Biochemical fractionation of nuclear extracts and reconstitution experiments show that these five GTFs play indispensable roles in the formation and function of RNA polymerase II complexes (83–85).

The polymerase cycle (Fig. 8) involves a progressive series of changes in association among pol II, the GTFs, and the template (1). Following recruitment

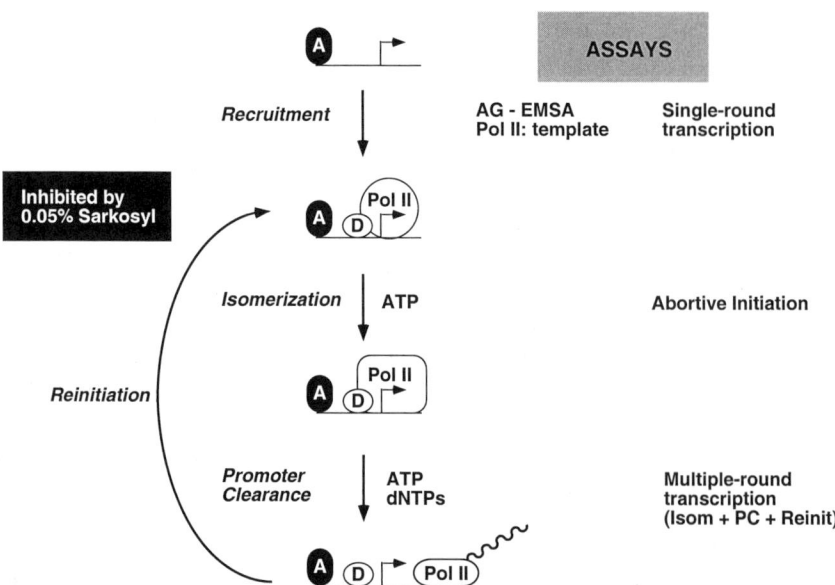

FIG. 8. The polymerase cycle and assessment of individual steps in the reaction mechanism. Discrete steps in transcription initiation are indicated to the left of the arrows and the assays used to assess the contributions of CREB domains to regulation of these processes are shown to the right of the arrows. As indicated, reinitiation is inhibited by 0.05% Sarkosyl.

of the GTFs to the promoter, isomerization of the complex and hydrolysis of ATP are required to melt the template at the start site to form an open complex capable of abortive synthesis of short RNAs. The polymerase then escapes the complex to engage in productive transcription in a reaction called promoter clearance (86–89). In addition, recent evidence indicates that many of the GTFs remain associated with the promoter in a scaffold complex that facilitates efficient reinitiation (90).

Recruitment of the GTFs is required to establish a preinitiation complex. An important concept to emerge from early studies with GTFs was that of template commitment (91). Preincubation of templates with various nuclear extract fractions demonstrated that all but one of the GTFs were freely exchangeable and suggested that association of IID with the template is a prerequisite for further complex formation. Subsequent reconstitution studies showed a preferred order of addition of GTFs, whether assessed by *in vitro* transcription or by mobility shift assay (92, 93, 1). The model for preinitiation complex formation to emerge from these studies was that the template recognition factor IID binds first, followed by IIB, which allows the binding of a pol II:IIF complex formed in solution, followed by the binding of late factors, IIE and IIH. In many cases, the GTFs may be assembled with pol II and other regulatory factors making up the mediator complex in a megadalton complex termed holoenzyme (94, 95). Regardless of whether the remaining GTFs are separate entities or present in a holoenzyme complex, recruitment of IID is an obligate first step in complex assembly. The CREB CAD interacts with dTAF$_{II}$110 and its human homolog, hTAF$_{II}$135, which is functionally equivalent (66, 65, 71, 70). As described below, we have shown that the CAD:TAF interaction is necessary and sufficient to recruit a polymerase complex and establish basal transcription (70, 68, 69, 72).

Melting of the template is required following assembly of GTFs into a preinitiation complex. A progressive series of changes in protein:DNA and protein:protein interactions, accompanied by ATP hydrolysis, is required to overcome energy barriers in two subsequent steps requiring melting of the template: isomerization and promoter clearance (86, 96, 88). Isomerization is a conformational change in the pol II complex that allows it to melt the template in an ~10-nucleotide (nt) bubble at the start site of transcription, but not to elongate (87). An isomerized polymerase complex is capable of synthesizing abortive transcripts of 3–10 nt (96, 87). This abortive initiation can be readily measured by providing a short primer and assaying the addition of a labeled nucleotide, which can only occur if the complex has isomerized and melted the template at the initiation site (97, 98). The open complex is inherently unstable and collapses back to a closed complex in the absence of continual ATP hydrolysis (96). Promoter clearance involves the input of additional energy for further DNA unwinding and disassembly of the complex to allow escape of the polymerase, which assumes a tighter, more stable conformation when elongating (80, 99, 82).

All the GTFs are required for efficient isomerization of the polymerase complex (*96*) as well as for promoter clearance (*100*). An important point is that RNA polymerase II remains in close association with the GTFs and with transcription activators during both of these crucial steps in transcription initiation. As a result, both isomerization and promoter clearance, in addition to recruitment, can be regulated by transcription factors, such as CREB. Little attention has been paid to regulation of postrecruitment steps in transcription activation until very recently, and our work shows that phosphorylation of CREB provides an important model for study of these processes.

Our major hypothesis regarding the mechanism of action of CREB, that the constitutive and kinase-inducible activation domains in CREB stimulate sequential steps in the transcription initiation reaction, has been tested and is supported by several different experiments. Our data, which are summarized below, suggests that the CAD in CREB is responsible for binding IID through hTAF$_{II}$135 and that this interaction is necessary and sufficient for the recruitment of a polymerase complex and the establishment of basal transcription. Polymerase recruitment was unaffected by phosphorylation of the KID in CREB or by mutation of the kinase-regulated site, indicating that phosphorylation acts at a subsequent step in the pathway to enhance transcription activation. In contrast, phosphorylation of the KID in CRG enhanced late steps in the reaction. This was determined by examining the separate and combined effects of CAD and KID on abortive initiation, a measure of isomerization, and single- versus multiple-round transcription, measures of recruitment versus late steps (isomerization, promoter clearance, and reinitiation), respectively. During the course of this work, we developed a new method for the direct assessment of recruitment of a polymerase complex, which also allows us to determine what specific components are included in the recruited complex and the contributions of each activation domain in CREB.

We used a minimal promoter (5XGT) containing the TATA region footprint through the start site of the PEPCK promoter under control of five Gal4 sites as a template for transcription and as a probe for binding (recruitment) studies. Recombinant CREB-Gal4 proteins containing either, both, or neither of the activation domains were expressed in insect cells using baculovirus and purified by affinity chromatography. CRG or KID-G4 protein was prephosphorylated with PKAc (Fig. 9A) and the reaction was stopped by the addition of the potent and specific PKA inhibitor PKI to avoid direct addition of PKAc to the nuclear extract, because the phosphorylation of other factors could complicate the interpretation of these experiments. Incorporation of isotope showed that CRG and KID-G4 are phosphorylated stoichiometrically under these conditions, and western blotting with anti-P-CREB antibody showed that phosphorylation was maintained throughout the course of these reactions (*69*). We used rat liver nuclear extracts that supported CRG-dependent basal and kinase-inducible

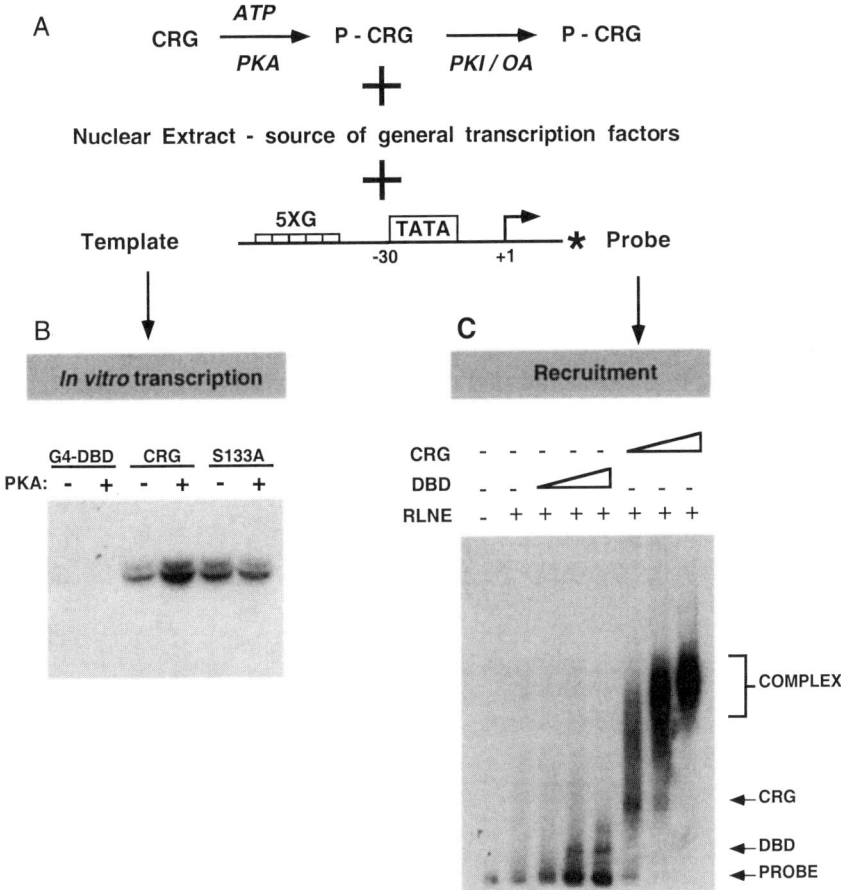

FIG. 9. Establishment of an in vitro transcription system. Recombinant CREB–G4 proteins, containing either (CAD-G4, KID-G4), both (CRG), or neither (G4$_{DBD}$) of the CREB activation domains, were introduced into insect cells in baculovirus and purified by affinity chromatography. Native or phosphorylated CRG (preincubated with PKA, which was then inhibited with PKI) were combined with rat liver nuclear extracts (RLNE), which was the source of general transcription factors, and the 5XGT promoter used as template or probe. Identical conditions were used for determinations of transcription and recruitment activity.

transcription activation (Fig. 9B) in vitro as a source of polymerase and GTFs (Fig. 9A), because this makes no assumptions about which factors are required for complex formation. Transcription (Fig. 9B) and binding (Fig. 9C) studies were done under identical conditions (molar ratio of factor : template : extract), except that only 1/10 as much material is required for the sensitive binding assay.

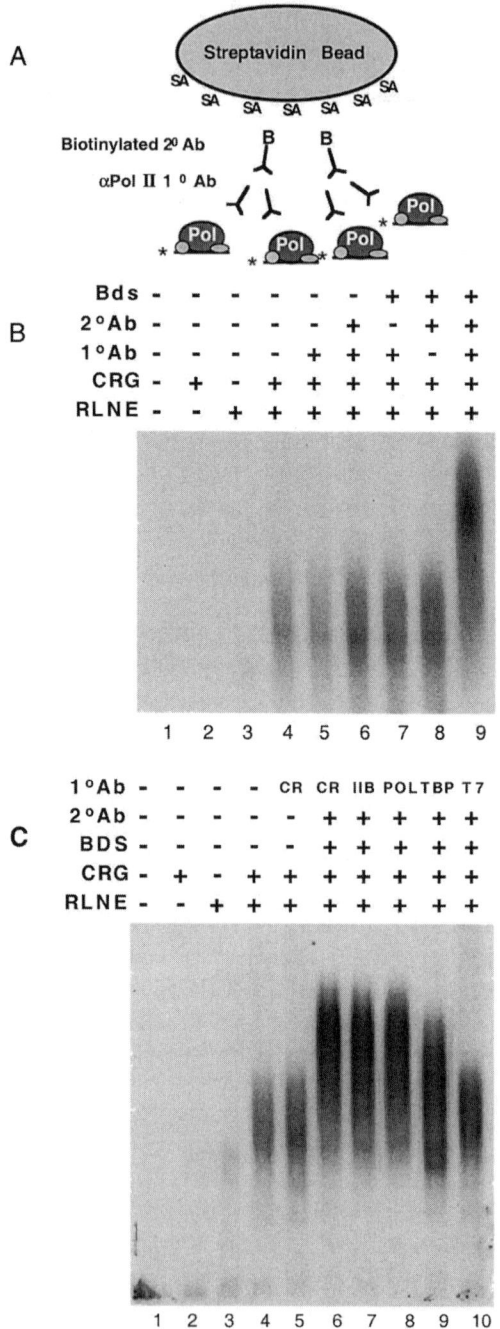

FIG. 10. The complex recruited to the promoter by CRG contains components of an RNA polymerase II complex, as indicated by an antibody supershift assay. (A) The antibody supershifting method uses rabbit primary antibodies specific for components of the preinitiation complex, a

A. Role of the CAD in the Recruitment of an RNA Polymerase II Complex

The most likely step to be affected by an interaction between CREB and IID is recruitment. Early work showed that binding of IID led to template commitment (91) and that inclusion of CRE/ATF sites upstream of the AdML promoter enhanced the binding of IID and other GTFs (2, 3). However, most of the available assays do not directly measure recruitment of polymerase complexes to an activator-associated template. We developed an assay for the direct measurement of recruitment of a polymerase complex to the 5XGT promoter and we employed a single-round *in vitro* transcription assay, which provides a functional test of polymerase recruitment, to assess the contributions of the CREB activation domains to recruitment.

1. AN ASSAY FOR DIRECT MEASUREMENT OF POLYMERASE COMPLEX RECRUITMENT

Lieberman and Berk (101) had developed an Mg^{2+}–agarose gel system for examining the recruitment of TBP to a labeled template by the activator Zta. We extended this observation to develop a method for examining recruitment of a polymerase complex to a template from crude extracts. The large pore size of agarose gels allows resolution of megadalton-sized complexes expected for a pol II complex. Initial recruitment experiments showed that CRG, but not $G4_{DBD}$, stimulated formation of a large complex in a nuclear extract- and dose-dependent manner (Fig. 9C).

In order to determine whether RNA polymerase or other GTFs were a part of the complex forming on the promoter probe, we had to develop a method for examining specific components of the complex (69). In mobility shift analysis of transcription factor:DNA interactions, addition of antibody recognizing a DNA-binding protein will ablate or supershift binding due to interference with binding or an increase in molecular size of the complex, respectively. However, in the case of a polymerase complex in the megadalton size range, addition of antibody would not be expected to affect the molecular size or migration of the complex. We tested different antibody-binding reagents to identify conditions where we could detect an increase in the size of the complex formed on the promoter probe, as illustrated for the use of a pol II primary antibody in Fig. 10A.

---

biotinylated goat anti-rabbit secondary antibody, and streptavidin-conjugated Dynabeads to bind the complexes into larger "supercomplexes." (B) Binding reactions contained 1° Ab directed against CREB, 2° Ab and/or beads, as indicated, with 5 fmol of probe, 10 fmol of purified CRG, and 3 μg of rat liver nuclear extract (RLNE). (C) Primary antibodies specific for CREB, TFIIB, the large subunit of RNA polymerse II, TBP, or an irrelevant T7 antibody were included with secondary antibody and streptavidin beads where indicated in agarose–electrophoretic mobility shift assay (Ag–EMSA) reactions that contained 5 fmol of probe DNA, 10 fmol of purified CRG, and 3 μg of RLNE where shown. Adapted from Ref. 69.

CREB antibody was used to validate the method (Fig. 10B), because CRG was required for complex formation (Fig. 9C). As expected, direct addition of anti-CREB antibody, with or without a secondary antibody to crosslink it, had no effect on the already large molecular weight of the complex (lanes 5 and 6, Fig. 10B). We found that a combination of specific primary antibody with biotinylated secondary antibody and streptavidin-coated beads provided further retardation of complex mobility (lane 9, Fig. 10B). Omission of any of these components abrogated supershifting (lanes 6–8, Fig. 10B).

Having established a method that allowed detection of specific components of the complex, we tested for components expected to be present in a minimal preinitiation complex, pol II, IIB, and TBP. Our results show that antibodies directed against CREB, pol II, or the GTFs (IIB, TBP) produce a supershift, whereas T7 antibody had no effect (Fig. 10C). Notably, retardation was less dramatic for antibody directed against TBP, perhaps because of steric hindrance by the antibody, which inhibits further complex formation. Thus, essential components of an RNA polymerase II preinitiation complex are present in the complex recruited to the promoter by the CREB activation domain.

Having demonstrated that CREB stimulated recruitment of a complex containing pol II and GTFs to the promoter, we assessed the independent and combined contributions of the CAD and KID to recruitment. Both the CAD and CRG were effective in mediating recruitment (Fig. 11). KID did not mediate recruitment, even when phosphorylated by PKAc (Fig. 11). CRG mediated recruitment more effectively than CAD alone, as has been seen in transcription assays

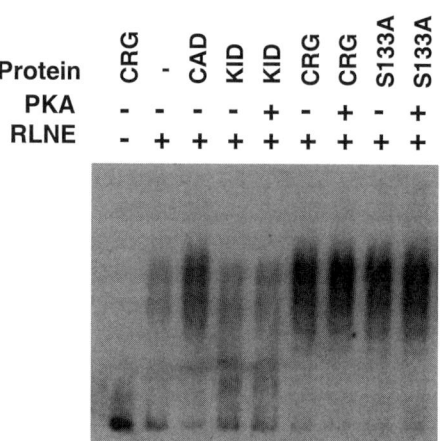

FIG. 11. The CAD, but not the KID, in CREB is sufficient to recruit a complex to the promoter. Ag–EMSA binding reactions included 5 fmol of probe DNA, 10 fmol of purified G4-DBD, KID-G4, CAD-G4, or CRG, and 3 μg of RLNE where indicated.

in vitro and *in vivo* (*68*). This enhanced activity of CAD within CRG is probably attributable to a more stable conformation of the activation domain, since no other activation domain contributing to basal transcription can be mapped (*8*). In support of that interpretation, mutation of the PKA phosphorylation site in CREB (S133A) abolished induction of transcription (Figs. 4, 9B), but did not diminish recruitment, and phosphorylation of CRG did not enhance recruitment (Fig. 11). Thus, all the binding data are consistent with recruitment being mediated by the CAD, independently of phosphorylation of the KID.

2. ROLE OF THE CAD:TAF INTERACTION IN TRANSCRIPTION ACTIVATION

In more recent studies (*72*), we analyzed the TAF interaction-defective mutants of CAD for recruitment of a polymerase complex and augmentation of transcription in transfected mammalian cells. CAD mutants defective for interaction with $dTAF_{II}110$ in the yeast two-hybrid assays were unable to recruit a polymerase complex to the promoter or to augment transcription activation in the presence of $dTAF_{II}110$ (*72*). These data provide further support for the idea that interaction between CAD and $dTAF_{II}110/hTAF_{II}135$ leads to recruitment of IID and the polymerase complex to the promoter and suggested that the TAF acts as a coactivator, binding the CAD and IID, but not DNA, to enhance transcription activation.

Finally, we tested the role of dTAFII110 to serve as a coactivator in mammalian cells in an experiment modeled after that used to determine whether the TAF served as a coactivator for bicoid in *Drosophila* extracts (*102, 103*). Deletion of either the CREB- or TBP-binding domains of the TAF is expected to abolish its ability to serve as a coactivator. $dTAF_{II}110$ interacts with the CREB CAD through its N-terminus (amino acids 1–308) (*66*) and binds to the scaffolding $hTAF_{II}250$ through its C-terminus (amino acids 796–921), integrating it into TFIID (*104*) (Fig. 11A). To determine whether $dTAF_{II}110$ served as a coactivator for CREB, we tested $dTAF_{II}110$ constructs deleted of either domain for potentiation of CAD activation in mammalian cells. Full-length $dTAF_{II}110$ potentiated transcription activation by the CAD *in vivo*. However, neither the N-terminal truncation (deleted of amino acids 1–308) nor the C-terminal truncation (deleted of amino acids 796–921) mutant of $dTAF_{II}110$ produced levels of luciferase distinguishable from samples in which no $dTAF_{II}110$ expression vector was added. This result demonstrates that the domains of $dTAF_{II}110$ that interact with the CAD and integrate it into the TFIID complex are both required for activation, as would be expected for a coactivator of the CREB CAD.

Although the biochemical studies above strongly support the hypothesis that the CAD:TAF interaction is required for recruitment and transcription activation, such a point can only be proven genetically. The reason that it was possible to use yeast two-hybrid assays to map the CAD:TAF interaction is that

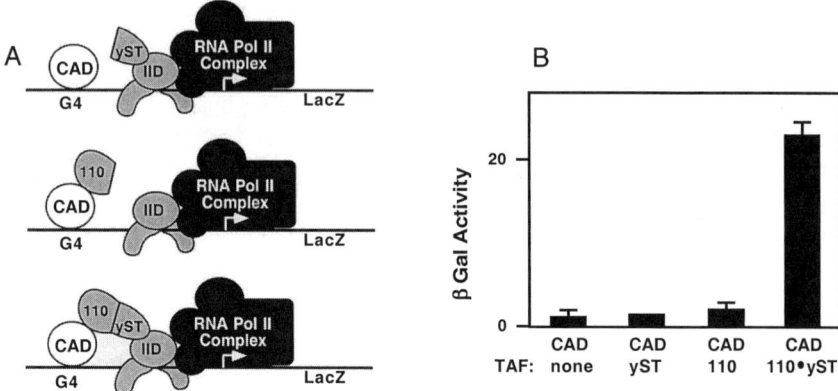

FIG. 12. A hybrid TAF recognizing the CAD and yeast TBP is sufficient to mediate activation in yeast. (A) Schematic diagram depicting the components (110 and yST) of the hybrid TAF (110-yST) and their interaction partners, CAD and TBP of IID. (B) Expression vectors for the hybrid TAF (110-yST) fusion protein, the single TAF domain control proteins (110 or yST), and CAD-G4 were transformed into the yeast strain Y-153 where indicated. Colonies were grown in liquid culture, protein lysates made, and β-galactosidase activity was determined using the O-nitrophenyl-βD-galactopyranoside (ONPG) synthetic substrate assay. Data represents three independent transformations with three colonies assayed from each. From Ref. 72.

the CREB CAD fails to activate transcription in yeast. As the various subunits of IID have been cloned from different organisms, it has become clear that TBP and the TAFs are highly conserved among yeast, *Drosophila*, and humans, where they serve similar functions (*105*). An important exception to this general rule is the absence of a yeast homolog for dTAF$_{II}$110/hTAF$_{II}$135, which mediates activation by CREB, Sp1, and others. This suggested to us that a hybrid TAF targeting the CREB-interacting domain of dTAF$_{II}$110 to yeast IID may mediate activation by CREB in yeast. To test this idea, we fused the CREB-interacting domain of dTAF$_{II}$110 (110) to the TBP-binding domain of the yeast scaffolding TAF (yST) to produce a hybrid TAF (110-yST) (Fig. 12A) (*72*). Transformation of yeast with the hybrid TAF 110-yST allowed the CAD to activate transcription in yeast, whereas neither component alone had any effect, indicating that formation of a ternary complex is required (Fig. 12B). We also tested G4–VP16 and G4–VDR, both of which interact with a IIB-Gal4 activation domain fusion (IIB-AD) to stimulate transcription in yeast (Fig. 13A) (*106, 107*). Both G4-VP16 and G4-VDR interacted with IIB-AD, but not 110–yST, whereas CAD-G4 interacted with 110-yST, but not IIB-AD, demonstrating that these interactions are specific for different activation domains (Fig. 13B) (*72*). Together with the biochemical experiments described above, this genetic experiment indicates that the CAD:TAF interaction is necessary and sufficient for recruitment of a polymerase complex and activation of basal transcription.

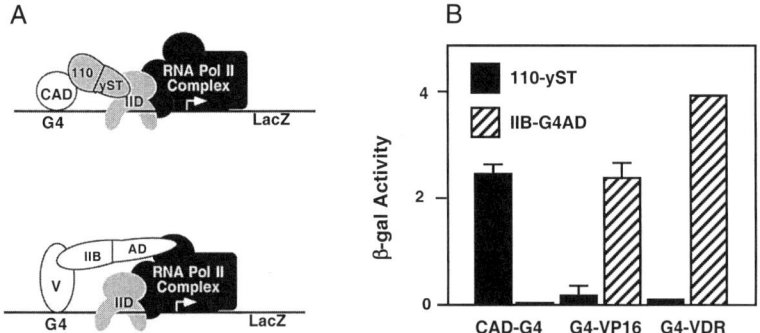

FIG. 13. Activator-specific interactions mediate activation by hybrid TAF (110–yST) and a IIB–AD fusion protein. (A) The two hybrids tested are depicted with their interaction partners, V = VP16 or VDR. (B) Expression vectors for CAD-G4, G4-VP16, and G4-VDR were transformed into Y-153 with the hybrid TAF protein or with the TFIIB–AD two-hybrid expression vector as indicated. β-Galactosidase activity was determined using the ONPG synthetic substrate assay. Data represent three independent transformations with three colonies assayed from each. From Ref. 72.

B. Role of Phosphorylation of the KID in Activation of the RNA Polymerase II Complex

The data described thus far provide strong evidence for a role of the CAD in recruitment of a polymerase complex and establishment of basal transcription activation and indicate that phosphorylation of the KID is not involved in recruitment. Thus, we hypothesized that phosphorylation of the KID may affect one or more postrecruitment steps in transcription activation, which would be in keeping with the very rapid transcriptional response to activation of protein kinases by hormones and neurotransmitters.

To provide a functional correlate for the recruitment studies and to examine the role of phosphorylation of CREB in transcription activation, we examined the ability of each domain to facilitate single-round transcription (a functional assay of recruitment), abortive initiation (an assay of isomerization), and multiple-round transcription (an assay of the combined effects of isomerization, promoter clearance, and reinitiation) (Figures 8, 14A) (68). An important feature of the experimental design for these experiments (Fig. 14A) is that the same reaction was split in two to assay single-round transcription and multiple-round transcription, so that the results would be directly comparable. After formation of a preinitiation complex, ATP and nucleoside triphosphates (NTPs) were added to initiate transcription, which was measured by primer extension. Transcription was limited to a single round by inclusion of only two of the NTPs during the first 2 min to allow initiation, but limit synthesis, after which further initiation was inhibited by the inclusion of 0.05% Sarkosyl and the remaining NTPs were

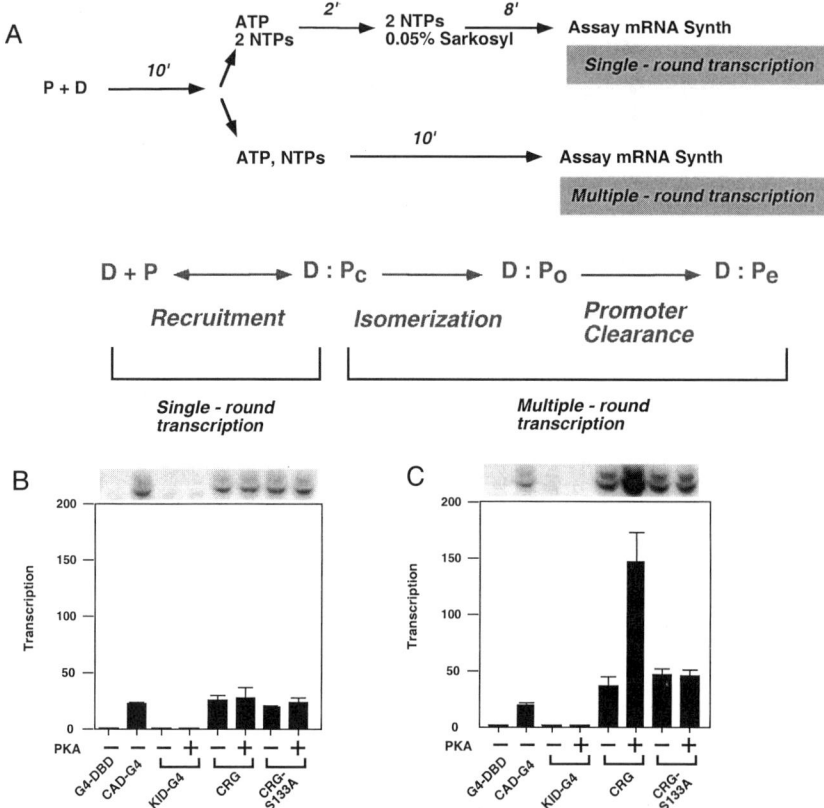

FIG. 14. The CAD, but not the KID, stimulates functional recruitment of a polymerase complex and phosphorylation of the KID in CRG stimulates activation of the recruited complex. (A) The reactions for single- and multiple-round transcription were done in parallel with the same template, nuclear extract, and CRG proteins. (B) Single-round transcription was measured by omission of two nucleotides until reinitiation was inhibited with 0.05% Sarkosyl after a 2-min preincubation. A representative experiment and a graphical representation of the results ($x \pm$ SEM) of four experiments are shown. Adapted from Ref. 68.

added to allow synthesis of the initiated transcript. All four NTPs were added with ATP to measure multiple-round transcription.

CAD-G4, CRG, and S133A were able to recruit a functional polymerase complex to the promoter, as judged by synthesis in the single-round assay (Fig. 14B). As in the binding studies of recruitment, phosphorylation of KID or CRG had no effect upon recruitment, and mutation of the PKA site (S133A) did not diminish recruitment, indicating that the CAD alone is sufficient to recruit a polymerase complex and establish basal transcription. In contrast, the order of potency in stimulating multiple-round transcription (P-CRG ≫ CRG = S133A > CAD) indicates that phosphorylation of the KID in CREB most effectively stimulates postrecruitment steps in transcription initiation (Fig. 14C). We attribute the lack of effect of P-KID in these assays to its inability to effectively recruit a polymerase complex, with the result that no polymerase complex is present to be regulated by phosphorylation of the activator.

To determine whether phosphorylation of KID stimulated the step following recruitment, we assayed abortive initiation, which is a measure of promoter melting during the isomerization step. Abortive initiation was measured by assessing the ability of the polymerase to add a radiolabeled NTP to a dinucleotide substrate, which can only bind if the polymerase complex has isomerized and melted the promoter at the start site (Fig. 15A). The substrate was separated from the product by electrophoresis (Fig. 15B) and quantitated to determine

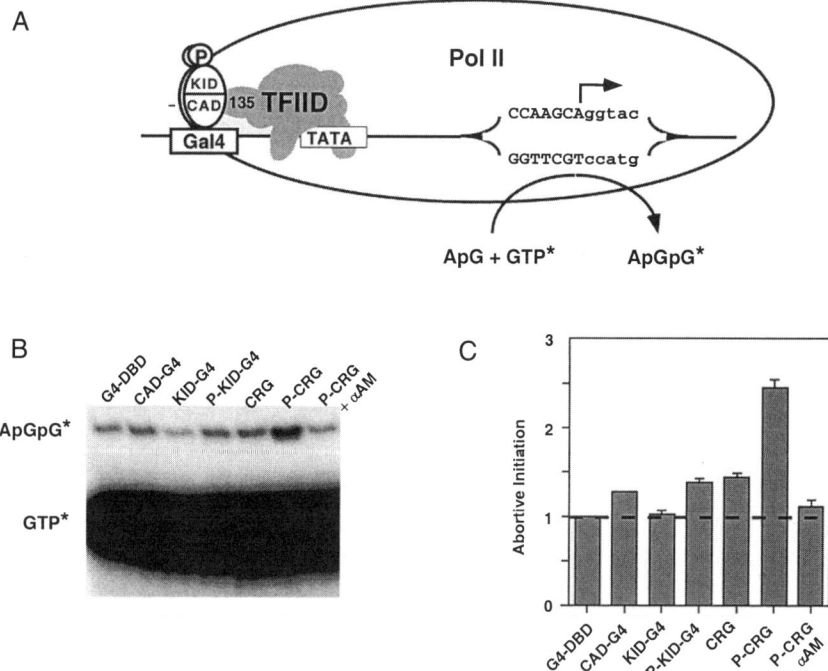

FIG. 15. Phosphorylation of the KID in CRG stimulates abortive initiation. (A) Abortive initiation was assayed by addition of labeled nucleotide to a dinucleotide substrate complementary to the transcription start site. (B) The products of abortive initiation reactions were separated by electrophoresis. Nonspecific reaction was assessed by inclusion of 1 μg/ml α-amanitin to inhibit RNA polymerase II. (C) The combined results ($x \pm$ SEM) of four experiments are shown. Adapted from Ref. 68.

the extent of stimulation of this reaction (Fig. 15C). Phosphorylation of the KID in CRG was the strongest stimulus of abortive initiation, which was stimulated only weakly by CAD-G4 or unphosphorylated CRG, consistent with their limited ability to enhance transcription activation in the absence of phosphorylation of the KID. Stimulation of abortive initiation by P-CRG was abrogated by inhibition of RNA polymerase II activity with α-amanitin. These results indicate that phosphorylation of the KID in CREB stimulates the isomerization reaction. The observation that the extent of stimulation of multiple-round transcription by P-CRG exceeds the extent of stimulation of abortive initiation may suggest that phosphorylation of KID also stimulates other postrecruitment steps in transcription activation, such as polymerase escape or reinitiation. However, more direct comparison of these activities in comparable assays is required to establish the extent to which different reaction steps are affected by phosphorylation of the CREB activation domain.

C. A Model for a Concerted Mechanism of Transcription Activation by Phosphorylated CREB

Taken together, our data provide evidence for a concerted mechanism of transcription activation (Fig. 16). Recruitment of IID, mediated by interaction of the CAD with hTAF$_{II}$135, is both necessary and sufficient to mediate recruitment of a functional polymerase complex to the promoter in a sequential mechanism. However, the CAD is relatively ineffective at stimulating later steps in transcription initiation, which limits basal transcription to low levels. Phosphorylation of the KID in CREB then modifies the activity of the polymerase complex recruited by the CAD (or other factors, as in the case of G4-PEPCK, Fig. 4A) to

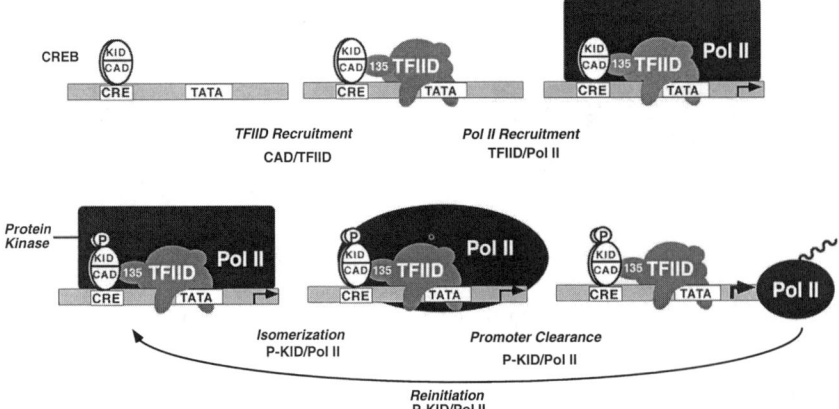

FIG. 16. Model for a concerted mechanism of activation of transcription by phosphorylated CREB.

provide rapid enhancement of transcription activation in response to extracellular signals. Coactivators, such as CREB-binding protein (CBP) are likely to be recruited by phosphorylation of CREB as well (reviewed in Ref. *108*). We found that immunodepletion of CBP with a combination of antibodies directed against amino- and carboxy-terminal epitopes had no effect on transcription, but we were unable to deplete more than 75% of the CBP present in our extracts, regardless of how much antibody we used. The antibody epitopes of the remaining CBP are most likely masked by its inclusion in a multiprotein complex. The question of whether this is a transcriptionally active complex and what it regulates will need to be addressed in a reconstituted transcription system. It is possible that the role of CBP is primarily in facilitating chromatin remodeling in reactions that would precede the polymerase recruitment and activation steps described here (*109, 108*). In addition, other coactivators continue to be discovered (*110, 111*) and one or more of these may be important in mediating induction by phosphorylated CREB. These components could be recruited in a CREB phosphorylation-dependent manner or they could be recruited independently of phosphorylation and have their interaction and activity modified by phosphorylation of CREB.

This concerted model is consistent with a number of observations of CREB-regulated gene transcription in biology. It provides a mechanistic explanation for how genes can exhibit measurable levels of basal transcription, which are reduced by mutation of the CRE (*34, 33, 112, 113*). It also accounts for how these genes can be rapidly induced in response to hormones and neurotransmitters activating a variety of protein kinases that phosphorylate Ser133 in CREB in different cells and tissues (*114, 115, 32*). This concerted model, in which each activation domain contributes to different steps in transcription initiation, also provides an explanation for why deletion of the CAD (naturally occurring in CREMα) or mutation of the kinase-regulated Ser133 in CREB results in dominant negative molecules interfering with induction by protein kinases (*61, 116*). The importance of recruitment by the CAD is also evident in an intriguing recent paper which suggests a role for the polyglutamine stretches found in the CREB CAD and Sp1 and also in huntingtin and other proteins involved in neurodegenerative disease (*117*). Overexpression of polyglutamine stretches interfered with CREB-dependent transcription activation in neurons in a dose-dependent manner and enhanced apoptosis (*117*). Both of these effects were diminished by overexpression of hTAF$_{II}$135, suggesting that these polyglutamine proteins interfere with the CAD:TAF interaction and prevent activation of survival genes by CREB. Finally, our model for a concerted mechanism of activation is in agreement with experiments in which either the IID or holoenzyme complexes are tethered to a promoter region by linking various components to a DNA-binding domain. Dorris and Struhl (*118*) found that tethering of IID to a promoter mediated transcription activation in mammalian cells, whereas tethering of various components of the holoenzyme complex did not result in transcription activation.

That result is consistent with the finding by Nakajima *et al.* (*74*) that, even when P-CREB was associated with holoenzyme, CAD-mediated recruitment of IID was required for transcription, and with our demonstration that the CAD:TAF interaction is necessary and sufficient for recruitment of a polymerase complex and transcription activation (*68, 72*). The mechanism of activation described here is likely used by all three cAMP-regulated CREB/ATF factors, CREB, CREM, and ATF-1. In addition, a similar mechanism has been described for stimulation of recruitment and isomerization/promoter clearance by different activation domains within the estrogen receptor (*119, 120*). It will be interesting to see how widely a concerted reaction mechanism is utilized to regulate gene transcription, either by employing distinct domains within a single factor or by employing distinct factors with different properties that are bound to the same promoter.

IV. Perspectives

This model for a concerted mechanism of activation provides a solid framework for further investigation of the mechanism of transcription activation by the CREB protein and raises a number of interesting questions. Are both isomerization and promoter clearance regulated by phosphorylation of the KID in CREB? What is the function of CBP in kinase-induced transcription activation? What general transcription factors interact to overcome these energy barriers in the transition steps to elongation by pol II? How does phosphorylation of the KID in CREB change its ability to interact with the general transcription machinery to facilitate these transitions? Does P-KID in CREB interact directly with general transcription factors? Are mediators required for these interactions? If so, which interactions are regulated as a consequence of phosphorylation? Which are required, but not directly regulated? Ultimately, what is needed to understand the mechanism of transcription activation by phosphorylated CREB is independent assessment of isomerization, promoter clearance, and reinitiation in different model systems, together with analyses of interactions between P-CREB and general transcription factors and genetic experiments to examine the consequences of mutations in interacting proteins to specific steps in transcription initiation.

ACKNOWLEDGMENTS

The work reported here was supported by grants from the NIH, the Juvenile Diabetes Research Foundation, and the American Diabetes Association. The author thanks Lorraine Koncar, Lianping Xing, Venkatesh Gopal, Joyce Agati, David Yeagley, Edward Felinski, Jingfang Liu, and Jeonga Kim, who contributed to this work; James Hopper and David Spector, whose discussions advanced the work; and all of our colleagues who contributed the valuable reagents acknowledged in our primary publications.

REFERENCES

1. L. Zawel and D. Reinberg, Common themes in assembly and function of eukaryotic transcription complexes. *Annu. Rev. Biochem.* **64,** 533–561 (1995).
2. T. W. Hai, M. Horikoshi, R. G. Roeder, and M. R. Green, Analysis of the role of the transcription factor ATF in the assembly of a functional preinitiation complex. *Cell* **54,** 1043–1051 (1988).
3. M. Horikoshi, T. Hai, Y. S. Lin, M. R. Green, and R. G. Roeder, Transcription factor ATF interacts with the TATA factor to facilitate establishment of a preinitiation complex. *Cell* **54,** 1033–1042 (1988).
4. S. Narayan, F. He, and S. H. Wilson, Activation of the human DNA polymerase beta promoter by a DNA alkylating agent through induced phosphorylation of cAMP response element binding protein-1. *J. Biol. Chem.* **271,** 18508–18513 (1996).
5. B. S. Wolner and J. D. Gralla, Promoter activation via a cyclic AMP response element in vitro. *J. Biol. Chem.* **272,** 32301–32307 (1997).
6. S. Narayan, W. A. Beard, and S. H. Wilson, DNA damage induced transcriptional activation of a human DNA polymerase beta chimeric promoter: recruitment of preinitiation complex in vitro by ATE/CREB. *Biochemistry* **34,** 73–80 (1995).
7. P. Brindle, S. Linke, and M. Montminy, Protein kinase A dependent activator in transcription factor CREB reveals new role for CREM repressors. *Nature* **364,** 821–824 (1993).
8. P. G. Quinn, Distinct activation domains within cAMP response element binding protein (CREB) mediate basal and cAMP-stimulated transcription. *J. Biol. Chem.* **268,** 16999–17009 (1993).
9. R. A. Maurer, Transcriptional regulation of the prolactin gene by ergocryptine and cyclic AMP. *Nature* **294,** 94–97 (1981).
10. M. R. Montminy, K. A. Sevarino, J. A. Wagner, G. Mandel, and R. H. Goodman, Identification of a cyclic AMP-responsive element within the rat somatostatin gane. *Proc. Natl. Acad. Sci. USA* **83,** 6682–6686 (1986).
11. J. M. Short, B. A. Wynshaw, H. P. Short, and R. W. Hanson, Characterization of the phosphoenolpyruvate carboxykinase (GTP) promoter-regulatory region. II. Identification of cAMP and glucocorticoid regulatory domains. *J. Biol. Chem.* **261,** 9721–9726 (1986).
12. W. J. Roesler, G. R. Vandenbark, and R. W. Hanson, Cyclic AMP and the induction of eukaryotic gene transcription. *J. Biol. Chem.* **263,** 9063–9066 (1988).
13. K. A. Lee, T. Y. Hai, L. SivaRaman, B. Thimmappaya, H. C. Hurst, N. C. Jones, and M. R. Green, A cellular protein, activating transcription factor, activates transcription of multiple E1A inducible adenovirus early promoters. *Proc. Natl. Acad. Sci. USA* **84,** 8355–8359 (1987).
14. Y. S. Lin and M. R. Green, Interaction of a common cellular transcription factor, ATF, with regulatory elements in both E1a and cyclic AMP-inducible promoters. *Proc. Natl. Acad. Sci. USA* **85,** 3396–3400 (1988).
15. T. Hai and M. G. Hartman, The molecular biology and nomenclature of the activating transcription factor/cAMP responsive element binding family of transcription factors: activating transcription factor proteins and homeostasis. *Gene* **273,** 1–11 (2001).
16. T. Hai, F. Liu, E. A. Allegretto, M. Karin, and M. R. Green, A family of immunologically related transcription factors that includes multiple froms of ATF and AP-1. *Genes Dev.* **2,** 1216–1226 (1988).

17. T. W. Hai, F. Liu, W. J. Coukos, and M. R. Green, Transcription factor ATF cDNA clones: an extensive family of leucine zipper proteins able to selectively form DNA-binding heterodimers. *Genes Dev.* **3,** 2083–2090 (1989) [Erratum, Genes Dev. 4(4), 682 (1990)].
18. K. J. Flint and N. C. Jones, Differential regulation of three members of the ATF/CREB family of DNA-binding proteins. *Oncogene* **6,** 2019–2026 (1991).
19. N. S. Foulkes, B. Mellstrom, E. Benusiglio, and P. Sassone-Corsi, Developmental switch of CREM function during spermatogenisis: from antagonist to activator. *Nature* **355,** 80–84 (1992).
20. J. Corbin, P. Sugden, L. West, D. Flockhart, T. Lincoln, and D. McCarthy, Studies on the properties and mode of action of the purified regulatory subunit of bovine heart adenosine 3′:5′-monophosphate. *J. Biol. Chem.* **253,** 3997–4003 (1978).
21. J. D. Corbin, C. E. Cobb, S. J. Beebe, D. K. Granner, S. R. Koch, T. W. Gettys, P. F. Blackmore, S. H. Francis, and J. N. Wells, Mechanism and function of cAMP- and cGMP-dependent protein-kinases. *Adv. Second Messenger Phosphoprotein Res.* **21,** 75–86 (1988).
22. M. Hagiwara, P. Brindle, A. Harootunian, R. Armstrong, J. Rivier, W. Vale, R. Tsien, and M. R. Montminy, Coupling of hormonal stimulation and transcription via the cyclic AMP-responsive factor CREB is rate limited by nuclear entry of protein-kinase A. *Mol. Cell. Biol.* **13,** 4852–4859 (1993).
23. K. T. Riabowol, J. S. Fink, M. Z. Gilman, D. A. Walsh, R. H. Goodman, and J. R. Feramisco, The catalytic subunit of cAMP-dependent protein kinase induces expression of genes containing cAMP-responsive enhancer elements. *Nature* **336,** 83–86 (1988).
24. R. A. Maurer, Both isoforms of the cAMP dependent protein kinase catalytic subunit can activate transcription of the prolactin gene. *J. Biol. Chem.* **264,** 6870–6873 (1989).
25. P. L. Mellon, C. H. Clegg, L. A. Correll, and G. S. McKnight, Regulation of transcription by cyclic AMP dependent protein kinase. *Proc. Natl. Acad. Sci. USA* **86,** 4887–4891 (1989).
26. J. Nakagawa, D. von der Ahe, D. Pearson, B. A. Hemmings, S. Shibahara, and Y. Nagamine, Transcriptional regulation of a plasminogen activator gene by cyclic AMP in a homologous cell free system. Involvement of cyclic AMP dependent protein kinase in transcriptional control. *J. Biol. Chem.* **263,** 2460–2468 (1988).
27. K. K. Yamamoto, G. A. Gonzalez, W. I. Biggs, and M. R. Montminy, Phosphorylation induced binding and trancriptional efficacy of nuclear factor CREB. *Nature* **334,** 494–498 (1988).
28. D. J. Klemm *et al.*, In vitro analysis of promoter elements regulating transcription of the phosphoenolpyruvate carboxykinase (GTP) gene. *Mol. Cell. Biol.* **10,** 480–485 (1990).
29. P. G. Quinn and D. K. Granner, Cyclic AMP dependent protein kinase regulates transcription of the phosphoenolpyruvate carboxykinase gene but not binding of nuclear factors to the cyclic AMP regulatory element. *Mol. Cell. Biol.* **10,** 3357–3364 (1990).
30. G. A. Gonzalez and M. R. Montminy, Cyclic AMP stimulates somatostatin gene transcription by phosphorylation of CREB at serine 133. *Cell* **59,** 675–680 (1989).
31. K. Du and M. Montminy, CREB is a regulatory target for the protein kinase Akt/PKB. *J. Biol. Chem.* **273,** 32377–32379 (1998).
32. D. De Cesare, G. M. Fimia, and P. Sassone-Corsi, Signaling routes to CREM and CREB: plasticity in transcriptional activation. *Trends Biochem. Sci.* **24,** 281–285 (1999).
33. P. G. Quinn, T. W. Wong, M. A. Magnuson, J. B. Shabb, and D. K. Granner, Identification of basal and cyclic AMP regulatory elements in the promoter of the phosphoenolpyruvate carboxykinase gene. *Mol. Cell. Biol.* **8,** 3467–3475 (1988).
34. A. M. Delegeane, L. H. Ferland, and P. L. Mellon, Tissue specific enhancer of the human glycoprotein hormone α-subunit gene: dependence on cyclic AMP-inducible elements. *Mol. Cell. Biol.* **7,** 3994–4002 (1987).

35. P. J. Deutsch, J. L. Jameson, and J. F. Habener, Cyclic AMP responsiveness of human gonadotropin alpha gene transcription is directed by a repeated 18 base pair enhancer. Alpha promoter receptivity to the enhancer confers cell preferential expression. *J. Biol. Chem.* **262**, 12169–12174 (1987).
36. O. M. Andrisani, Z. N. Zhu, D. A. Pot, and J. E. Dixon, In vitro transcription directed from the somatostatin promoter in dependent upon a purified 43-kDa-DNA-binding protein. *Proc. Natl. Acad. Sci. USA* **86**, 2181–2185 (1989).
37. G. A. Gonzalez, K. K. Yamamoto, W. H. Fischer, D. Karr, P. Menzel, W. I. Biggs, W. W. Vale, and M. R. Montminy, A cluster of phosphorylation site on the cyclic AMP regulated nuclear factor CREB predicted by its sequence. *Nature* **337**, 749–752 (1989).
38. J. P. Hoeffler, T. E. Meyer, Y. Yun, J. L. Jameson, and J. F. Habener, Cyclic AMP-responsive DNA-binding protein: structure based on a cloned placental cDNA. *Science* **242**, 1430–1433 (1988).
39. L. A. Berkowitz and M. Z. Gilman, Two distinct forms of active transcription factor CREB (cAMP response element binding protein). *Proc. Natl. Acad. Sci. USA* **87**, 5258–5262 (1990).
40. L. Raycroft, H. Wu, and G. Lozano, Transcription activation by wild-type but not transforming mutants of the p53 anti oncogene. *Science* **249**, 1049–1051 (1990).
41. C. Q. Lee, Y. D. Yun, J. P. Hoeffler, and J. F. Habener, Cyclic AMP responsive transcriptional activation of CREB-327 involves interdependent phosphorylated subdomains. *EMBO J.* **9**, 4455–4465 (1990).
42. N. S. Foulkes and P. Sassone-Corsi, More is better: activators and repressors from the same gene. *Cell* **68**, 411–414 (1992).
43. H. C. Cheng, B. E. Kemp, R. B. Pearson, A. J. Smith, L. Misconi, S. M. Van Patten, and D. A. Walsh, A potent synthetic peptide inhibitor of the cAMP-dependent protein kinase. *J. Biol. Chem.* **261**, 989–992 (1986).
44. R. N. Day, J. A. Walder, and R. A. Maurer, A protein kinase inhibitor gene reduces both basal and multihormone stimulated prolactin gene transcription. *J. Biol. Chem.* **264**, 431–436 (1989).
45. J. R. Grove, P. J. Deutsch, D. J. Price, J. F. Habener, and J. Avruch, Plasmids encoding PKI(1-31), a specific inhibitor of cAMP stimulated gene expression, inhibit the basal transcriptional activity of some but not all cAMP regulated DNA response elements in JEG-3 cells. *J. Biol. Chem.* **264**, 19506–19513 (1989).
46. S. R. Olsen and M. D. Uhler, Inhibition of protein kinase A by overexpression of the cloned human protein kinase inhibitor. *Mol. Endocrinol.* **5**, 1246–1256 (1991).
47. M. Vallejo, L. Penchuk, and J. F. Habener, Somatostatin gene upstream enhancer element activated by a protein complex consisting of CREB, Isl-1-like, and alpha-CBF-like transcription factors. *J. Biol. Chem.* **267**, 12876–12884 (1992).
48. W. J. Roesler et al., Evidence for the involvement of at least two distinct transcription factors, one of which is liver-enriched, for the activation of the phosphoenolpyruvate carboxykinase gene promoter by cAMP. *J. Biol. Chem.* **268**, 3791–3796 (1993).
49. P. Budworth, P. Quinn, and J. Nilson, Multiple characteristics of a pentameric regulatory array endow the human α subunit glycoprotein hormone promoter with trophoblast specificity and maximal activity. *Mol. Endocrinol.* **11**, 1669–1680 (1997).
50. D. Yeagley, J. Agati, and P. Quinn, A tripartite array of transcription factor binding sites mediates cAMP induction of PEPCK gene transcription and its inhibition by insulin. *J. Biol. Chem.* **273**, 18743–18750 (1998).
51. T. W. Fawcett, J. L. Martindale, K. Z. Guyton, T. Hai, and N. J. Holbrook, Complexes containing activating transcription factor (ATF)/cAMP responsive element binding protein (CREB) interact with the CCAAT/enhancer binding protein (C/EBP)-ATF composite site to regulate Gadd153 expression during the stress response. *Biochem. J.* **339**, 135–141 (1999).

52. D. Yeagley, J. Moll, C. A. Vinson, and P. G. Quinn, Characterization of elements mediating regulation of phosphoenolpyruvate carboxykinase gene transcription by protein kinase A and insulin. Identification of a distinct complex formed in cells that mediate insulin inhibition. *J. Biol. Chem.* **275**, 17814–17820 (2000).
53. K. K. Yamamoto, G. A. Gonzalez, P. Menzel, J. Rivier, and M. R. Montminy, Characterization of a bipartite activator domain in transcription factor CREB. *Cell* **60**, 611–617 (1990).
54. M. R. Montminy and L. M. Bilezikjian, Binding of a nuclear protein to the cyclic-AMP response element of the somatostatin gene. *Nature* **328**, 175–178 (1987).
55. R. P. Rehfuss, K. M. Walton, M. M. Loriaux, and R. H. Goodman, The cAMP-regulated enhancer-binding protein ATF-1 activates transcription in response to cAMP dependent protein kinase A. *J. Biol. Chem.* **266**, 18431–18434 (1991).
56. P. J. Mitchell and R. Tjian, Transcriptional regulation in mammalian cells by sequence-specific DNA binding protein. *Science* **245**, 371–378 (1989).
57. J. P. Hoeffler, T. E. Meyer, G. Waeber, and J. F. Habener, Multiple adenosine 3′,5′-cyclic [corrected] monophosphate response element DNA binding proteins generated by gene diversification and alternative exon splicing. *Mol. Endocrinol.* **4**, 920–930 (1990) [Erratum, *Mol. Endocrinol.* **4**(7), 1016 (1990)].
58. S. Ruppert, T. J. Cole, M. Boshart, E. Schmid, and G. Schutz, Multiple mRNA isoforms of the transcription activator protein CREB. Generation by alternative splicing and specific expression in primary spermatocytes. *EMBO J.* **11**, 1503–1512 (1992).
59. I. Radhakrishnan, G. C. Perez-Alvarado, D. Parker, H. J. Dyson, M. R. Montminy, and P. E. Wright, Solution structure of the KIX domain of CBP bound to the transactivation domain of CREB: a model for activator: coactivator interactions. *Cell* **91**, 741–752 (1997).
60. D. Parker, M. Rivera, T. Zor, A. Henrion-Caude, I. Radhakrishnan, A. Kumar, L. H. Shapiro, P. E. Wright, M. Montminy, and P. K. Brindle, Role of secondary structure in discrimination between constitutive and inducible activators. *Mol. Cell. Biol.* **19**, 5601–5607 (1999).
61. N. S. Foulkes, B. M. Laoide, F. Schlotter, and C. P. Sassone, Transcriptional antagonist cAMP-responsive element modulator (CREM) down regulates e-fos cAMP-induced expression. *Proc. Natl. Acad. Sci. USA* **88**, 5448–5452 (1991).
62. L. P. Xing and P. G. Quinn, Three distinct regions within the constitutive activation domain of CREB are required for transcription activation. *J. Biol. Chem.* **269**, 28732–28736 (1994).
63. B. Pugh and R. Tjian, Mechanism of transcriptional activation by Sp1: Evidence for coactivators. *Cell* **61**, 1187–1197 (1990).
64. G. Gill, I. Sadowski, and M. Ptashne, Mutations that increase the activity of a transcriptional activator in yeast and mammalian cells. *Proc. Natl. Acad. Sci. USA* **87**, 2127–2131 (1990).
65. G. Gill, E. Pascal, Z. H. Tseng, and R. Tjian, A glutamine rich hydrophobic patch in transcription factor SP1 contacts the dTAFII 110 component of the *drosophila* TFIID complex and mediates transcriptional activation. *Proc. Natl. Acad. Sci. USA* **91**, 192–196 (1994).
66. K. Ferreri, G. Gill, and M. Montminy, The cAMP-regulated transcription factor CREB interacts with a component of the TFIID complex. *Proc. Natl. Acad. Sci. USA* **91**, 1210–1213 (1994).
67. L. Xing, V. K. Gopal, and P. G. Quinn, cAMP response element binding protein (CREB) interacts with transcription factors IIB and IID. *J. Biol. Chem.* **270**, 17488–17493 (1995).
68. J. Kim, J.-F. Lu, and P. Quinn, Distinct cAMP response element binding protein (CREB) domains stimulate different steps in a concerted mechanism of transcription activation. *Proc. Natl. Acad. Sci. USA* **97**, 11292–11296 (2000).
69. E. A. Felinski, J. Kim, J. Lu, and P. G. Quinn, Recruitment of an RNA polymerase II complex is mediated by the constitutive activation domain in CREB, independently of CREB phosphorylation. *Mol. Cell. Biol.* **21**, 1001–1010 (2001).

CREB TRANSCRIPTION ACTIVATION MECHANISMS 303

70. E. A. Felinski and P. G. Quinn, The CREB constitutive activation domain interacts with TATA-binding protein-associated factor 110 (TAF110) through specific hydrophobic residues in one of the three subdomains required for both activation and TAF110 binding. *J. Biol. Chem.* **274**, 11672–11678 (1999).
71. D. Saluja, M. Vassallo, and N. Tanese, Distinct subdomains of human TAFII130 are required for interactions with glutamine rich transcriptional activators. *Mol. Cell. Biol.* **18**, 5734–5743 (1998).
72. E. A. Felinski and P. G. Quinn, The coactivator dTAFII110/hTAFII135 is sufficient to recruit a polymerase complex and activate basal transcription mediated by CREB. *Proc. Natl. Acad. Sci. USA* **98**, 13078–13083 (2001).
73. M. W. Van Dyke, R. G. Roeder, and M. Sawadogo, Physical analysis of transcription preinitiation complex assembly on a class II gene promoter. *Science* **241**, 1335–1338 (1988).
74. T. Nakajima, C. Uchida, S. F. Anderson, C. G. Lee, J. Hurwitz, J. D. Parvin, and M. Montminy, RNA helicase A mediates association of CBP with RNA polymerase II. *Cell* **90**, 1107–1112 (1997).
75. T. Nakajima, C. Uchida, S. F. Anderson, J. D. Parvin, and M. Montminy, Analysis of a cAMP-responsive activator reveals a two component mechanism for transcriptional induction via signal dependent factors. *Genes Dev.* **11**, 738–747 (1997).
76. G. Orphanides, T. Lagrange, and D. Reinberg, The general transcription factors of RNA polymerase II. *Genes Dev.* **10**, 2657–2683 (1996).
77. S. K. Burley, K. L. Clark, D. A. A. Ferré, J. L. Kim, and D. B. Nikolov, X-ray crystallographic studies of eukaryotic transcription factors. *Cold Spring Harb. Symp. Quant. Biol.* **58**, 123–132 (1993).
78. F. Andel, 3rd, A. G. Ladurner, C. Inouye, R. Tjian, and E. Nogales, Three dimensional structure of the human TFIID-IIA IIB complex. *Science* **286**, 2153–2156 (1999).
79. F. J. Asturias, Y. W. Jiang, L. C. Myers, C. M. Gustafsson, and R. D. Kornberg, Conserved structures of mediator and RNA polymerase II holoenzyme. *Science* **283**, 985–987 (1999).
80. R. A. Mooney and R. Landick, RNA polymerase unveiled. *Cell* **98**, 687–690 (1999).
81. P. Schultz, S. Fribourg, A. Poterszman, V. Mallouh, D. Moras, and J. M. Egly, Molecular structure of human TFIIH. *Cell* **102**, 599–607 (2000).
82. A. L. Gnatt, P. Cramer, J. Fu, D. A. Bushnell, and R. D. Kornberg, Structural Basis of Transcription: An RNA Polymerase II Elongation Complex at 33 A Resolution. *Science* **292**, 1863–1876 (2001).
83. R. Conaway and J. Conaway, General initiation factors for RNA polymerase II. *Annu. Rev. Biochem.* **62**, 161–190 (1993).
84. H. Ge, E. Martinez, C. M. Chiang, and R. G. Roeder, Activator dependent transcription by mammalian RNA polymerase II: in vitro reconstitution with general transcription factors and cofactors. *Meth. Enzymol.* **274**, 57–71 (1996).
85. L. C. Myers, K. Leuther, D. A. Bushnell, C. M. Gustafsson, and R. D. Kornberg, Yeast RNA polymerase II transcription reconstituted with purified proteins. *Methods* **12**, 212–216 (1997).
86. A. Dvir, R. C. Conaway, and J. W. Conaway, Promoter escape by RNA polymerase II. A role for an ATP cofactor in suppression of arrest by polymerase at promoter-proximal sites. *J. Biol. Chem.* **271**, 23352–23356 (1996).
87. F. C. Holstege, P. C. van der Vliet, and H. T. Timmers, Opening of an RNA polymerase II promoter occurs in two distinct steps and requires the basal transcription factors IIE and IIH. *EMBO J.* **15**, 1666–1677 (1996).
88. A. Dvir, S. Tan, J. W. Conaway, and R. C. Conaway, Promoter escape by RNA polymerase II. Formation of an escape-competent transcriptional intermediate is a prerequisite for exit of polymerase from the promoter. *J. Biol. Chem.* **272**, 28175–28178 (1997).

89. J. W. Conaway, A. Dvir, R. J. Moreland, Q. Yan, B. J. Elmendorf, S. Tan, and R. C. Conaway, Mechanism of promoter escape by RNA polymerase II. *Cold Spring Harb. Symp. Quant. Biol.* **63**, 357–364 (1998).
90. N. Yudkovsky, J. A. Ranish, and S. Hahn, A transcription reinitiation intermediate that is stabilized by activator. *Nature* **408**, 225–229 (2000).
91. M. W. Van Dyke, M. Sawadogo, and R. G. Roeder, Stability of transcription complexes on class II genes. *Mol. Cell. Biol.* **9**, 342–344 (1989).
92. S. Buratkowski, S. Hahn, L. Guarente, and P. A. Sharp, Five intermediate complexes in transcription initiation by RNA polymerase II. *Cell* **56**, 549–561 (1989).
93. J. W. Conaway, J. N. Bradsher, and R. C. Conaway, Mechanism of assembly of the RNA polymerase II preinitiation complex. Transcription factors delta and epsilon promote stable binding of the transcription apparatus to the initiator element. *J. Biol. Chem.* **267**, 10142–10148 (1992).
94. A. J. Koleske and R. A. Young, An RNA polymerase II holoenzyme responsive to activators. *Nature* **368**, 466–469 (1994).
95. V. Ossipow, J. P. Tassan, E. A. Nigg, and U. Schibler, A mammalian RNA polymerase II holoenzyme containing all components required for promoter-specific transcription initiation. *Cell* **83**, 137–146 (1995).
96. A. Dvir, K. P. Garrett, C. Chalut, J. M. Egly, J. W. Conaway, and R. C. Conaway, A role for ATP and TFIIH in activation of the RNA polymerase II preinitiation complex prior to transcription initiation. *J. Biol. Chem.* **271**, 7245–7248 (1996).
97. D. S. Luse and G. A. Jacob, Abortive initiation by RNA polymerase II *in vitro* at the adenovirus major late promoter. *J. Biol. Chem.* **262**, 14990–14997 (1987).
98. J. A. Goodrich and R. Tjian, Transcription factors IIE and IIH and ATP hydrolysis direct promoter clearance by RNA polymerase II. *Cell* **77**, 145–156 (1994).
99. P. Cramer, D. A. Bushnell, and R. D. Kornberg, Structural Basis of Transcription: RNA Polymerase II at 2.8 A Resolution. *Science* **292**, 1876–1882 (2001).
100. A. Dvir, R. C. Conaway, and J. W. Conaway, A role for TFIIH in controlling the activity of early RNA polymerase II elongation complexes. *Proc. Natl. Acad. Sci. USA* **94**, 9006–9010 (1997).
101. P. Lieberman and A. Berk, The Zta trans activator protein stabilizes TFIID association with promoter DNA by direct protein-protein interaction. *Genes Dev.* **5**, 2441–2454 (1991).
102. T. Hoey, R. O. J. Weinzierl, G. Gill, J. L. Chen, B. D. Dynlacht, and R. Tjian, Molecular cloning and functional analysis of Drosophila TAF110 reveal properties expected of coactivators. *Cell* **72**, 247–260 (1993).
103. F. Sauer, S. K. Hansen, and R. Tjian, DNA template and activator coactivator requirements for transcriptional synergism by Drosophila bicoid. *Science* **270**, 1825–1828 (1995).
104. K. Yokomori, J. L. Chen, A. Admon, S. Zhou, and R. Tjian, Molecular cloning and characterization of dTAFII30α and dTAFII30β. two small subunits of Drosophila TFIID. *Genes Dev.* **7**, 2587–2597 (1993).
105. S. K. Burley and R. G. Roeder, Biochemistry and structural biology of transcription factor IID (TFIID). *Annu. Rev. Biochem.* **65**, 769–799 (1996).
106. J. A. Goodrich, T. Hoey, C. J. Thut, A. Admon, and R. Tjian, Drosophila TAFII40 interacts with both a VP16 activation domain and the basal transcription factor TFIIB. *Cell* **75**, 519–530 (1993).
107. P. R. MacDonald, D. R. Sherman, D. R. Dowd, S. C. J. Jefcoat, and R. K. DeLisle, The vitamin D receptor interacts with general transcription factor IIB. *J. Biol. Chem.* **270**, 4748–4752 (1995).
108. N. Vo and R. Goodman, CREB Binding Protein and p300 in Transcriptional Regulation. *J. Biol. Chem.* **276**, 13505–13508 (2001).
109. H. Asahara, B. Santoso, E. Guzman, K. Du, P. A. Cole, I. Davidson, and M. Montminy, Chromatin Dependent Cooperativity between Constitutive and Inducible Activation Domains in CREB. *Mol. Cell. Biol.* **21**, 7892–7900 (2001).

110. D. Knutti and A. Kralli, PGC-1, a versatile coactivator. *Trends Endocrinol. Metab.* **12,** 360–365 (2001).
111. J. C. Yoon, P. Puigserver, G. Chen, J. Donovan, Z. Wu, J. Rhee, G. Adelmant, J. Stafford, C. R. Kahn, D. K. Granner, C. B. Newgard, and B. M. Spiegelman, Control of hepatic gluconeogenesis through the transcriptional coactivator PGC-1. *Nature* **413,** 131–138 (2001).
112. K. S. Kim, M. K. Lee, J. Carroll, and T. H. Joh, Both the basal and inducible transcription of the tyrosine hydroxylase gene are dependent upon a cAMP response element. *J. Biol. Chem.* **268,** 15689–15695 (1993).
113. M. Lazaroff, S. Patankar, S. O. Yoon, and D. M. Chikaraishi, The cyclic AMP response element directs tyrosine hydroxylase expression in catecholaminergic central and peripheral nervous system cell lines from transgenic mice. *J. Biol. Chem.* **270,** 21579–21589 (1995).
114. K. Sasaki, T. P. Cripe, S. R. Koch, T. L. Andreone, D. D. Petersen, E. G. Beale, and D. K. Granner, Multihormonal regulation of phosphoenolpyruvate carboxykinase gene transcription. The dominant role of insulin. *J. Biol. Chem.* **259,** 15242–15251 (1984).
115. P. Brindle, T. Nakajima, and M. Montminy, Multiple protein kinase A regulated events are required for transcriptional induction by cAMP. *Proc. Nat. Acad. Sci. USA* **92,** 10521–10525 (1995).
116. K. Barton, N. Muthusamy, C. Fischer, C. Clendenin, and J. Leiden, Defective thymocyte proliferation and IL-2 production in transgenic mice expressing a dominant negative form of CREB. *Nature* **379,** 81–85 (1996).
117. T. Shimohata, T. Nakajima, M. Yamada, C. Uchida, O. Onodera, S. Naruse, T. Kimura, R. Koide, K. Nozaki, Y. Sano, H. Ishiguro, K. Sakoe, T. Ooshima, A. Sato, T. Ikeuchi, M. Oyake, T. Sato, Y. Aoyagi, I. Hozumi, T. Nagatsu, Y. Takiyama, M. Nishizawa, J. Goto, I. Kanazawa, I. Davidson, N. Tanese, H. Takahashi, and S. Tsuji, Expanded polyglutamine stretches interact with TAFII130, interfering with CREB dependent transcription. *Nat. Genet.* **26,** 29–36 (2000).
118. D. R. Dorris and K. Struhl, Artificial recruitment of TFIID, but not RNA polymerase II holoenzyme, activates transcription in mammalian cells. *Mol. Cell. Biol.* **20,** 4350–4358 (2000).
119. D. Chen, T. Riedl, E. Washbrook, P. E. Pace, R. C. Coombes, J. M. Egly, and S. Ali, Activation of estrogen receptor alpha by S118 phosphorylation involves a ligand dependent interaction with TFIIH and participation of CDK7. *Mol. Cell.* **6,** 127–137 (2000).
120. X. Wu, H. Li, and J. D. Chen, The human homologue of the yeast DNA repair and TFIIH regulator MMS19 is an AF1 specific coactivator of estrogen receptor. *J. Biol. Chem.* **276,** 23962–23968 (2001).

eIF4A: The Godfather of the DEAD Box Helicases

GEORGE W. ROGERS, JR.,*
ANTON A. KOMAR, AND
WILLIAM C. MERRICK

Department of Biochemistry
School of Medicine
Case Western Reserve University
Cleveland, Ohio 44106-4935

I. eIF4A: The Protein... 308
II. The Biology of eIF4A... 312
III. eIF4A: Biochemical Properties 313
IV. Influence of Other Proteins on eIF4A Function..................... 319
V. Does Any of This Biochemistry Make Sense or Is There a
   Contradiction Somewhere?..................................... 321
VI. The Role of the DEAD Box Sequences........................... 323
VII. The Function of the Helicase Activity of eIF4A..................... 324
VIII. eIF4A and the 80S Initiation Pathway............................ 326
IX. Lessons Learned from eIF4A.................................... 328
    References.. 329

eIF4A has long been considered the "gold standard" for DEAD box helicases. In large measure, this reflected two items: first, the role of eIF4A in protein synthesis initiation was relatively well established. Second, a wide variety of biochemical studies had established the ability of eIF4A to bind nucleic acids in an ATP-dependent manner, to hydrolyze ATP in an RNA-dependent manner, and to unwind RNA duplexes in an ATP-dependent manner. In this article, these basic observations are reviewed for biochemical consistency and also interpreted in light of the available crystal structures for DEAD box proteins. The role of nonprocessive vs. processive helicase activity in protein synthesis is discussed. Also examined is the influence of ancillary protein factors (eIF4B, eIF4G, and eIF4H) on this activity. Finally, the "real" role(s) for eIF4A helicase activity in protein synthesis is discussed and related to other circumstances that likely also involve the use of non-processive or slightly processive DEAD box helicases (ribosome biosynthesis, RNA splicing). © 2002, Elsevier Science (USA).

*Current address: The Scripps Research Institute, Department of Neurobiology, La Jolla, California 92037.

I. eIF4A: The Protein

The original work on the purification of eIF4A was accomplished in the mid-1970s by a variety of research groups working with either plant or animal systems. From a number of studies, it was concluded that eIF4A was functional as a single polypeptide chain about 45,000 Da in size and generally spherical in shape. In the late-1980s, as cloning became a more routine process, the amino acid sequence of eIF4A was determined by cloning and partial amino acid sequencing (1–3). At the same time, Linder and colleagues noted that a number of proteins appeared to share common sequence motifs with eIF4A (4). They termed the family of proteins the "DEAD box" proteins, referring to one of the conserved motifs, the amino acid sequence aspartic acid–glutamic acid–alanine–aspartic acid (DEAD in the single-letter code). This group identified seven conserved sequence motifs with approximately, but not exactly, the same spacing between the motifs. They concluded that "These motifs are probably involved in the ATP binding and ATPase activity and in the interaction with nucleic acids" (4). Today, this family and closely related families number over 200 proteins and where these proteins have been examined, they generally all have the ability to bind both ATP and nucleic acid (for up-to-the minute compilations, see the web sites www.expasy.ch/linder/RNA—helicases.html and www.helicase.net). Although more difficult to test in some instances, these proteins often have demonstrable RNA-dependent ATPase activity and ATP-dependent duplex RNA unwinding activity and generally fall into the group of proteins referred to as helicases (see below).

Figure 1 is a representation of a number of different DEAD box proteins. Two things should be evident. First, it appears as if eIF4A is the smallest member of the DEAD box family of proteins. Second, the larger DEAD box proteins have extensions at the amino terminus, the carboxy terminus, or at both ends. However, the cores (containing the conserved sequence motifs) are accurately represented by just eIF4A. This has led to the idea that perhaps eIF4A is a common core (the "motor") for each of the DEAD box proteins (P. Linder, personal communication). However, efforts to take the entire eIF4A sequence and attach flanking sequences from other DEAD box proteins have not resulted in the generation of active proteins, either *in vitro* or *in vivo*. This result has been taken to suggest that the folding and interaction of the extra domains is specific for each core, even though the function of each of the cores may be the same as that of eIF4A.

Although the conservation of the linear amino acid sequence demonstrates the family identity, this is seen in even greater (and biologically, more important) detail when one compares the three-dimensional structure. Although limited structures are available, especially in comparison to the number of gene

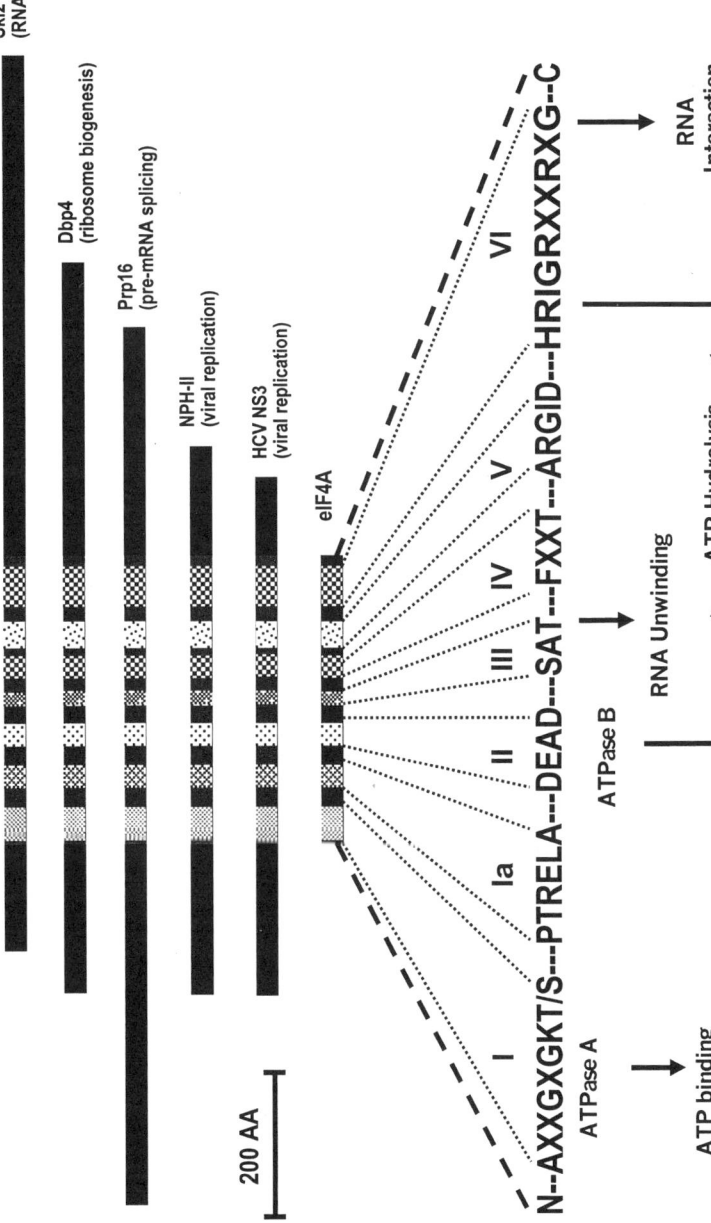

FIG. 1. Schematic alignment of the conserved motifs of the DEAD box family of proteins. Shown in the figure are a representative number of DEAD box proteins aligned to the conserved motifs seen in eIF4A. Note that although there is considerable variation at both the amino and carboxy termini, there is little or no variation in the sequence of the conserved DEAD box motifs (additionally, there is little variation in the spacing between the DEAD box elements).

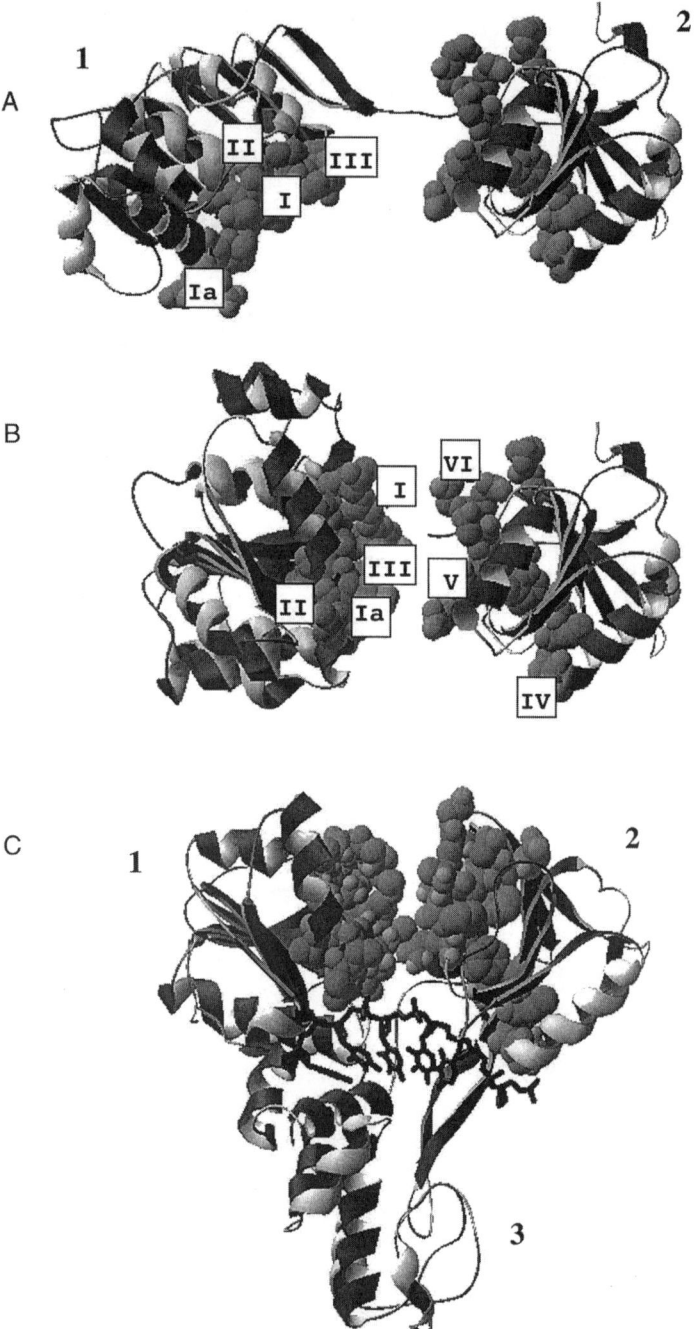

sequences, those of both eIF4A and the hepatitus C virus (HCV) NS3 proteins have been determined (5, 6) and are shown in Fig. 2. eIF4A exists as a "dumbbell"-shaped molecule with two structural domains connected by an 11-amino-acid linker (6). The structure of eIF4A can be superimposed upon that of a number of helicases, including HCV NS3 helicase. These superimpositions (see Fig. 3 in Ref. 6) indicate a general conservation of structure for the two domains of eIF4A with regard to the other helicases and allow one to infer where the RNA-binding site for eIF4A is likely to be. The N-terminal domain of eIF4A contains the ATP-binding domain and contains the classical Walker B motif, whereas the C-terminal domain has a parallel $\alpha-\beta$ structure. Assuming the eIF4A binds RNA in a manner similar to HCV NS3 helicase, it has been suggested that a number of conserved arginine residues in eIF4A could be responsible for generating the RNA-binding site (arginine residues 98, 148, 269, 270, 298, and perhaps 321; Ref. 6).

For the HCV NS3 helicase (and presumably other DEAD box proteins with amino- or carboxy-terminal extensions), an additional domain assists in binding the RNA. The presence of this additional domain may explain why eIF4A appears to be both a rather poor RNA-binding protein and a nonprocessive helicase, whereas helicases like HCV NS3 are capable of processively unwinding duplex nucleic acid substrates with high affinity (see below). It should be noted that the structure of eIF4A in Fig. 2A may be a bit misleading, as it has been shown that eIF4A undergoes several conformational changes upon the binding of ATP, RNA, or both ATP and RNA (7). In this context, the crystal structure for eIF4A (Fig. 2A) does not show the conserved DEAD box motifs in the same relative position as noted in HCV NS3 (Fig. 2C). By fixing the position of domain 2 and rotating domain 1, it is possible to achieve the same orientation of conserved motifs as seen in the HCV NS3 helicase (Fig. 2B). It should be noted that although this rotation is not possible without severing a bond in the connecting linker region, it is possible that the ligand-driven conformation changes noted by Lorsch and Herschlag (7) may allow for such a reorientation in the native eIF4A.

FIG. 2. Structural features of eIF4A and HCV NS3 helicase. Shown above are ribbon diagrams of eIF4A (A) and HCV NS3 helicase (C) obtained from the primary coordinates deposited in the PDB protein structure database by Caruthers *et al.* (6) and Kim *et al.* (5). The structure of the HCV NS3 helicase has been positioned to show the binding of the single strand of DNA across the face of the helicase. The structure of eIF4A has been oriented to match domain 2 of the HCV NS3 helicase. Note that domain 1 appears to be twisted relative to the equivalent orientation in domain 1 of the HCV NS3 helicase. The middle part of figure (B) shows the eIF4A structure after severing the linking peptide (amino acids 222–232) that was necessary to orient eIF4A domain 1, as is seen for domain 1 in the HCV NS3 helicase. The conserved DEAD box sequence motifs are represented by spheres and labeled as shown in Fig. 1.

A second feature that has emerged from the structure for the HCV NS3 helicase is that the conserved motifs are in close proximity to one another, even though they are separated by many amino acids in the linear sequence. It is presumed that it is this orientation and the preservation of the catalytic residues that yield the conserved spacing (see Fig. 1) between the DEAD box sequence motifs and the conserved folding pattern of domains 1 and 2 (Fig. 2). It will be interesting to see if future crystal structures will lead to distinctions that might explain why some helicases are nonprocessive and others are processive. At the same time, it may also be possible to begin to define the requirements of complementary factors to convert nonprocessive helicases to slightly or fully processive helicases.

II. The Biology of eIF4A

In most eukaryotic organisms, there are two distinct eIF4A genes, most commonly termed eIF4A1 and eIF4A2 (8). The sequence identity of the two proteins is generally in the 90–95% range. Where these isoforms have been tested, each species appears to have the same biological activity in *in vitro* assays (3, 8, 9). Although this would appear to indicate that the two isoforms are functionally indistinguishable, there is a distinctly different pattern of expression, either between tissues or as a function of development (2, 10). In particular, it was noted that the eIF4A1 mRNA was synthesized and translated most efficiently in those cells that were actively growing, whereas the eIF4A2 mRNA was synthesized and translated during growth-arrested conditions (i.e., the quiescent state; 10). However, the aggregate amount of the two mRNAs appears to be relatively constant.

Curiously, a third isoform of eIF4A, eIF4A3, does not appear to function in protein synthesis (11). This isoform, which is 65% similar to the other isoforms eIF4A1 and eIF4A2, has been shown to have several of the eIF4A activities (RNA-dependent ATP hydrolysis, ATP-dependent RNA duplex unwinding), but when added to reticulocytes lysates, inhibited translation (11). However, as the relative abundance of eIF4A3 to eIF4A1 is about 1:10, it is not clear how effective eIF4A3 might be as a negative regulator of translation.

Given eIF4A's relative abundance to most other translation factors at about three copies per ribosome (12), it would not be surprising that the activity of eIF4A might be directly regulated. To date, there is no evidence for regulated posttranslational modifications of mammalian eIF4A, although the only modification well examined has been phosphorylation, which is technically the easiest modification to monitor. In contrast, in plants eIF4A can exist as a phosphoprotein. The phosphorylation state is altered either during development or in response to heat shock (13–15). The increase in eIF4A phosphorylation with heat shock would suggest that the phosphorylated form of eIF4A would be less active (13). However, during early germination and during pollen tube germination, there appears to be extensive eIF4A phosphorylation, which would suggest a correlation with increased protein synthetic activity (14, 15). Thus, it is currently uncertain whether phosphorylation directly alters eIF4A activity, or whether it may play some role in the mRNA selectivity of the eIF4 group of factors in the process of mRNA activation and binding of the mRNA to 40S subunits. At present, there are no *in vitro* data to directly address this question.

Currently three different proteins (i.e., non-translation factors) have been shown to bind to eIF4A, although their mechanism of action is unknown. Paip1, the poly(A)-binding protein interacting protein 1, contains an amino acid sequence that shares homology with the middle region of eIF4G and thus can bind both poly(A)-binding protein (PABP) and eIF4A (16). The role of Paip1 appears to be as a positive regulator, as its expression in COS cells has been shown to stimulate the translation of a reporter mRNA (16).

Two other proteins have been shown to be inhibitors of translation, p97 (or NAT1/DAP-5) and Pdcd4 (17–19; Colburn, N. H. *et al.*, personal communication). p97, which displays some similarity with the carboxyl two-thirds of eIF4G, is capable of binding to both eIF3 and eIF4A, and has been proposed to act as a dominant negative inhibitor of both cap-dependent and cap-independent translation (17–19). Pdcd4 was initially characterized as a putative tumor suppressor (20). By use of yeast two-hybrid analyses and coimmunoprecipitation experiments, Pdcd4 has been shown to directly bind eIF4A and, as monitored using the ATP-dependent unwinding assay, block the ability of eIF4A to unwind an RNA duplex (Colburn, N. H. *et al.*, personal communication). The full appreciation of the role of these two proteins in regulating translation awaits further experiments which will detail the absolute amount of these regulators and their possible regulation as well.

III. eIF4A: Biochemical Properties

As is true for all of the translation initiation factors, one would like to know exactly what the role of each protein is, when it is functional in the initiation pathway, and what reaction/rearrangement it has caused. As the introduction of the ribosome often complicates interpretations, a number of studies have been performed to determine what eIF4A is capable of doing in the absence of other translation factors. This not only gives insight into the function of eIF4A in the initiation process, but also provides insight into the possible functions/mechanisms of action of other DEAD box proteins. As will be noted later, the presence of other proteins (or the ribosome) might easily alter the characteristics to be described below.

TABLE I
BINDING OF RNA BY eIF4A[a]

	eIF4A	eIF4B	eIF4A + eIF4B
[³H]Poly(A) binding (cpm)			
Minus ATP	20	210	530
Plus ATP	470	120	5090
[³H]Globin mRNA binding (cpm)			
Minus ATP	130	230	150
Plus ATP	1800	350	5170

[a]Data taken from Tables 4 and 5 of Ref. 24. Binding of RNA by eIF4A or eIF4A + eIF4B was measured by retention of radiolabeled RNA on nitrocellulose filters.

Initial studies with cell-free systems indicated that ATP was required for the optimal binding of mRNA to ribosomes (21–23). Subsequent studies revealed that eIF4A, eIF4B, and, later, eIF4F were the primary proteins associated with steps in initiation that led to the binding of mRNA to either 40S subunits or the formation of 80S initiation complexes with mRNA. These studies then led to the following three assays: ATP-dependent retention of mRNA on nitrocellulose filters; RNA-dependent ATP hydrolysis; and, finally, ATP-dependent duplex unwinding.

Sample data on the eIF4A-dependent retention of RNA on nitrocellulose filters is presented in Table I. These data, and similar data from a number of laboratories, indicated that eIF4A was capable of binding RNA in an ATP-dependent manner and that this binding could be synergistically increased by the presence of eIF4B. One curious feature noted in these assays was that the binding was specific for ATP (or dATP) and that the nonhydrolyzable ATP analog 5′-adenyl-β,γ-imidodiphosphate (ADPNP) would not substitute for ATP.

More sophisticated analyses aimed at determining various binding constants for ligands (ATP or RNA) were developed, the most useful being the RNA-dependent ATPase assay. This assay in particular has been most extensively used in the preliminary biochemical characterization of most of the DEAD box proteins. An example of the early data is presented in Table II. These data provided initial insights into the eIF4A interaction with RNA. It appears that single-stranded RNAs optimally activate the hydrolysis of ATP, but that neither single-stranded DNA nor double-stranded RNA [as either poly(I–C), poly(G), tRNA, or globin mRNA] is an effective activator of the eIF4A ATPase activity. In a separate experiment, it was shown that eIF4B enhanced this ATP hydrolysis activity of eIF4A, primarily by improving the apparent affinity of eIF4A for the RNA activator (25).

In a much more sophisticated analysis, Lorsch, Peck and Herschlag examined both the kinetic parameters and substrate specificity for the

TABLE II
RNA-DEPENDENT ATP HYDROLYSIS
BY eIF4A[a]

Activator	P_i released (fmol/μg sec)
None	3.4
Poly(C)	47.4
Poly(A)	37.4
Poly(U)	35.4
Poly(I)	31.6
Poly(G)	5.7
tRNA	13.2
Globin mRNA	11.2
Poly(I–C)	6.2
Poly(dA)	5.8

[a]Data from Table 1 in Ref. 24; note that poly(G) tends to exist as a triple helix under the conditions used for this assay. RNA-dependent ATP hydrolysis was measured as phosphate released from [γ-^{32}P]ATP and extracted into isobutanol:benzene (1:1) as a phosphomolybdate complex.

RNA-dependent ATP hydrolysis assay (7, 26, 27) using single turnover kinetics and an altered buffer system {essentially a lower pH of 6.0 and a lower salt concentration of 30 mM [2-(N-morpholino)ethanesulfonic acid (MES)–KOH + KCH$_3$COO)}. These conditions allowed for a much greater formation of either eIF4A•ATP, eIF4A•RNA, or eIF4A•ATP•RNA complexes than could be achieved using buffer conditions more compatible with total protein synthesis (pH of 7.3–7.6, salt concentration of 100–150 mM KCl). Under these conditions, the K_m for ATP was determined to be 80 μM and the K_{act} for poly(U) was determined to be 5 μM (20-mer units; 26). Most interestingly, it was determined that the sequence of binding ligands was random, not ordered, as had been suggested from studies done under the more physiological conditions. Additionally, the binding of the ligands was coupled; that is, the binding of ATP enhanced the binding of RNA and visa versa. Also, the relatively poor binding of RNA by eIF4A was determined to be due to the rapid off-rate of eIF4A from the eIF4A•RNA complex (and the inability to even detect complexes under physiological conditions suggested that here the off-rates were even greater).

Perhaps the most surprising result stemmed from the determination of the time required for the "chemical step," the hydrolysis of ATP (26). The k_{cat} value was around 3 min^{-1}, or about 20 s was required for the hydrolysis of eIF4A-bound ATP. Given that the estimated time for protein synthesis initiation *in vivo* is 30 s, the step(s) requiring eIF4A-mediated ATP hydrolysis would seem to be rate-limiting in this process. As will be discussed later, this does not appear to be true.

Substrate specificity for eIF4A was assessed both for length and nucleic acid composition. In agreement with previous data (28), Peck and Herschlag found that nucleic acids with either a 2'-H group or a 2'-OCH$_3$ group would not activate the ATPase activity of eIF4A (27). Unexpectedly, single-stranded nucleic acids with either the 2'-H or a 2'-OCH$_3$ group were inhibitory, with K_i values about two-fold greater than the K_{act} for RNA (27). Effective/optimal stimulation of the ATPase activity is accomplished with oligonucleotides containing 15 nucleotides (27, 28). Substitution of the oxygen/phosphorus backbone with a phosphothioate backbone resulted in oligonucleotides that yielded a two- to three-fold increase in (k_{cat}/K_m) for ATP and decreased the K_{act} value at least 50-fold (27). Thus, whereas eIF4A appeared to be able to interact with most single-stranded nucleic acids, the binding constants as well as the ability of the oligonucleotide to stimulate eIF4A-dependent ATP hydrolysis varied considerably. In short, an unstructured RNA would appear to be the biological substrate.

The third assay to examine the activity of eIF4A was the ATP-dependent unwinding of duplex nucleic acids. The original assay was developed by Thach and colleagues (29) and was moved into a more convenient format by Sonenberg and colleagues (30). A better assay system that derived from both of the above was developed by Rogers et al. (31, 32). In general terms, there were two specific elements that were examined in establishing the assay. The first was the use of a family of relatively unstable duplexes (ΔG from −16 to −25 kcal/mol). The second was monitoring of the helicase activity kinetically. The general methodology is represented in Figs. 3 and 4.

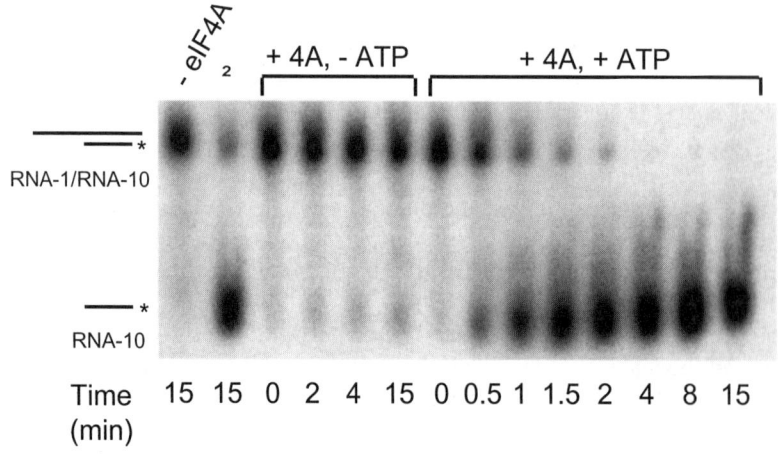

FIG. 3. The eIF4A-dependent unwinding of an RNA/RNA duplex where the long, unlabeled strand is about 50 nucleotides and the short, [^{32}P]-radiolabeled strand is 10 nucleotides (ΔG = −16.1 kcal/mol) and is marked in the figure with an asterisk to indicate radioactivity. Following incubation, duplex and monomer were resolved by native gel electrophoresis and quantitated using an Ambis radioanalytic scanner and autoradiography. Figure taken from Ref. 31, Fig. 1A.

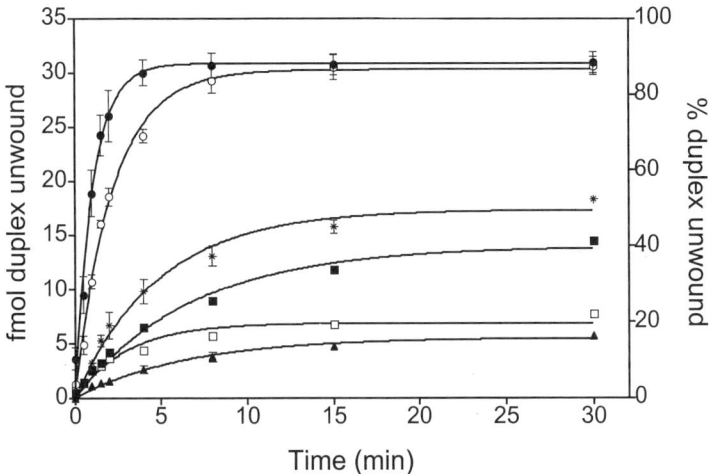

FIG. 4. The kinetics of duplex unwinding by eIF4A. The duplexes unwound were: 10 base pairs, filled circles; 11 base pairs, open circles; 12 base pairs, asterisks; 13 base pairs, filled squares; 14 base pairs, open squares; 15 base pairs, filled triangles. The ΔG values for these duplexes range from -16.1 to -24.7 kcal/mol (for the 10- and 15-base-pair duplexes, respectively). Analysis of the unwinding was as described in the legend to Fig. 3. Figure taken from Ref. *31*, Fig. 3.

As is evident from Fig. 3, unwinding of the duplex is dependent on both added protein and ATP. The reaction also occurs as a function of time (and as a function of eIF4A concentration, not shown) with complete unwinding occurring in roughly 4 min. When the extent of unwinding for a variety of RNA duplexes is plotted as a function of time, Fig. 4 is obtained. As can be seen, the linear portion of the curves is obtained within the first 3 min, where product formation (free monomer) is linear with time. From these observations, it is clear that the initial rate of unwinding decreases with increasing stability of the duplexes.

From general thermodynamic considerations, it was anticipated that $\Delta G = -RT \ln(K_{eq}) = -RT \ln(k_1) + RT \ln(k_2)$, where k_1 is the kinetic rate constant for unwinding and k_2 is the kinetic rate constant for hybridization. For the experimental conditions used, there is essentially little change in k_2 and thus the equation can be considered to be $\Delta G = -RT \ln(k_1) + \text{const}$ (*33*). This predicts that a plot of ln(initial rate of unwinding) versus ΔG should be a straight line and this is what was observed.

At the same time two controls were performed. The first was to check the rate of reannealing under the conditions used and it was found that this was insignificant. The second was to check the rate of unwinding due to temperature (i.e., eIF4A-independent unwinding, which will be referred to as "thermal melting"). For the fewest base pairs (10 or 11 base pairs), the thermal melting rate was appreciable and for the 10-base-pair duplex was about 10–15% of the total rate of unwinding. Curiously, when the rate of thermal melting was plotted as ln(initial rate of thermal melting) versus ΔG, this yielded a straight line with a slope equal to that observed with eIF4A (Fig. 4 in ref. 31). This suggested to the authors that eIF4A functioned to partially unwind the duplexes and that according to the stability of the remaining duplex, either melting would occur (ΔG less than -19 kcal) or the eIF4A would be released and the duplex would reanneal (ΔG greater than -22 kcal).

Although suggested as an initial hypothesis based upon the more exacting studies done with DNA helicases (i.e., Ref. 34) it was proposed that eIF4A unwound two to four base pairs in a single ATP-hydrolyzing event and was then released from the duplex. The hydrolysis event was predicted to be slow (20 s) in keeping with the slow step in the RNA-dependent ATPase assay (26). During this slow, chemical step when there is partial unwinding of the duplex, the duplex behaves as if it is shorter by these two to four base pairs and if sufficiently unstable, the remaining portion of the duplex undergoes thermal melting. If, on the other hand, eIF4A was as processive as are the DNA helicases, then one would anticipate that it would be difficult to observe any difference in the initial rate of unwinding of the different duplexes, and that the plot of ln(initial rate of unwinding) versus ΔG would have a slope of zero. As was evident from the model and the experimental data, eIF4A appears to be a nonprocessive helicase.

Having established the basic characteristic parameters of the eIF4A-dependent unwinding assay (duplex stability, ionic strength, [Mg^{2+}], pH, etc.), the next study was to better characterize the specificity of eIF4A for duplex substrates. The general "dogma" for helicases was that they bound to single-stranded regions and then unwound the adjacent duplex region. When both single-stranded 5' and 3' extensions were examined, it was observed that optimal unwinding was achieved when the single-stranded region was 6–10 nucleotides in length (32). This was a bit smaller than the suggested site size for eIF4A of about 15 nucleotides (24, 27, 35), but if one includes the size of the duplex (13 nucleotides), the resulting substrate is much larger than the suggested site size (20–25 nucleotides rather than 13 nucleotides).

However, the most striking feature was that the presence of a single-stranded region only increased the rate of unwinding by about 50% when compared to the blunt-ended duplex (32). Thus, eIF4A is capable of rather efficiently unwinding a blunt-ended duplex, in contrast to the expectations cited above. The second unusual feature was that for the increase in rate noted, it did not appear to matter whether the single-stranded region was 5' or 3' of the RNA duplex. Thus, eIF4A is a bidirectional helicase. This observation of duplex binding to eIF4A with subsequent unwinding was confirmed using a stable double-stranded RNA (17 base pairs, $\Delta G > -25$ kcal/mol), which served as an effective inhibitor of eIF4A in an assay with a 13-base-pair duplex substrate ($\Delta G = -21.1$ kcal/mol; 36).

Subsequent tests of possible duplex substrates indicated that RNA/RNA (the first nucleic acid indicates the long strand, the second nucleic acid the radiolabeled short strand), RNA/DNA, DNA/RNA, but not DNA/DNA duplexes were unwound. The rate of unwinding of all duplexes was strictly related to the ΔG value of the duplex in question, and did not vary depending on the number of base pairs, the sequence involved, or whether one of the two strands was DNA. In this same light, the changing of the sugar–phosphate backbone of the short strand to a phosphothioate linkage did not affect the rate of unwinding either. However, a change of the 2'-hydroxyl group to a 2'-methoxyethyl (2'-MOE) group yielded a duplex substrate that was not unwound. This suggested that perhaps the RNA/RNA-2'-MOE duplex was too bulky to fit the presumed binding pocket and thus this substrate was not unwound.

To test this suggestion, RNA–2'-MOE was tested as a single strand for its ability to inhibit the eIF4A-dependent unwinding of a 13-base-pair duplex. It was ineffective (single-stranded DNA or single-stranded DNA with phosphothioate linkages were also ineffective; 32). Thus, whereas there may be some importance to how much a duplex is A form, B form or a form in between, it still would appear that to observe unwinding activity, eIF4A must contact an RNA strand.

IV. Influence of Other Proteins on eIF4A Function

The primary role of eIF4A appears to be in the activation of an mRNA and its attachment to the 43S ribosomal subunit complex (40S•eIF2•GTP•Met-tRNA$_i$•eIF1•eIF3; Ref. 8 and see below). In this activation, three other proteins also participate, eIF4B, eIF4F, and eIF4H. Thus, it was of interest to see if the duplex-unwinding properties of eIF4A would be altered by the presence of these other proteins (36). When eIF4F was tested by itself (this initiation factor contains three subunits: eIF4A, eIF4E, and eIF4G), it had the same properties as eIF4A, although it was more active per mole of eIF4A. That is, eIF4F is also a nonprocessive helicase. In contrast, the presence of either eIF4B or eIF4H caused the eIF4A activity to become slightly processive (36, 37) with the estimation that two or three rounds of ATP hydrolysis might be coupled to each unwinding event as was suggested by a change in the slope of the plot of ln(initial rate of unwinding) versus ΔG from -0.25 to -0.13.

In concert with this finding, either eIF4F or a combination of eIF4F and eIF4B could form a stable enough complex with an RNA duplex to demonstrate a gel mobility shift (30). eIF4A, even in the presence of eIF4B, was unable to do so and the addition of ATP to the eIF4F + eIF4B mixture resulted in the loss of the gel-mobility-shifted complex. This suggested that the helicase activity of eIF4F had been activated leading to the release of the RNA substrate. These experiments provided additional weight to the general interpretation that the binding of nucleic acids by eIF4A was quite poor and that to obtain realistic binding (i.e., under concentrations of RNA *in vivo*), eIF4A would need to work in conjunction with other translation factors.

A second change in the character of the eIF4A-dependent unwinding assay was the influence of a 5′ or 3′ single-stranded region (37). In the presence of eIF4H, a duplex with 10–12 nucleotides of single-stranded region adjacent to the duplex resulted in a 2- to 3-fold increase in eIF4A-dependent unwinding activity (compared to just a 1.5-fold increase with eIF4A alone; see above). With added eIF4B, the presence of a 24 nucleotide single-stranded region resulted in about a 3.5-fold increase in the rate of unwinding, whereas eIF4F by itself showed the greatest stimulation with a 24-nucleotide single-stranded region, leading to an 8-fold increase in unwinding rate relative to the rate of unwinding the blunt duplex. In all combinations, the rates were essentially identical independent of whether the single-stranded region was 5′ or 3′ of the duplex. This finding is consistent with previous reports that suggested that the helicase activity of eIF4A/eIF4F was bidirectional (38).

When the substrate specificity of eIF4A was examined in the presence of eIF4B or eIF4H, it was the same as seen with eIF4A alone (i.e., RNA/RNA, RNA/DNA, DNA/RNA, but not DNA/DNA duplexes were unwound). However, in sharp contrast, eIF4F would only unwind an RNA/RNA duplex. Thus, the presence of the eIF4G 170-kDa subunit (and perhaps the eIF4E subunit as well) somehow influences the binding site to exclude the other eIF4A substrates, RNA/DNA and DNA/RNA duplexes (37).

Thus the unusual features of the eIF4A-dependent unwinding of RNA duplexes are (relative to REAL helicases, such as those associated with DNA replication):

1. eIF4A (and eIF4F) is a nonprocessive helicase and appears to have only a single ATP hydrolytic event associated with its unwinding.
2. eIF4A will unwind blunt-ended duplexes (although when associated with eIF4B, eIF4H, or in the eIF4F complex, this ability is impaired).
3. eIF4A is a bidirectional helicase
4. eIF4A will unwind RNA/RNA, RNA/DNA, DNA/RNA, but not DNA/DNA duplexes.

The general characteristics change slightly when eIF4B and eIF4H are present such that:

1. Using either eIF4A or eIF4F, the presence of either eIF4H or eIF4B causes the unwinding to become slightly processive.
2. Using eIF4F and either eIF4H or eIF4B, the unwinding reaction has a much more restrictive substrate, one with at least 20 single-stranded

nucleotides adjacent to the duplex and a duplex substrate which must be an RNA/RNA duplex.

Perhaps the most insightful article on eIF4A function was that of Pause et al. (39). Using a mutation in one of the DEAD box motifs (HRIGRXXRXG ⟶ HRIGQXXRXG; R362Q), Pause et al. were able to show that when added to a reticulocyte extract, this eIF4A mutant displayed characteristics of a dominant-negative mutant (as did other mutations as well). An examination of the biochemical properties of this mutant indicated that it would inhibit the helicase activity of eIF4F (presumably through eIF4A subunit exchange; 9) in proportion to the amount of mutant eIF4A present (that from eIF4F and that from the R362Q mutant added). However, the eIF4A mutant caused no inhibition of the (eIF4A + eIF4B)-mediated unwinding of duplex RNA. These results suggested that the mutant eIF4A exhibited its inhibition through a complex of eIF4F, not as free eIF4A. The simplest mechanism that would account for this discrepancy (i.e., the dominant-negative behavior) was that there must be more than a single round of utilization of eIF4F for each mRNA initiated (the math suggested that about four eIF4Fs might be used per mRNA activated).

Although the above studies do not directly address the molecular mechanism of how eIF4A unwinds duplex structures, it is anticipated that the mechanism is likely to be similar to the "inchworm" mechanism as proposed by Velankar et al. (40). The main features consistent with such a model would be (1) the function of eIF4A as a monomer, whereas the "rolling circle" model would necessitate eIF4A functioning as an aggregate (dimer or higher ordered structure), and (2) the step size (number of base pairs unwound per ATP hydrolysis event) could be quite small, as little as one or two base pairs (although larger step sizes could be possible). The main feature that would appear to be at odds with this model is that eIF4A is a bidirectional helicase. This, combined with the ability of eIF4A to unwind blunt-ended duplexes, would suggest the direct prying apart of double strands rather than the directed translocation and unwinding as suggested in the inchworm model. It is anticipated that both a crystal structure of eIF4A with bound substrates (especially the RNA duplex) and additional biochemical characterization will be required to more accurately assess the molecular details of eIF4A unwinding.

V. Does Any of This Biochemistry Make Sense or Is There a Contradiction Somewhere?

The very astute reader (and one who keeps up with all the literature) will recognize there are more than a few inconsistencies with the above data derived from the RNA-binding, RNA-dependent ATP hydrolysis, and RNA duplex-unwinding assays. Several of these are discussed below:

1. Binding of double-stranded RNA by eIF4A: Although not tested directly in the RNA-binding assay, eIF4A clearly binds (and unwinds) short RNA duplexes. Therefore, why doesn't double-stranded RNA stimulate the RNA-dependent hydrolysis of ATP by eIF4A? A simple explanation is that it does. In Table II, the stimulation by double-stranded RNA is small, but above background. It could be that in hydrolyzing ATP with double-stranded RNA as the activator, work needs to be attempted, and in being forced to couple hydrolysis with work, the rate of ATP hydrolysis is reduced (i.e., there would be no work done when single-stranded RNA stimulated the ATP hydrolysis).

If this is true, then double-stranded RNA (especially as a stable RNA duplex; $\Delta G > -25$ kcal/mol) ought to be able to serve as an inhibitor. When tested in the helicase assay, this inhibition was observed, although the apparent K_i was about two to three times larger than that for single-stranded RNA (32). In contrast, efforts to observe inhibition in the ATPase assay proved negative (26). It is not clear if this failure represents the use of lowered pH and ionic strength in the ATPase assay. Consistent with this possibility is the result that the ligand-dependent conformational changes observed with eIF4A are pH-dependent (7).

2. Binding of DNA by eIF4A: Previous studies have shown that DNA did not stimulate the nucleic acid-dependent hydrolysis of ATP. Inhibition studies, however, yielded conflicting results. In the helicase assay, neither single-stranded nor double-stranded DNA inhibited the unwinding of an RNA duplex (32). In contrast, Peck and Herschlag reported that single-stranded DNA effectively bound to eIF4A, but did not activate the ATPase activity (27). Again, this may reflect the use of nonphysiological buffers.

3. Binding by eIF4A of nucleic acids with modified 2'-OH groups: In the unwinding assay, the uniform addition of a 2'-methoxyethyl (2'-MOE) group to one of the duplexed strands of RNA resulted in a substrate that was either barely unwound or not unwound at all (32). At the same time, a single-stranded RNA with 2'-MOE modifications was unable to function as an inhibitor in the helicase assay. In contrast, 2'-O-methyl-substituted RNA strands bound well to eIF4A, as judged by inhibition of the RNA-dependent ATPase assay, although they did not stimulate ATP hydrolysis (27).

The three examples of apparent inconsistencies cited above are not likely due to an error in the experimentation. Although a full explanation will require further study, it is suggested that the conformations assumed by eIF4A upon binding the various analogs of duplexed RNA (single-stranded DNA, single-stranded RNAs with 2'-substitutions, or double-stranded RNA) are different and thus the influence of these molecules as either inhibitors of the helicase assay or stimulators of the nucleic acid-dependent ATPase assay is also different. The authors postulate that the actual binding site is undoubtedly the same for either single- or double-stranded RNAs, although the conformations of eIF4A with bound ligand may depend on the ligand.

It is curious to note that similar inconsistencies are apparent in other DEAD box proteins as well. In studies with the hepatitis C virus (HCV) NS3/4A helicase, it was observed that an RNA with a 2'-O-methyl modification was just as effective as RNA in stimulating the RNA-dependent ATPase activity (41). In contrast, single-stranded 2'-O-methyl RNA was a poor inhibitor of the unwinding assay and when this RNA was the 3' strand (defined as the strand providing the 3' overhang), the resulting duplex was not unwound. Again one must assume that the initial binding conformations must differ depending on the ligand.

A somewhat different inconsistency is the amount of stimulation observed when the other mRNA-specific translation initiation factors are added. In general, the data obtained with the binding assay and the helicase assay yield relatively similar numbers (i.e., the addition of eIF4B stimulates the eIF4A activity about 3- to 6-fold). In contrast, eIF4B enhances the binding of RNA by eIF4A by approximately 100-fold when measured as K_{act} in the ATPase assay (25). In a similar vein, the K_{act} for eIF4F for RNA in the ATPase assay was also observed to be about 100-fold lower than that for eIF4A, whereas the ability of eIF4F to either bind or unwind RNA is only about 3- to 5-fold greater (mostly in binding, not catalytic turnover; 25). The one characteristic shared by the RNA-binding and the duplex-unwinding assays is that in both instances the concentration of initiation factor is greater than the concentration of ligand/substrate. In contrast, the RNA-dependent ATP hydrolysis assay is performed under conditions where the RNA is in excess of the initiation factor.

VI. The Role of the DEAD Box Sequences

The most complete analyses of the DEAD box sequence elements of eIF4A were conducted by Pause and Sonenberg (39, 42, 43). Shown in Table III are some of the data they obtained. As is obvious from the data in Table III, most of the substitutions of the wild-type eIF4A DEAD box motifs result in an eIF4A with reduced activity. Alteration of either the K or T in the AXXXXGKT motif is characteristic of inactivation of a number of proteins that bind ATP, as this is the Walker A motif.

In a similar fashion, alteration of the Walker B motif (DEAD) is also deleterious. There are, however, two notable mutants. The EEAD mutant is capable of hydrolyzing ATP in an RNA-dependent manner and yet fails to unwind RNA duplex structures. The DEAH mutant has activities that are significantly better than the wild type, except that duplex unwinding is only 10% of wild type (note:

TABLE III
BIOCHEMICAL ACTIVITIES OF eIF4A MUTANTS (WILD TYPE = 100)[a]

Motif	ATP cross linking	ATPase	RNA cross linking	RNA helicase activity
AXXXXGKT	100	100	100	100
SXXXXGKT	220	110		110
VXXXXGKT	7	0	0	0
AXXXXGNT	2	0		0
DEAD	100	100	100	100
NEAD	150	0		0
DQAD	60	0		0
EEAD	110	80		0
DEAH	410	340	123	10
SAT	100	100	100	100
AAA	340	240	67	0
HRIGRGGR	100	100	100	100
QRIGRGGR	60	30	4	0
HQIGRGGR	10	0	5	0
HRIGQGGR	22	15	0	0

[a]Data in this table are from Table 2 of Ref. 42 and Table 2 of Ref. 43. Radiolabeled ATP and radiolabeled RNA crosslinking were achieved by irradiation with ultraviolet light and subsequent analysis by sodium dodecyl sulfate gel electrophoresis and autoradiography. RNA-dependent ATP hydrolysis was monitored as described in the note to Table II. RNA helicase activity was determined as described in the legend to Fig. 3.

this is of particular interest, since DEAH is a signature motif for one of the subfamilies of the DEAD box proteins). Thus, although there is sufficient structure maintained to effect efficient ATP hydrolysis in an RNA-dependent manner, the coupling to unwinding is lost. Given the observations of Lorsch and Herschlag (7, 26), it is tempting to assume that this loss in the ability to unwind secondary structures reflects an increased off-rate of eIF4A from the duplex. This may be complicated in the DEAH mutant, where the fourfold increase in V_{max} for ATP hydrolysis might suggest that the time required for the chemical step has been shortened, which would also be predicted to reduce duplex-unwinding activity (note: for this and other studies, it is assumed that all factor preparations are 100% active; therefore an increase in activity above wild type should not reflect a preparation which contains a higher percentage of active molecules).

VII. The Function of the Helicase Activity of eIF4A

As noted above, the most obvious function of the helicase activity of eIF4A (either as eIF4A or as eIF4F) would be to unwind RNA duplex structures at the 5' end of eukaryotic mRNAs. This activity would appear to be necessary for threading the mRNA onto the 40S subunit. A working example of this is the correlation between the degree of secondary structure in an mRNA and the amount of eIF4A required for optimal translation (44). However, there are several other very real possibilities. The first is the use of the helicase activity to remove proteins from the mRNA. As the mRNA exits the nucleus, it is coated with protein, and heterogeneous nuclear ribonucleoproteins (hnRNPs) (or even mRNPs) are estimated to be about 50% protein and 50% nucleic acid (based upon analyses using density gradients). How is this protein removed? Although there is no specific answer, it appears likely that the helicase activity of eIF4A could also be used to remove protein from the mRNA just as it does to remove secondary structure (perhaps requiring the help of the other eIF4 group of translation factors). An example of protein displacement has been shown for the NPH-II helicase in displacing the U1A protein from RNA (45) and one assumes that this property may apply to other DEAD box helicases as well.

A second possible use of the helicase activity of eIF4A is in the process of mRNA scanning, the ATP-dependent movement of the 40S subunit from the cap structure toward the initiating AUG (8). In mammalian translation systems, the only protein that has been identified that binds or hydrolyzes ATP is eIF4A. This includes analyses using a wide variety of assays such as mRNA binding to 40S subunits, toeprinting, or 80S complex formation. One can envision the helicase activity being converted into a motor in which the position of the eIF4A/ribosome complex is fixed and the mRNA is translocated along the complex as ATP is hydrolyzed. There is no proof of this idea, only the lack of identification of any other protein which might be capable of performing this task.

The third possibility is that the helicase activity is required for RNA structural rearrangements to allow the association of other translation factors with either the mRNA or the ribosomal RNA. Such activities have been demonstrated for the ATP-dependent, eIF4A-directed binding of eIF4B to RNA or to ribosomes (46) and may apply to other factors as well. This activity could be associated with either the initial binding of the translation factors to the mRNA or the binding of the translation factor•mRNA complex to the 40S subunit, or both.

Finally, the hydrolysis of ATP by eIF4A may lead to the dissociation of factors from either the mRNA or the 40S subunit. Although there are many possibilities, the best-documented one is the release of eIF4F from mRNAs upon ATP hydrolysis (24, 29). In this case, it appears that the hydrolysis of ATP drives a conformational change in the eIF4E subunit of eIF4G such that it releases from the m^7G cap structure, thus allowing for the recycling of eIF4F. This observation is confirmed by the finding that eIF4F molecules that lack the eIF4A subunit do not undergo this recycling (29). This nucleotide-hydrolysis-driven recycling is similar in nature to that seen with the hydrolysis of the GTP in the eIF2 • GTP • Met-tRNA$_i$ complex, the GTP in the eEF1A • GTP•aminoacyl-tRNA complex, the GTP in the eEF2 • GTP complex, or the GTP in the eRF1 • eRF3 • GTP complex (8).

VIII. eIF4A and the 80S Initiation Pathway

Shown in Fig. 5 is a model for the formation of 80S initiation complexes. Although there is experimental evidence to support this pathway, there are several parts that require further experimentation to delineate more clearly the precise times at which factors enter and leave the pathway, their precise function, and the stoichiometry of their utilization. If this pathway were exactly correct as shown, there would be utilization of each of the factors only once. If this were true, then one might expect to find the levels of the factors on the ribosomes in roughly a one-to-one ratio. This is clearly not the case. At the upper end, eIF4A appears to be present in the highest amount, at roughly 3.0 copies per ribosome, eIF2 is present at 0.5 copy per ribosome, and eIF5B is present at 0.05 copy per ribosome (12; W. C. Merrick, unpublished results). Thus, there is at least a 20- to 40-fold difference in the levels of the different factors. As most of the ribosomes exist in polysomes and there is only a limited amount of subunits, the 0.5 copy of eIF2 per ribosome in reality probably represents close to a one-to-one ratio of ternary complexes with 40S subunits (either free or as attached to polysomes). In this light, it should be noted that most (about 80–90% of the total) of the initiation factors are associated with 40S subunits/polysomes, with the remaining small percentage free in solution.

How does one explain the range of factor concentrations on the ribosome? Those proteins needed for all steps (eIF2 and eIF3, for example) are likely to be the most abundant, as they are not released until just prior to or concomitant with subunit joining. Proteins like eIF5 and eIF5B function late in the initiation process and their activities are associated with GTP hydrolysis. These factors appear to have greater turnover numbers and can act more catalytically than eIF2 or eIF3 and thus are present in reduced amounts. The real problem is eIF4A. Clearly this protein is present in considerable excess over eIF2 and eIF3. How might this happen?

There are several possible explanations for this observation. The first is that since eIF4A can bind to RNA in the absence of other proteins, perhaps its abundance in polysomes reflects in part the direct binding of eIF4A to the mRNA or to the RNA in the 40S subunit. Although this is possible, it is considered unlikely, as eIF4A has a poor intrinsic affinity for RNA and one would suspect that such binding would be nonspecific and perhaps not useful for translation. An alternate mechanism has been suggested based upon results using a dominant-negative mutant of eIF4A (39). In this model, it is proposed that the initiation of each mRNA requires multiple bindings of eIF4F (which contains the subunits eIF4E, eIF4A, and eIF4G) and that with each round of recycling, eIF4A molecules might be left behind or "deposited" on the mRNA (or on the ribosome). In this scheme the "extra" eIF4A molecules could participate

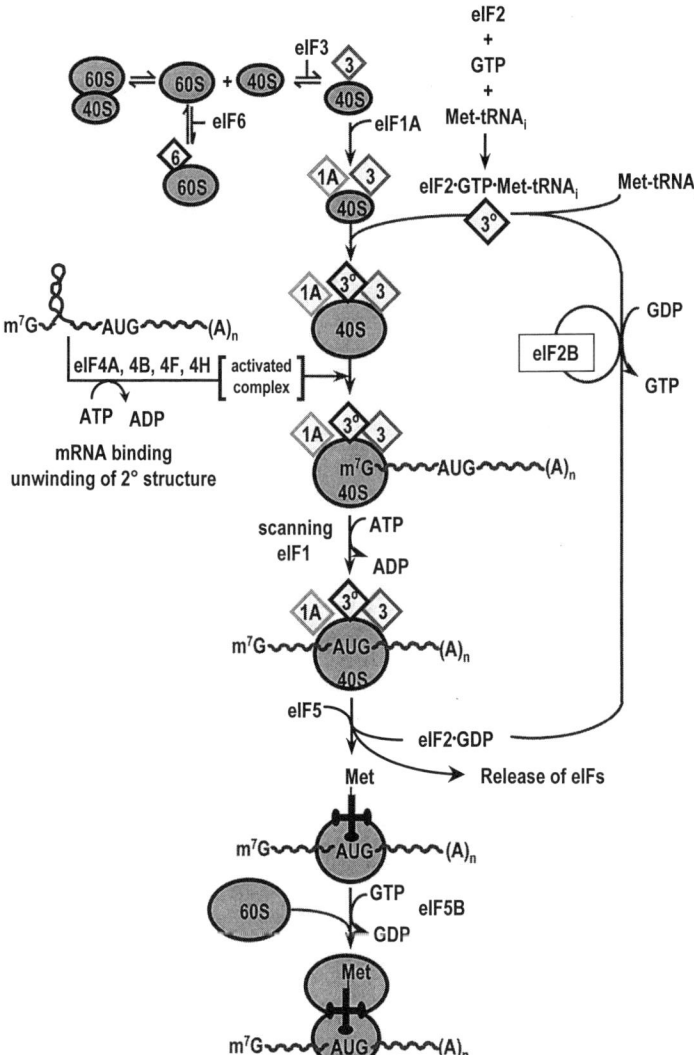

FIG. 5. The 80S initiation pathway. This flow scheme attempts to place the translation initiation factors in the appropriate context for their utilization in the formation of an 80S initiation complex. It should be noted that for some of the factors, the precise points at which they enter the pathway and at which they exit the pathway is uncertain or unknown. The basic steps in this pathway are (1) the formation of a small subunit pool (which is facilitated by eIF3 and eIF1A), (2) the binding of the ternary complex of eIF2•GTP•Met-tRNA$_i$, (3) the activation of mRNA and its binding to the 40S subunit (which requires ATP, eIF4A, eIF4B, eIF4F, and eIF4H), (4) scanning of the mRNA to reach the initiating AUG codon (a process that requires ATP), and (5) subunit joining (which appears to require eIF5, eIF5B, and GTP). For further details see Ref. 8.

either in the unwinding of mRNA secondary structure or, as noted above, scanning of the mRNA. Alternatively, it could be that eIF4A is required for each of the "possible" functions and thus the apparent excess of eIF4A reflects multiple functions rather than multiple eIF4A molecules required for a single function.

As the 80S pathway scheme in Fig. 5 does not adequately account for the different levels of translation factors, as noted above for eIF4A, it should not be taken as necessarily the correct flow scheme. However, in lieu of a more precise pathway, it is useful for orienting the apparent temporal use of the translation factors, defining some of the functions of the factors, and providing a reference base for keeping track of which factor does what and when it might do it. This pathway also allows for the planning of experiments to test the validity of certain aspects of the pathway or to expand specific regions for further definition. Thus, although likely to be incorrect in some respects, the pathway is still useful.

IX. Lessons Learned from eIF4A

At the start of the studies with eIF4A, especially with respect to its helicase activity, one already had in mind the DNA helicases, which unwind thousands of base pairs in a nucleotide-dependent, highly processive manner. It has become clear that eIF4A is not this type of helicase. It is a nonprocessive helicase. However, other DEAD box proteins have been shown to be much more processive (45). Thus, one cannot decide *a priori* how processive a DEAD box helicase might be. Second, eIF4A was found to be a bidirectional helicase, in contrast to the other helicases studied, which are unidirectional. This raises the question, Are all/most of the nonprocessive helicases also bidirectional? Clearly more studies of nonprocessive helicases will need to be performed before the answer is known.

As is likely the case for most, if not all, of the DEAD box helicases, the specificity of the helicase activity is very likely to be influenced by auxiliary factors. The presence of eIF4B or eIF4H is sufficient to convert eIF4A to a slightly processive helicase. It is possible that the presence of the ribosome will allow for even greater processivity, allowing eIF4A to become the motor that drives mRNA scanning. The presence of both eIF4G and eIF4E along with eIF4A in the complex that is eIF4F changed the substrate specificity dramatically. While still nonprocessive, eIF4F could only unwind RNA/RNA duplexes and these duplexes had to have a single-stranded region of about 30 nucleotides for optimal activity. Given that many of the DEAD box helicases described to date are part of other macromolecular synthetic pathways (most especially ribosome biogenesis and mRNA processing), it is likely that they, too, will have altered functions in the presence of auxiliary factors and in the presence of other components bound to the RNA target (i.e., the ribosome, the splicesome, etc.). The consequence of this is that it is likely that the complete unraveling of the specificities and functions of the DEAD box helicases will be an interesting and challenging biochemical problem that will require years of continued experimental study.

ACKNOWLEDGMENTS

The authors thank Dr. Raman Bhasker for his comments and critical reading of the manuscript. This work was supported in part by a grant from the National Institutes for General Medical Sciences, NIH (GM26796).

REFERENCES

1. J. P. Nielsen, G. K. McMaster, and H. Trachsel, Cloning of eukaryotic protein synthesis initiation factor genes: isolation and characterization of cDNA clones encoding factor eIF4A. *Nucleic Acid Res.* **13,** 6867–6880 (1985).
2. J. P. Nielsen and H. Trachsel, The mouse protein synthesis initiation factor 4A gene family includes two related functional genes which are differentially expressed. *EMBO J.* **7,** 2097–2105 (1988).
3. S. C. Conroy, T. E. Dever, C. L. Owens, and W. C. Merrick, Characterization of the 46,000 dalton subunit of eIF4F. *Arch. Biochem. Biophys. Acta* **282,** 363–371 (1990).
4. P. Linder, P. F. Lasko, M. Ashburner, P. Leroy, P. J. Nielsen, K. Nishi, J. Schnier, and P. P. Slonimski, Birth of the D-E-A-D box. *Nature* **337,** 121–122 (1989).
5. J. L. Kim, K. A. Morgenstern, J. P. Griffith, M. D. Dwyer, J. A. Thomson, M. A. Murcko, C. Lin, and P. R. Caron, Hepatitus C virus NS3 RNA helicase domain with a bound nucleotide: the crystal structure provides insights into the mode of unwinding. *Structure* **6,** 89–100 (1998).
6. J. M. Caruthers, E. R. Johnson, and D. B. McKay, Crystal structure of yeast initiation factor 4A, a DEAD-box RNA helicase. *Proc. Natl. Acad. Sci. USA* **97,** 13080–13085 (2000).
7. J. R. Lorsch and D. Herschlag, The DEAD box protein eIF4A: 2, a cycle of nucleotide and RNA-dependent conformational changes. *Biochemistry* **37,** 2194–2206 (1998).
8. W. C. Merrick and J. W. B. Hershey, The pathway and mechanism of eukaryotic protein synthesis. in "Translational Control of Eukaryotic Gene Expression" (J. W. B. Hershey, M. B. Mathews, and N. Sonenberg, eds.), pp. 33–88, Cold Spring Harbor Press, Cold Spring Harbor, New York, 2000.
9. J. Yoder-Hill, A. Pause, N. Sonenberg, and W. C. Merrick, The p46 subunit of eukaryotic initiation factor (eIF)-4F exchanges with eIF4A. *J. Biol. Chem.* **268,** 5566–5573 (1993).
10. D. M. Williams,-Hill, R. F. Duncan, P. J. Nielsen, and S. M. Tahara, Differential expression of the murine eukaryotic translation initiation factor isogenes eIF4A1 and eIF4A2 is dependent upon cellular growth status. *Arch. Biochem. Biophys.* **338,** 111–120 (1997).
11. Q. Li, H. Imataka, S. Morino, G. W. Rogers, Jr., N. J. Richter, W. C. Merrick, and N. Sonenberg, Eukaryotic translation initiation factor eIF4AIII is functionally distinct from eIF4AI and eIF4AII. *Mol. Cell. Biol.* **19,** 7336–7346 (1999).
12. R. Duncan and J. W. B. Hershey, Identification and quantitation of levels of protein synthesis initiation factors in crude HeLa cell lysates by two-dimensional polyacrylamide gel electrophoresis. *J. Biol. Chem.* **258,** 7228–7235 (1983).

13. D. R. Gallie, H. Le, C. Caldwell, R. T. Tanguay, N. X. Hoang, and K. S. Browning, The phosphorylation state of translation initiation factors is regulated developmentally and following heat shock in wheat. *J. Biol. Chem.* **272**, 1046–1053 (1997).
14. H. Le, K. S. Browning, and D. R. Gallie, The phosphorylation state of the wheat translation initiation factors eIF4B, eIF4A and eIF2 is differentially regulated during seed development and germination. *J. Biol. Chem.* **273**, 20084–20089 (1998).
15. R. G. L. op den Camp and C. Kuhlemeier, Phosphorylation of tobacco eukaryotic translation initiation factor 4A upon pollen tube germination. *Nucleic Acids Res.* **26**, 2058–2062 (1998).
16. A. W. Craig, A. Haghighat, A. T. Yu, and N. Sonenberg, Interaction of polyadenylate-binding protein with the eIF4G homologue PAIP enhances translation. *Nature* **392**, 520–523 (1998).
17. H. Imataka and N. Sonenberg, Human eukaryotic translation initiation factor 4G (eIF4G) possesses two separate and independent binding sites for eIF4A. *Mol. Cell. Biol.* **17**, 6940–6947 (1997).
18. H. Imakata, H. S. Olsen, and N. Sonenberg, A new translational regulator with homology to eukaryotic translation initiation factor 4G. *EMBO J.* **16**, 817–825 (1997).
19. S. Yamanaka, K. S. Poksay, K. S. Arnold, and T. L. Innerarity, A novel translational repressor mRNA is edited extensively in livers containing tumors caused by the transgene expression of the apoB mRNA-editing enzyme. *Genes Dev.* **11**, 321–333 (1997).
20. J. L. Cmarik, H. Min, C. Hegamyer, S. Zhan, M. Kulesz-Martin, H. Yoshinaga, S. Matsuhashi, and N. H. Colburn, Differentially expressed protein Pdcd4 inhibts tumor promoter-induced neoplastic translfromation. *Proc. Natl. Acd. Sci. USA* **96**, 14037–14042 (1999).
21. M. Giesen, R. Roman, S. N. Seal, and A. Marcus, Formation of an 80S methionyl-tRNA initiation complex with soluble factors from wheat germ. *J. Biol. Chem.* **251**, 6075–6081 (1976).
22. H. Trachsel, B. Erni, M. H. Schreier, and T. Staehelin, Initiation of mammalian protein synthesis. II. The assembly of the initiation complex with purified initiation factors. *J. Mol. Biol.* **116**, 755–767 (1977).
23. R. Benne and J. W. B. Hershey, The mechanism of action of protein synthesis initiation factors from rabbit reticulocytes. *J. Biol. Chem.* **253**, 3078–3087 (1978).
24. R. D. Abramson, T. E. Dever, T. G. Lawson, B. K. Ray, R. E. Thach, and W. C. Merrick, The ATP-dependent interaction of eukaryotic initiation factors with mRNA. *J. Biol. Chem.* **262**, 3826–3832 (1987).
25. R. D. Abramson, T. E. Dever, and W. C. Merrick, Biochemical evidence supporting a mechanism for cap-independent and internal initiation of eukaryotic mRNA. *J. Biol. Chem.* **263**, 6016–6019 (1988).
26. J. R. Lorsch and D. Herschlag, The DEAD box protein eIF4A: 1. A minimal kinetic and thermodynamic framework reveals coupled binding of RNA and nucleotide. *Biochemistry* **37**, 2180–2193 (1998).
27. M. L. Peck and D. Herschlag, Effects of oligonucleotide length and atomic composition on stimulation of the ATPase activity of translation initiation factor eIF4A. *RNA* **5**, 1210–1221 (1999).
28. J. A. Grifo, R. D. Abramson, C. A. Satler, and W. C. Merrick, RNA-stimulated ATPase activity of eukaryotic initiation factors. *J. Biol. Chem.* **259**, 8648–8654 (1984).
29. B. K. Ray, T. G. Lawson, J. C. Kramer, M. H. Cladaras, J. A. Grifo, R. D. Abramson, W. C. Merrick, and R. E. Thach, ATP-dependent unwinding of messenger RNA structure by eukaryotic initiation factors. *J. Biol. Chem.* **260**, 7651–7658 (1985).
30. M. Jaramillo, T. E. Dever, W. C. Merrick, and N. Sonenberg, RNA unwinding in translation: Assembly of helicase complex intermediates comprising eukaryotic initiation factors eIF4F and eIF4B. *Mol. Cell. Biol.* **11**, 5992–5997 (1991).

31. G. W. Rogers, Jr., N. J. Richter, and W. C. Merrick, Biochemical and kinetic characterization of the RNA helicase activity of eukaryotic initiation factor 4A. *J. Biol. Chem.* **274**, 12236–12244 (1999).
32. G. W. Rogers, Jr., W. F. Lima, and W. C. Merrick, Further characterization of the helicase activity of eIF4A: Substrate specificity. *J. Biol. Chem.* **276**, 12598–12608 (2001).
33. J. G. Wetmur, DNA probes: Applications of the principals of nucleic acid hybridization. *Crit. Rev. Biochem. Mol. Biol.* **26**, 227–259 (1991).
34. J. A. Ali and T. M. Lohman, Kinetic measurement of the step size of DNA unwinding by *Escherichia coli* UvrD helicase. *Science* **275**, 377–380 (1997).
35. D. J. Goss, C. L. Woodley, and A. J. Wahba, A fluorescence study of the binding of eukaryotic initiation factors to messenger RNA and messenger RNA analogues. *Biochemistry* **26**, 1551–1556 (1987).
36. G. W. Rogers, Jr., N. J. Richter, W. Lima, and W. C. Merrick, Modulation of the helicase activity of eIF4A by eIF4B, eIF4H and eIF4F. *J. Biol. Chem.* **276**, 30914–30922 (2001).
37. N. J. Richter, G. W. Rogers, Jr., J. O. Hensold, and W. C. Merrick, Further biochemical kinetic characterization of human eukaryotic initiation factor 4H. *J. Biol. Chem.* **274**, 35415–35424 (1999).
38. F. Rozen, I. Edery, K. Meerovitch, T. E. Dever, W. C. Merrick, and N. Sonenberg, Bidirectional RNA helicase activity of eukaryotic translation initiation factors 4A and 4F. *Mol. Cell. Biol.* **10**, 1134–1144 (1990).
39. A. Pause, N. Methot, Y. Svitkin, W. C. Merrick, and N. Sonenberg, Dominant negative mutants of mammalian translation initiation factor eIF4A define a critical role for eIF4F in cap-dependent and cap-independent initiation of translation. *EMBO J.* **13**, 1205–1215 (1994).
40. S. S. Velankar, P. Soultanas, M. S. Dillingham, H. S. Subramanya, and D. B. Wigley, Crystal structure of complexes of PcrA DNA helicase with a DNA substrate indicate an inchworm mechanism. *Cell* **97**, 75–84 (1999).
41. T. Hesson, A. Mannarino, and M. Cable, Probing the relationship between RNA-stimulated ATPase and helicase activities of HCV NS3 using 2′-O-methyl RNA substrates. *Biochemistry* **39**, 2619–2625 (2000).
42. A. Pause and N. Sonenberg, Mutational analysis of a DEAD box RNA helicase: the mammalian translation initiation factor eIF4A. *EMBO J.* **11**, 2643–2654 (1992).
43. A. Pause, N. Methot, and N. Sonenberg, The HRIGRXXR region of the DEAD box RNA helicase eukaryotic translation initiation factor 4A is required for RNA binding and ATP hydrolysis. *Mol. Cell. Biol.* **13**, 6789–6798 (1993).
44. V. Y. Svitkin, A. Pause, A. Haghighat, S. Pyronnet, G. Witherell, G. J. Belsham, and N. Sonenberg, The requirement for eukaryotic initiation factor 4A (eIF4A) in translation is in direct proportion to the degree of mRNA 5′ secondary structure. *RNA* **7**, 382–394 (2001).
45. E. Jankowsky, C. H. Gross, S. Shuman, and A. M. Pyle, Active disruption of an RNA-protein interaction by a DexH/D RNA helicase. *Science* **291**, 121–125 (2001).
46. D. L. Hughes, T. E. Dever, and W. C. Merrick, Further biochemical characterization of rabbit reticulocytes eIF4B. *Arch. Biochem. Biophys.* **301**, 311–319 (1993).

CTD Phosphatase: Role in RNA Polymerase II Cycling and the Regulation of Transcript Elongation

PATRICK S. LIN,
NICHOLAS F. MARSHALL,
AND MICHAEL E. DAHMUS

Section of Molecular and Cellular Biology
Division of Biological Sciences
University of California
Davis, California 95616

I. Historical Overview.. 334
   A. Dynamics of RNAP II Phosphorylation........................... 334
   B. Discovery of CTD Phosphatase....................................... 335
II. General Properties of CTD Phosphatase................................ 337
   A. Purification and Molecular Characterization of Yeast and Mammalian CTD Phosphatase.. 337
   B. CTD Phosphatase Regulation by TFIIF and TFIIB............. 341
   C. Molecular Cloning and Expression of Recombinant CTD Phosphatase 342
III. RNAP II Recycling Mediated by CTD Phosphatase.................. 346
   A. The Role of CTD Phosphatase during the Transcription Cycle........ 346
   B. The Role of CTD Phosphatase in the Mobilization of Stored RNAP II 347
IV. Involvement of CTD Phosphatase in the Regulation of Transcript Elongation.. 349
   A. Interplay of Positive and Negative Factors That Determine the Fate of Early Elongation Complexes... 349
   B. The Link between CTD Phosphorylation and Processive Transcript Elongation.. 350
   C. Phosphate Turnover during Transcript Elongation................. 351
   D. A Possible Role of CTD Phosphatase in the Regulation of mRNA Processing... 353
   E. Potential Coupling between CTD Dephosphorylation and Transcript Termination.. 354
   F. Sensitivity of RNAP II in Elongation Complexes to CTD Phosphatase.. 355
V. Perspectives and Future Directions.. 357
   References.. 359

Abbreviations: RNAP, RNA polymerase; CTD, C-terminal domain; Ad2-MLP, adenovirus 2 major late promoter; CK, casein kinase; SDS–PAGE, sodium dodecyl sulfate–polyacrylamide gel electrophoresis; P-TEFb, positive transcription elongation factor b.

The repetitive C-terminal domain (CTD) of the largest RNA polymerase II subunit plays a critical role in the regulation of gene expression. The activity of the CTD is dependent on its state of phosphorylation. A variety of CTD kinases act on RNA polymerase II at specific steps in the transcription cycle and preferentially phosphorylate distinct positions within the CTD consensus repeat. A single CTD phosphatase has been identified and characterized that in concert with CTD kinases establishes the level of CTD phosphorylation. The involvement of CTD phosphatase in controlling the progression of RNAP II around the transcription cycle, the mobilization of stored RNAP IIO, and the regulation of transcript elongation and RNA processing is discussed. © 2002, Elsevier Science (USA).

Gene expression in eukaryotic cells is regulated by a variety of factors that interact with DNA, either directly or indirectly, to modify the chromatin template and/or to influence assembly of a preinitiation complex at the promoter (for recent reviews see Refs. *1–4*). Clearly, the initiation phase of transcription is a target for multiple regulators. It has become increasingly clear that transcript elongation and termination are also important targets in the regulation of gene expression.

The central enzyme driving transcription of protein-coding genes is RNA polymerase (RNAP) II. RNAP II is unique in that it contains at the C-terminus of its largest subunit an unusual domain composed of tandem repeats of the consensus sequence $Tyr^1-Ser^2-Pro^3-Thr^4-Ser^5-Pro^6-Ser^7$ (for review see Refs. 5, 6). This domain, designated the CTD for C-terminal domain, is essential for cell viability. Although the number of repeats varies from 26–27 in yeast to 52 in mammals, the CTD consensus repeat is conserved in evolution. A variety of studies have established that the CTD can be extensively phosphorylated and that this phosphorylation is highly dynamic (*7–10*). Because the hypophosphorylated and hyperphosphorylated forms of RNAP II have distinct functions in the transcription cycle, CTD kinases and CTD phosphatase(s), which control the level of CTD phosphorylation, can play important roles in the regulation of gene expression. A multiplicity of CTD kinases have been described, whereas a single CTD phosphatase has been reported. This review will focus on CTD phosphatase, including the general properties of the enzyme and its role in transcription.

I. Historical Overview
A. Dynamics of RNAP II Phosphorylation

A striking feature of RNAP II is the massive amount of phosphorylation that occurs within the CTD. Phosphorylation of the CTD defines two distinct conformational states of RNAP II. The hyperphosphorylated form is designated

RNAP IIO and the hypophosphorylated form is designated RNAP IIA. The largest subunits of RNAPs IIA and IIO are referred to as IIa and IIo, respectively. RNAP IIA is the form of the enzyme recruited to the promoter and assembled into the preinitiation complex (*11–13*). In general, phosphorylation of the CTD is concomitant with transcript initiation and facilitates promoter clearance (*14, 15*). Transcript elongation is catalyzed by RNAP IIO (*7, 15, 16*). At some point prior to the next round of transcription, RNAP IIO must be dephosphorylated to regenerate RNAP IIA, thereby completing the transcription cycle. Accordingly, each round of transcription is associated with phosphorylation and dephosphorylation of the CTD. Thus, CTD kinases and CTD phosphatase(s) play essential roles in the progression of RNAP II through the transcription cycle.

Apart from the dynamics of phosphorylation associated with each transcription cycle, increasing evidence supports the idea that phosphate turnover occurs during the elongation phase of transcription (*8–10*). Because of the apparent requirement for a fully phosphorylated CTD for processive elongation, CTD kinase(s) and CTD phosphatase(s) that act on RNAP II in an elongation complex can behave as either positive or negative regulators of transcription, respectively. Although we are far from understanding the role of phosphate turnover during elongation, certain model systems have been examined in detail and establish this as an important regulatory mechanism (*9, 10*).

The CTD appears to mediate the interaction of RNAP II with a complex array of proteins during the transcription process (*5, 17, 18*). Reversible phosphorylation generates different conformational states of the CTD, which in turn enables the CTD to interact with a discrete complement of proteins depending on its position in the transcription cycle. Several distinct protein kinases can phosphorylate the CTD at different steps in the cycle. Furthermore, these CTD kinases differ in their preferred phosphorylation site within the consensus repeat. CTD phosphatase was discovered several years after the first CTD kinase was reported (*13*). Less is known about its role at specific steps in the transcription cycle and how its activity is regulated.

B. Discovery of CTD Phosphatase

The finding that the hypophosphorylated form of RNAP II is recruited to the promoter, whereas transcript elongation is catalyzed by the hyperphosphorylated form led to two specific predictions. First, a CTD kinase(s) must act on RNAP IIA after its association with the promoter to generate the elongation-competent form of the enzyme, RNAP IIO. Second, a protein phosphatase(s) must exist to reverse this reaction, thereby regenerating RNAP IIA for a subsequent round of transcription.

CTD phosphatase activity was first observed in an experiment designed to examine the role of CTD phosphorylation on the recruitment of RNAP II to

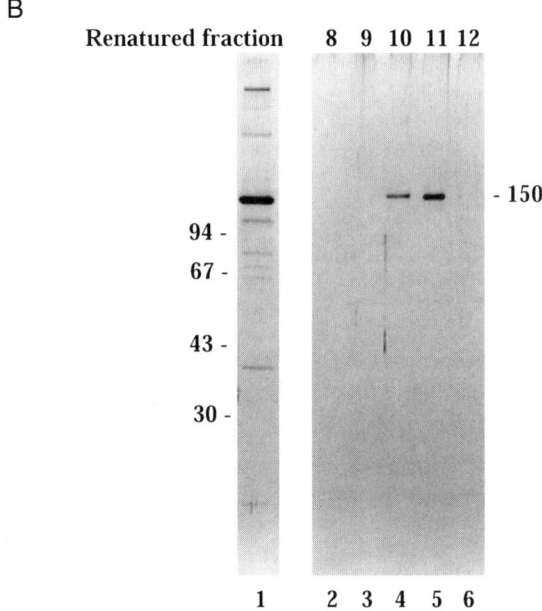

FIG. 1. Renaturation of CTD phosphatase activity from the SDS–PAGE purified 150-kDa polypeptide. Highly purified CTD phosphatase (B, lane 1) was resolved by SDS–PAGE. Proteins were eluted and renatured from individual gel slices as previously described (21). (A) CTD phosphatase assay. Lanes 1 and 11 contain marker RNAP IIO; lane 10 contains marker RNAP IIA. Lanes 2–4, RNAP IIO incubated with 1, 0.1, and 0.01 unit CTD phosphatase, respectively. Lanes 5–9, RNAP IIO incubated with renatured protein from fractions 8–12, respectively. (B) Silver-stained SDS–PAGE. Renatured CTD phosphatase samples assayed in A are analyzed for protein content. Lane 1 contains 30 units of input CTD phosphatase. The M_r markers ($\times 10^{-3}$) are indicated on the left. From Marshall et al. (21).

the adenovirus 2 major late promoter (Ad2-MLP) (13). ^{32}P-labeled RNAPs IIA and IIO were prepared, as described below, and incubated in the presence of a HeLa cell extract to assay the ability of each to be assembled into preinitiation complexes. Apart from the observation that RNAP IIA was assembled into complexes, whereas RNAP IIO was not, a small fraction of the RNAP IIO in the reaction was converted to RNAP IIA. Furthermore, the RNAP IIA generated was functional in that it assembled into preinitiation complexes. This experiment established that the HeLa cell extract contained CTD phosphatase(s) and formed a basis for the subsequent purification.

The key to the discovery of CTD phosphatase and the development of a convenient assay system was the preparation of an appropriate substrate. The most C-terminal serine in the largest subunit of mammalian RNAP II is flanked by acidic residues and can be selectively phosphorylated by casein kinase (CK) II. This serine lies outside the last consensus repeat and phosphorylation of this site does not appear to influence RNAP II activity. The *in vitro* phosphorylation of purified mammalian RNAP IIA by CKII in the presence of [γ-^{32}P]ATP has been used extensively to prepare ^{32}P-labeled RNAP II for a variety of *in vitro* studies (12, 13, 15, 19–24).

A substrate for CTD phosphatase can be prepared by the phosphorylation of CKII-labeled RNAP IIA with CTD kinase in the presence of excess unlabeled ATP (13, 25, 26). This generates a substrate in which the consensus CTD repeats are phosphorylated with unlabeled phosphates, whereas only the terminal serine is labeled. Therefore, dephosphorylation by a specific CTD phosphatase removes the unlabeled phosphates, but not ^{32}P at the CKII site. Because phosphorylation of the CTD results in a marked mobility shift of the largest subunit in sodium dodecyl sulfate–polyacrylamide gel electrophoresis (SDS–PAGE), a shift in mobility from the position of subunit IIo to the position of subunit IIa is a direct measure of phosphatase activity (Fig. 1A; compare lanes 10 and 11). CTD phosphatase does not utilize as substrate either phosphorylated recombinant CTD (rCTD) or phosphorylated synthetic peptides composed of multiple copies of the consensus repeat. Accordingly, phosphorylated rCTD and synthetic peptides were not useful reagents in assays for CTD phosphatase.

II. General Properties of CTD Phosphatase

A. Purification and Molecular Characterization of Yeast and Mammalian CTD Phosphatase

Since each round of transcription is associated with reversible phosphorylation of the CTD, RNAP IIO must be dephosphorylated concomitant with or following termination to regenerate RNAP IIA that can reinitiate transcription. As a critical test of this idea, an assay was developed for the purification of an enzyme(s) capable of converting RNAP IIO to RNAP IIA (13). As discussed in Section I.B, ^{32}P-labeled RNAP IIO was used as substrate in mobility shift assays for the purification and characterization of mammalian CTD phosphatase. Human CTD phosphatase was purified from HeLa cell extracts (25, 26) and shown by SDS–PAGE to contain a prominent polypeptide of 150 kDa (Fig. 1B, lane 1) (21). To confirm that the 150-kDa protein is necessary and sufficient for CTD phosphatase activity, highly purified CTD phosphatase was denatured, fractionated by SDS–PAGE, and protein renatured from individual gel slices (21). Results presented in Fig. 1 show that CTD phosphatase activity was recovered following renaturation of the 150-kDa protein.

Mammalian CTD phosphatase requires Mg^{2+} for activity and is resistant to okadaic acid, a potent inhibitor of several families of protein phosphatases (25). The Mg^{2+} dependence is consistent with the observation that high concentrations of EDTA and EGTA are necessary to preserve RNAP IIO in mammalian cell extracts (27). The dephosphorylation of RNAP IIO by CTD phosphatase has the appearance of a processive reaction as reflected by the absence of partially dephosphorylated species with mobilities intermediate between that of subunits IIo and IIa (Fig. 1A, lanes 7 and 8). Accordingly, CTD phosphatase dephosphorylates RNAP IIO in a manner similar to the processive reaction catalyzed by several CTD kinases.

There are several aspects of the specificity of CTD phosphatase that are of interest. First, CTD phosphatase appears to be highly specific for the dephosphorylation of the CTD. A variety of other phosphoproteins, such as phosphorylase a or the α and β subunits of phosphorylase kinase, are not substrates of CTD phosphatase (25). It remains to be established if CTD phosphatase can dephosphorylate proteins such as Spt5 that contain CTD-like repeats (28, 29), which likewise appear to be phosphorylated (30). Second, CTD phosphatase dephosphorylates RNAP IIO generated by serine/threonine CTD kinases (25). A subset of RNAP IIO in vivo is known to be phosphorylated on tyrosine (31). RNAP IIO prepared by the in vitro phosphorylation of mammalian RNAP IIA with c-Abl tyrosine kinase cannot be dephosphorylated by the 150-kDa CTD phosphatase (R. S. Chambers and M. E. Dahmus, unpublished results). Furthermore, CTD phosphatase is not sensitive to vanadate, a known tyrosine phosphatase inhibitor. Accordingly, a second CTD phosphatase specific for the dephosphorylation of phosphotyrosine may exist. These observations indicate that CTD phosphatase is a serine/threonine protein phosphatase that is highly specific toward the CTD of RNAP II.

To more critically assess the specificity of CTD phosphatase with respect to its ability to dephosphorylate phosphoserines at different positions within the consensus repeat, it is necessary to briefly discuss the specificity of different CTD kinases. TFIIH is targeted to the preinitiation complex and phosphorylates the CTD of RNAP II concomitant with transcript initiation (32–34). Phosphorylation by TFIIH may disrupt protein–protein interactions that mediate the recruitment of RNAP II to the preinitiation complex, thereby facilitating promoter clearance. Positive transcription elongation factor b (P-TEFb) plays an important role in rescuing paused elongation complexes and promotes processive transcript elongation (35, 36). P-TEFb can render early elongation complexes resistant to negative regulation by factors such as 5,6-dichloro-1-β-D-ribofuranosyl benzimidazole (DRB) sensitivity-inducing factor (DSIF) and negative elongation factor (NELF) (37, 38). Homologous CTD kinases such as Kin28 (33, 34, 39), Ctk1 (10, 40), and Bur1 (41) appear to play analogous roles in *Saccharomyces cerevisiae*.

Besides TFIIH and P-TEFb, other, more general protein kinases have been found to phosphorylate the CTD as one of their many targets. Mitogen-activated protein kinase (MAPK) phosphorylates the CTD of mammalian RNAP II in response to environmental stress and during oocyte maturation (42–44). Cdc2 kinase, a regulatory enzyme in the cell cycle that triggers the entry of mammalian cells into mitosis, has also been implicated as a CTD kinase (45) that can inhibit transcription *in vitro* (46).

Apart from acting at distinct steps in the transcription cycle, different CTD kinases preferentially phosphorylate different positions within the consensus repeat. A number of studies using synthetic peptides have established that both TFIIH and MAPK phosphorylate serines at position 5 within the consensus repeat (47–49), whereas Cdc2 kinase phosphorylates serines at positions 2 and 5 (50). An analysis of preinitiation complexes using anti-CTD monoclonal antibodies confirms that TFIIH phosphorylates serine 5, whereas P-TEFb phosphorylates serine 2 (51). Interestingly, the specificity of P-TEFb appears to be altered in the presence of HIV-1 Tat protein in that under these conditions P-TEFb phosphorylates serines at positions 2 and 5. This finding suggests that the specificity of CTD kinases with respect to the phosphorylation of different positions within the consensus repeat may be context-dependent.

A recent study has shown that the mammalian CTD phosphatase is able to dephosphorylate RNAP IIO generated *in vitro* by the phosphorylation of RNAP IIA with TFIIH, P-TEFb, MAPK, and Cdc2 kinase (P. S. Lin and M. E. Dahmus, unpublished results). This finding establishes that CTD phosphatase can act on RNAP IIO generated by a variety of serine/threonine CTD kinases. Future studies are necessary to establish whether CTD phosphatase can regulate cellular events activated by the phosphorylation of either serine 2 and/or serine 5 within the consensus repeat (see Sections III and IV).

A striking feature of CTD phosphatase is that it will not dephosphorylate rCTDo prepared by phosphorylation of rCTDa with the same CTD kinases used to prepare the RNAP IIO substrates (26). Purified RNAP subunit IIo is also not a substrate. Interestingly, RNAP IIB, which lacks a CTD, and RNAP IIA are competitive inhibitors of CTD phosphatase. These results support the idea that RNAP II contains a docking site that is distinct from the CTD and that the catalytic activity of CTD phosphatase is dependent on its interaction with that site. Although the mechanism by which this docking site stimulates phosphatase activity is not known, one possibility is that the interaction of CTD phosphatase with RNAP II generates an active conformation of the enzyme. Alternatively, this interaction may serve to simply localize CTD phosphatase adjacent to the CTD to promote processive dephosphorylation. In this regard, it would be of interest to establish whether dephosphorylation proceeds in an orderly fashion from either the N- or C-terminus of the CTD or if it is random.

The observation that CTD phosphatase must interact with the docking site in order to dephosphorylate the CTD opens the possibility that CTD phosphatase activity could in part be regulated by controlling its interaction with the docking site. Accordingly, it is important to map the CTD phosphatase-binding site on RNAP II and to determine if and how access to this site is altered during the transcription cycle. Further studies are necessary to establish whether proteins that associate with RNAP II directly influence the binding of CTD phosphatase and whether phosphate turnover during transcript elongation is mediated by the recruitment of CTD phosphatase.

CTD phosphatase purified from *S. cerevisiae* has properties similar to the mammalian homolog (52). The 100-kDa yeast CTD phosphatase is highly specific for dephosphorylating the largest subunit of yeast RNAP II. Like the mammalian enzyme, yeast CTD phosphatase requires Mg^{2+} for activity and is resistant to okadaic acid. Genetic evidence demonstrates that CTD phosphatase is required for dephosphorylation of the CTD *in vivo* (53). This finding is supported by the observation that RNAP IIO accumulates at the nonpermissive temperature in yeast harboring a CTD phosphatase temperature-sensitive mutation. Furthermore, a genome-wide expression profile generated by high-density oligonucleotide arrays demonstrates that the expression of nearly 70% of all yeast genes is significantly reduced in a CTD phosphatase temperature-sensitive strain (53). This is similar to the level of reduced expression observed in a *rpb1-1* mutant strain at the nonpermissive temperature. This observation establishes that CTD phosphatase is required for transcription *in vivo* and its inactivation leads to a global defect in mRNA synthesis.

CTD phosphatase purified from yeast extracts appears to have a unique requirement for an additional factor(s) for the *in vitro* dephosphorylation of RNAP IIO (52). This is supported by the observation that CTD phosphatase activity is reconstituted in the presence of partially purified TFIIF, but not recombinant TFIIF. As described below in the Section II.B, the general transcription factor IIF stimulates the activity of both human and yeast CTD phosphatase.

The nature of the additional factor requirement, apart from the stimulatory activity of TFIIF, remains to be established. Interestingly, recombinant yeast CTD phosphatase (Fcp1)[1] is able to dephosphorylate RNAP IIO *in vitro* in the absence of extrinsic factors (53). Differences in posttranslational modification and/or the concentration of enzyme being assayed could in principle account for the differential requirement.

The finding that excess RNAP II inhibits the activity of yeast CTD phosphatase suggests that yeast CTD phosphatase also stably interacts with a site on RNAP II distinct from the CTD (52). Interestingly, there appears to be a strict species specificity for both the mammalian and yeast CTD phosphatase. Neither the HeLa CTD phosphatase nor the yeast CTD phosphatase dephosphorylates RNAP II from the other species, despite the fact that the heptapeptide repeat is identical. The species specificity may reflect divergent protein structures at the CTD phosphatase-binding sites between the mammalian and yeast enzymes.

B. CTD Phosphatase Regulation by TFIIF and TFIIB

Given the potential significance of CTD phosphatase in the regulation of transcription, it is important to know whether the activity of CTD phosphatase is modulated by any of the general transcription factors. A systematic analysis of TFIIB, E, F, H, and TBP demonstrates that CTD phosphatase is stimulated by TFIIF and that this stimulation is inhibited by TFIIB (26). Human TFIIF is a heterodimeric factor consisting of 26-kDa (RAP30) and 58-kDa (RAP74) subunits (54, 55). TFIIF, through the action of RAP30, can escort RNAP II to the promoter to initiate transcription (56, 57). However, it is the RAP74 subunit that stimulates CTD phosphatase activity *in vitro* (26). Experiments involving RAP74 deletion constructs establish that the minimal region sufficient for stimulatory activity corresponds to C-terminal residues 358–517. TFIIB abrogates the stimulatory activity of TFIIF, but has no influence on CTD phosphatase activity in the absence of TFIIF. Not surprisingly, species specificity also exists at the level of CTD phosphatase regulation. Although the yeast and HeLa CTD phosphatase can interchangeably bind RAP74, each CTD phosphatase is activated only by its cognate TFIIF (52). This suggests that the requirements for enzyme activity are more complex than the simple binding assays would indicate.

Regulation of CTD phosphatase by TFIIF and TFIIB is consistent with the role that these transcription factors play in the transcription cycle. During initiation, TFIIB is recruited to the promoter subsequent to the binding of TFIIA

---

[1]The yeast protein corresponding to the 100-kDa CTD phosphatase is designated Fcp1 for TFIIF-associating component of CTD phosphatase (62). The yeast gene is designated *FCP1*, and alleles of that gene are designated *fcp1*. The human and *Xenopus* homologs to the yeast protein are designated hFCP1 (59) and xFCP1 (24), respectively.

and TFIID and facilitates the association of the RNAP IIA/TFIIF complex. Its primary role is to specify the start site of transcription for RNAP II (58). After TFIIH phosphorylates the CTD and RNAP II clears the promoter, TFIIB remains at the promoter. Therefore, TFIIB localized at the promoter may prohibit CTD phosphatase from acting on RNAP II during initiation. Furthermore, since CTD phosphatase appears to assemble in the preinitiation complex (59), its activity must be tightly regulated to prevent a futile cycle of CTD phosphorylation and dephosphorylation. Once RNAP II leaves the promoter, TFIIF is free to modulate CTD phosphatase activity in the absence of TFIIB.

Although the precise mechanism of CTD phosphatase regulation by transcription factors has not been established, several studies have contributed insights concerning the multiple interactions of CTD phosphatase, TFIIF, and TFIIB. The interaction between TFIIF and TFIIB appears to be mediated by the C-terminus of RAP74, the same protein region required for the stimulation of CTD phosphatase (60). Accordingly, TFIIB may inhibit the stimulatory activity of TFIIF by binding to the C-terminus of RAP74, thereby masking the critical region of RAP74 required for CTD phosphatase activation.

Recent studies have established that in *S. cerevisiae*, TFIIB, like TFIIF, can bind CTD phosphatase (61). A KEFGK motif shared by TFIIF and TFIIB has been shown to mediate their interactions with CTD phosphatase. This motif is present in the TFIIB and RAP74 proteins found in human, *Xenopus laevis*, *Drosophila melanogaster*, and *Caenorhabditis elegans*. More importantly, TFIIB and RAP74 each binds adjacent and possibly overlapping regions on the C-terminus of CTD phosphatase. Therefore, it is possible that TFIIB may compete with RAP74 for its binding to CTD phosphatase. Repression of CTD phosphatase activity during initiation may result from the preferential association of TFIIB. It will be important to determine whether the interactions between TFIIB and TFIIF with CTD phosphatase are mutually exclusive or whether these proteins can form a ternary complex. Clearly, additional studies are necessary to establish the molecular mechanism by which TFIIB and TFIIF regulate the activity of CTD phosphatase.

C. Molecular Cloning and Expression of Recombinant CTD Phosphatase

Soon after the discovery of both mammalian and yeast CTD phosphatase, efforts were initiated to clone the cDNAs encoding these phosphatases and express the recombinant enzymes. Purification of yeast CTD phosphatase from extracts of *S. cerevisiae* led to the resolution of two component fractions essential for phosphatase activity (52). One of these components was yeast TFIIF, and the other contained a polypeptide with an apparent molecular mass of 100 kDa which copurified with the activity. The 100-kDa polypeptide was excised from SDS gels, renatured, and shown to contain CTD phosphatase activity in the presence of the TFIIF-containing fraction. The 100-kDa polypeptide was digested with protease lysC and the sequences of several peptides were determined. These peptides were contained in a 732-amino-acid open reading frame on chromosome XIII of *S. cerevisiae* (*62*). Expression of this yeast gene in *Escherichia coli* resulted in the production of a protein with CTD phosphatase activity. Since the sequence of the yeast protein did not resemble any known phosphatases and was stimulated by TFIIF, the gene was designated *FCP1* for TFIIF-associating-component of CTD phosphatase (*62*). The *FCP1* gene is essential for yeast cell growth and its gene product was later confirmed to be the major CTD phosphatase activity in *S. cerevisiae* (*53*).

The mammalian CTD phosphatase cDNA was first identified in a yeast two-hybrid screen used to characterize RAP74-interacting proteins in humans (*59*). The relationship of this clone to the previously characterized human CTD phosphatase was established by immunochemical analysis utilizing antibodies prepared against a recombinant peptide. Furthermore, five tryptic peptides recovered from purified mammalian CTD phosphatase were localized in the hFCP1-encoded sequence (M. E. Dahmus, unpublished results). Two forms of hFCP1 were isolated, hFCP1a, which encodes an 842-amino-acid polypeptide, and a shorter splicing variant called hFCP1b (*59*). Northern blot analysis indicates that the amounts of hFCP1 mRNA are similar in various tissues examined. The deduced amino acid sequence of hFCP1a was similar to the yeast Fcp1 sequence (*59, 62*). Both hFCP1a and hFCP1b exhibited very low activity and were eventually found to be missing the N-terminus required for full catalytic activity (*63*). The full-length human cDNA encoding CTD phosphatase was isolated by the Reinberg laboratory and shown to have catalytic activity capable of recycling RNAP II (*63*). This functional hFCP1 is a polypeptide of 961 amino acids, which is larger than hFCP1a at the N-terminus by 119 amino acids. The difference between hFCP1a and hFCP1 is due to a different assignment of the translation initiation codon.

The cDNA encoding the full-length FCP1 in *Xenopus laevis* has recently been cloned (*24*). xFCP1 is a homolog of the human and yeast FCP1 CTD phosphatase. Immunodepletion with antibody directed against hFCP1 establishes that xFCP1 is the major CTD phosphatase activity in *Xenopus* egg and cell extracts. Interestingly, RNAP II appears to be fully phosphorylated when *Xenopus* oocytes enter meiosis and upon fertilization undergoes massive dephosphorylation prior to the onset of zygotic gene expression. The CTD dephosphorylation triggered by fertilization is completely attributed to xFCP1 activity. This observation indicates that CTD phosphatase plays a key role in mobilizing RNAP IIO during early development.

Close examination of aligned FCP1 sequences (Fig. 2) from various organisms reveals two interesting functional domains in CTD phosphatase.

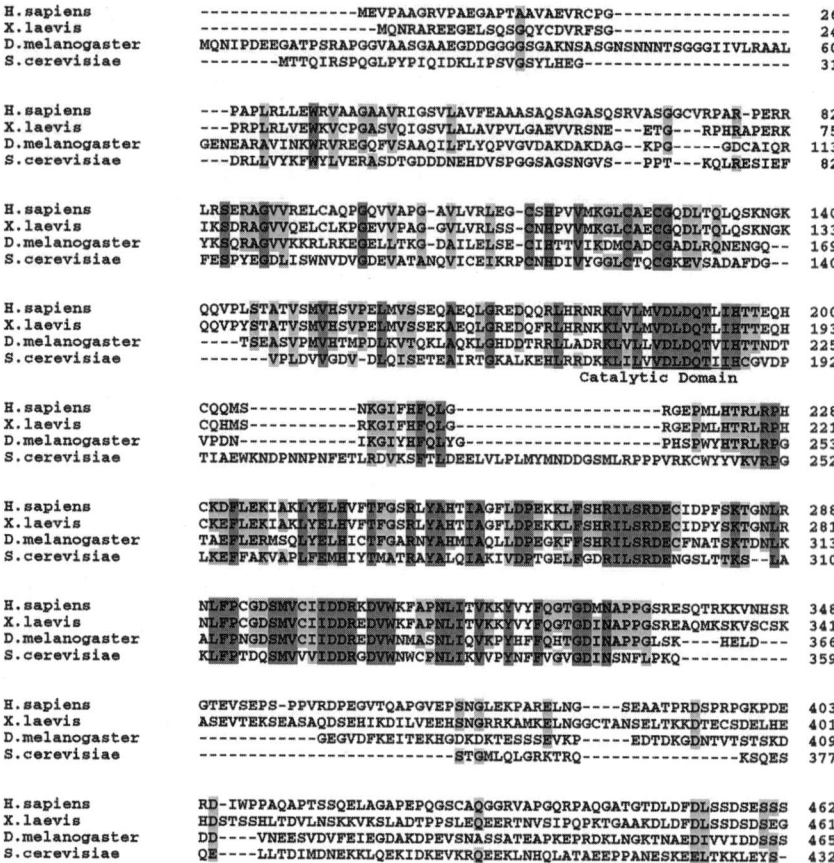

FIG. 2. Alignment of FCP1 protein sequences from human, *Xenopus*, *Drosophila*, and yeast. Fcp1 sequences were obtained from the GenBank database and aligned using the ClustalX sequence alignment program (University of Strasbourg, France). The shading assigned to selective amino acid residues depends on its identity/similarity relative to the other aligned sequences. Light gray is assigned to three identical amino acid residues in a given position or three identical and a fourth which represents a conservative change. Dark gray is assigned to amino acid residues in a given position where all four are identical. The underlined motif ΨΨΨDXDX(T/V)ΨΨ(Ψ represents a hydrophobic residue) is the catalytic domain and is present in the FCP1 sequences from all four organisms (amino acid residues 185–194 in human, 178–187 in *Xenopus*, 210–219 in *Drosophila*, 177–186 in yeast).

The *Drosophila* sequence was recovered from the GenBank Database (CG12252, map position 60D5). Based on sequence similarity, this gene appears homologous to FCP1, although the activity of the gene product has not been reported. Analysis of the N-terminal sequences suggests that the catalytic domain of FCP1 resembles similar domains found in a number of proteins of unknown function

```
H.sapiens        ESEGTKSSSSASDGESEGKRGRQKPKAAPEGAGALAQGSSLEPGRPAAPSLPGEAEPGAH  522
X.laevis         ETRKISSPSSASGSGENECKRSWRKSNKKDEDCIASQELCTDDDSKKARPENHSNLERPIF 521
D.melanogaster   GSP--DAEKAASDGED----------------VVVIDDNSKESTKAEVPPTPAEKNEVVA 507
S.cerevisiae     ---------ASLEVQQ--------------------------------------------  439


H.sapiens        APDKEPELGGQ----EEGERDGLCGLGNGCADRKEAETESQNSELSGVTAGESLDQSMEE  578
X.laevis         ESKDTLPVEDDEMEVQSAEQDSLCDLGNGCTGKKEVETESQNSEQSGITVGESLDQSMEE  581
D.melanogaster   SSTTSPDEKRP----------------------------SADADVATTSKTPSLRAPLEG 539
S.cerevisiae     -------QNRP--------------------------------------LAKLQKHLHD  453


H.sapiens        EEE--EDTDEDDHLIYLEEILVRVHTDYYAKYDRYLNKEIEEAPDIRKIVPELKSKVLAD  636
X.laevis         EDEDSEDTDEDDHLIYLEEILVRVHTDYYAKYDRYLKKEVDSVPDIRKIVPELKSKVLEN  641
D.melanogaster   QKQ-IEIEDPDDYLYLYLEVILRNIHKRFYSIYD-----ETTEIPDLKVIVPKIRSEVLRG  593
S.cerevisiae     QKL---LVDDDDELYYLMGTLSNIHKTYYDMLS----QQNEPEPNLMEIIPSLKQKVFQN  506


H.sapiens        VAIIFSGLHPTNFPIEKTREHYHATALGAKILTRLVLSPDAPDRATHLIAARAGTEKVLQ  696
X.laevis         VIISFSGLHPTNFPIERTREFYHARALGASIHKNLILKPDDPDRTTHLIAARAGTEVRK   701
D.melanogaster   KNLVFSGLVPTQMKLEQSRAYFIAKSLGAEVK------PNIDKEITHLVAVNAGTYKVNA  647
S.cerevisiae     CYFVFSGLILGTDIQRSDIVIWTSTFGATST------PDIDYLTTHLITKNPSTYKARL  560


H.sapiens        AQEC-GHLHVVNPDWLWSCLERWDKVEEQLFPLRDDHTKAQRENSPAAFPDREGVPPTAL  755
X.laevis         AQNC-KHLHVVNPEWLWSCLERWEEKVEEQLFPLKDDYMKSQRTISPTTFPDVQSAFQTPL  760
D.melanogaster   AKKE-PAIKVVNANWLWTCAERWEEVEEKLFPLDRKVRNKGRQPPAHCHSPEHVVNYSER  706
S.cerevisiae     AKKFNPQIKIVHPDWIFECLVNWKKVDEKPYTLIVDSP---------ISDEELQNFQTQ  610


H.sapiens        FHPMPVLPKAQPGPEVRIYDSNTGKLIRTGARGP------PAPS-SSLPIRQEPSSFRAV 808
X.laevis         FHPSPIHPNTQPGFNVTYDGKGLIRKGSQASRESPYIQAPSPSVTPVHGEHSSFRVV   820
D.melanogaster   SEISPSSSKQQEEQSGNFRETLNPLLVFTNADIES-------------MNKDYETFFES  752
S.cerevisiae     LQKRQEYLEETQEQQHMLTSQENLNLFAAGTSWLN---------------NDDDEDIPDT 655


H.sapiens        PPPQPQMFG-EELPDAQ-DGEQPGPSRRKRQPSMSETMPLYTLCKEDLESMDKEVDDILG  866
X.laevis         QPHQEQLFDDEELATANPDEEQSGPSRRKRQPSMSETMPLYTLCKEDLESMDKEVDDILG  880
D.melanogaster   DSSSD--EGPVNFENPP---MDKKLLKRKREDDNSRAHDFFTRSDDIMIGAPNLVEVDI  807
S.cerevisiae     ASDDD---EDDDHDDESDDENNSEGIDRKR----S--I------EDN------------  687


H.sapiens        EGSDDSDSEKRRPEEQEEEPQPRKPGTRRGADAR--APASSERSAAGGRGPRGHKRKLNE  924
X.laevis         EGSDDSDSEKKKTIKIKKSQIAAQGNKLKNPEERNESSSSSERSLSGSR-PRGHKRKLEE  939
D.melanogaster   SSNEEADDNNEKEDDDDEMP---SAKFRRGEDLP------SDLELG--------------  844
S.cerevisiae     --HDDTSQKKTKAEPSQDGP---------------------VQHK--------------  709


H.sapiens        E---DAASESSRESSNEDE-GSSSEADEMAKALEAELNDLM----  961
X.laevis         EEEEDAESEISKESSNEDEEGSSSEADEMAAAIEAELNDFI---- 980
D.melanogaster   -----SESNSEKEPEDEDD----GEWNMMGAALEREFLGLEDFDM  880
S.cerevisiae     ------------GEGDDN-------EDSDSQLEEELMDMLDD--  732
```

FIG. 2. (continued)

and has now been designated the FCP homology (FCPH) domain (53, 62). This domain, which encompasses amino acid residues 181–335 in the human sequence, contains the conserved motif KLΨLΨVDLDQTΨIH (Ψ designates a hydrophobic residue). This region includes the general phosphatase motif ΨΨΨDXDX(T/V)ΨΨ at its catalytic center, which is found in a large family of proteins consisting of phosphotransferases and phosphohydrolases (64). Since the catalytic domain in FCP1 is different from the motifs found in major classes of known phosphatases, FCP1 may be the founding member of a new class of eukaryotic protein phosphatases. Site-directed mutagenesis has established that both aspartate residues in the catalytic center are essential for CTD phosphatase function *in vivo* (53). Yeast *fcp1* alleles carrying mutations in these critical residues fail to complement a chromosomal deletion of the *FCP1* gene.

Furthermore, replacement of the aspartate (D181 or D183) by glutamate or asparagine in the xFCP1 sequence suppressed the CTD phosphatase activity of the mutant recombinant GST fusion protein (24).

Analysis of the C-terminal sequence of Fcp1 reveals the presence of a BRCT domain (62). This domain, which encompasses amino acid residues 628–728 in the human sequence and 499–593 in the yeast sequence, is found in a number of DNA damage-responsive cell cycle checkpoint proteins (65) and can mediate biologically important protein–protein interactions (66). Genetic analysis indicates that the BRCT domain, although not the catalytic center, is essential for Fcp1 activity *in vivo* (61). A point mutation (W575A) in the BRCT domain results in the accumulation of RNAP IIO and a reduction in mRNA synthesis. This observation may be partly explained by the fact that the BRCT domain overlaps one of two RAP74-binding sites. Since the precise role of the BRCT domain in Fcp1 has not been established, it will be important to identify the binding partners of this domain. Although more domains may be uncovered in the future, it appears that CTD phosphatase consists of two prominent functional modules both of which are required for activity.

III. RNAP II Recycling Mediated by CTD Phosphatase

A. The Role of CTD Phosphatase during the Transcription Cycle

Protein kinases and phosphatases that alter the state of CTD phosphorylation can serve as transcriptional activators or inhibitors, depending on the point in the transcription cycle at which they function. There is evidence that some CTD kinases function as activators of transcription, whereas others function as inhibitors. For example, TFIIH phosphorylates the CTD of RNAP II in the preinitiation complex concomitant with initiation of the transcript (32–34). Likewise, P-TEFb can serve as a transcriptional activator by protecting early elongation complexes against the negative effects of DSIF and NELF (37, 38). However, Srb10/Srb11 contained in the yeast mediator converts free RNAP IIA to RNAP IIO, thereby reducing the level of RNAP IIA available for transcriptional initiation (34, 67). In this case, Srb10/Srb11 would be considered a transcriptional inhibitor.

Although much less is known about the role of CTD phosphatase in transcription, CTD phosphatase, like CTD kinases, can in principle serve as both an activator and inhibitor of transcription. RNAP IIO in early elongation complexes is a substrate for CTD phosphatase (22, 23, 63). Since RNAP IIO is required for processive elongation, CTD phosphatase acting on transcribing polymerase would be expected to function as an inhibitor of mRNA synthesis. Recent studies suggest that CTD phosphatase can also enhance RNAP II elongation independent of its phosphatase activity (63). Most importantly, CTD phosphatase is capable of dephosphorylating free RNAP IIO, thereby generating RNAP IIA, which can be efficiently incorporated into preinitiation complexes for another round of transcription (13, 63). The ability to regenerate RNAP IIA to support multiple rounds of transcription is an essential activity of CTD phosphatase. It is not clear whether the conversion of RNAP IIO to IIA is coupled to the termination process or whether RNAP IIO is released from the template and subsequently dephosphorylated to generate RNAP IIA.

B. The Role of CTD Phosphatase in the Mobilization of Stored RNAP II

At specific times during growth and development, in response to the appropriate signal(s), cells can undergo major changes in the level of transcriptional activity. An example of such a change is the onset of transcription during early embryogenesis. A plausible model to account for major alterations in the level of gene expression involves the storage and release of sequestered RNAP II. This has been studied in detail in early *Xenopus* and mammalian development (24, 44, 68). Transcriptionally quiescent *Xenopus* and mammalian oocytes contain almost exclusively RNAP IIO. Since transcription in early embryonic development must utilize this maternal store of RNAP II, it must be dephosphorylated to generate RNAP IIA. This is consistent with the observation that oocytes rapidly dephosphorylate endogenous RNAP IIO upon fertilization (44). The rapid dephosphorylation of RNAP IIO at fertilization can be reproduced *in vitro* by calcium activation of *Xenopus* egg extracts (24).

RNAP IIO dephosphorylation in egg extracts is dramatically inhibited by immunodepletion of CTD phosphatase with antibodies directed against hFCP1. Interestingly, in xFCP1-depleted extracts, the steady-state level of RNAP IIO is elevated, indicating that CTD phosphatase is also involved in maintaining the RNAP IIO/IIA balance under conditions where transcription is not taking place (24). These observations support the idea that the mobilization of RNAP IIO is catalyzed by CTD phosphatase (FCP1) (24) and demonstrate directly the rapid turnover of CTD phosphates in egg extracts in the absence of transcription. Accordingly, this activity of CTD phosphatase during early development is distinct from the involvement of CTD phosphatase in the recycling of RNAP II and in transcript elongation (24, 44).

Apart from the storage and subsequent mobilization of RNAP II during early development, there is some indication that RNAP II can be localized at discrete sites within the nucleus in a transcriptionally quiescent state in differentiated cells. The subnuclear distribution of RNAP II has been determined in transcriptionally active and repressed cells by immunofluorescence microscopy (69). These studies utilized monoclonal antibodies (H5 and H14) that distinguish between different patterns of CTD phosphorylation.

In transcriptionally active cells, RNAP II is distributed in irregularly shaped speckle domains, which appear to be interconnected via a reticular network. The inhibition of transcription results in the redistribution of RNAP IIO into enlarged, dotlike speckle domains. Upon release from transcriptional inhibition, RNAP II is redistributed back to the interconnected speckle pattern. Interestingly, H5, which recognizes phosphoserine 2, stains a diffuse nucleoplasmic fraction of RNAP II as well as RNAP II contained in irregularly shaped speckle domains. H14, which recognizes phosphoserine 5, reacts with the diffusely distributed pattern of RNAP II in transcriptionally active cells and with speckle domains in transcriptionally inhibited cells. The speckle domains, defined by morphology and immunological methods, are also enriched for splicing and transcription factors (69, 70). Accordingly, RNAP IIO within the nucleus changes its localization and pattern of phosphorylation within the CTD in response to changes in transcriptional activity. It is important to remember that even though transcript elongation is catalyzed exclusively by RNAP IIO, not all RNAP IIO is engaged in transcription. Since RNAP IIO does not assemble directly into preinitiation complexes, the activation of transcription requires the conversion of sequestered RNAP IIO to RNAP IIA. This directly implicates the involvement of CTD phosphatase in the mobilization of stored RNAP II.

RNAP II can also be localized to subnuclear structures referred to as Cajal or coiled bodies (for a review see Ref. 71). Cajal bodies are stained by H14 monoclonal antibody and this staining is reversibly inhibited by DRB, indicating that the level of RNAP IIO is maintained by the action of a DRB-sensitive CTD kinase(s). Recent results suggest that newly translated RNAP II subunits are targeted to Cajal bodies (72). Immunostaining has also demonstrated the presence of a large number of proteins involved in the transcription and processing of RNA products of the three nuclear polymerases (71, 73). Although there is evidence that Cajal bodies can be specifically targeted to sites near histone mRNA transcription, the majority are free in the nucleoplasm.

Nuclear localization studies utilizing monoclonal antibodies of known specificity can provide important information relating the pattern of CTD phosphorylation to nuclear function. However, caution must be exercised in the interpretation of these results, because the reactivity of antibodies such as H5 and H14 are not rigorously defined and the possibility that other epitopes are recognized in the staining of fixed cells cannot be eliminated. For example, human Spt5 contains a serine/threonine-rich domain that is phosphorylated by P-TEFb and is reminiscent of the consensus CTD repeat (30).

IV. Involvement of CTD Phosphatase in the Regulation of Transcript Elongation

In addition to being required for the progression of RNAP II through the transcription cycle and the mobilization of sequestered RNAP IIO, CTD phosphatase appears to play a role in the regulation of transcript elongation. The progression of RNAP II along the template is positively influenced by factors that maintain a high level of CTD phosphorylation, such as P-TEFb, and negatively influenced by factors that act on RNAP IIA to inhibit transcript elongation, such as DSIF and NELF. During transcript elongation the phosphorylated CTD serves to couple the RNA processing machinery directly to RNAP II (5, 74). Furthermore, the association of factors involved in specific processing events can be sensitive to the pattern of CTD phosphorylation (75–77). Accordingly, the sequential activation of specific CTD kinases and CTD phosphatase(s) may be necessary for the correct processing of the primary transcript. Finally, CTD phosphatase acting on elongation complexes may play an integral part in the termination process. These areas of potential involvement of CTD phosphatase in the elongation phase of transcription will be discussed in this section.

A. Interplay of Positive and Negative Factors That Determine the Fate of Early Elongation Complexes

The fate of each transcription initiation event is controlled by the interplay of a host of positive and negative transcription elongation factors generally termed P-TEFs and N-TEFs, respectively (9, 78). Positively acting factors include, but are not limited to, proteins such as SII, SIII, TFIIF, ELL, and similar proteins (79–81). These factors influence the ability of elongating RNAP II to move through specific pause sites and/or enhance the inherent rate of elongation. Factor 2 is an important ATP-dependent N-TEF that causes termination of paused RNAP II (82, 83).

In a parallel situation, there is a group of P-TEFs and N-TEFs whose activities require the CTD and that work in opposition to control transcription. The first of these factors to be discovered was P-TEFb. P-TEFb was initially identified and purified from *Drosophila* Kc cell nuclear extracts (35). P-TEFb works in a DRB-sensitive manner to promote processive elongation and limit premature termination. P-TEFb is a CTD kinase composed of Cdk9 (previously PITALRE) and one of four cyclin subunits, cyclin T1, T2a, T2b, or K (84, 85). Shortly after the discovery of P-TEFb, a negative factor was identified and named DRB-sensitivity inducing factor (DSIF) (29, 86). DSIF, which is the human homolog of the yeast Spt4/Spt5 complex, is required for reconstruction of DRB-sensitive transcription *in vitro*. DSIF, along with the more recently described negative elongation factor (NELF), act in a CTD-dependent manner to inhibit the transition into productive elongation (37, 38). A multitude of positive and negative factors work in concert to regulate transcript elongation in a manner that is sensitive to the phosphorylation state of the CTD.

B. The Link between CTD Phosphorylation and Processive Transcript Elongation

The finding that transcript elongation is catalyzed almost exclusively by RNAP IIO led to the hypothesis that CTD kinases and CTD phosphatase may play either positive or negative roles in transcription, depending on their temporal regulation. This is supported by the observation that P-TEFb, initially identified as a factor required for productive elongation *in vitro* (35, 36), is a CTD kinase. The inverse correlation, that the dephosphorylation of RNAP IIO in an elongation complex leads to pausing, is not rigorously established. However, a number of specific cases establish a clear correlation between the absence of CTD phosphorylation and a paused elongation complex.

The heat shock genes of *Drosophila* are a well-studied case (87). In the uninduced state, RNAP IIA is paused about 25 nucleotides downstream of the transcriptional start site of the hsp70 gene (88). Heat shock results in the activation of transcription and the conversion of RNAP IIA to IIO. Similarly, paused RNAP II complexes are found on several other genes, including the heat-shock gene hsp26 and non-heat-shock genes such as GAPDH and β-tubulin (87). This supports the idea that paused polymerases near the promoter might be a common feature of transcription *in vivo*. In principle, paused RNAP IIA could arise from the failure to fully phosphorylate RNAP II at the time of transcript initiation or from the action of CTD phosphatase on early elongation complexes.

In what is likely a related phenomenon, a number of cellular genes, as well as certain viruses, generate short abortive transcripts in the absence of induction (89). Induction results in the establishment of a processive elongation complex and expression of the gene or viral genome. An example of such a transcriptional activator is the human immunodeficiency virus 1 (HIV-1)-encoded protein Tat. Transcription of the HIV genome from the long terminal repeater (LTR) promoter is a particularly interesting model for the control of transcript elongation.

Tat is the only viral protein required to regulate transcription from the LTR and controls transcription of the HIV template by stimulating the formation of highly processive elongation complexes. The activity of Tat is dependent on a *cis*-acting element, TAR, present near the 5′ end of the nascent transcript. As well as binding to TAR, Tat requires the CTD and host proteins to carry out its function (90–93). P-TEFb has been shown to be critical for Tat transactivation and inhibitors of P-TEFb such as DRB, H-8, or flavopiridol (94) are potent inhibitors of Tat transactivation. Tat binds tightly and specifically to P-TEFb through the cyclin T1 subunit (95, 96). Whereas any of the cyclin subunits can partner with Cdk9 and form a transcriptionally active P-TEFb, only cyclin T1 can mediate Tat transactivation (9, 97). Tat also activates the CTD kinase activity of P-TEFb (98) and is reported to shift the substrate specificity of P-TEFb from serine 2 to the phosphorylation of both serine 2 and serine 5 when assayed with partially purified early elongation complexes (51).

In addition to binding P-TEFb, Tat can also associate with CTD phosphatase (59) and inhibit its activity *in vitro* (21). Mutant Tat proteins with reduced transactivation potential inhibit CTD phosphatase less effectively, thus establishing a correlation between the inhibition of CTD phosphatase and the ability to transactivate. As Tat can bind to and influence the activity of multiple factors that together determine the state of CTD phosphorylation, it was proposed that the regulation of both P-TEFb and CTD phosphatase is integral to Tat transactivation. Recent support for this idea comes from the observation that overexpression of CTD phosphatase *in vivo* inhibits Tat transactivation (99).

A rationale for why mechanisms might have evolved to terminate transcription catalyzed by RNAP IIA comes from a consideration of RNA processing. Both capping enzymes and some splicing factors bind specifically to RNAP IIO. The finding that the phosphorylated CTD provides an interaction site that physically couples the RNA processing machinery to elongating RNA polymerase II suggests that dephosphorylation would result in uncoupling RNA processing from transcription. Accordingly, there may exist a feedback loop that shuts down transcript elongation if the primary transcript is not being properly processed. The negative regulators of transcript elongation may be a part of such a putative signal transduction pathway.

C. Phosphate Turnover during Transcript Elongation

Although a variety of studies suggest that CTD phosphates turn over during transcript elongation, direct confirmation of this idea is lacking. Most studies examining the level of CTD phosphorylation are based on the electrophoretic mobility difference between subunits IIa and IIo. This provides an estimate of the steady-state level of phosphorylation, but provides no information concerning turnover. The recent observation that the pattern of CTD phosphorylation changes as RNAP II moves down the gene provides the best evidence of turnover (8, 10, 77).

As noted previously, RNA processing factors that catalyze capping, the removal of introns, and polyadenylation couple to elongating RNAP II via the CTD. In an effort to determine whether these activities simultaneously associate with the CTD or interact transiently as needed, chromatin immunoprecipitation assays were performed in yeast (8, 10). RNAP II phosphorylated at serine 5 is found near the promoter on several genes. This modification coincides with the localization of capping enzyme, which preferentially associates with the CTD phosphorylated at position 5 (8, 77). Although mammalian capping enzyme will bind CTD repeats phosphorylated on either serine 2 or 5 *in vitro*, guanylyltransferase activity is only stimulated by serine 5 phosphates (75). In yeast, CTD mutants in which half of all serine 5 positions are changed to alanine are synthetically lethal in combination with *ceg1* (guanylyltransferase) capping enzyme mutants (76). The association of capping enzyme with RNAP IIO near the promoter is also in agreement with the observation that TFIIH, which likely phosphorylates RNAP II at the promoter, specifically targets serine 5 (47–49, 100).

The relative amount of serine 5 phosphorylation and colocalized capping enzyme decrease as RNAP II moves downstream. The removal of serine 5 phosphates by the time RNAP II elongates to about position 200 suggests that a CTD phosphatase acts on RNAP II contained in early elongation complexes. The possibility cannot be eliminated that the H14 monoclonal antibody used to monitor phosphoserine 5 no longer reacts due to further modification of the CTD, as opposed to the loss of serine 5 phosphates. In one case, mutation of the *FCP1* gene caused a reduction in the 5′-to-3′ polarity of serine 5 phosphorylation, which correlated with aberrant retention of capping enzyme (77). However, in a separate study, there appeared to be no specific loss of serine 5 phosphorylation in *fcp1* mutant strains (10). This study differs from earlier work in that chromatin immunoprecipitations were carried out in cells grown at a semipermissive temperature. Furthermore, results were normalized to a negative RNAP II internal control. These findings suggest that the decrease in serine 5 phosphorylation at promoter distal sites is not mediated by Fcp1. Accordingly, a distinct CTD phosphatase may exist that is responsible for serine 5 phosphate turnover.

RNAP II phosphorylated on serine 2 is observed throughout the coding region, but is present in increased amounts toward the 3′-end of genes (8). This observation is also in agreement with the specificity of P-TEFb, which acts after initiation and phosphorylates serine 2. Furthermore, the amount of serine 2 phosphorylation increases dramatically at the semipermissive temperature in a yeast *fcp1* mutant (10). This finding suggests that CTD phosphatase is involved in maintaining the appropriate level of CTD phosphorylation during elongation over the body of the gene. It is likely that the level of CTD phosphorylation in elongation complexes is maintained at its appropriate level by the continual action of P-TEFb (Ctk1) and CTD phosphatase (10). Precedence exists for this idea in the *Xenopus* oocyte, where phosphorylation of nontranscribing RNAP IIO is apparently maintained by the competing actions of MAP kinase Xp42 and CTD phosphatase (24). These studies suggest a dynamic interaction between CTD kinases and CTD phosphatase(s) in generating distinct conformations of the CTD during the elongation phase of transcription. This in turn enables the association of the appropriate RNA processing machinery.

D. A Possible Role of CTD Phosphatase in the Regulation of mRNA Processing

The evidence suggesting a link between CTD phosphatase and RNA processing is twofold. First, results described in the preceding section suggest that CTD phosphatase is involved in modulating the phosphorylation pattern of elongating RNAP II. Second, RNA processing factors have been shown to bind the CTD and in the case of capping, this binding is sensitive to the pattern of CTD phosphorylation. Therefore, it is conceivable that CTD phosphatase activity on elongation complexes could play a role in regulating pre-mRNA processing.

In mammals, the capping enzyme (*101–103*), and in yeast, both guanylyltransferase and methyltransferase (*101, 104*), bind specifically to the phosphorylated CTD of RNAP IIO. In mammals, the capping enzyme is stimulated by phosphorylated, but not unphosphorylated CTD peptides. As indicated in the preceding section, the serine 5 phosphorylation that correlates with retention of the capping enzyme in yeast displays 5′–3′ polarity on the gene, which means it is prevalent near the promoter and is progressively lost downstream. The reduction of phosphoserine 5 correlates with reduced levels of yeast capping enzymes Abd1 (methylase) and Ceg1 (guanylyltransferase) (*77*). Interestingly, in control experiments the gradient for Abd1 is not as steep as that for Ceg1, indicating that the methyltransferase remains bound to elongating RNAP II for longer periods of time (*8, 77*).

Splicing factors, like the capping enzymes, specifically associate with RNAP IIO (*70, 105, 106*). RNAP IIO, but not RNAP IIA, can stimulate the splicing reaction, apparently by stimulating the formation of spliceosomal complexes (*74*). It would be of interest to determine whether the splicing machinery selectively associates with serine 2 phosphorylated CTD. CTD phosphatase could conceivably affect the course of the splicing reaction by dephosphorylating the CTD at specific times/locations, causing the temporary dissociation of splicing factors.

In the case of 3′-end formation, evidence for the potential involvement of CTD phosphatase is somewhat less compelling. Two factors involved in cleavage and polyadenylation of mRNAs, CPSF and CstF, can bind to the CTD (*101*). This binding does not appear to be sensitive to the state of CTD phosphorylation. RNAP II can greatly stimulate the cleavage reaction in both crude and purified *in vitro* reactions; however, both RNAP IIA and IIO can function in this role (*107*).

The CTD may mediate 3′-end formation through the enzyme peptidyl prolyl isomerase (PPIase), an enzyme that catalyzes the *cis–trans* isomerization of the peptide bond of proline residues. PPIase, encoded by the *ESS1* gene in yeast and PIN1 in humans, is specific for phosphorylated Ser–Pro or Thr–Pro bonds in model peptides (*108, 109*). Most importantly, *ESS1* has been shown genetically to be important for 3′-end formation (*108*). Furthermore, a PPIase of the cyclophilin family has been found to interact with the CTD of mammalian RNAP II in a yeast two-hybrid screen (*110*). Ess1/Pin1 binds to phosphorylated, but not unphosphorylated CTD peptides and can bind to the phosphorylated form of the largest subunit of RNAP II in a far-Western analysis (*111*). This supports the idea that CTD phosphatase activity would be antagonistic to Ess1/Pin1 activity. Although the exact pattern of CTD phosphorylation required to induce Ess1/Pin1 binding is not known, it is possible that CTD phosphatase acts to eliminate the critical phosphate(s), thereby preventing the binding of Ess1/Pin1 until the appropriate 3′-end formation signals are received. Alternatively, Ess1 might act to promote CTD dephosphorylation (*112*). This is supported by the observation that overexpression of Fcp1 can suppress certain temperature-sensitive Ess1 mutations. Other hypotheses are certainly possible, but the potential interaction between Ess1/Pin1 activity and CTD phosphatase activity is an exciting area for future studies. Finally, 3′-end formation and termination are tightly coupled in yeast. Accordingly, in the studies discussed above, it might be difficult to distinguish between a primary defect in 3′-end formation and a defect in termination.

E. Potential Coupling between CTD Dephosphorylation and Transcript Termination

Although neither the factors nor the mechanism required for termination by RNAP II are clearly defined, it would not be surprising to find that this step in transcription is also in part mediated by the CTD. As discussed above, the CTD-interacting protein PPIase has been implicated in 3′-end formation/termination in yeast. One favored model for termination is that once an elongating RNAP II passes a functional polyadenylation signal in the template, RNAP IIO is dephosphorylated and thereby converted to a nonprocessive form. Such elongation complexes, now containing RNAP IIA, are more prone to termination facilitated by NELF/DSIF negative factors. Therefore, it is possible that CTD phosphatase acts on elongating RNAP II to promote termination. There is evidence that a functional polyadenylation signal in the template and 3′-end cleavage factors are required for termination (*113*). Recently, nascent transcript cleavage at the poly(A) site has been shown to precede the dissociation of RNAP II from template (*113–115*). Accordingly, if the 3′-end formation is a signal for termination and CTD phosphatase is part of that signal transduction pathway, CTD phosphatase activity might be induced in complexes downstream of the transcript cleavage site. This activation may be mediated by a direct interaction with factors, such as Pin1, which are part of the normal cleavage/polyadenylation process. In this way, the process of termination can be temporally linked to 3′-end formation.

F. Sensitivity of RNAP II in Elongation Complexes to CTD Phosphatase

Results summarized above establish that phosphates within the CTD turn over during transcript elongation and that the concerted action of CTD kinases and CTD phosphatase determine the level and pattern of CTD phosphorylation. The Reinberg laboratory first demonstrated that CTD phosphatase can act on RNAP IIO in transcription complexes formed in a defined *in vitro* reaction (63). RNAP IIO contained in preinitiation complexes as well as complexes at 392 nucleotides downstream from the start site were found to be sensitive to dephosphorylation by CTD phosphatase. These studies relied on Western blots to detect RNAPs IIA and IIO. In experiments of this type it is difficult to know the fraction of bound RNAP II that is in functional complexes. Unless nearly all the RNAP II bound is transcriptionally active, there is no assurance that the sensitivity determined accurately reflects the sensitivity of RNAP II in functional complexes. It is important to design experiments in a way that enables quantification of the molar amount of RNAP II bound and the molar amount of transcript produced. To more clearly understand the role of CTD phosphatase during elongation, it is necessary to understand what controls the recruitment and activation of CTD phosphatase in the context of the elongation complex.

Using ^{32}P-labeled RNAP II, CKII-labeled as described in Section I.B, immobilized templates, and an RNAP II-dependent transcription extract system, the CTD phosphatase sensitivity of elongation complexes was examined under conditions where the RNAP IIO to transcript ratio was one (23). These conditions require the use of a 1% Sarkosyl wash to remove a preponderance of nonengaged RNAP II. As a reference point, the CTD phosphatase sensitivity of free RNAP IIO and RNAP IIO transcriptionally engaged on dC-tailed templates was determined. RNAP IIO contained in transcription complexes on the dC-tailed template exhibited the same sensitivity as free RNAP IIO. Accordingly, the formation of an elongation complex does not in itself result in protection of the CTD against dephosphorylation. RNAP IIO in early elongation complexes generated by initiation on the Ad2-MLP was approximately 50 times more resistant to dephosphorylation than control RNAP IIO. Furthermore, the sensitivity of RNAP IIO to dephosphorylation was in part dependent on its position relative to the start site of transcription. RNAP IIO contained in downstream complexes was more resistant to dephosphorylation (23).

To more precisely define CTD phosphatase sensitivity of RNAP IIO in elongation complexes as a function of their position from the transcriptional start site, the immobilized template protocol was adapted for use in HeLa nuclear extracts that are capable of generating highly synchronous transcribing RNAP II (22). Sarkosyl-washed early elongation complexes were stalled at the earliest stable

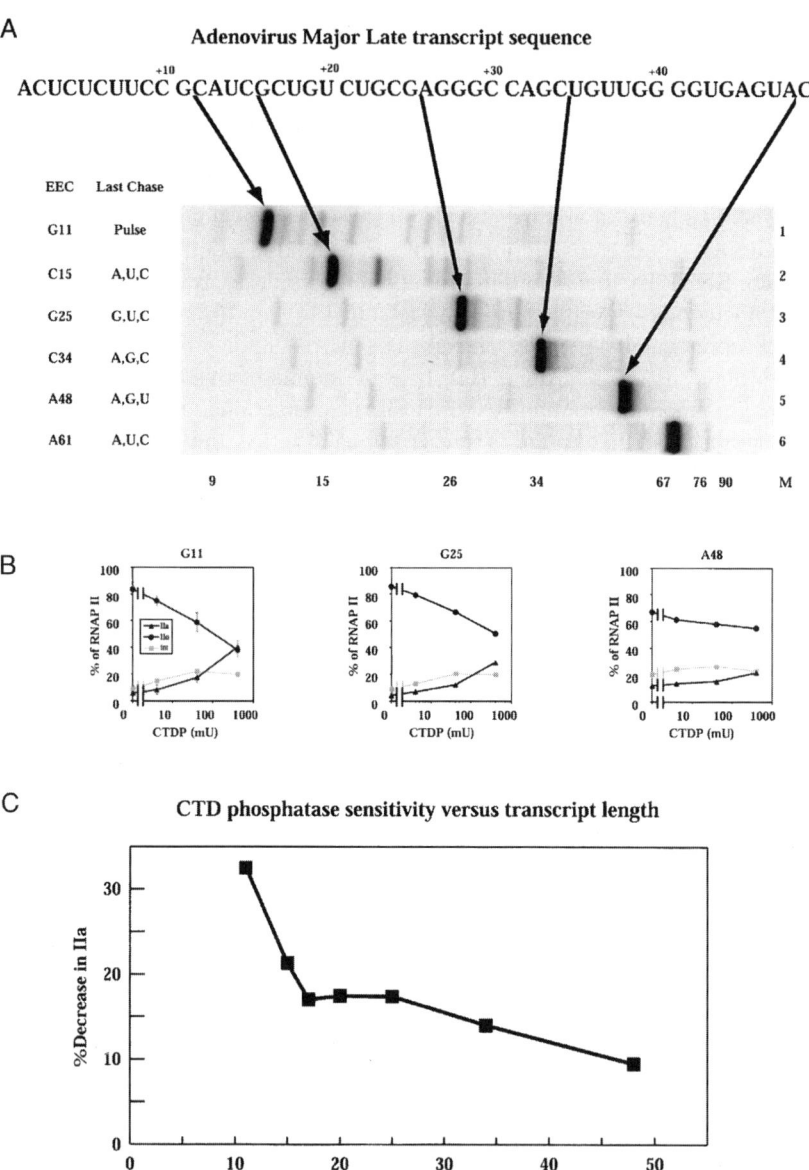

FIG. 3. CTD phosphatase sensitivity of RNAP IIO in elongation complexes on the Ad2-ML template as a function of their position from the transcriptional start site. (A) Sequence of the first 50 nucleotides of the Ad2-ML transcript. The lower part of A is a urea–PAGE gel of RNA generated by transcription of immobilized template DNA. Labeled nucleotides are present only during the

point after initiation on both HIV (U14/G16) and Ad2-ML (G11) templates and were found to be approximately equal in sensitivity to CTD phosphatase (22) (Fig. 3 shows results from the Ad2-ML template). These stringently washed complexes were "walked" to various positions on each template and assayed for CTD phosphatase sensitivity. The synchronous population of complexes stalled at specific sites on the Ad2-ML template is shown in Fig. 3A. On both the HIV and Ad2-ML templates, complexes that stalled further from the promoter were less sensitive to CTD phosphatase (Fig. 3C). Sarkosyl-treated complexes that have moved past position 20 are highly resistant to dephosphorylation by CTD phosphatase. The position-dependent change in CTD phosphatase sensitivity likely results from conformational changes in the elongation complex, as RNAP II progression down the template is accomplished entirely by the addition of different nucleotide mixtures.

Subsequent transcript elongation by Sarkosyl-washed early elongation complexes was unaffected by pretreatment with CTD phosphatase (22). This is not surprising, as the Sarkosyl treatment likely removes the negative elongation factors which would be required to prematurely terminate elongating RNAP IIA. This system provides an appropriate assay for examining promoter-specific effects as well the characterization of extrinsic factors that influence the CTD phosphatase sensitivity of elongation complexes.

V. Perspectives and Future Directions

The remarkable structure of the CTD and its dynamic phosphorylation have intrigued biologists since its discovery more than 15 years ago. Clear evidence for the involvement of the CTD at multiple levels of transcription has been established. Most importantly, the activity of the CTD is regulated by reversible phosphorylation. Considerable progress has been made in the identification and characterization of CTD kinases that function at different steps in the transcription cycle and, by virtue of their selectivity for the phosphorylation of certain positions within the consensus repeat, can generate distinct CTD conformations. Although the importance of CTD phosphatase in reversing this reaction and in establishing a balance between RNAPs IIA and IIO is recognized, much

---

initial pulse. Complexes were washed with 1% Sarkosyl after pulse labeling and buffer-washed between each successive triple NTP chase (22). (B) Quantitation of CTD phosphatase assays. Early elongation complexes shown in A were analyzed for sensitivity to increasing amounts of CTD phosphatase. (C) CTD phosphatase sensitivity versus transcript length. The decrease in RNAP IIA at 400 mU of CTD phosphatase from analyses similar to those in B is plotted against length of transcript. Adapted from Marshall and Dahmus (22).

remains to be done to clarify our understanding of how CTD phosphatase is regulated or its precise function at different stages in the transcription cycle.

Areas of special interest for future investigation include the involvement of CTD phosphatase in (1) the mobilization of stored RNAP IIO, (2) the cycling of RNAP II during the transcription cycle, and (3) phosphate turnover during the elongation phase of transcription. A technical challenge in pursuing these studies is to establish methods to more accurately assess the nature of CTD modifications that occur *in vivo*. Although monoclonal antibodies that preferentially react with phosphoserines 2 or 5 have proven to be valuable reagents, their specificity is likely influenced by other modifications that can occur within the CTD. For example, the consequence of phosphorylation of multiple sites within a single repeat is not clearly understood. An additional complication is that the specificity of certain kinases appears to be context-dependent. Accordingly, the specificity established utilizing synthetic peptides may not hold true for the phosphorylation of RNAP II in the context of a transcriptional complex. Given the different conformations generated by distinct CTD kinases, it is important to know whether these are differentially dephosphorylated by CTD phosphatase.

CTD phosphatase is known to be regulated by TFIIB, TFIIF, and the HIV-1 transcriptional activator, Tat. It is important to know whether other cellular proteins function in the regulation of CTD phosphatase activity and whether they do so in a gene-specific manner. Furthermore, to understand the mechanism by which such factors regulate CTD phosphatase activity, it will be necessary to localize the CTD phosphatase-binding site on RNAP II and to determine whether the accessibility of this site is altered during the transcription cycle.

Finally, although FCP1 appears to be the major CTD phosphatase in eukaryotic cells, other CTD phosphatases may exist and may play important roles in the regulation of RNAP II activity. Possibilities include a CTD phosphatase specific for phosphoserine 5 that may mediate the changeover from serine 5 to serine 2 phosphorylation during early transcript elongation. Furthermore, based on the assumption that tyrosine phosphates in the CTD also turn over, a specific tyrosine phosphatase may exist. Methods for the preparation of specific RNAP II substrates for the identification and characterization of such putative CTD phosphatases are available.

Acknowledgments

We thank Alexandre Tremeau-Bravard for his critical review of the manuscript and Stephen Buratowski for making his results available prior to publication. We apologize to those who have contributed to our understanding of CTD phosphatase and CTD function, but whose work was not included due to space limitations. Research in the laboratory during the preparation of this manuscript was supported by National Institutes of Health Grant GM33300.

References

1. B. Lemon and R. Tjian, Orchestrated response: A symphony of transcription factors for gene control. *Genes Dev.* **14**, 2551–2569 (2000).
2. D. V. Fyodorov and J. T. Kadonaga, The many faces of chromatin remodeling: SWItching beyond transcription. *Cell* **106**, 523–525 (2001).
3. P. Cheung, C. D. Allis, and P. Sassone-Corsi, Signaling to chromatin through histone modifications. *Cell* **103**, 263–271 (2000).
4. A. M. Naar, B. D. Lemon, and R. Tjian, Transcriptional coactivator complexes. *Annu. Rev. Biochem.* **70**, 475–501 (2001).
5. Y. Hirose and J. L. Manley, RNA polymerase II and the integration of nuclear events. *Genes Dev.* **14**, 1415–1429 (2000).
6. M. E. Dahmus, Reversible phosphorylation of the C-terminal domain of RNA polymerase II. *J. Biol. Chem.* **271**, 19009–19012 (1996).
7. D. L. Cadena and M. E. Dahmus, Messenger RNA synthesis in mammalian cells is catalyzed by the phosphorylated form of RNA polymerase II. *J. Biol. Chem.* **262**, 12468–12474 (1987).
8. P. Komarnitsky, E. J. Cho, and S. Buratowski, Different phosphorylated forms of RNA polymerase II and associated mRNA processing factors during transcription. *Genes Dev.* **14**, 2452–2460 (2000).
9. D. H. Price, P-TEFb, a cyclin-dependent kinase controlling elongation by RNA polymerase II. *Mol. Cell. Biol.* **20**, 2629–2634 (2000).
10. E.-J. Cho, M. S. Kobor, M. Kim, J. Greenblatt, and S. Buratowski, Opposing effects of Ctk1 kinase and Fcp1 phosphatase at serine 2 of the RNA polymerase II C-terminal domain. *Genes Dev.*, **15**, 3319–3329 (2001).
11. H. Lu, O. Flores, R. Weinmann, and D. Reinberg, The nonphosphorylated form of RNA polymerase II preferentially associates with the preinitiation complex. *Proc. Natl. Acad. Sci. USA* **88**, 10004–10008 (1991).
12. P. J. Laybourn and M. E. Dahmus, Phosphorylation of RNA polymerase IIA occurs subsequent to interaction with the promoter and before the initiation of transcription. *J. Biol. Chem.* **265**, 13165–13173 (1990).
13. J. D. Chesnut, J. H. Stephens, and M. E. Dahmus, The interaction of RNA polymerase II with the adenovirus-2 major late promoter is precluded by phosphorylation of the C-terminal domain of subunit IIa. *J. Biol. Chem.* **267**, 10500–10506 (1992).
14. M. E. Maxon, J. A. Goodrich, and R. Tjian, Transcription factor IIE binds preferentially to RNA polymerase IIa and recruits TFIIH: a model for promoter clearance. *Genes Dev.* **8**, 515–524 (1994).
15. J. M. Payne, P. J. Laybourn, and M. E. Dahmus, The transition of RNA polymerase II from initiation to elongation is associated with phosphorylation of the carboxyl-terminal domain of subunit IIa. *J. Biol. Chem.* **264**, 19621–19629 (1989).
16. J. R. Weeks, S. E. Hardin, J. Shen, J. M. Lee, and A. L. Greenleaf, Locus-specific variation in phosphorylation state of RNA polymerase II in vivo: correlations with gene activity and transcript processing. *Genes Dev.* **7**, 2329–2344 (1993).
17. N. Fong and D. L. Bentley, Capping, splicing, and 3' processing are independently stimulated by RNA polymerase II: different functions for different segments of the CTD. *Genes Dev.* **15**, 1783–1795 (2001).
18. D. Bentley, Coupling RNA polymerase II transcription with pre-mRNA processing. *Curr. Opin. Cell Biol.* **11**, 347–351 (1999).
19. J. M. Payne and M. E. Dahmus, Partial purification and characterization of two distinct protein kinases that differentially phosphorylate the carboxyl-terminal domain of RNA polymerase subunit IIa. *J. Biol. Chem.* **268**, 80–87 (1993).

20. M. E. Kang and M. E. Dahmus, RNA polymerases IIA and IIO have distinct roles during transcription from the TATA-less murine dihydrofolate reductase promoter. *J. Biol. Chem.* **268**, 25033–25040 (1993).
21. N. F. Marshall, G. K. Dahmus, and M. E. Dahmus, Regulation of carboxyl-terminal domain phosphatase by HIV-1 tat protein. *J. Biol. Chem.* **273**, 31726–31730 (1998).
22. N. F. Marshall and M. E. Dahmus, C-terminal domain phosphatase sensitivity of RNA polymerase II in early elongation complexes on the HIV-1 and adenovirus 2 major late templates. *J. Biol. Chem.* **275**, 32430–32437 (2000).
23. A. L. Lehman and M. E. Dahmus, The sensitivity of RNA polymerase II in elongation complexes to C-terminal domain phosphatase. *J. Biol. Chem.* **275**, 14923–14932 (2000).
24. B. Palancade, M. F. Dubois, M. E. Dahmus, and O. Bensaude, Transcription-independent RNA polymerase II dephosphorylation by the FCP1 carboxy-terminal domain phosphatase in Xenopus laevis early embryos. *Mol. Cell. Biol.* **21**, 6359–6368 (2001).
25. R. S. Chambers and M. E. Dahmus, Purification and characterization of a phosphatase from HeLa cells which dephosphorylates the C-terminal domain of RNA Polymerase II. *J. Biol. Chem.* **269**, 26243–26248 (1994).
26. R. S. Chambers, B. Q. Wang, Z. F. Burton, and M. E. Dahmus, The activity of COOH-terminal domain phosphatase is regulated by a docking site on RNA polymerase II and by the general transcription factors IIF and IIB. *J. Biol. Chem.* **270**, 14962–14969 (1995).
27. W. Y. Kim and M. E. Dahmus, Immunochemical analysis of mammalian RNA polymerase II subspecies. Stability and relative in vivo concentration. *J. Biol. Chem.* **261**, 14219–14225 (1986).
28. M. S. Swanson, E. A. Malone, and F. Winston, SPT5, an essential gene important for normal transcription in Saccharomyces cerevisiae, encodes an acidic nuclear protein with a carboxy-terminal repeat. *Mol. Cell. Biol.* **11**, 3009–3019 (1991).
29. Y. Yamaguchi, T. Wada, D. Watanabe, T. Takagi, J. Hasegawa, and H. Handa, Structure and function of the human transcription elongation factor DSIF. *J. Biol. Chem.* **274**, 8085–8092 (1999).
30. J. B. Kim and P. A. Sharp, Positive transcription elongation factor b phosphorylates hSPT5 and RNA polymerase II carboxyl-terminal domain independently of cyclin-dependent kinase-activating kinase. *J. Biol. Chem.* **276**, 12317–12323 (2001).
31. R. Baskaran, M. E. Dahmus, and J. Y. Wang, Tyrosine phosphorylation of mammalian RNA polymerase II carboxyl-terminal domain. *Proc. Natl. Acad. Sci. USA* **90**, 11167–11171 (1993).
32. H. Lu, L. Zawel, L. Fisher, J. M. Egly, and D. Reinberg, Human general transcription factor IIH phosphorylates the C-terminal domain of RNA polymerase II. *Nature* **358**, 641–645 (1992).
33. W. J. Feaver, J. Q. Svejstrup, N. L. Henry, and R. D. Kornberg, Relationship of CDK-activating kinase and RNA polymerase II CTD kinase TFIIH/TFIIK. *Cell* **79**, 1103–1109 (1994).
34. C. J. Hengartner, V. E. Myer, S. M. Liao, C. J. Wilson, S. S. Koh, and R. A. Young, Temporal regulation of RNA polymerase II by Srb10 and Kin28 cyclin-dependent kinases. *Mol. Cells* **2**, 43–53 (1998).
35. N. F. Marshall and D. H. Price, Purification of P-TEFb, a transcription factor required for the transition into productive elongation. *J. Biol. Chem.* **270**, 12335–12338 (1995).
36. N. F. Marshall, J. Peng, Z. Xie, and D. H. Price, Control of RNA polymerase II elongation potential by a novel carboxyl-terminal domain kinase. *J. Biol. Chem.* **271**, 27176–27183 (1996).
37. T. Wada, T. Takagi, Y. Yamaguchi, D. Watanabe, and H. Handa, Evidence that P-TEFb alleviates the negative effect of DSIF on RNA polymerase II-dependent transcription in vitro. *EMBO J.* **17**, 7395–7403 (1998).
38. Y. Yamaguchi, T. Takagi, T. Wada, K. Yano, A. Furuya, S. Sugimoto, J. Hasegawa, and H. Handa, NELF, a multisubunit complex containing RD, cooperates with DSIF to repress RNA polymerase II elongation. *Cell* **97**, 41–51 (1999).

39. M. J. Cismowski, G. M. Laff, M. J. Solomon, and S. I. Reed, KIN28 encodes a C-terminal domain kinase that controls mRNA transcription in Saccharomyces cerevisiae but lacks cyclin-dependent kinase-activating kinase (CAK) activity. *Mol. Cell. Biol.* **15**, 2983–2992 (1995).
40. J. M. Lee and A. L. Greenleaf, CTD kinase large subunit is encoded by CTK1, a gene required for normal growth of Saccharomyces cerevisiae. *Gene Expr.* **1**, 149–167 (1991).
41. S. Murray, R. Udupa, S. Yao, G. Hartzog, and G. Prelich, Phosphorylation of the RNA polymerase II carboxy-terminal domain by the Bur1 cyclin-dependent kinase. *Mol. Cell. Biol.* **21**, 4089–4096 (2001).
42. M. F. Dubois, V. T. Nguyen, M. E. Dahmus, G. Pages, J. Pouyssegur, and O. Bensaude, Enhanced phosphorylation of the C-terminal domain of RNA polymerase II upon serum stimulation of quiescent cells: possible involvement of MAP kinases. *EMBO J.* **13**, 4787–4797 (1994).
43. A. Venetianer, M.-F. Dubois, V. T. Nguyen, S. Bellier, S.-J. Seo, and O. Bensaude, Phosphorylation state of the RNA polymerase II C-terminal domain (CTD) in heat-shocked cells: Possible involvement of the stress-activated mitogen-activated protein (MAP) kinases. *Eur. J. Biochem.* **233**, 83–92 (1995).
44. S. Bellier, M. F. Dubois, E. Nishida, G. Almouzni, and O. Bensaude, Phosphorylation of the RNA polymerase II largest subunit during Xenopus laevis oocyte maturation. *Mol. Cell. Biol.* **17**, 1434–1440 (1997).
45. L. J. Cisek and J. L. Corden, Phosphorylation of RNA polymerase by the murine homologue of the cell-cycle control protein cdc2. *Nature* **339**, 679–684 (1989).
46. M. M. Gebara, M. H. Sayre, and J. L. Corden, Phosphorylation of the carboxy-terminal repeat domain in RNA polymerase II by cyclin-dependent kinases is sufficient to inhibit transcription. *J. Cell. Biochem.* **64**, 390–402 (1997).
47. R. Roy, J. P. Adamczewski, T. Seroz, W. Vermeulen, J.-P. Tassan, L. Schaeffer, E. A. Nigg, J. H. J. Hoeijmakers, and J.-M. Egly, The MO15 cell cycle kinase is associated with the TFIIH transcription-DNA repair factor. *Cell* **79**, 1093–1101 (1994).
48. S. Trigon, H. Serizawa, J. W. Conaway, R. C. Conaway, S. P. Jackson, and M. Morange, Characterization of the residues phosphorylated in vitro by different C-terminal domain kinases. *J. Biol. Chem.* **273**, 6769–6775 (1998).
49. Y. Ramanathan, S. M. Rajpara, S. M. Reza, E. Lees, S. Shuman, M. B. Mathews, and T. Pe'ery, Three RNA polymerase II carboxyl-terminal domain kinases display distinct substrate preferences. *J. Biol. Chem.* **276**, 10913–10920 (2001).
50. J. Zhang and J. L. Corden, Identification of phosphorylation sites in the repetitive carboxyl-terminal domain of the mouse RNA polymerase II largest subunit. *J. Biol. Chem.* **266**, 2290–2296 (1991).
51. M. Zhou, M. A. Halanski, M. F. Radonovich, F. Kashanchi, J. Peng, D. H. Price, and J. N. Brady, Tat modifies the activity of CDK9 to phosphorylate serine 5 of the RNA polymerase II carboxyl-terminal domain during human immunodeficiency virus type 1 transcription. *Mol. Cell. Biol.* **20**, 5077–5086 (2000).
52. R. S. Chambers and C. M. Kane, Purification and characterization of an RNA polymerase II phosphatase from yeast. *J. Biol. Chem.* **271**, 24498–24504 (1996).
53. M. S. Kobor, J. Archambault, W. Lester, F. C. Holstege, O. Gileadi, D. B. Jansma, E. G. Jennings, F. Kouyoumdjian, A. R. Davidson, R. A. Young, and J. Greenblatt, An unusual eukaryotic protein phosphatase required for transcription by RNA polymerase II and CTD dephosphorylation in S. cerevisiae. *Mol. Cell.* **4**, 55–62 (1999).
54. O. Flores, I. Ha, and D. Reinberg, Factors involved in specific transcription by mammalian RNA polymerase II. Purification and subunit composition of transcription factor IIF. *J. Biol. Chem.* **265**, 5629–5634 (1990).

55. J. W. Conaway and R. C. Conaway, A multisubunit transcription factor essential for accurate initiation by RNA polymerase II. *J. Biol. Chem.* **264**, 2357–2362 (1989).
56. O. Flores, H. Lu, M. Killeen, J. Greenblatt, Z. F. Burton, and D. Reinberg, The small subunit of transcription factor IIF recruits RNA polymerase II into the preinitiation complex. *Proc. Natl. Acad. Sci. USA* **88**, 9999–10003 (1991).
57. M. Killeen, B. Coulombe, and J. Greenblatt, Recombinant TBP, transcription factor IIB, and RAP30 are sufficient for promoter recognition by mammalian RNA polymerase II. *J. Biol. Chem.* **267**, 9463–9466 (1992).
58. Y. Li, P. M. Flanagan, H. Tschochner, and R. D. Kornberg, RNA polymerase II initiation factor interactions and transcription start site selection. *Science* **263**, 805–807 (1994).
59. J. Archambault, G. Pan, G. K. Dahmus, M. Cartier, N. Marshall, S. Zhang, M. E. Dahmus, and J. Greenblatt, FCP1, the RAP74-interacting subunit of a human protein phosphatase that dephosphorylates the carboxyl-terminal domain of RNA polymerase IIO. *J. Biol. Chem.* **273**, 27593–27601 (1998).
60. S. M. Fang and Z. F. Burton, RNA polymerase II-associated protein (RAP) 74 binds transcription factor (TF) IIB and blocks TFIIB-RAP30 binding. *J. Biol. Chem.* **271**, 11703–11709 (1996).
61. M. S. Kobor, L. D. Simon, J. Omichinski, G. Zhong, J. Archambault, and J. Greenblatt, A motif shared by TFIIF and TFIIB mediates their interaction with the RNA polymerase II carboxy-terminal domain phosphatase Fcp1p in Saccharomyces cerevisiae. *Mol. Cell. Biol.* **20**, 7438–7449 (2000).
62. J. Archambault, R. S. Chambers, M. S. Kobor, Y. Ho, M. Cartier, D. Bolotin, B. Andrews, C. M. Kane, and J. Greenblatt, An essential component of a C-terminal domain phosphatase that interacts with transcription factor IIF in Saccharomyces cerevisiae. *Proc. Natl. Acad. Sci. USA* **94**, 14300–14305 (1997).
63. H. Cho, T. K. Kim, H. Mancebo, W. S. Lane, O. Flores, and D. Reinberg, A protein phosphatase functions to recycle RNA polymerase II. *Genes Dev.* **13**, 1540–1552 (1999).
64. J. F. Collet, V. Stroobant, M. Pirard, G. Delpierre, and E. Van Schaftingen, A new class of phosphotransferases phosphorylated on an aspartate residue in an amino-terminal DXDX(T/V) motif. *J. Biol. Chem.* **273**, 14107–14112 (1998).
65. P. Bork, K. Hofmann, P. Bucher, A. F. Neuwald, S. F. Altschul, and E. V. Koonin, A superfamily of conserved domains in DNA damage-responsive cell cycle checkpoint proteins. *FASEB J.* **11**, 68–76 (1997).
66. R. M. Taylor, B. Wickstead, S. Cronin, and K. W. Caldecott, Role of a BRCT domain in the interaction of DNA ligase III-alpha with the DNA repair protein XRCC1. *Curr. Biol.* **8**, 877–880 (1998).
67. S. M. Liao, J. Zhang, D. A. Jeffery, A. J. Koleske, C. M. Thompson, D. M. Chao, M. Viljoen, H. J. J. van Vuuren, and R. A. Young, A kinase-cyclin pair in the RNA polymerase II holoenzyme. *Nature* **374**, 193–196 (1995).
68. S. Bellier, S. Chastant, P. Adenot, M. Vincent, J. P. Renard, and O. Bensaude, Nuclear translocation and carboxyl-terminal domain phosphorylation of RNA polymerase II delineate the two phases of zygotic gene activation in mammalian embryos. *EMBO J.* **16**, 6250–6262 (1997).
69. D. B. Bregman, L. Du, S. van der Zee, and S. L. Warren, Transcription-dependent redistribution of the large subunit of RNA polymerase II to discrete nuclear domains. *J. Cell. Biol.* **129**, 287–298 (1995).
70. M. J. Mortillaro, B. J. Blencowe, X. Wei, H. Nakayasu, L. Du, S. L. Warren, P. A. Sharp, and R. Berezney, A hyperphosphorylated form of the large subunit of RNA polymerase II is associated with splicing complexes and the nuclear matrix. *Proc. Natl. Acad. Sci. USA* **93**, 8253–8257 (1996).

71. J. G. Gall, Cajal bodies: the first 100 years. *Annu. Rev. Cell Dev. Biol.* **16**, 273–300 (2000).
72. G. T. Morgan, O. Doyle, C. Murphy, and J. G. Gall, RNA polymerase II in Cajal bodies of amphibian oocytes. *J. Struct. Biol.* **129**, 258–268 (2000).
73. J. G. Gall, A role for Cajal bodies in assembly of the nuclear transcription machinery. *FEBS Lett.* **498**, 164–167 (2001).
74. Y. Hirose, R. Tacke, and J. L. Manley, Phosphorylated RNA polymerase II stimulates pre-mRNA splicing. *Genes Dev.* **13**, 1234–1239 (1999).
75. C. K. Ho and S. Shuman, Distinct roles for CTD Ser-2 and Ser-5 phosphorylation in the recruitment and allosteric activation of mammalian mRNA capping enzyme. *Mol. Cell.* **3**, 405–411 (1999).
76. C. R. Rodriguez, E.-J. Cho, M.-C. Keogh, C. L. Moore, A. L. Greenleaf, and S. Buratowski, Kin28, the TFIIH-associated carboxy-terminal domain kinase, facilitates the recruitment of mRNA processing machinery to RNA polymerase II. *Mol. Cell. Biol.* **20**, 104–112 (2000).
77. S. C. Schroeder, B. Schwer, S. Shuman, and D. Bentley, Dynamic association of capping enzymes with transcribing RNA polymerase II. *Genes Dev.* **14**, 2435–2440 (2000).
78. N. F. Marshall and D. H. Price, Control of formation of two distinct classes of RNA polymerase II elongation complexes. *Mol. Cell. Biol.* **12**, 2078–2090 (1992).
79. A. Shilatifard, The RNA polymerase II general elongation complex. *Biol. Chem.* **379**, 27–31 (1998).
80. D. Reines, R. C. Conaway, and J. W. Conaway, Mechanism and regulation of transcriptional elongation by RNA polymerase II. *Curr. Opin. Cell Biol.* **11**, 342–346 (1999).
81. J. W. Conaway and R. C. Conaway, Transcription elongation and human disease. *Annu. Rev. Biochem.* **68**, 301–319 (1999).
82. Z. Xie and D. Price, Drosophila factor 2, an RNA polymerase II transcript release factor, has DNA-dependent ATPase activity. *J. Biol. Chem.* **272**, 31902–31907 (1997).
83. Z. Xie and D. H. Price, Unusual nucleic acid binding properties of factor 2, an RNA polymerase II transcript release factor. *J. Biol. Chem.* **273**, 3771–3777 (1998).
84. T. J. Fu, J. Peng, G. Lee, D. H. Price, and O. Flores, Cyclin K functions as a CDK9 regulatory subunit and participates in RNA polymerase II transcription. *J. Biol. Chem.* **274**, 34527–34530 (1999).
85. J. Peng, M. Liu, J. Marion, Y. Zhu, and D. H. Price, RNA polymerase II elongation control. *Cold Spring Harb. Symp. Quant. Biol.* **63**, 365–370 (1998).
86. T. Wada, T. Takagi, Y. Yamaguchi, A. Ferdous, T. Imai, S. Hirose, S. Sugimoto, K. Yano, G. A. Hartzog, F. Winston, S. Buratowski, and H. Handa, DSIF, a novel transcription elongation factor that regulates RNA polymerase II processivity, is composed of human Spt4 and Spt5 homologs. *Genes Dev.* **12**, 343–356 (1998).
87. T. O'Brien, S. Hardin, A. Greenleaf, and J. T. Lis, Phosphorylation of RNA polymerase II C-terminal domain and transcriptional elongation. *Nature* **370**, 75–77 (1994).
88. A. E. Rougvie and J. T. Lis, The RNA polymerase II molecule at the 5' end of the uninduced hsp70 gene of D. melanogaster is transcriptionally engaged. *Cell* **54**, 795–804 (1988).
89. D. L. Bentley, Regulation of transcriptional elongation by RNA polymerase II. *Curr. Opin. Genet. Dev.* **5**, 210–216 (1995).
90. R. F. Chun and K. T. Jeang, Requirements for RNA polymerase II carboxyl-terminal domain for activated transcription of human retroviruses human T-cell lymphotropic virus I and HIV-1. *J. Biol. Chem.* **271**, 27888–27894 (1996).
91. H. Okamoto, C. T. Sheline, J. L. Corden, K. A. Jones, and B. M. Peterlin, Trans-activation by human immunodeficiency virus Tat protein requires the C-terminal domain of RNA polymerase II. *Proc. Natl. Acad. Sci. USA* **93**, 11575–11579 (1996).
92. C. A. Parada and R. G. Roeder, Enhanced processivity of RNA polymerase II triggered by Tat-induced phosphorylation of its carboxy-terminal domain. *Nature* **384**, 375–378 (1996).

93. X. Yang, C. H. Herrmann, and A. P. Rice, The human immunodeficiency virus Tat proteins specifically associate with TAK in vivo and require the carboxyl-terminal domain of RNA polymerase II for function. *J. Virol.* **70,** 4576–4584 (1996).
94. S. H. Chao, K. Fujinaga, J. E. Marion, R. Taube, E. A. Sausville, A. M. Senderowicz, B. M. Peterlin, and D. H. Price, Flavopiridol inhibits P-TEFb and blocks HIV-1 replication. *J. Biol. Chem.* **275,** 28345–28348 (2000).
95. J. Peng, Y. Zhu, J. T. Milton, and D. H. Price, Identification of multiple cyclin subunits of human P-TEFb. *Genes Dev.* **12,** 755–762 (1998).
96. P. Wei, M. E. Garber, S. M. Fang, W. H. Fischer, and K. A. Jones, A novel CDK9-associated C-type cyclin interacts directly with HIV-1 Tat and mediates its high-affinity, loop-specific binding to TAR RNA. *Cell* **92,** 451–462 (1998).
97. J. Wimmer, K. Fujinaga, R. Taube, T. P. Cujec, Y. Zhu, J. Peng, D. H. Price, and B. M. Peterlin, Interactions between Tat and TAR and human immunodeficiency virus replication are facilitated by human cyclin T1 but not cyclins T2a or T2b. *Virology* **255,** 182–189 (1999).
98. Y. H. Ping and T. M. Rana, DSIF and NELF interact with RNA polymerase II elongation complex and HIV-1 Tat stimulates P-TEFb-mediated phosphorylation of RNA polymerase II and DSIF during transcription elongation. *J. Biol. Chem.* **276,** 12951–12958 (2001).
99. P. Licciardo, G. Napolitano, B. Majello, and L. Lania, Inhibition of Tat transactivation by the RNA polymerase II CTD-phosphatase FCP1. *AIDS* **15,** 301–307 (2001).
100. P. Rickert, J. L. Corden, and E. Lees, Cyclin C/CDK8 and cyclin H/CDK7/p36 are biochemically distinct CTD kinases. *Oncogene* **18,** 1093–1102 (1999).
101. S. McCracken, N. Fong, E. Rosonina, K. Yankulov, G. Brothers, D. Siderovski, A. Hessel, S. Foster, S. Shuman, and D. L. Bentley, 5′-Capping enzymes are targeted to pre-mRNA by binding to the phosphorylated carboxy-terminal domain of RNA polymerase II. *Genes Dev.* **11,** 3306–3318 (1997).
102. C. K. Ho, V. Sriskanda, S. McCracken, D. Bentley, B. Schwer, and S. Shuman, The guanylyltransferase domain of mammalian mRNA capping enzyme binds to the phosphorylated carboxyl-terminal domain of RNA polymerase II. *J. Biol. Chem.* **273,** 9577–9585 (1998).
103. Z. Yue, E. Maldonado, R. Pillutla, H. Cho, D. Reinberg, and A. J. Shatkin, Mammalian capping enzyme complements mutant Saccharomyces cerevisiae lacking mRNA guanylyltransferase and selectively binds the elongating form of RNA polymerase II. *Proc. Natl. Acad. Sci. USA* **94,** 12898–12903 (1997).
104. E. J. Cho, T. Takagi, C. R. Moore, and S. Buratowski, mRNA capping enzyme is recruited to the transcription complex by phosphorylation of the RNA polymerase II carboxy-terminal domain. *Genes Dev.* **11,** 3319–3326 (1997).
105. A. Yuryev, M. Patturajan, Y. Litingtung, R. V. Joshi, C. Gentile, M. Gebara, and J. L. Corden, The C-terminal domain of the largest subunit of RNA polymerase II interacts with a novel set of serine/arginine-rich proteins. *Proc. Natl. Acad. Sci. USA* **93,** 6975–6980 (1996).
106. E. Kim, L. Du, D. B. Bregman, and S. L. Warren, Splicing factors associate with hyperphosphorylated RNA polymerase II in the absence of pre-mRNA. *J. Cell Biol.* **136,** 19–28 (1997).
107. Y. Hirose and J. L. Manley, RNA polymerase II is an essential mRNA polyadenylation factor. *Nature* **395,** 93–96 (1998).
108. J. Hani, B. Schelbert, A. Bernhardt, H. Domdey, G. Fischer, K. Wiebauer, and J.-U. Rahfeld, Mutations in a peptidylprolyl-cis/trans-isomerase gene lead to a defect in 3′-end formation of a pre-mRNA in Saccharomyces cerevisiae. *J. Biol. Chem.* **274,** 108–116 (1999).
109. M. Schutkowski, A. Bernhardt, X. Z. Zhou, M. Shen, U. Reimer, J.-U. Rahfeld, K. P. Lu, and G. Fischer, Role of phosphorylation in determining the backbone dynamics of the serine/threonine-proline motif and Pin1 substrate recognition. *Biochemistry* **37,** 5566–5575 (1998).

110. J.-P. Bourquin, I. Stagljar, P. Meier, P. Moosmann, J. Silke, T. Baechi, O. Georgiev, and W. Schaffner, A serine/arginine-rich nuclear matrix cyclophilin interacts with the C-terminal domain of RNA polymerase II. *Nucleic Acids Res.* **25,** 2055–2061 (1997).
111. D. P. Morris, H. P. Phatnani, and A. L. Greenleaf, Phospho-carboxyl-terminal domain binding and the role of a prolyl isomerase in pre-mRNA 3′-end formation. *J. Biol. Chem.* **274,** 31583–31587 (1999).
112. X. Wu, C. B. Wilcox, G. Devasahayam, R. L. Hackett, M. Arévalo-Rodríguez, M. E. Cardenas, J. Heitman, and S. D. Hanes, The Ess1 prolyl isomerase is linked to chromatin remodeling complexes and the general transcription machinery. *EMBO J.* **19,** 3727–3738 (2000).
113. C. E. Birse, L. Minvielle-Sebastia, B. A. Lee, W. Keller, and N. J. Proudfoot, Coupling termination of transcription to messenger RNA maturation in yeast. *Science* **280,** 298–301 (1998).
114. M. J. Dye and N. J. Proudfoot, Terminal exon definition occurs cotranscriptionally and promotes termination of RNA polymerase II. *Mol. Cell.* **3,** 371–378 (1999).
115. M. J. Dye and N. J. Proudfoot, Multiple transcript cleavage precedes polymerase releasein termination by RNA polymerase II. *Cell* **105,** 669–681 (2001).

Translational Control of Gene Expression: Role of IRESs and Consequences for Cell Transformation and Angiogenesis

ANNE-CATHERINE PRATS
AND HERVÉ PRATS

*Institut National de la Santé et de la
Recherche Médicale U397
Endocrinologie et Communication
Cellulaire
Institut Fédératif de Recherche Louis
Bugnard
C.H.U. Rangueil
31403 Toulouse Cedex 04, France*

I. Introduction 369
II. Background: Translation Initiation in Eukaryotes 371
   A. The Scanning Model of Translation Initiation 371
   B. The Alternatives to the Scanning Mechanism 372
   C. Messengers with Alternative Initiations of Translation 377
III. The Murine Leukemia Virus: The Retroviral Genomic mRNA Codes for
   Two Gag–Pol Polyproteins from Two CUG and AUG Initiation Codons .. 378
   A. Gross Cell Surface Antigen Synthesis Is Initiated at a CUG
      Initiation Codon 378
   B. The Function of an Alternative Initiation Process: The GCSA
      Is Required for MuLV Dissemination and Pathogenicity 380
   C. The MuLV Genomic mRNA Contains an IRES Located between
      the CUG and AUG Start Codons 380
   D. Identification of PTB as a Putative Mo-MuLV IRES *Trans*-Acting
      Factor.. 382
   E. Generalization of IRES-Dependent Translation to Other
      Oncoretroviruses and Lentiviruses............................. 382

Abbreviations: ATR, alternatively translated region; CAT, chloramphenicol acetyl transferase; CrPV, cricket paralysis virus; eIF, eukaryotic initiation factor; EMCV, encephalomyocarditis virus; FGF, fibroblast growth factor; FMDV, foot-and-mouth disease virus; F-MuLV, Friend murine leukemia virus; GCSA, gross cell surface antigen; HIV, human immunodeficiency virus; HMW, high molecular weight; HTLVI, human T lymphocyte leukemia virus; IRES, internal ribosome entry site; ITAF, IRES *trans*-acting factor; Mo-MuLV, Moloney murine leukemia virus; NLS, nuclear localization sequence; ORF, open reading frame; PABP, poly(A)-binding protein; pCBP, poly(C)-binding protein; PDGF, platelet-derived growth factor; PTB, pyrimidine-tract-binding protein; RRL, rabbit reticulocyte lysate; SIV, simian immunodeficiency virus; TE, translational enhancer; TGF, transforming growth factor; UTR, untranslated region; VEGF, vascular endothelial growth factor; WGE, wheat germ extract; XIAP, X-linked inhibitor of apoptosis.

IV. The FGF-2 mRNA: How Cap-Dependent and IRES-Dependent
   Translation of a Single mRNA Leads to Expression of Five Isoforms
   with Distinct Localizations and Functions......................... 383
   A. A Process of Alternative Initiation of Translation Using Three CUG
      Start Codons in the FGF-2 mRNA Leads to the Synthesis of Several
      Isoforms.. 383
   B. The FGF-2 Isoforms Have Different Localizations and Functions.... 385
   C. Different Modes of Action of the FGF-2 Isoforms................. 385
   D. A Fifth FGF-2 Isoform, Synthesized from an Upstream CUG
      Codon on the Same mRNA, Behaves as a Survival Factor........... 386
   E. Translational Regulation of FGF-2 Expression by *Cis*-Acting
      Elements in the mRNA 5'-UTR................................ 387
   F. Identification of an IRES in the FGF-2 mRNA.................... 387
   G. Regulation of Expression of the Different FGF-2 Isoforms
      in Transformed and Stressed Cells............................ 390
   H. Characterization of FGF-2 IRES-Binding *Trans*-Acting Factors...... 391
   I. FGF-2 mRNA Expression Is Controlled by Alternative Poly(A)
      Sites, an RNA Destabilization Element, and a Translational
      Enhancer Present in the 3'-UTR.............................. 392
V. VEGF mRNA: Two Distinct IRESs Control Alternative Initiation of the
   Translation of Two Isoforms...................................... 393
   A. Identification of Two Distinct IRESs in the VEGF mRNA.......... 393
   B. Identification of a New VEGF Isoform, L-VEGF, Initiated at a
      CUG Codon.. 394
VI. C-*myc* mRNA: A Natural Multicistronic Messenger.................. 396
   A. Identification of an IRES in the P2 C-*myc* mRNA................ 397
   B. Identification of a Second C-*myc* IRES in the P0 Leader.......... 397
VII. IRES-Dependent Translational Control: IRES Activity Enhancement
    in Transformed Cells... 398
VIII. IRES Tissue Specificity *in Vivo*................................. 399
IX. FGF-2 Translational Silencing by Tumor Suppressor p53............. 400
X. Concluding Remarks... 403
   References.. 405

Translational control of gene expression has, over the last 10 years, become appreciated as an important process in its regulation in eukaryotes. Among a series of control mechanisms exerted at the translational level, the use of alternative codons provides a very subtle means of increasing gene diversity by expressing several proteins from a single mRNA. The internal ribosome entry sites (IRESs) act as specific translational enhancers that allow translation initiation to occur independently of the classic cap-dependent mechanism, in response to specific stimuli and under the control of different *trans*-acting factors. It is striking to observe that the two processes mostly concern genes coding for control proteins such as growth factors, protooncogenes, angiogenesis factors, and apoptosis regulators. Here, we focus on the translational regulation of four mRNAs, with both IRESs and alternative initiation codons, which are the messengers of retroviral murine leukemia virus, fibroblast growth factor 2, vascular endothelial growth factor, and protooncogene c-*myc*. Four of them are involved in cell transformation and/or angiogenesis, with important consequences for such translation regulations in these pathophysiological processes. © 2002, Elsevier Science (USA).

I. Introduction

For decades scientific knowledge about gene expression was dominated by two main concepts. The first was that eukaryotic gene expression followed the rule "one gene, one protein," in that one transcriptional unit coded for a single open reading frame, resulting in a single gene product. The second important concept was that the regulation of gene expression essentially occurred at the transcriptional level, as documented by numerous studies of the regulation of gene transcription.

Today we know that neither of these notions is the absolute truth. On one hand the idea of one gene encoding only one protein has been disproven in many cases where a single gene is able to generate several protein products. This was first discovered for viruses such as the Sendai virus and the murine leukemia virus (MuLV) (1, 2). The use of noncanonical translation initiation codons ACG or CUG present on a single mRNA was shown to be responsible for the synthesis of several distinct proteins, either in overlapping reading frames, as in the Sendai virus, or in the same reading frame, as shown for MuLV *gag* gene products (1, 2). In both cases the process of alternative initiation of translation gave rise to gene products having distinct functions in the viral replication cycle, demonstrating that a process that was obviously translational rather than transcriptional had important consequences for gene expression.

Such a process might have been of limited interest if it had been restricted to viruses. However, it appeared that the use of noncanonical codons also occurred in cellular genes: The first examples of alternative initiations of translation using noncanonical codons were the c-*myc* protooncogene and the fibroblast growth factor 2 (3–5). In both cases the noncanonical CUG codon(s) were shown to produce protein isoforms with additional N-terminal domains conferring on these isoforms new functions, distinct from those of the AUG-initiated gene products. The process of alternative initiation of translation was shown later to extend to many cellular genes, using either noncanonical codons, mainly CUG and ACG, or alternative AUG codons. Overlapping genes, well described in several viruses, were also discovered in cellular genes (6, 7).

The main consequence of the discovery of an alternative initiation of the translation process, and to a lesser extent overlapping genes, was an unexpected increase in gene diversity that had not been predicted by computer analysis, which was only able to detect AUG-initiated open reading frames in messenger

RNAs. Although we are going to focus here on translational regulation, we must mention that such an increase in gene diversity has also been observed for other posttranscriptional gene regulation processes, the most striking being splicing. Many examples of alternative splicing have been discovered, giving rise to alternative mRNA isoforms resulting in synthesis of protein isoforms with distinct functions. The posttranscriptional control of gene expression is thus a major cause of gene diversity increase and provides part of the explanation of how the human organism is able to be so complex with only 30,000 genes. In eukaryotic organisms the most important aspect is probably not the number of genes, but their complexity, providing several gene products from a single gene.

The discovery of posttranscriptional gene complexity also had important consequences for the concept of mainly transcriptional gene expression regulation. Whereas regulation of transcription remains the major mechanism governing gene expression in prokaryotes, it has appeared during the last decade that posttranscriptional regulations play a pivotal role in the control of gene expression in eukaryotes. First, the processes of alternative initiation of translation and alternative splicing generate a gene diversity that is completely independent of the transcriptional process. Second, eukaryotic mRNAs have been shown to contain untranslated regions able to control the efficiency of protein synthesis. The efficiency of gene expression appeared to be controlled at different posttranscriptional levels, including mRNA polyadenylation, stability, and translation. The *cis*-acting mRNA elements controlling such regulations are mainly located in the 5' untranslated or alternatively translated region (UTR or ATR, respectively) and in the 3'-UTR, whereas few examples of regulation of mRNA stability by coding regions have been described (for a review, see Ref. 8).

Strikingly, the first examples of posttranscriptionally controlled genes in eukaryotic cells were genes for transcription factors, cytokines, and growth factors. Cytokine mRNAs, as well as protooncogene mRNAs such as c-*myc* or c-*fos*, were shown to contain AU-rich elements in their mRNA 3'-UTRs, which were responsible for RNA destabilization or translation blockade (8, 9). On the other hand, the mRNAs of c-*myc*, transforming growth factor-β (TGF-β), fibroblast growth factor 2 (FGF-2), and others were shown to contain translational regulatory elements in their 5'-UTRs or ATR (in the case of FGF-2). In yeast, a subtle translation regulation mechanism involving several upstream μORFs (open reading frames) was described for transcription factor GCN4 (*10*).

In addition, an important event in the understanding of translational control was the discovery of the so-called internal ribosome entry sites (IRESs). Until the end of the 1980s, the unchallenged model was that eukaryotic mRNAs could only be translated with the so-called scanning mechanism (*11*). The hypothesis was that ribosome recruitment could only occur upon recognition of the capped 5' mRNA end by the initiation factor eIF4F, followed by the ribosome scanning along the RNA molecule until it encountered an initiation codon. This dogma was in fact simultaneously discredited in two laboratories when it was found that poliovirus and encephalomyocarditis virus (EMCV) mRNAs were translated in a cap-independent manner by an internal ribosome entry mechanism involving an mRNA element called ribosome landing pad or internal ribosome entry site, respectively (12, 13). The scanning mechanism rule was also questioned in the case of cellular mRNAs following the discovery of an IRES in the Bip mRNA, which codes for an immunoglobulin chaperon of the heat shock protein family (14). This posed an interesting question: Why should a cellular capped mRNA be translated by a cap-independent mechanism? One hypothesis was that such a mechanism could allow the translation of specific cellular mRNAs in conditions where cap-dependent translation is blocked, as would occur under stress conditions and during the M phase of the cell cycle (15). This suggested that IRESs could be found in mRNAs coding for proteins involved in a stress response (like Bip) or in the control of cell growth. Several IRESs were indeed identified in cellular mRNAs over the following years, starting with *Drosophila* antennapedia transcription factor and human FGF-2 mRNAs (16, 17). Several cellular mRNA IRESs were found during the following years, always in mRNAs coding for regulatory proteins, including protooncogenes, growth factors, apoptosis factors, and angiogenesis factors. Initiation of IRES-dependent translation seems to concern, once again, genes involved in the control of cell proliferation and differentiation and whose aberrant expression is often related to cell transformation and cancerous tumor progression.

We shall focus here on the hypothesis that a link exists between the translational control of gene expression and the appearance of cancer. The hypothesis will be illustrated by the detailed analysis of several translation regulation model systems, one retroviral mRNA, the Moloney murine leukemia virus mRNA, and three cellular mRNAs, which have been studied in our laboratory for about 15 years. The cellular mRNAs code for two major angiogenic factors, FGF-2 and vascular endothelial growth factor (VEGF), and for the oncoprotein c-myc. Strikingly, all four mRNAs are subjected to alternative initiations of translation and contain IRESs.

II. Background: Translation Initiation in Eukaryotes

A. The Scanning Model of Translation Initiation

Until the end of the 1980s, the unchallenged model was that eukaryotic mRNAs could only be translated according to the so-called scanning mechanism (11, 18). This canonical model envisaged the recruitment at the capped 5' end of the mRNA of the 43S complex composed of a ribosome 40S subunit bound to eIF1A and eIF3 initiation factors and to eIF2 complexed to GTP and initiator Met-tRNA$_i$ (19, 20). 43S recruitment occurs upon recognition of the 5′ capped mRNA by the eIF4E subunit of the initiation factor eIF4F, a process that is greatly enhanced by the poly(A)-binding protein bound to the mRNA 3′ poly(A) tail (21). EIF4F is composed of three subunits: (1) eIF4E, the cap-binding protein, (2) eIF4A, an RNA helicase, and (3) eIF4G, a large polypeptide acting as a molecular scaffold binding and coordinating the activity of eIF3, eIF4A, eIF4E, and PABP. The attachment stage is followed by ribosome scanning in the 5′-to-3′ direction upon the synergistic action of eIF1 and eIF1A, along the RNA molecule, which is unwound by eIF4A and eIF4B, until it encounters an initiation codon. The initiation codon recognition, corresponding to a base pairing of the mRNA initiation codon with the Met-tRNA$_i$ anticodon, is mediated by eIF1, eIF2, and eIF5, whose binding results in the 48S ribosome complex (20). Finally, the ribosome 60S subunit is recruited upon the action of eIF5B, resulting in formation of the 80S ribosome (22). This is the end of the translation initiation process and the beginning of protein synthesis.

The classical scanning model also predicts that the AUG triplet that will be used is the most proximal to the 5′ end of the mRNA, if this AUG is located in a favorable context, that is, ACCAUGG (23). However, if the AUG is in an unfavorable context (pyrimidine at positions −3 and +4), ribosome scanning can continue until the next downstream AUG. This process is called leaky scanning.

B. The Alternatives to the Scanning Mechanism

1. THE DISCOVERY OF INTERNAL RIBOSOME ENTRY SITES IN VIRAL mRNAs

Whereas it was claimed that the only mechanism was cap-dependent scanning, as it had been reported that a ribosome was not able to bind to a circular RNA (24), such a mechanism could not explain the efficient translation initiation of picornavirus mRNAs, which are uncapped and contain 800- to 1000-nucleotide (nt)-long 5′ untranslated regions with numerous AUG start codons upstream of the authentic start codon (12, 13). Furthermore, such viruses produce a protease that cleaves an eIF4F subunit, the initiation factor 4G, thereby generating a shut-off of the cellular cap-dependent translation while the synthesis of viral proteins remains unaffected. This suggested, at the end of the 1980s, that the cap-dependent scanning mechanism was not a general rule for translation initiation. The scanning mechanism rule was broken when it was found that poliovirus and EMCV mRNAs were translated in a cap-independent manner by an internal ribosome entry mechanism involving an mRNA element called IRES (12, 13).

Strikingly, internal initiation of translation of these viral mRNAs was shown to involve cellular proteins. The first found to be involved in IRES activity were p52 and p57, identified later as the La autoantigen and the splicing factor pyrimidine tract-binding protein (PTB), for poliovirus and EMCV, respectively (*25, 26*). Whereas the role of La autoantigen appears to be nonspecific and thus remains unclear, PTB has been cited for its activating effect on EMCV IRES activity (*27*). This protein is also able to bind to other IRESs or 5′-UTRs, but this does not seem to be related to an intrinsic ability of the protein to activate IRES-dependent translation. In human T-lymphocyte leukemia virus (HTLV-I) IRES, this can be explained by a very different dissociation constant, as shown by surface plasmon resonance measurement and/or by the requirement for additional IRES *trans*-acting factors (ITAF) (*28*). This has been reported in foot-and-mouse disease virus (FMDV) IRES, which requires both PTB and ITAF45, a protein expressed in proliferating cells (*29*). As regards poliovirus, its IRES is highly activated by both PTB and the poly(rC)-binding protein 2 (pCPB-2) (*30*). In the case of human rhinovirus IRES, a biochemical study identified that the IRES was synergistically activated by PTB and by a complex between the cold-shock-domain protein *unr* (upstream of N-*ras*) and the WD-repeat protein *unrip* (*31*).

Although most IRESs require ITAFs for optimal activity, canonical initiation factors are also required. The minimum set of factors required for recruitment of the 40S subunit of the EMCV initiation codon has been determined by biochemical reconstitution of the initiation complex *in vitro* (*32*). This complex formation requires the same factors as classical initiation except for eIF4E, PABP, or large N-terminal and C-terminal fragments of eIF4G, and is ATP-dependent (for a review see Ref. *33*). Furthermore, initiation factor requirement has been shown to differ from one IRES to the another. Toeprinting experiments have even shown, in hepatitis C and classical swine fever virus IRESs, that AUG initiation factor is positioned at or near the P site of the bound 40S subunits in the absence of any eIF. In such IRESs resembling prokaryotic translation initiation sites, the RNA structure may form a scaffold presenting conserved unpaired regions as ribosome-binding sites to assemble functional initiation complexes.

Translation initiation by an IRES-dependent mechanism seems to occur quite frequently in the viral world: Numerous IRESs have been found not only in all picornaviruses, flaviviruses, and pestiviruses studied, but also in herpesvirus and retroviruses, including murine, avian, and human oncoretro- or lentiviruses (*34–42*). Mutagenesis and comparison of the different picornavirus IRESs provided better insight into the internal ribosome entry mechanism. IRES involves the secondary and/or tertiary structure of RNA rather than the primary sequence (*34*). This structural element allows ribosome recruitment at an AUG triplet located in the 3′ end of the structure. The picornavirus IRESs are divided into three groups: in cardiovirus (EMCV), the ribosome recruiting codon is used as a start codon; in entero/rhinovirus (poliovirus), ribosome entry is followed by scanning and initiation occurs at an AUG downstream; whereas in aphtovirus, initiation occurs both at the 3′ end of the IRES and at an AUG downstream. A fourth class of IRES is found with hepatitis C and pestivirus IRESs, which require a portion of the coding sequence in order to function (43).

A fifth class of IRES has been recently described in the insect cricket paralysis-like virus (CrPV). CrPV belongs to a picorna-like insect virus family having RNA genomes, which have been revealed to be naturally dicistronic, translation of the two ORFs being mediated by two distinct IRESs (44–46). The striking feature of the CrPV IRES is that it promotes translation initiation in the absence of any cognate or weak cognate start codon (such as CUG, GUG, etc.). In fact, translation starts at CAA or GCU triplets and does not require initiator Met-tRNA. Thus, the new concept has been proposed that, in the case of such an IRES, translation initiation starts from the A site of the ribosome (47).

2. The Discovery of Internal Ribosome Entry Sites in Cellular mRNAs

Although cellular mRNAs are capped and thus assumed to be translated by a cap-dependent mechanism, the involvement of cellular proteins in the regulation of picornavirus IRES activity suggests that certain cellular mRNAs could be translated by an IRES-dependent mechanism. Actually, the first IRES element discovered in cellular mRNA was in Bip mRNA, which codes for an immunoglobulin chaperone showing a continuous activity in poliovirus-infected cells despite the strong inhibition of the host cell mRNA translation (14). The first hypothesis to be proposed was that such a mechanism could allow mRNA translation in response to stress, viral infection, G2/M transition in the cell cycle, etc. Such conditions are associated with the blockade of cap-dependent translation (48–50). Evidence of IRES-regulated gene expression during development has been provided by the discovery of IRES in *Drosophila* antennapedia mRNA, coding for a developmentally regulated transcription factor (16).

The next mammalian gene reported to contain an IRES in its mRNA was FGF-2 (17), which, in addition to its major angiogenic role, is more generally involved in the control of cell growth and differentiation and in the repair of tissue lesions. This discovery was followed by the description of IRESs in several other growth factors and transcription factors: platelet-derived growth factor 2 (PDGF2) (c-*sis*), c-*myc*, and VEGF-A (51–57). More recently, IRESs have been discovered in numerous cellular mRNAs (for a review, see Ref. 58 and our web site, http://www.rangueil.inserm.fr/iresdatabase), mostly including genes exhibiting a control function in the cell, primarily the control of gene expression, but also cell communication, cell cycle, or signal transduction. The important function of the genes containing an IRES suggests a physiological role of these translational regulatory elements, even though such a role remains to be demonstrated for most IRES-containing messengers. However, a physiological function for IRESs has been reported for a small number of messengers (Fig. 1). PDGF

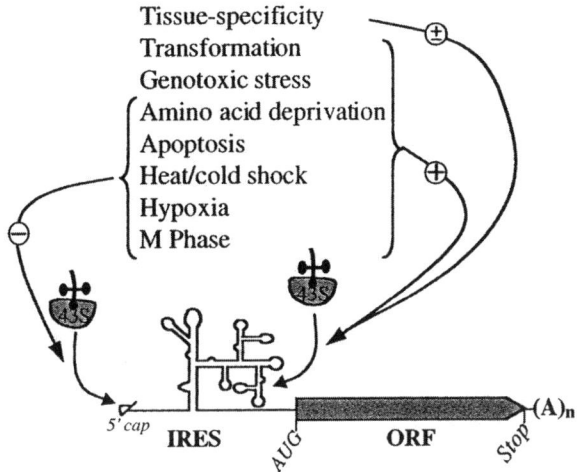

FIG. 1. Cap- and IRES-dependent translation initiation in response to different stimuli. A capped IRES-containing cellular mRNA is represented, with different stimuli reported to activate cellular IRESes while inhibiting cap-dependent translation. These stimuli include a series of stresses as well as the G2/M transition during the cell cycle (references are given in the text). Tissue specificity can result either in IRES activation or in IRES silencing. As regards cell transformation, the literature indicates that this process is accompanied by activation of both cap-dependent and IRES-dependent translation initiation. The increase or decrease of ribosome recruitment is represented by arrows marked with plus or minus signs.

IRES is activated during megakaryocytic differentiation (51); ornithine decarboxylase and the kinase p58PITSLRE mRNAs are translationally regulated in an IRES-dependent manner during the cell cycle (59, 60); translation of several messengers is also activated or maintained in response to stress in an IRES-dependent manner: the VEGF IRES allows translation of VEGF mRNA after hypoxia stimulation (57); FGF-2 and BAG-1 mRNAs are translationally activated upon heat shock (61, 62); the cold-stress induced mRNA Rbm3 exhibits a cold-shock-sensitive IRES (63); the IRES of the cat-1 arginine transporter mRNA is activated upon amino acid starvation (64); and several pro- or antiapoptotic mRNAs exhibit an enhanced IRES-dependent translation upon apoptosis induction: this is the case of the antiapoptotic X-linked inhibitor of apoptosis (XIAP) gene (65) and the proapoptotic genes DAP5 (belonging to the eIF4G family) and c-myc (66, 67). Recently, the c-myc IRES was also shown to mediate translation following genotoxic stress (68). These different observations indicate that IRESs are translation regulatory elements that permit the cell to provide a rapid answer to various exogenous stimuli. This also prompts us to state the hypothesis that a given stimulus might be accompanied by the coordinate stimulation of a family of messengers having a related or complementary function.

Another argument favoring a crucial role of IRESs in the control of gene expression has been provided by the observation that IRES-dependent translation is deregulated in transformed cells. This has been shown for FGF-2, where translation regulation was found to be cell-density-dependent in normal cells, but not in transformed cells (61, 69). Interestingly, c-*myc* has also been reported as translationally upregulated in Bloom's syndrome cells, where the c-*myc* IRES contains a natural point mutation (70, 71). This naturally mutated IRES exhibits an enhanced activity, apparently resulting from the increased binding of a cellular protein to the c-*myc* mRNA (72). Actually, most cellular IRESs possess a significant activity, albeit with different efficiencies, in transiently transfected cell lines that mostly correspond to transformed or at least immortalized cells (73, 74) (L. Creancier and A. C. Prats, unpublished). In contrast, the *in vivo* behavior of cellular IRESs is completely different: In the FGF-2 and c-*myc* IRESs, whose activities have been tested in transgenic mice, a stringent regulation is observed, as these IRESs are essentially active in the embryo, whereas they are silenced or highly tissue-specific in adult mice (73, 74).

3. VIRAL AND CELLULAR IRESs EXHIBIT SIGNIFICANTLY DIFFERENT PHYSIOLOGICAL FUNCTIONS

Recent data obtained *in vivo* have revealed strong differences between cellular IRESs and the picornaviral EMCV IRES, which showed a high activity in embryonic as well as in adult organs (73, 74). This may reflect a strong difference between cellular and viral IRESs, related to the function of the genes controlled by the IRES-dependent translation. The IRES of a pleiotropic virus such as poliovirus or EMCV needs to be highly active in most cells, independently of their proliferating or quiescent, differentiated, or undifferentiated state (the result will be cell death in all cases), in order to obtain efficient viral replication. In contrast, in addition to other transcriptional or posttranscriptional regulatory processes, the FGF-2 or c-*myc* IRESs must control the expression of genes whose deregulation will have drastic pathophysiological consequences. An important question that remains concerning the control of IRES tissue specificity is whether the IRESs need specific factors to be activated, as shown for the FMDV IRES, which is activated by the proliferating cell-specific factor ITAF 45 (29), or whether IRESs are silenced by the presence of inhibitory factors, as was recently shown for the FGF-2 IRES, which is downregulated by p53 (75, 76).

4. TRANSLATION INITIATION BY RIBOSOME SHUNT

Another important process breaking the rule of solely ribosome scanning is ribosome shunting. This was first shown to occur with virus mRNAs, the cauliflower mosaic virus 35S mRNA and the adenovirus late mRNA (77, 78). Stable RNA structures were shown to (1) prevent ribosome scanning from the mRNA 5' end and (2) promote ribosome shunting to a downstream landing site involving, at least in adenovirus, sequences complementary to the 3' end of the ribosome 18S RNA (79). As was the case of other unusual regulatory processes described in viruses, ribosome shunting has also been observed in the cellular mRNA encoding heat shock protein 70 (79). We can assume that the future will reveal additional examples of ribosome shunting in cellular mRNAs and that the RNA structures involved in the ribosome jump will provide new elements for the translational regulation of cellular gene expression.

C. Messengers with Alternative Initiations of Translation

The combination of both cap-dependent leaky scanning and IRES-dependent internal entry processes results in increased possibilities of translation initiation when different initiation codons are present. Whereas few examples of overlapping genes have been described, the most frequent phenomenon corresponds to the use of several initiation codons in the same reading frame. This results in proteins exhibiting a common C-terminal part with different N-terminal domains. Strikingly, most of the studied mRNAs with alternative initiation codons have revealed the presence of an IRES. However, the position of the IRES with respect to the start codons can vary. In the Moloney murine leukemia virus mRNA, the IRES is positioned between the CUG and AUG start codons, giving rise to the two different *gag* polyproteins (36). Here the IRES activity can regulate the ratio of the two CUG- and AUG-initiated proteins. Actually, the IRES of the kinase PITSLRE is also located between the two alternative AUG codons (59). One might expect that the IRES must be systematically located between the initiation codons, but reality is different: for instance, in the c-*myc* mRNA the IRES is located upstream from the two CUG and AUG start codons (52, 53). The FGF-2 mRNA, with its five initiation codons (four CUGs and one AUG), is more complex (80). However the IRES is located between the first and the second CUG (called CUG0 and CUG1, respectively) (61). Consequently, CUG0 is the only codon used by the cap-dependent mechanism exclusively (80). A second IRES has been suspected between the CUGs and the AUG codons, but has been impossible to identify, probably because of an overlap of two (putative) IRES structures (L. Creancier *et al.*, unpublished). The VEGF mRNA, which is simpler as regards the number of start codons, has two IRESs, each controlling the synthesis of a CUG- or AUG-initiated isoform (81).

The list of the cellular mRNAs exhibiting alternative initiations of translation in the same reading frame for which an IRES has been identified includes the estrogen receptor, neuroblastoma specific N-*myc*, and antiapoptotic protein BAG-1 mRNAs (62, 82, 83). The list is expected to increase in the near future. It can be noticed that all of these mRNAs code for regulatory proteins. It is probably not a coincidence that the two features are often found in a single mRNA. Combining the two features of containing one or several IRESs and several initiation codons provides subtle possibilities for the regulation of gene expression at the translational level. This is precisely the aspect that we develop below by detailing how translational regulation occurs for four mRNAs that we have studied for the last 15 years. These messengers are Moloney murine leukemia virus, FGF-2, VEGF-A, and c-*myc* mRNAs. All four are related to cancerogenesis, by acting either on the cell transformation process or on tumoral angiogenesis or both. The description of our studies of the translational regulation of these mRNAs will support the crucial role of translational control of gene expression in such pathophysiological processes.

III. The Murine Leukemia Virus: The Retroviral Genomic mRNA Codes for Two Gag–Pol Polyproteins from Two CUG and AUG Initiation Codons

The RNA genome of retroviruses is also a premessenger and a messenger (*84*). These different functions are controlled by regulatory elements located in the RNA 5′ untranslated region (5′-UTR), a sequence extending to several hundred nucleotides. These regulatory elements control crucial steps of the viral cycle, including reverse transcription, splicing, RNA dimerization, encapsidation, and translation. The murine leukemia virus (MuLV) genomic mRNA has the interesting feature of coding for two precursors of the group-specific antigen (gag), Pr65gag and Pr75gag, *in vivo*. Pr65gag is the precursor of the virion structural proteins found in all retroviruses. In contrast, Pr75gag is not involved in virus assembly, but is glycosylated and anchored in the infected cell membrane, where it is known as the gross cell surface antigen (GCSA) (*85*). The existence of such a second gag precursor, unique among retroviruses, is shared by the MuLV and the feline leukemia virus (*86*). In the middle of the 1980s, nothing was known about the translation initiation site or the function of this protein.

A. Gross Cell Surface Antigen Synthesis Is Initiated at a CUG Initiation Codon

The 8-kb Moloney murine leukemia virus (Mo-MuLV) genomic mRNA exhibits a 621-nt-long 5′-UTR upstream from the Pr65gag AUG initiation codon (Fig. 2). The GCSA has been described as resulting from an upstream initiation of translation, although no in-frame AUG codon is present to explain such an initiation. At that time translation initiation at non-AUG codons had been described only in two cases, adenovirus-associated virus 2 and the Sendai virus. In both cases the noncanonical codon was an ACG (*1, 87*). Such a codon was not present in-frame in the Mo-MuLV 5′-UTR (*84*). However, analysis of the 5′-UTR sequence revealed a CUG codon in a favorable context, ACCUGG

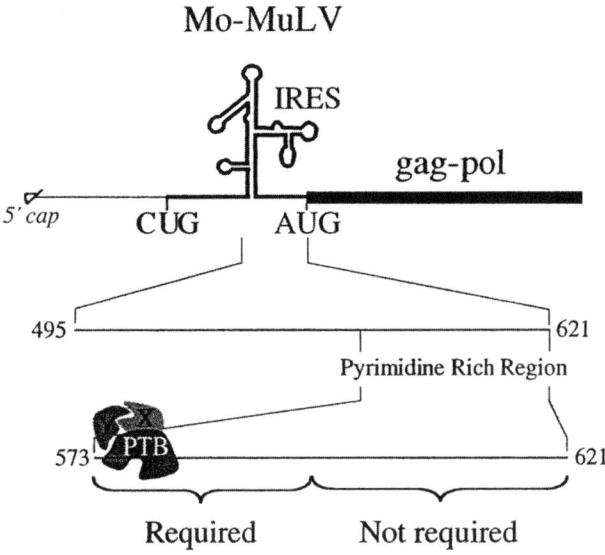

FIG. 2. Alternative initiations of translation in the Mo-MuLV genomic mRNA. The 5′ part of the Mo-MuLV genomic mRNA is represented with the gag–pol ORF. The AUG codon is the start codon of the Pr65gag precursor and of the gag–pol precursor. CUG is the start codon of the Pr75gag, precursor of the GCSA, which forms a major antigen on the membrane of infected cells. The IRES located between the two initiation codons is shown. Below, the minimal IRES fragment (nt 495–621) is enlarged to show the pyrimidine-rich region located just upstream from the AUG codon: The 5′ part of this pyrimidine tract contains the PTB-binding site and is required for IRES activity, whereas the downstream part is not required (36).

(23), at position 357 from the mRNA 5′ end. Using a site-directed mutagenesis approach, we were able to demonstrate for the first time, by translation *in vitro* in rabbit reticulocyte lysate, that a CUG codon is used for translation initiation of GCSA synthesis (Fig. 2) (2, 88). This discovery was just concomitant with the publication of c-*myc* alternative initiation at a non-AUG codon (4), which proved to be the first of a whole list of cellular mRNAs subjected to the alternative translation initiation process.

The following years revealed that the CUG codon, used in several cellular mRNAs including FGF-2, FGF-3, VEGF, tumor suppressor WT1, protooncogenes c-*myc*, N-*myc*, L-*myc*, and *pim1*, antiapoptotic gene *BAG-1*, and tyrosine kinase *hck*, is actually the most frequently found noncanonical start codon in eukaryotes (3, 5, 55, 81, 89–96). The discovery of the CUG start codon of the MoMuLV Pr75gag (GCSA) played a pivotal role in the emergence of a new concept in gene expression control (2). Translational control of gene expression by the use of alternative start codons increases gene diversity and maintains an appropriate ratio of two functionally different proteins coded by a single mRNA.

B. The Function of an Alternative Initiation Process: The GCSA Is Required for MuLV Dissemination and Pathogenicity

We have also shown that the CUG start codon is conserved in several MuLV strains, including the Friend MuLV (F-MuLV) (2). Thus, in order to study the function of the CUG-initiated GCSA, we used the F-MuLV retrovirus, known to induce erythroleukemia preceded by severe hemolytic anemia when administrated to new-born mice.

The CUG start codon was mutated into a noninitiator CGG codon and a complete provirus bearing this mutation was reconstituted and used for NIH 3T3 cell transfection and analysis of virus dissemination. We showed that the infectivity of the GCSA mutant virus was 100-fold lower than that of the wild type when analyzed 5 days after cell transfection with the proviral plasmid (97). However, this difference disappeared after 7 days. Suspecting that the *in vitro* analysis did not allow a correct evaluation of the mutant virus behavior, probably because of the appearance of revertant viruses, we infected newborn mice with the GCSA-mutant F-MuLV and analyzed both viral dissemination and appearance of the pathological effects. These experiments showed clearly that the mutant F-MuLV not only exhibited a 1-week-delayed spread in animals, but in addition did not induce hemolytic anemia, whereas the latency of erythroleukemia was significantly increased (97). Furthermore, analysis of the virus present in the erythroleukemic mice revealed the presence of a revertant virus able to produce the GCSA (97). This was the demonstration of the GCSA requirement for the F-MuLV pathological effects and spread, and of the functional significance of the CUG alternative initiation codon in viral replication and pathogenesis.

C. The MuLV Genomic mRNA Contains an IRES Located between the CUG and AUG Start Codons

We looked for the presence of an IRES in the Mo-MuLV start codon by two different strategies expected to block the cap-dependent translation initiation. The first approach consisted in the addition of a stable hairpin at the mRNA 5' end to block ribosome binding and scanning; the second approach corresponded to the bicistronic vector strategy previously described for the identification of the picornavirus and Bip IRESs (Fig. 3) (14). The principle of this strategy is that the cap-dependent mechanism of translation initiation only allows expression of the first cistron of a bicistronic mRNA: The second cistron is expressed only if an IRES is present in the intercistronic region. To avoid the possibility of ribosome reinitiation at the end of the first cistron, we used the system described by Macejak and Sarnow (14) of adding a stable hairpin in the 5' end of the mRNA. Such a hairpin is expected to downregulate expression of the first

A   Monocistronic mRNA with a 5′ hairpin

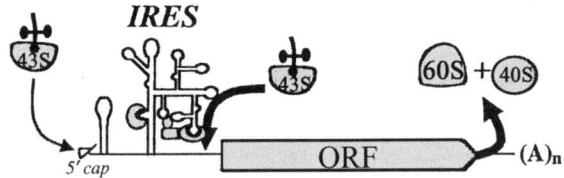

B   Bicistronic mRNA with or without a 5′ hairpin

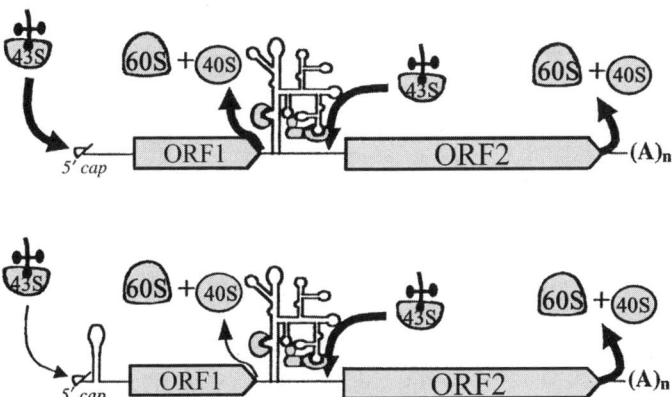

FIG. 3. Two approaches used for IRES characterization. (A) Monocistronic vector with a 5′ hairpin: The principle is to insert a stable hairpin at the 5′ end of the mRNA, in order to block the cap-dependent translation initiation. The remaining expression of the mRNA results from IRES-dependent initiation. This approach requires the precise calibration of the ratio of protein expression to mRNA level. (B) Bicistronic vectors with or without a 5′ hairpin: The principle of this approach is to express two ORFs from a single mRNA. The cap-dependent mechanism drives the translation of the first ORF, but not of the second one, unless there is a reinitiation process. To evaluate the possibility of reinitiation, a hairpin is inserted 5′ of the bicistronic mRNA expected to decrease the level of cap-dependent translation: IRES-dependent expression of the second cistron must not be affected by this hairpin. The advantage of the system is that IRES activity can be evaluated independently of the mRNA level in the cells, by calculating the ratio of second ORF versus first ORF expression.

cistron, which is cap-dependent, but not that of the second cistron if it is IRES-dependent (Fig. 3). In both cases (mono- and bicistronic vectors) the 5′ first 671 nucleotides of the MuLV sequence were fused to a chloramphenicol acetyl-transferase (CAT) coding sequence, under the control of the cytomegalovirus promoter (36). Such a construct was expected to express both the CUG and the AUG-initiated MuLV–CAT fusion isoforms. In the bicistronic construct, the CAT gene was introduced as the first (cap-dependent) cistron and the MuLV–CAT fusion was the second (IRES-dependent) cistron (36).

The two approaches (mono- and bicistronic) gave the same conclusion: the Mo-MuLV leader is able to promote internal ribosome entry. However, this IRES allows initiation of translation exclusively from the AUG start codon, but not from the CUG. Further deletions and site-directed mutagenesis in the leader sequence confirmed that the IRES is indeed located between the two initiation codons, more precisely, within the 126-nt fragment located upstream from the AUG codon (36). This region is pyrimidine-rich and at that time it had been reported that the IRES of several picornaviruses involved an oligopyrimidine tract, starting with UUUUC, located about 25 nt upstream from the start codon (98). Actually, an oligopyrimidine tract was present at this position in the Mo-MuLV leader, but was shown to be dispensable for IRES activity. In contrast, another pyrimidine-rich tract present at about 50 nt upstream from the AUG codon was shown to be required for the IRES to function (Fig. 3).

D. Identification of PTB as a Putative Mo-MuLV IRES Trans-Acting Factor

The identification of the pyrimidine tract-binding protein (PTB, also called hnRNPI) and La autoantigen involved in the internal entry process of EMCV and poliovirus (25, 26) prompted us to look for *trans*-acting factors controlling the MuLV IRES activity. By ultraviolet (UV) crosslinking experiments we detected six proteins able to bind to the MuLV IRES, migrating at 63, 57, 55, 52, 44, and 38 kDa, respectively (36). Interestingly, the 57-kDa protein was displaced by addition of EMCV IRES and did not bind to the inactive IRES mutated for the −50 oligopyrimidine tract. This favored a role of PTB in the Mo-MuLV IRES activity.

E. Generalization of IRES-Dependent Translation to Other Oncoretroviruses and Lentiviruses

We found an IRES between the two start codons of the Mo-MuLV genomic mRNA. However, another study reported that F-MuLV genomic mRNA contains an IRES located upstream from the CUG codon (35). Such a difference could result from different properties of the F-MuLV and Mo-MuLV 5'-UTRs. However, we consider that it is highly improbable, given the strong homology between the 5'-UTRs of these two retroviruses. It seems more likely that both initiations occur in an IRES-dependent manner, and that the MuLV mRNA contains two different IRESs to control the alternative translation initiation processes. The differences between the two reports describing MuLV IRESs could reflect some specificity of these IRES activities or the translation system used for the experiments (35, 36).

In any event, the discovery of an IRES in Mo-MuLV and F-MuLV suggested that such a process could be generalized to other retroviruses. Thus, with the same approach, we looked for such a possibility in several primate retroviruses, including an oncoretrovirus, human T-lymphocyte leukemia virus (HTLV-I), and three lentiviruses, human immunodeficiency virus 1 and 2 (HIV-1 and HIV-2) and simian immunodeficiency virus (SIV). The bicistronic approach allowed us to conclude that all these viruses contain an IRES in their mRNA 5'-UTR (*42*), which was confirmed by several reports (*37, 40, 41*).

We can conclude from these different studies that the retroviruses, which all have a long 5'-UTR expected to prevent an efficient cap-dependent translation, have their genomic mRNA translated (exclusively or not) by an IRES-dependent mechanism.

In an attempt to determine the nature of the retroviral ITAFs we analyzed the binding of PTB and La autoantigen to these different retroviral mRNA IRESs in comparison to picornaviral IRESs (*28*). In addition to UV crosslinking experiments, which are not really quantitative, we developed a new method for calculating the affinity of RNA–protein interactions, based on the BIAcore technology (surface plasmon resonance). This allowed us to demonstrate that PTB exhibits affinity only for oncoretroviral, but not for lentiviral IRESs. The PTB dissociation constant was 1000-fold greater for oncoretrovirus than for picornavirus IRESs, a feature that was not detectable from the UV crosslinking experiments. Furthermore, addition of PTB in *in vitro* translation experiments weakly enhanced translation initiation of mRNAs containing the HTLV-I IRES. This suggests that PTB might be an oncoretrovirus ITAF, but probably needs other, coworking ITAFs to promote internal initiation of translation. In contrast, we can clearly conclude that PTB is not an ITAF for lentiviral IRESs (*28*).

IV. The FGF-2 mRNA: How Cap-Dependent and IRES-Dependent Translation of a Single mRNA Leads to Expression of Five Isoforms with Distinct Localizations and Functions

A. A Process of Alternative Initiation of Translation Using Three CUG Start Codons in the FGF-2 mRNA Leads to the Synthesis of Several Isoforms

Fibroblast growth factor 2 (FGF-2, also known as basic FGF) is a prototype member of an ever-increasing family that now includes 22 genes involved in the control of cell proliferation and differentiation, in particular during embryogenesis (*99, 100*; for review see Ref. *101*). FGF-2 is produced by numerous cell types and has a pleiotropic role. It stimulates endothelial cell growth and migration *in vitro*, whereas it promotes angiogenesis *in vivo*. This angiogenic factor is involved in wound-healing processes, but also in tumor vascularization (*102–104*). FGF-2 is also a neurotrophic factor involved in nerve cell survival, neuron differentiation, and regeneration of central nervous system injuries (*105–107*).

The various physiological roles of FGF-2 can be partly explained by the existence of different isoforms of the factor, which display different modes of action. Indeed, in the 1980s, FGF-2 was called basic FGF and was known as a single 18-kDa protein. However, when sequencing the FGF-2 polypeptide, one of the sequenced peptides corresponded to the predicted protein sequence deduced from a nucleotide located upstream from the 18-kDa AUG start codon (*5*). Furthermore, in looking at the FGF-2 mRNA nucleotide sequence, we observed the presence of a very long open reading frame in the 484-nt-long 5' untranslated region, with no in-frame AUG, but with three in-frame CUG codons located in a favorable translation initiation context. It turned out that *in vitro* transcription and translation of the FGF-2 mRNA led to three bands migrating more slowly than the expected 18-kDa isoform, that is, at 22, 22.5, and 24 kDa, called high molecular weight (HMW) FGF-2 (Fig. 4). As we had recently identified a CUG initiating codon in the Mo-MuLV genomic mRNA (*2*),

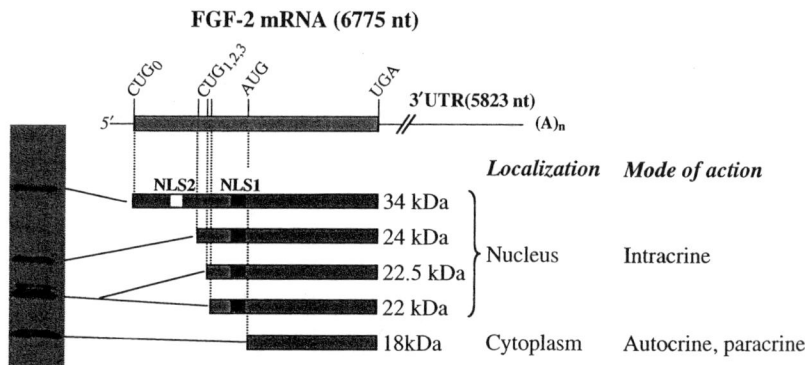

FIG. 4. FGF-2 isoform expression from alternative initations of translation. FGF-2 mRNA is represented with the five in-frame initiation codons: CUG0 (position 86), CUG1 (position 319), CUG2 (position 346), CUG3 (position 361), and AUG (position 484). The stop codon is also shown (UGA at position 950). The 3'-UTR maximal length is indicated (5823 nt). It can exhibit different sizes depending on the use of eight alternative poly(A) sites. The gray box represents the FGF-2 ORF. Beneath, the FGF-2 isoforms are schematized as boxes, and the presence of one or two NLS is shown. On the left, a Western immunoblot of endogenous FGF-2 (using anti-FGF-2 antibody) in SK-Hep-1 cells is shown, and each band is linked to the corresponding isoform. Shown on the right are the molecular weights of the five isoforms, their intracellular localizations, and their modes of action.

we performed site-directed mutagenesis of the first of these three CUGs and demonstrated by *in vitro* transcription and translation that this codon was responsible for translation initiation and synthesis of the 24-kDa isoform (5). This was confirmed by other authors, who showed, in addition, that the two other CUGs present in the cDNA corresponded to the start codons of the 22- and 22.5-kDa isoforms (3) (Fig. 4).

B. The FGF-2 Isoforms Have Different Localizations and Functions

At this stage, the main issue was to determine whether these different isoforms had distinct functions, which could provide a physiological basis for the presence of the three alternative CUG start codons. The first element which suggested the existence of different functions was the discovery of a nuclear localization of HMW FGF-2, resulting from the presence of a nuclear localization sequence (NLS) in their additional N-terminal domain (108, 109) (Fig. 4). In parallel, the FGF-2 isoforms were expressed separately in primary adult bovine endothelial cells (ABAE), using retroviral vectors. This revealed that the constitutive expression of such proteins in ABAE generated different phenotypes: Expression of the 18-kDa FGF-2 resulted in cell transformation (rendered them able to grow in soft agar), whereas that of the 24-kDa isoform generated an immortalized phenotype in ABAE (110). Interestingly, ABAE cells coexpressing all isoforms were not only both immortalized and transformed, but had a tumorigenic effect when injected in nude mice. Such a tumorigenic effect was observed only when all isoforms were coexpressed and not with an FGF-2 isoform alone, recalling an oncogene cooperation process. The original feature is, in this case, that the two cooperating gene products are coded by the same mRNA. The transforming power of FGF-2 18-kDa was shown in parallel by other authors, who overexpressed FGF-2 isoforms in NIH 3T3 mouse fibroblasts (111). In addition, it was shown a few years later that the 18-kDa FGF-2 was responsible for cell migration, whereas the 24-kDa FGF-2 presented the interesting feature of promoting cell radioresistance (112, 113).

C. Different Modes of Action of the FGF-2 Isoforms

The FGF-2 signal pathway was known to involve recognition of high- and low-affinity receptors by the growth factor, followed by classical signal transduction mediated by second messengers. The high-affinity receptors are tyrosine kinase receptors, whereas the low-affinity receptors are heparan sulfate proteoglycans (114, 115). When we started to study the mode of action of this growth factor by using iodinated recombinant FGF-2, we made the surprising discovery that the growth factor was translocated to the nucleolus, where it was shown to activate, by a direct interaction with the ribosomal DNA promoter, the synthesis of ribosomal RNA (*116, 117*). This was the first time that such a mode of action was described for a growth factor.

However, we had not finished with the noncanonical mode of action of FGF-2: It was shown by other authors, using dominant-negative FGF-2 receptors, that, as expected, the 18-kDa molecule had a paracrine and an autocrine mode of action (*118*). In contrast, the same report showed that HMW FGF-2 acts independently of receptor recognition, creating the new concept of an intracrine mode of action for a growth factor (Fig. 4). This could explain the different functions of FGF-2 isoforms and suggested that HMW FGF-2, instead of acting by a receptor-mediated pathway, could act by direct interaction with nuclear targets.

D. A Fifth FGF-2 Isoform, Synthesized from an Upstream CUG Codon on the Same mRNA, Behaves as a Survival Factor

We believed we had finished with possibilities of new translation initiations in FGF-2 mRNA, but further studies of its expression revealed that this was not the case: By *in vitro* translation (in rabbit reticulocyte lyase, RRL) of FGF-2 mRNA we detected a faint band migrating at 34 kDa, a band which disappeared upon deletion of the 5' part of the mRNA leader sequence (*80*). Constructs expressing FGF–CAT protein fusions allowed us to confirm, in transiently transfected cells, that an upstream translation initiation was indeed occurring in the FGF-2 mRNA 5' region. A detailed analysis of endogenous FGF-2 expression in different cell types revealed the presence of a band migrating at 34 kDa. By using a polyclonal antibody directed against the expected additional N-terminal domain of the putative 34-kDa FGF-2, we demonstrated that such an isoform was indeed present endogenously in cells (*80*). Site-directed mutagenesis of putative start codons in the FGF-2 mRNA 5' led us to identify the 34-kDa FGF-2 initiation codon: a CUG codon located 86 nt from the mRNA 5' end, which was called CUG0 (Fig. 4).

The localization and function of the new FGF-2 isoform were investigated and revealed interesting features. Using a chimeric construct of the 34-kDa FGF-2 N-terminal domain with CAT, we demonstrated, by immunolocation and subcellular fractionation, that this domain contains a NLS, in addition to that present in the HMW FGF-2 (initiated at CUGs 1, 2, and 3; Fig. 4). Furthermore, mutagenesis of the nucleotide sequence allowed the new NLS to be characterized as a noncanonical, Arg-rich NLS resembling that of HIV *rev* protein. Such an NLS has been described as directly interacting with importin-β and could account for a regulation of 34-kDa FGF-2 nuclearization distinct from that of the other FGF-2 nuclear isoforms (*119*). This Arg-rich domain could also be responsible for interaction of FGF-2 with nuclear targets that may differ from the target of the so-called HMW FGF-2 of 22, 22.5, and 24 kDa.

TRANSLATIONAL CONTROL OF GENE EXPRESSION   387

Does this additional domain confer a new function to the 34-kDa FGF-2? The question was addressed by transfecting NIH 3T3 fibroblasts with a series of vectors expressing each of the different FGF-2 isoforms. Permanent clones were analyzed to determine their ability to proliferate in low-serum conditions. Results showed that only the clones expressing the 34-kDa FGF-2 were able to survive and grow in such culture conditions, whereas the clones expressing either the 18- or the 24-kDa FGF-2 did not. Thus, the 34-kDa isoform exhibits a specific cell survival activity (*80*).

E. Translational Regulation of FGF-2 Expression by *Cis*-Acting Elements in the mRNA 5'-UTR

At the beginning of the 1990s, very little information was available as regards the translational regulation of gene expression, but the presence of long 5' and 3' untranslated regions in many messengers posed the question of the function of such noncoding RNA elements. The features of human FGF-2 mRNA were particularly striking, with a 484-nt-long leader region (mostly composed of alternatively translated regions rather than untranslated regions) and a 5823-nt-long 3'-UTR (Fig. 4) (*5*). In total more than 90% of the mRNA was untranslated or alternatively translated. This suggested the presence of *cis*-acting regulatory elements. Using a deletion approach, we analyzed the presence of such regulatory elements in the 5' region of the FGF-2 mRNA, both *in vitro*, by translation in RRL and wheat germ extract (WGE), and in transiently transfected COS-7 cells. The results revealed five regulatory elements, one of which (position 192–256) acted as an inhibitory element in the wheat germ extract specifically (Fig. 5, W). Another element (A), located just upstream from the CUG1 codon, had an effect in RRL and COS-7, but not in WGE (*120*). Three other elements, B, C and D, were involved in the control of alternative initiation of translation in the three systems (*120*). Interestingly, these elements corresponded to features of the predicted secondary structure of the FGF-2 mRNA leader (Fig. 5). The different activities of the regulatory elements depending on the translation system suggested the presence of cell-specific *trans*-acting factors able to regulate translation.

F. Identification of an IRES in the FGF-2 mRNA

According to Kozak's model of the cap-dependent ribosome scanning mechanism, it was expected that the GC-rich leader would have an inhibitory effect on translation. However, such an effect of the 319-nt-long region located upstream from the CUG1 codon was observed only in WGE, but not in RRL or in COS-7 cells (*120*). This suggested the existence of another mechanism to account for this translation in the two latter systems. As the mechanism of ribosome internal entry had been recently described for picornavirus, we looked at

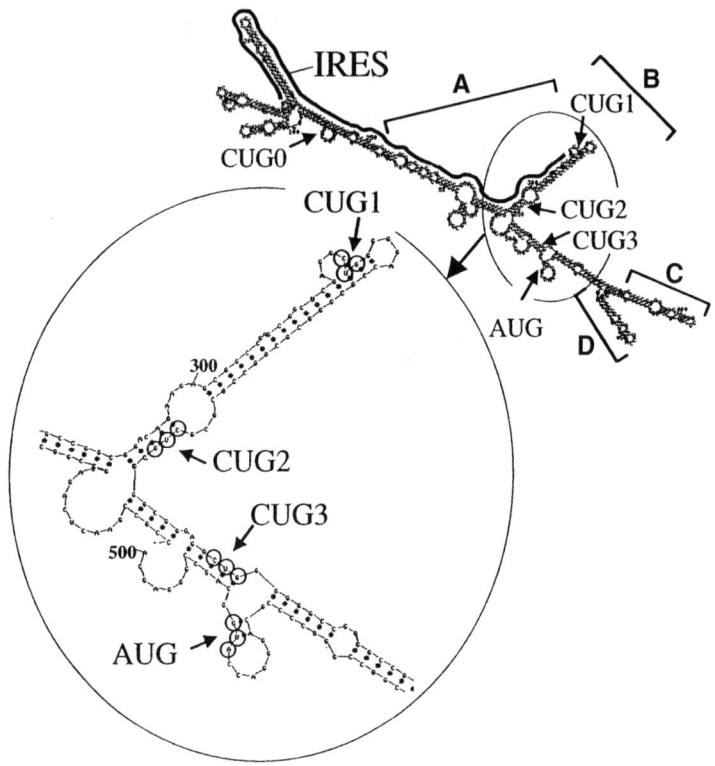

FIG. 5. *In silico* prediction of the FGF-2 mRNA leader secondary structure. Nucleotides 1–500 of the FGF-2 mRNA have been folded using the Zuker RNA folding program. The five initiation codons are indicated: CUG0 (position 86), CUG1 (position 320), CUG2 (position 347), CUG3 (position 362), and AUG (position 485) (*80*). The IRES, located between nt 154 and 319, is represented by a bold line. The domains A, B, C, and D, shown to be involved in the control of the alternative initiation of translation are shown: A, nt 257–313; B, nt 313–342 ; C, nt 386–412; D, nt 412–473 (*120*). W (nt 193–257) is not shown, as it is located within the IRES (*17, 120*). The circle represents an enlargement of the structural domain containing the four initiations codons (CUG1–3 and AUG) that are used by the IRES-dependent mechanism. CUG0 is used only by the cap-dependent mechanism.

the possibility of cap-independent translation and showed that the W element allows cap-independent translation in RRL (*17*). Thus, as identification of an IRES in BiP mRNA had just been published, we used bicistronic vectors to characterize the presence of an IRES in FGF-2 mRNA. The principle of the bicistronic approach has been described above (Fig. 3). It allowed us to identify an IRES within the 150-nt fragment located upstream from CUG-1 (Fig. 6) (*17*). The IRES coincides with the W element that was inhibitory in the WGE,

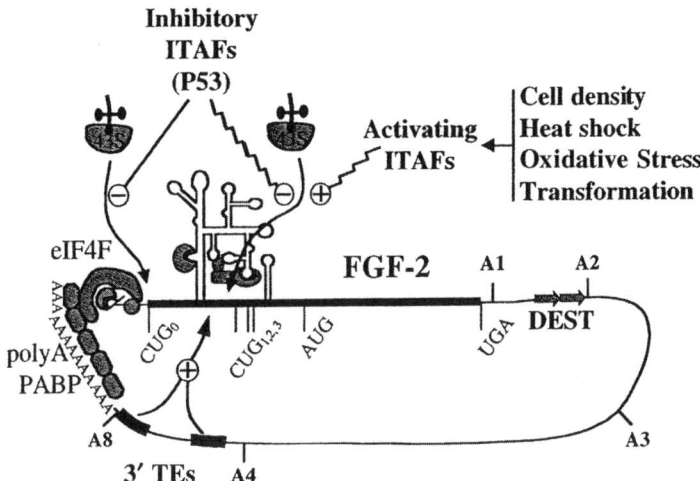

FIG. 6. Posttranscriptional regulation of FGF-2 expression. FGF-2 mRNA is represented with the different cis-acting elements present in the 5′-UTR and ATR as well as the 3′-UTR, which are responsible for posttranscriptional regulation of expression. In the 5′ region, these elements include five initiation codons responsible for the synthesis of different isoforms, the IRES allowing expression of FGF-2 from CUG1–3 and AUG, but not from CUG0 (which is cap-dependent). In the 3′-UTR, elements regulate expression at the three levels of RNA polyadenylation, stability, and translation, due to the presence of eight alternative poly(A) sites (represented by A1, A2, . . .), a destabilization element (DEST) corresponding to an AU-rich tandem repeat between A1 and A2, and a translational enhancer (3′ TE) composed of two elements between A4 and A8, which specifically activates initiation at the CUG1–3 codons. Translation initiation can occur by two alternative mechanisms: (1) the cap-dependent mechanism, as shown by the interaction of eIF4F with cap. Interaction between PABP and eIF4G (a subunit of eIF4F) results in cross-talk between the 3′ polyadenylated end of the mRNA and the 5′ cap. This cross-talk also exists with IRESs (but has not yet been demonstrated for FGF-2 IRES). The cap-dependent mechanism drives translation initiation at CUG0, but can also be responsible for initiation at the other CUGs. (2) The IRES-dependent mechanism, mediated by the IRES present upstream from the CUG1 codon. IRES activity is regulated by positive or negative ITAFs, according to tissue specificity and to exogenous stimuli (heat shock, oxidative stress, cell transformation, tissue specificity). P53 has been identified as a negative ITAF. P60 and P110, specifically binding to FGF-2 mRNA 5′, are putative positive ITAFs that remain to be identified.

providing an explanation for the absence of translation inhibition in RRL and COS-7, where translation occurs in a cap-independent manner (Fig. 5).

While attempting to pinpoint the role of the FGF-2 IRES in the control of alternative initiation of translation, we found that this IRES (located upstream from the CUG1 codon, in contrast to that of Mo-MuLV located between the CUG and the AUG codons) could not directly control the ratio of CUG- versus AUG-initiated forms. We showed that initiation of translation of the CUG1–3- and AUG-initiated forms is IRES-dependent, whereas that of the

CUG0-initiated form (34-kDa FGF-2) is strictly cap-dependent (80). Such data suggest that the presence of an IRES has a role in maintaining a favorable ratio of FGF-2 HMW 18-kDa isoforms (initiated at CUG1–3 and AUG) versus the 34-kDa FGF-2.

G. Regulation of Expression of the Different FGF-2 Isoforms in Transformed and Stressed Cells

Our approach to looking for a possible regulation of expression of FGF-2 isoforms was to analyze the endogenous status of different cell types for FGF-2 isoforms by Western immunoblotting using FGF-2 antibodies. This allowed us to detect different profiles of the four isoforms, depending on the cell type. The CUG-initiated forms were detected in most human transformed cell lines, whereas different normal cell types (human skin fibroblasts, retinal pigmentary epithelial cells, human aorta endothelial cells) mostly expressed the AUG-initiated form (61). Transfection of these cells by constructs expressing FGF–CAT fusion proteins showed a similar difference of isoform expression according to the cell type, indicating that this regulation is mediated by the 5′ region (containing the IRES) of the FGF-2 mRNA. Interestingly, both the CUG0- and CUG1–3-initiated forms were detected in several transformed cells including HeLa, SK-Hep-1, and A431. This suggested that both the cap-dependent translation and the IRES-dependent translation were aberrantly upregulated in transformed cells. Actually, it had been published that overexpression of the cap-binding protein eIF4E leads to cell transformation (121). In contrast, increased IRES activity in transformed cells provided a new concept.

This hypothesis was addressed by looking at the regulation of HMW FGF-2 expression in primary and transformed human skin fibroblasts. Indeed, we observed that expression of the CUG-initiated forms is regulated in normal skin fibroblasts according to cell density. HMW FGF-2 expression is detected at low cell density in proliferating cells and decreases at high cell density (69). In contrast, such expression is constitutive in transformed cell types. Thus, we transformed the primary skin fibroblasts by transfecting them with the large T SV40 antigen. This experiment resulted in the disappearance of density-dependent, CUG-initiated-form regulation and in their constitutive expression independently of the cell density (69).

We also looked for stress conditions able to affect FGF-2 isoform expression, given the fact that cap-dependent initiation of translation is inhibited in stress conditions (48–50). Human primary skin fibroblasts were subjected to heat shock at 45°C for 15–60 min. This resulted in induction of HMW (CUG-initiated) FGF isoform expression in a time-dependent manner (61). Furthermore, we showed that induction of CUG-initiated forms also occurs in response to oxidative stress (61). Although we have not directly shown that the CUG-initiated-form induction involves the IRES, such a hypothesis is most probable, as it has long been known that the heat shock response is accompanied by a blockade of cap-dependent translation (*48*) (Fig. 6).

H. Characterization of FGF-2 IRES-Binding *Trans*-Acting Factors

A small number of IRES *trans*-acting factors (ITAFs) has been characterized in recent years as regulating the viral IRES activities. These factors include PTB, La autoantigen, Unr, pCBP, and ITAF45 (*27, 29, 31, 34, 122*). However, very little information is available as regards ITAFs controlling the cellular mRNA IRESes: hnRNPC was reported to regulate the PDGF2 IRES, whereas it has been recently shown that PTB inhibits the translation of Bip mRNA (*123, 124*). Recent reports indicate that DAP-5 is able to activate its own IRES activity and that APAF-1 IRES is activated by the synergic activity of PTB and Unr (*66, 125*).

When we analyzed the binding of cellular proteins to the FGF-2 mRNA 5' by UV crosslinking experiments, the results indicated that several proteins were able to specifically bind to the FGF-2 mRNA 5' region, migrating at 110, 75, 60, 50, and 40 kDa, respectively (*61*). The comparison of the binding profile of cell proteins from different cell types indicated that all the transformed cell line proteins had similar binding profiles, whereas the profile was drastically different using proteins from human primary skin fibroblasts, RRL, or WGE. This suggested that cell transformation is accompanied by the binding of *trans*-acting factors to the FGF-2 mRNA leader. To determine whether this difference is indeed related to cell transformation and not just to a difference of cell type or origin, we analyzed the protein binding in the large T Antigen-transformed skin fibroblasts. This resulted in a change in the binding profile compared to normal skin fibroblasts, the new profile being similar to that of previously analyzed transformed cell types (*69*). This definitively demonstrated that the binding of several proteins (putative ITAFs) is linked to cell transformation and also correlated to constitutive expression of CUG-initiated forms.

Further analysis of ITAF binding was performed using normal skin fibroblasts subjected to cell-density variations, heat shock, or oxidative stress. In both cell-density variations and stress treatment, the protein binding profile changed. However, at low cell density, the profile resembled that obtained in transformed cells, but this was not the case after stress treatment. Heat shock treatment, in particular, is accompanied by the major binding of a 60-kDa protein (*61, 69*). The first hypothesis about the identity of this 60-kDa protein was that it could correspond to PTB. However, immunoprecipitation experiments and two-dimensional gels showed that this protein is different from PTB, both in stressed skin fibroblasts and in transformed cells (*61*) (S. Audigier *et al.*, unpublished). The FGF-2 IRES is thus probably regulated by specific ITAFs (Fig. 6).

I. FGF-2 mRNA Expression Is Controlled by Alternative Poly(A) Sites, an RNA Destabilization Element, and a Translational Enhancer Present in the 3'-UTR

The FGF-2 mRNA is 6775 nt long (5, 80). However, most of these sequences are noncoding and contained in the 3'-UTR, which is 5823 nt long. Furthermore, different species of FGF-2 RNA are expressed from a single promoter: They vary by the length of their 3'-UTR, due to the presence of eight alternative poly(A) sites (Fig. 6). By analyzing the length of the FGF-2 mRNA by quantitative reverse transcription-polymerase chain reaction (RT-PCR), we showed that the poly(A) used differed according to the cell type considered (126). In transformed cells such as COS-7, HeLa, or SK-Hep-1 cells, most FGF-2 mRNA was cleaved at the first or second poly(A) site (A1 and A2). In contrast, in human primary skin fibroblasts, the longest form of FGF-2 mRNA, cleaved at the eight poly(A) site, was the major species detected (126). We also showed that the length of 3'-UTR remained unchanged in tumoral cell lines, but did vary in skin fibroblasts subjected to cell-density variations or to heat shock (127). Such observations suggested that the 3'-UTR of FGF-2 mRNA, as was shown for the 5'-UTR, could be responsible for regulating the mRNA translation and/or stability.

In order to study the putative regulatory role of *cis*-acting elements in the 3'-UTR, we used an RNA transfection approach, which had the advantage of analyzing the expression of a single RNA species, which was impossible when using a DNA transfection approach. *In vitro*-transcribed CAT mRNAs to the different alternative FGF-2 3'-UTRs were introduced in different cell types by RNA transfection; measurement of CAT activity and of mRNA half-life revealed the presence of an RNA destabilization element (DEST; Fig. 6). The DEST element was mapped, by deletion analysis and by using antisense oligonucleotides, within an AU-rich direct repeat located just upstream of the second poly(A) site (126). This DEST element is unique among the destabilization elements previously published for protooncogene, cytokine, and growth factor mRNA: It does not contain the classic AUUUA tandem motifs or a long U-rich region described as an ARE-enhancing element (8). Evidence was provided by these findings that AU-rich elements are not sufficient by themselves for mRNA destabilization: FGF-2 mRNA contains 15 AUUUA motifs that are outside the DEST element, and do not influence mRNA decay (126).

When we measured the CAT activity and RNA stability of a CAT mRNA fused to the complete 5823-nt-long 3'-UTR, we observed in human skin fibroblasts, but not in transformed cell lines, that CAT expression was high, despite the presence of the RNA destabilization element. This prompted us to look for a translational enhancer in the FGF-2 mRNA 3'-UTR (127). Such an element revealed its presence in the distal part of the long 3'-UTR, between the fourth and the eighth poly(A) sites. It was shown by deletion experiments that this translational element (called 3' TE) was composed of two segments each responsible for 50% of the enhancement. Furthermore, the 3′ TE effect was more efficient in the presence of the FGF-2 mRNA 5′ region. Western blots were performed to evaluate the effect of the 3′ TE on the control of alternative initiation of translation; this allowed us to show that the 3′ TE favors synthesis of the CUG1–3-initiated FGF–CAT isoforms (127) (Fig. 6).

In conclusion, our data indicate that the posttranscriptional regulation of FGF-2 expression is not only mediated by the IRES-containing 5′ mRNA region, but is also regulated by the 3′-UTR. This 3′-UTR-dependent regulation involves three posttranscriptional processes; alternative polyadenylation, RNA destabilization, and translation.

V. VEGF mRNA: Two Distinct IRESs Control Alternative Initiation of the Translation of Two Isoforms

The vascular endothelial cell growth factors (VEGFs) make up a family of genes containing five members (128–130). Among them, VEGF-A (often simply called VEGF) plays a pivotal role in the angiogenesis process. Whereas the VEGF gene is mostly transcribed from a single promoter (the presence of a second cryptic promoter has been reported), it is subjected to a process of alternative splicing giving rise to at least four subunits composed of 121–204 amino acids (131). The VEGF is expressed in the normal arterial wall (but not in endothelial cells) (132); however, the pathophysiological implications of the different isoforms generated by alternative splicing remain to be defined. Complete invalidation of the VEGF gene is lethal, but selective expression of the 121-amino-acid isoform is not sufficient for normal development (133). Angiogenesis stimulation by VEGF is mediated by two tyrosine kinase receptors, VEGFR1 and R2 (134, 135). VEGF endothelial tropism has indicated it for a long time as a potential therapeutic agent of atherosclerosis. However, it seems that its administration to hypercholesterolemic mice has an aggravating effect. VEGF also plays an important role in tumoral angiogenesis, as its expression is induced in response to hypoxia conditions at both transcriptional and posttranscriptional levels (136, 137).

A. Identification of Two Distinct IRESs in the VEGF mRNA

When we started studying the mechanism controlling human VEGF expression, the only information available was the identification of an RNA-destabilizing element in the VEGF mRNA 3′-UTR. Nothing had been reported about translational control of VEGF expression, despite the 1038-nt-long 5′-UTR of its mRNA (138). By using the bicistronic vector approach previously described for

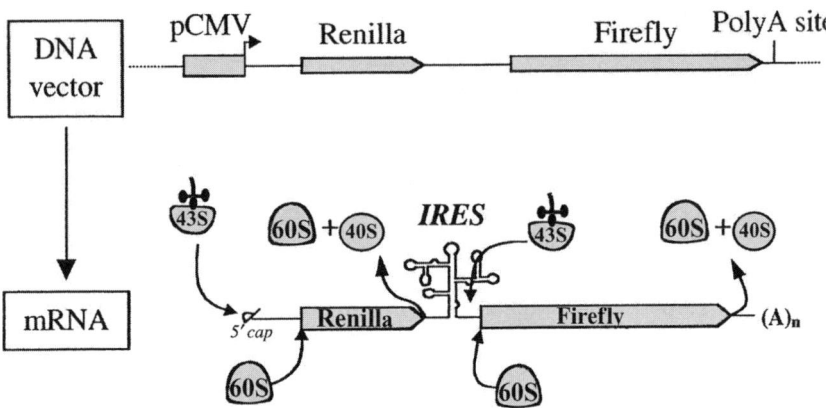

FIG. 7. The "Lucky Luke" bicistronic vectors. The "Lucky Luke" vectors express the two renilla (LucR) and firefly (LucF) luciferase genes, respectively, from a bicistronic mRNA transcribed at a single promoter (cytomegalovirus). These two reporter genes exhibit the advantage of high sensitivity and quantitative measurement of their expression. Thus, the IRES activity can be measured (*ex vivo* and *in vivo*) by calculating the LucF to LucR ratio.

the identification of FGF-2 and MuLV IRES (*17, 36*) (Fig. 3), we discovered, at the same time as other authors, that such a leader sequence is able to promote internal ribosome entry (*54–57*). We developed a new bicistronic vector, called "Lucky Luke," based on the expression of two renilla and firefly luciferase reporter genes (Fig. 7). This new vector provided highly sensitive and quantitative results. We succeeded in showing, by precise mapping of the IRES segment, that the VEGF mRNA 5′-UTR contains two distinct IRESes, called IRES A and IRES B (*55*) Fig. 8. UV crosslinking experiments allowed the detection of different factors selectively binding to IRES A or to IRES B: in particular, a 100-kDa protein specifically binds to IRES A, actually to a hairpin shown to be required for IRES A activity (*55*). This ITAF has not been identified yet. Several factors are also bound to IRES B; the major one, migrating at 60 kDa, has been identified as PTB (*55*). The PTB-binding site has been mapped, and is just upstream from the minimal segment necessary for the IRES B activity. This suggests that PTB more probably has a chaperone role, stabilizing the RNA structure in a conformation favorable for IRES activity (as proposed for picornavirus IRESs; *34*), rather than a direct role in ribosome recruitment (*55*).

B. Identification of a New VEGF Isoform, L-VEGF, Initiated at a CUG Codon

Two IRESs controlling the use of a single AUG initiation codon could allow fine tuning of VEGF expression at the translational level. However, the mechanism by which IRES B (located in a distal position as regards the AUG codon,

TRANSLATIONAL CONTROL OF GENE EXPRESSION

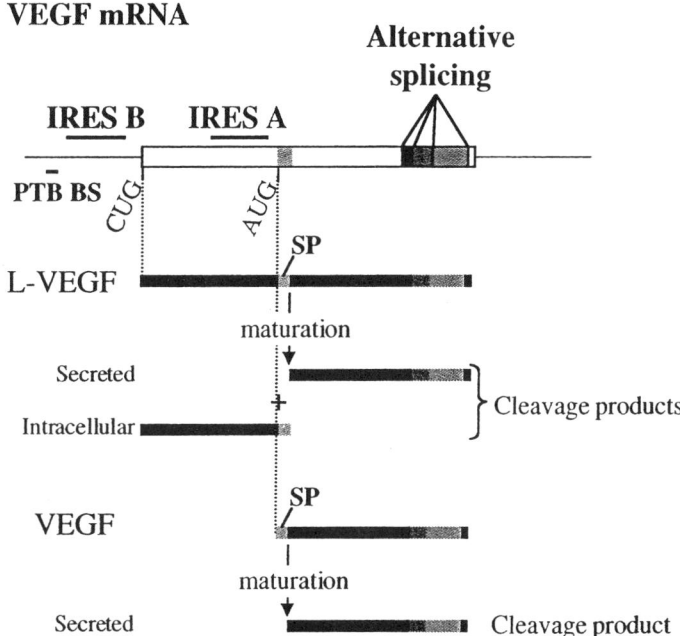

FIG. 8. Expression of two VEGF isoforms from alternative initations of translation. The VEGF mRNA is schematized with the VEGF ORF (empty box) and the two initiation codons CUG and AUG (their positions in the mRNA are 499 and 1039, respectively). The signal peptide is shown by a small box just downstream from the AUG, whereas the alternative exons are indicated by gray and black boxes. Positions of the two IRESs and of the PTB-binding sites are represented. Beneath the mRNA, the two translation products, L-VEGF and VEGF, are schematized, as well as their maturation products and their intracellular or secreted fate. The signal peptide is indicated by an arrow.

see Fig. 8) promotes translation initiation at this AUG remains unclear. The hypothesis of ribosomes scanning across IRES A to reach the AUG codon seems highly improbable. A more likely possibility was that in such a case IRES A and B are imbricated in a "super-IRES" structure, or that ribosomes could shunt from the end of IRES B up to the AUG codon. However, recent results in our laboratory as well as from other authors have revealed a simpler explanation (81, 96): A CUG codon in-frame with the VEGF open reading frame is located at the 3′ end of IRES B (Fig. 8). The synthesis of a new VEGF isoform, called L-VEGF, from the CUG codon has been demonstrated. Interestingly, this isoform shows a Golgi and endoplasmic reticulum localization (81). As shown by animal tissue analysis, this isoform might be, at least in some tissues, the major precursor of secreted VEGF.

The presence of distinct *trans*-acting factors bound to IRES A and B provides an interesting opportunity for controlling the relative level of CUG- versus

AUG-initiated isoforms. The physiological relevance of the presence of these two alternatively initiated VEGF precursors is under investigation.

VI. C-*myc* mRNA: A Natural Multicistronic Messenger

In mammals, the human protooncogene c-*myc* is transcribed from four alternative promoters (P0, P1, P2, and P3), among which P2 is the most frequently used, while P0 is human-specific (*139*). P0, P1, and P2 are located upstream from exon 1 and thus give rise to mRNAs with leader sequences of different lengths that encode the same c-Myc1 and c-Myc2 isoforms of 67 and 64 kDa, initiated at two alternative CUG and AUG codons, respectively (*4*) (Fig. 9). The ratio of these two proteins, two transcription factors with distinct DNA targets, is regulated at the translational level in response to methionine starvation (*93, 140*). In addition, the c-*myc* ORF leads to synthesis of downstream-initiated c-*myc* proteins, called c-Myc-S, which are dominant-negative inhibitors of transactivation by full-length c-*myc* proteins (*141*). P3 is located within intron 1 and gives rise to the c-Myc2 protein as well as to c-Myc-S isoforms.

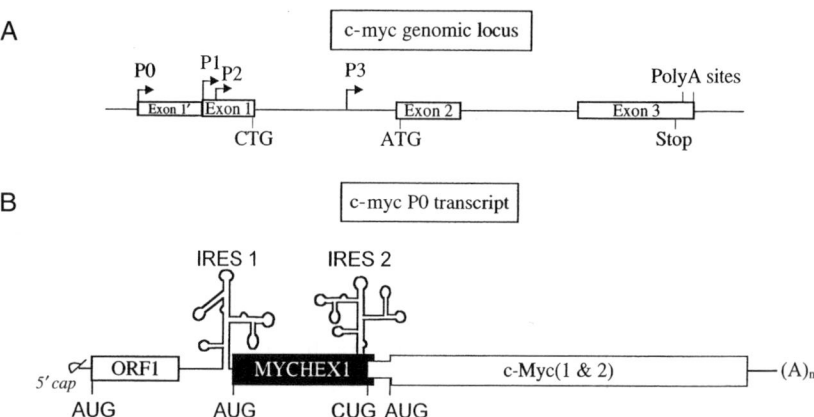

FIG. 9. The c-*myc* genomic locus and transcript. (A) Schema of the human c-*myc* genomic locus. The three exons are represented as empty boxes, with the initiation codon of c-Myc1 (CUG) in exon 1 and the initiation codon of c-Myc2 (AUG) in exon 2. The four alternative promoters are indicated by arrows. The promoter P3 is located within intron 1. (B) Schema of the c-*myc* P0 tricistronic mRNA. The three ORFs are ORF1 (nt 53–395, empty box), MYCHEX1 (nt 625–1188, black box), and c-*myc* (1173 to 2535). The c-Myc1 CUG codon is located at position 1173 and the c-Myc2 AUG codon, in the same frame, at position 1218 downstream from the 5' end of the c-*myc* P0 mRNA. IRES 1 has been localized just upstream from the MYCHEX1 ORF (nt 545–625) and IRES 2 upstream from the c-*myc* ORF (nt 811–1077).

The mRNA transcribed from promoter P0, with its leader sequence of 1172 upstream from the c-Myc1 CUG start codon, is a natural tricistronic mRNA: It contains, upstream from the c-myc open reading frame, two additional ORFs coding for polypeptides of 188 and 114 amino acids, respectively, which are both initiated upstream from the P1 promoter (Fig. 9). The middle ORF product (188 amino acids) has been characterized in human cells as MYCHEX1, but its function remains unknown (142).

A. Identification of an IRES in the P2 C-myc mRNA

The existence of such a large number of different proteins coded by the c-myc mRNA suggested by itself the presence of a translational control of their expression. In addition, the transcription start points of the P0, P1, and P2 c-myc mRNAs are located 1172, 524, and 363 nt upstream from the CUG initiation codon, respectively (Fig. 9). Such long leader sequences are likely to contain translation regulatory elements such as IRESs. The bicistronic approach was used to detect the possible presence of an IRES in the c-myc P2 mRNA by transient cell transfection. The second cistron of a bicistronic tandem CAT or double luciferase mRNA was efficiently expressed when the c-myc P2, P1, or P0 leader was introduced in the intercistronic region (52, 53). Deletion mutagenesis showed that the IRES is located between nucleotides 811 and 1077 downstream from P0 mRNA 5′ end, thus at about 100 nt upstream from the CUG start codon. This IRES is shared by the three leaders P0, P1, and P2 and is able to promote the synthesis of the two c-Myc1 and 2 isoforms (Fig. 9).

B. Identification of a Second C-myc IRES in the P0 Leader

Three ORFs are present in the P0 c-myc mRNA, but the c-myc IRES found in the P2 mRNA can drive initiation of the c-myc1 and 2 proteins only. What about the two upstream ORFs? This question was addressed by analyzing a putative IRES activity of the RNA leader segment located upstream from the P1 promoter position (Fig. 9). This segment, specific to the P0 mRNA, was introduced in the intercistronic region of the tandem CAT bicistronic vector. COS-7 and HeLa cell transfection clearly showed the presence of an IRES activity (143). This IRES, called IRES 1 (whereas the IRES common to the three P0, P1, and P2 mRNAs is now called IRES 2), was localized between nucleotides 545 and 625 downstream from the P0 mRNA 5′ end and shown to drive the translation of the second ORF, giving rise to the MYCHEX1 protein. As regards the third ORF located in the 5′ part of the mRNA, its expression was shown to be clearly cap-dependent in vitro (143).

In conclusion, the c-myc P0 mRNA is probably the first natural tricistronic mRNA described. The presence of the upstream ORFs may play a role in the control of the synthesis of the c-*myc* proteins, according to previous reports showing the regulatory role of upstream ORFs (*10*). However, in the other cases, the upstream ORFs are very short, whereas the c-*myc* P0 mRNA ORFs encode peptides of 114 and 188 amino acids (about 12 and 20 kDa, respectively). It is likely that such proteins might have a physiological function and that the presence of two IRESs in the mRNA provides the opportunity to express the three translational products of the c-*myc* P0 mRNA controlled by distinct factors. Finally, an additional question remains to be elucidated: Why is this P0 mRNA specific to the human c-*myc* gene?

VII. IRES-Dependent Translational Control: IRES Activity Enhancement in Transformed Cells

The first element suggesting that IRES-dependent translation might be enhanced in transformed cells has been reported for the c-*myc* gene (*70*): indeed, in Bloom's syndrome cells, c-*myc* expression is enhanced at a cap-independent translational level. Furthermore, in these cells, the c-*myc* gene contains a point mutation located about 100 nt upstream from the CUG codon (*144*). At this time the IRES had not been discovered. It appears that this mutation is located within the c-*myc* IRES 2 and that it leads to IRES activity enhancement, due to the higher binding affinity of an ITAF (as yet unidentified) (*72*).

We have also shown that FGF-2 isoform expression is constitutively activated in different transformed cell types (*69*). To confirm that this FGF-2 expression in transformed cells at confluence involves the IRES, the transformed and nontransformed cells have been transfected by bicistronic vectors. The second cistron expresses the FGF–CAT isoforms initiated at CUG1–3 in the transformed cells exclusively, demonstrating that expression of such isoforms is IRES-dependent in transformed cells (*69*). Interestingly, the AUG-initiated form is expressed in an IRES-dependent manner in transformed as well as in nontransformed cells, suggesting the presence of a second IRES responsible for initiation of translation at the AUG codon and subjected to a different regulation of its activity (*69*).

These two observations of a possible relationship between cell transformation and IRES activation prompted us to look at the IRES activity in different transformed and nontransformed cell types. This question was approached using the "Lucky Luke" vectors, given that luciferase expression measurement is more sensitive and more quantitative than that of CAT expression (Fig. 7). The transient transfection of 14 cell types was performed using the "Lucky Luke" vector containing any of the FGF-2, c-*myc*, VEGF, or EMCV IRES. The results showed that the four IRESs work in all the cell types, independently of the origin (human, simian, bovine, or murine). However, the levels of IRES activity vary:

The EMCV IRES has a high IRES or very high activity in transformed, immortalized, and primary cell types, except for HeLa and Jurkat, where the activity is moderate (74). The three cellular IRESs have a low activity in all the primary or immortalized cell types, whereas the activity is variable, depending on the IRES, in the transformed cell lines. The FGF-2 IRES activity is high in osteosarcoma Saos-2 cells and in neuroblastoma SK-N-BE and SK-N-AS cells (73); the c-*myc* IRES has a high activity in lymphocyte Jurkat cells, but also in the two neuroblastoma cells mentioned above (74); finally, VEGF IRES has a high activity in CHO, COS-7, ECV304, as well as the osteosarcoma and neuroblastoma cells mentioned above (L. Creancier *et al.*, unpublished results). These data indicate that, although there is no perfect correlation between IRES activation and cell transformation, the cellular IRESs show a low activity in all the nontransformed cell lines tested and have a high activity in several (but not all) transformed cell lines, with an IRES-dependent specificity.

VIII. IRES Tissue Specificity *in Vivo*

The analysis of IRES activity in cultured cells has provided much information about IRES regulation (see Section VI). However, we expected more drastic differences from one cell type to another. Significant activity was observed for all the tested IRESs in the different transformed and nontransformed cell types used in our experiments. One possible explanation for this IRES behavior is that the cells are not in a physiological situation: Indeed they are all in a proliferative state, in contrast to their normal situation in tissues, which is mostly quiescent. Furthermore, the transient transfection treatment generates a stress that could change the normal IRES activity of a given cell type. These two arguments prompted us to pursue the study of IRES regulation *in vivo*.

In order to study the tissue specificity of several IRESes *in vivo*, we developed several transgenic mice lines expressing "Lucky Luke" transgenes (Fig. 7), containing either the FGF-2, the c-*myc*, or the EMCV IRES (73, 74). Such a system has the advantage that the first cistron expression reflects the mRNA level and that the ratio of LucF to LucR activity provides an IRES activity value that is independent of the mRNA level. This was confirmed by the fact that similar IRES activities were measured from different organs of two independent mouse lines for each transgene (having different mRNA expression levels). EMCV, FGF-2, and c-*myc* IRES activities were measured in embryo and adult organs. In the 11-day-old embryo (E11), the three IRESes were active with similar efficiencies, but the two cellular IRESs showed no activity in placenta. Several organs of 16-day-old embryos were then analyzed, revealing a decrease of c-*myc* and FGF-2 IRES activities in the limb and in several organs, but not in brain and heart, in contrast to EMCV IRES, which shows a similar activity in

E16 organs to that measured in E11 (73, 74). The contrast between the cellular and viral IRESs increases in adults: The EMCV IRES continues to be active in all tissues (unfortunately, we obtained no data in the liver, where we could not detect the transgene). The FGF-2 IRES activity, however, is very low in lung, skin, muscle, heart, stomach, kidney, and intestine, and medium in ovary and tongue. In testis, the FGF-2 activity is similar to that of E11. The most unexpected result was a very high activity of FGF-2 IRES in brain: In this tissue, it is 20- to 40-fold more efficient than the EMCV IRES (which is already high) (73). These data revealed a high tissue specificity of the FGF-2 IRES, which may be related to the neurotrophic role of this growth factor (104, 106, 145, 146). As regards the role of FGF-2 in angiogenesis, it would have been interesting to determine the IRES activity in vessels, in particular in endothelial and smooth muscle cells of the vascular walls. Unfortunately, the detection system was not sensitive enough. The use of fluorescent proteins to analyze IRES activity *in situ* is currently being explored.

As regards the c-*myc* IRES, it is completely inactive in most adult organs (74). The highest activity is observed in brain, but is still very low: 5- to 10-fold less efficient than EMCV IRES, 100- to 200-fold less efficient than FGF-2 IRES. These results suggest that the c-*myc* IRES could play a role in c-*myc* expression in the embryo, in correlation with the indispensable function of c-*myc* in embryogenesis (147). The IRES could also be involved in the repression of c-*myc* expression in adult.

Taken together, these *in vivo* data show that the IRES-dependent translational regulation is likely to be more pronounced *in vivo*, where the cells are mostly in a nonproliferating state, than in cultured proliferating cells. The high-tissue-specificity features seem to mostly concern the cellular IRESs and to be related to the function of the corresponding gene, whereas the viral IRES has a large spectrum of activity. This suggests that the activity of cellular IRESes is strongly controlled by specific enhancing or silencing ITAFs. The concept of silencing ITAFs is new: Up to now, the different ITAFs involved in the regulation of viral IRES had been activators, but recent data from our laboratory have shown that proteins such as tumor suppressor p53 can act as an IRES silencer (75, 76).

IX. FGF-2 Translational Silencing by Tumor Suppressor p53

The observation that the FGF-2 IRES is upregulated or constitutively activated in several transformed, but not primary, cells incited us to look for the putative role of a tumor suppressor in the regulation of FGF-2 mRNA translation (69). As regards such a hypothesis, p53 was an interesting candidate: Although it is well known for its transcriptional transactivating effect, several reports indicate that p53 is also involved in translational processes. P53 is present in polysomes and has been reported as being covalently linked to the 5.8S ribosomal RNA and associated to the L5 ribosomal protein with mdm2 (*148–150*). It presents an RNA-binding and RNA-annealing activity and is involved in the repression of *cdk4* translation in G1-arrested osteosarcoma cells (*151, 152*). In addition, p53 is thought to inhibit the translation of its own mRNA *in vitro* (*153*).

To assess the possibility that p53 has an effect on FGF-2 expression, we constructed bigenic vectors containing both the effector and the target under the control of two distinct cytomegalovirus (CMV) promoters (*76*). Different cell types were transfected by bigenic vectors expressing wild-type or mutant p53 as effectors and, as target, a FGF–CAT fusion where the complete leader of FGF-2 mRNA or the complete leader plus the FGF-2 coding sequence had been fused to the CAT coding sequence. By this approach it has been clearly demonstrated that p53 is able to inhibit FGF-2 expression in a posttranscriptional manner (*75*). As regards the effect of the p53 mutants, all the mutants of the DNA-binding domain lose the ability to block FGF-2 expression, whereas a mutant of the N-terminal transactivating domain is fully able to promote the inhibitory effect. Furthermore, when the effector is p73, a p53-like protein sharing common transcriptional targets, the inhibition is not observed. Such data reveal that the observed inhibition by p53 does not involve its transactivating domain or its transcriptional targets. The effect no longer involves a change in the level of target mRNA, favoring the hypothesis of translational inhibition.

To determine whether the translational inhibition involved the FGF-2 IRES, a series of deletions was performed in the FGF-2 mRNA leader and their ability to mediate the translational inhibition was assessed. Surprisingly, at least three regions of the leader sequence are necessary for the effect to occur, one being in the 5' end of the leader (outside of the minimum IRES), another in the middle of the IRES, and a third in the region containing the CUG1–3 codons (*75*). This suggests that the effect of p53 probably involves a global RNA structure rather than a specific sequence.

We have demonstrated that this effect results from a direct interaction of p53 with the mRNA by using an *in vitro* approach where the FGF–CAT mRNA (transcribed *in vitro*) was preincubated with recombinant p53 before being used in a translation assay in rabbit reticulocyte lysate. This experiment results in a specific blockade of FGF–CAT mRNA translation in a dose-dependent manner (*75*). Furthermore, analysis of polysome formation in the presence or in the absence of p53 has clearly shown that the blockade occurs at the level of translation initiation (*75*). In an attempt to understand the mechanism of this inhibition, the affinity of p53 to FGF-2 mRNA was analyzed. We observed a significant RNA-binding activity, higher for the wild type than for the Ala143 mutant unable to promote the translational inhibition. However, this RNA-binding feature is not specific and thus cannot account for the specific translational blockade.

Another hypothesis was that p53 could induce a conformational change responsible for the translational inhibition. This was assessed using oligonucleotide probes complementary to the FGF-2 leader sequence, but unable to hybridize to RNA, because of the very stable structure of the leader. Preincubation of RNA and probe with p53 resulted in oligonucleotide hybridization, indicating that the protein is able to unwind the RNA structure to render it accessible to the probe. This led us to propose that p53 is able to induce a conformational change in FGF-2 mRNA. This results in the disappearance of the IRES structure, stopping IRES-dependent translation. Furthermore, the appearance of the new RNA structure also prevents ribosome attachment, thus blocking the cap-dependent translation (Fig. 10). As far as we know, p53 is the first ITAF discovered able to induce a specific blockade of IRES activity.

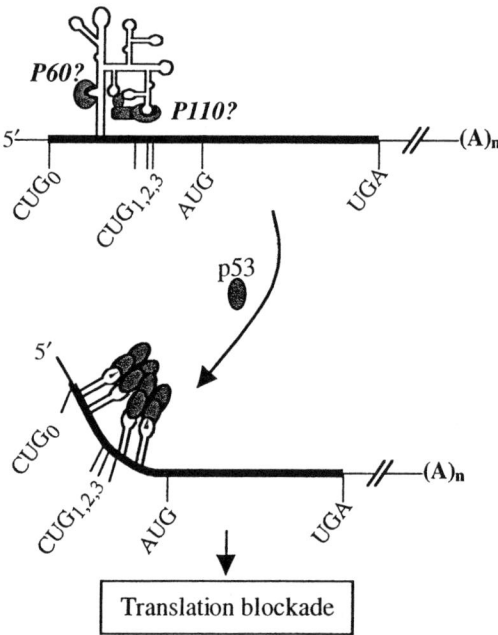

FIG. 10. Model for translation inhibition of FGF-2 mRNA expression by p53. In the absence of p53 binding, the FGF-2 mRNA is translated, as shown in Fig. 6, by both cap-dependent and IRES-dependent mechanisms, under the control of translation initiation factors (initiation from cap and IRES) and of positive ITAFs (only for initiation from the IRES), such as the putative ITAFs p60 or p110. Binding of p53 to the FGF-2 mRNA, resulting from stimuli that remain to be defined, induces an RNA conformational change. This destroys the IRES and furthermore results in a stable structure, which does not allow ribosome scanning. This new structure of FGF-2 mRNA is responsible for translational blockade of both cap- and IRES-dependent mechanisms.

The inhibitory effect of FGF-2 mRNA translation by p53 remains to be demonstrated in more physiological conditions. However, such an effect is physiologically very relevant, considering the antagonistic roles of p53 and FGF-2 in the control of cell proliferation, angiogenesis, and apoptosis. We propose that this translational inhibition, together with a transcriptional inhibition of FGF-2 expression by p53, participates in a feedback loop where p53 and FGF-2 both downregulate their own expression or activity. FGF-2 has indeed been shown to induce *mdm2* expression, which results in p53 destabilization (*154, 155*).

The process of translational inhibition by p53 raises the question of its effect on other mRNAs, especially IRES-containing mRNAs. We have shown that it cannot affect the translation of mRNAs controlled by the VEGF or the PDGF IRES (*76*). However, the strong inhibition of c-*myc* IRES observed in transgenic mice favors the involvement of an inhibitory ITAF, which could indeed be p53 (*74*). The role of other tumor suppressors in the control of IRES activity remains an interesting and innovative hypothesis.

X. Concluding Remarks

The data obtained during the last decade have not only revealed that translational mechanisms play a pivotal role in the control of gene expression, but they have led to the emergence of a new field in gene regulation, with the discovery that IRESs in messengers control functions in the cell. We have seen that IRESs allow translation initiation to be tissue-specific, so these translational elements may be compared to transcriptional enhancers. Such enhancers present in DNA promoters contain specific boxes able to bind transactivators, resulting in specific regulation of gene expression. At the translational level, IRESs could play a similar role, translational initiation thus being regulated by the binding of ITAFs. However, a major difference between the two mechanisms is that the DNA-binding sites are short nucleotide sequences, whereas at the translational level IRESs correspond to secondary or tertiary RNA structures. These features unfortunately prevent the detection of IRESs in databases. The same problem exists for ITAF-binding sites. Even though it may be known that a given ITAF binds a specific nucleotide stretch, the presence of such a stretch in the mRNA leader will not mean that it corresponds to a functional site. The accessibility of a nucleotide sequence assumed to bind an ITAF will always depend on the RNA structure. Furthermore, the question is rendered still more complex by the fact that ITAFs probably act not as single proteins, but as multiprotein complexes. This compounds the difficulty of studying IRES regulation, especially in cellular mRNAs.

Although complete elucidation of the physiological function of IRESs is still a long way off, the different data available from the literature either show, or strongly suggest, that these translational elements play a role in the control of gene expression in response to various pathophysiological stimuli including hypoxia, apoptosis, heat shock, amino acid starvation, and other stress. The simultaneous activation or inhibition of a group of IRESs regulated by the same ITAFs may also allow coordinated regulation of families of genes involved in a common process, such as angiogenesis or lesion repair. The identification of the ITAFs regulating different cellular mRNA IRESs, in particular among the angiogenic factors, will tell us whether such a hypothesis is true or not.

In addition to their role as translational enhancers allowing the cell to respond to exogenous stimuli very quickly and without requiring mRNA synthesis, IRESs clearly appear to be involved in the control of alternative initiation of translation. In the case of MuLV, FGF-2, and VEGF, the control of IRES activity by ITAFs will have consequences for the ratio of expression of the alternatively initiated isoforms coded by the mRNA. The isoforms synthesized from the different mRNAs mentioned above have different functions and/or localizations, so the control of IRES-dependent translation has important consequences for the function of the gene product.

As regards the role of IRESs in cell transformation, this is strongly suggested by a body of arguments, although not fully demonstrated. Several sources of data indicate constitutive activation of IRESs in transformed cells, especially in the case of c-*myc* and FGF-2, suggesting abnormal activation of ITAFs in such conditions. Furthermore, most of the gene products coded by IRES-containing mRNAs are related to the control of cell proliferation and differentiation, which means that aberrant or constitutive IRES activation will result in inadequate expression of such genes, leading to disorders in cell physiology. A striking feature of IRES-containing mRNAs is that they are often also regulated at other levels, transcriptionally and posttranscriptionally (splicing, RNA stability). A possible explanation is that the regulation of expression of such genes must be tightly controlled, and that different levels of control provide the multiple locking mechanism necessary to ensure adequate gene expression. An alternative (and not exclusive) possibility is that the different levels of regulation could be coupled and depend on common *trans*-acting factors. This is supported by the discovery that ITAFs are often nuclear proteins, or even transcription factors such as p53.

Finally, the discovery of numerous IRESs with either a large spectrum or with tissue specificity provides a big biotechnological potential. It is often necessary to express two or more genes from expression vectors, for instance, one or two genes of interest and a selection gene. The use of two independent transcription units in a given vector often creates a source of interference between the two promoter activities and can lead to the extinction of one of them. The only way to be sure that two proteins are synthesized together is to express them from the same mRNA. IRESs can thus be used to construct multicistronic expression cassettes where several genes are under the control of a single promoter. One important feature of such a vector is the possibility of controlling the stoechiometry, for example, in industrial production, of two enzymes in a given biosynthesis chain, or of two subunits of a heterodimeric enzyme. The multicistronic concept also allows coexpression of several therapeutic molecules in gene therapy programs.

ACKNOWLEDGMENT

We thank Peter Winterton for proofreading the manuscript.

REFERENCES

1. J. Curran and D. Kolakofsky, Ribosomal initiation from an ACG codon in the Sendai virus P/C mRNA. *EMBO J.* **7**, 245–251 (1988).
2. A. C. Prats, G. De Billy, P. Wang, and J. L. Darlix, CUG initiation codon used for the synthesis of a cell surface antigen coded by the murine leukemia virus. *J. Mol. Biol.* **205**, 363–372 (1989).
3. R. Z. Florkiewicz and A. Sommer, Human basic fibroblast growth factor gene encodes four polypeptides: three initiate translation from non-AUG codons. *Proc. Natl. Acad. Sci. USA* **86**, 3978–3981 (1989). [Erratum, *Proc. Natl. Acad. Sci. USA* **87**(5), 2045 (1990)].
4. S. R. Hann, M. W. King, D. L. Bentley, C. W. Anderson, and R. N. Eisenman, A non-AUG translational initiation in c-myc exon 1 generates an N-terminally distinct protein whose synthesis is disrupted in Burkitt's lymphomas. *Cell* **52**, 185–195 (1988).
5. H. Prats, M. Kaghad, A. C. Prats, M. Klagsbrun, J. M. Lelias, P. Liauzun, P. Chalon, J. P. Tauber, F. Amalric, J. A. Smith, and D. Caput, High molecular mass forms of basic fibroblast growth factor are initiated by alternative CUG codons. *Proc. Natl. Acad. Sci. USA* **86**, 1836–1840 (1989).
6. M. Guittaut, S. Charpentier, T. Normand, M. Dubois, J. Raimond, and A. Legrand, Identification of an internal gene to the human Galectin-3 gene with two different overlapping reading frames that do not encode Galectin-3. *J. Biol. Chem.* **276**, 2652–2657 (2001).
7. M. Klemke, R. H. Kehlenbach, and W. B. Huttner, Two overlapping reading frames in a single exon encode interacting proteins—a novel way of gene usage. *EMBO J.* **20**, 3849–3860 (2001).
8. J. A. Jarzembowski and J. S. Malter, Cytoplasmic fate of eukaryotic mRNA: identification and characterization of AU-binding proteins. *Prog. Mol. Subcell. Biol.* **18**, 141–172 (1997).
9. V. Kruys, B. Beutler, and G. Huez, Translational control mediated by UA-rich sequences. *Enzyme* **44**, 193–202 (1990).
10. A. G. Hinnebusch, Translational regulation of yeast GCN4. A window on factors that control initiator-trna binding to the ribosome. *J. Biol. Chem.* **272**, 21661–21664 (1997).
11. M. Kozak, Evaluation of the "scanning model" for initiation of protein synthesis in eucaryotes. *Cell* **22**, 7–8 (1980).
12. S. K. Jang, H. G. Krausslich, M. J. Nicklin, G. M. Duke, A. C. Palmenberg, and E. Wimmer, A segment of the 5' nontranslated region of encephalomyocarditis virus RNA directs internal entry of ribosomes during in vitro translation. *J. Virol.* **62**, 2636–2643 (1988).
13. J. Pelletier and N. Sonenberg, Internal initiation of translation of eukaryotic mRNA directed by a sequence derived from poliovirus RNA. *Nature* **334**, 320–325 (1988).
14. D. G. Macejak and P. Sarnow, Internal initiation of translation mediated by the 5' leader of a cellular mRNA [see comments]. *Nature* **353**, 90–94 (1991).

15. R. J. Jackson, mRNA translation Initiation without an end. *Nature* **353**, 14–15 (1991).
16. S. K. Oh, M. P. Scott, and P. Sarnow, Homeotic gene Antennapedia mRNA contains 5′-noncoding sequences that confer translational initiation by internal ribosome binding. *Genes Dev.* **6**, 1643–1653 (1992).
17. S. Vagner, M. C. Gensac, A. Maret, F. Bayard, F. Amalric, H. Prats, and A. C. Prats, Alternative translation of human fibroblast growth factor 2 mRNA occurs by internal entry of ribosomes. *Mol. Cell. Biol.* **15**, 35–44 (1995).
18. M. Kozak, The scanning model for translation: an update. *J. Cell Biol.* **108**, 229–241 (1989).
19. T. Preiss and M. W. Hentze, From factors to mechanisms: translation and translational control in eukaryotes. *Curr. Opin. Genet. Dev.* **9**, 515–521 (1999).
20. T. V. Pestova, V. G. Kolupaeva, I. B. Lomakin, E. V. Pilipenko, I. N. Shatsky, V. I. Agol, and C. U. Hellen, Molecular mechanisms of translation initiation in eukaryotes. *Proc. Natl. Acad. Sci. USA* **98**, 7029–7036 (2001).
21. A. B. Sachs and G. Varani, Eukaryotic translation initiation: there are (at least) two sides to every story. *Nat. Struct. Biol.* **7**, 356–361 (2000).
22. T. V. Pestova and C. U. Hellen, The structure and function of initiation factors in eukaryotic protein synthesis. *Cell. Mol. Life. Sci.* **57**, 651–674 (2000).
23. M. Kozak, Point mutations define a sequence flanking the AUG initiator codon that modulates translation by eukaryotic ribosomes. *Cell* **44**, 283–292 (1986).
24. M. Kozak, Binding of wheat germ ribosomes to bisulfite-modified reovirus messenger RNA: evidence for a scanning mechanism. *J. Mol. Biol.* **144**, 291–304 (1980).
25. S. K. Jang and E. Wimmer, Cap-independent translation of encephalomyocarditis virus RNA: structural elements of the internal ribosomal entry site and involvement of a cellular 57-kD RNA-binding protein. *Genes Dev.* **4**, 1560–1572 (1990).
26. K. Meerovitch, J. Pelletier, and N. Sonenberg, A cellular protein that binds to the 5′-noncoding region of poliovirus RNA: implications for internal translation initiation. *Genes Dev.* **3**, 1026–1034 (1989).
27. A. Kaminski and R. J. Jackson, The polypyrimidine tract binding protein (PTB) requirement for internal initiation of translation of cardiovirus RNAs is conditional rather than absolute. *RNA* **4**, 626–638 (1998).
28. A. Waysbort, S. Bonnal, S. Audigier, J. P. Esteve, and A. C. Prats, Pyrimidine tract binding protein and La autoantigen interact differently with the 5′ untranslated regions of lentiviruses and oncoretrovirus mRNAs. *FEBS Lett.* **490**, 54–58 (2001).
29. E. V. Pilipenko, T. V. Pestova, V. G. Kolupaeva, E. V. Khitrina, A. N. Poperechnaya, V. I. Agol, and C. U. Hellen, A cell cycle-dependent protein serves as a template-specific translation initiation factor. *Genes Dev.* **14**, 2028–2045 (2000).
30. L. B. Blyn, J. S. Towner, B. L. Semler, and E. Ehrenfeld, Requirement of poly(rC) binding protein 2 for translation of poliovirus RNA. *J. Virol.* **71**, 6243–6246 (1997).
31. S. L. Hunt, J. J. Hsuan, N. Totty, and R. J. Jackson, unr, a cellular cytoplasmic RNA-binding protein with five cold-shock domains, is required for internal initiation of translation of human rhinovirus RNA. *Genes Dev.* **13**, 437–448 (1999).
32. T. V. Pestova, C. U. Hellen, and I. N. Shatsky, Canonical eukaryotic initiation factors determine initiation of translation by internal ribosomal entry. *Mol. Cell. Biol.* **16**, 6859–6869 (1996).
33. C. U. Hellen and P. Sarnow, Internal ribosome entry sites in eukaryotic mRNA molecules. *Genes Dev.* **15**, 1593–1612 (2001).
34. R. J. Jackson and A. Kaminski, Internal initiation of translation in eukaryotes: the picornavirus paradigm and beyond. *RNA* **1**, 985–1000 (1995).
35. C. Berlioz and J. L. Darlix, An internal ribosomal entry mechanism promotes translation of murine leukemia virus gag polyprotein precursors. *J. Virol.* **69**, 2214–2222 (1995).
36. S. Vagner, A. Waysbort, M. Marenda, M. C. Gensac, F. Amalric, and A. C. Prats, Alternative

translation initiation of the Moloney murine leukemia virus mRNA controlled by internal ribosome entry involving the p57/PTB splicing factor. *J. Biol. Chem.* **270**, 20376–20383 (1995).
37. J. Attal, M. C. Theron, F. Taboit, M. Cajero-Juarez, G. Kann, P. Bolifraud, and L. M. Houdebine, The RU5 ('R') region from human leukaemia viruses (HTLV-1) contains an internal ribosome entry site (IRES)-like sequence. *FEBS Lett.* **392**, 220–224 (1996).
38. M. Lopez-Lastra, C. Gabus, and J. L. Darlix, Characterization of an internal ribosomal entry segment within the 5′ leader of avian reticuloendotheliosis virus type A RNA and development of novel MLV-REV-based retroviral vectors. *Hum. Gene. Ther.* **8**, 1855–1865 (1997).
39. C. Deffaud and J. L. Darlix, Rous sarcoma virus translation revisited: characterization of an internal ribosome entry segment in the 5′ leader of the genomic RNA. *J. Virol.* **74**, 11581–11588 (2000).
40. T. Ohlmann, M. Lopez-Lastra, and J. L. Darlix, An internal ribosome entry segment promotes translation of the simian immunodeficiency virus genomic RNA. *J. Biol. Chem.* **275**, 11899–11906 (2000).
41. C. B. Buck, X. Shen, M. A. Egan, T. C. Pierson, C. M. Walker, and R. F. Siliciano, The human immunodeficiency virus type 1 gag gene encodes an internal ribosome entry site. *J. Virol.* **75**, 181–191 (2001).
42. A. Waysbort, *Contrôle de la traduction chez les rétrovirus primates,* Sciences thesis. Université Paul Sabatier, Toulouse, France (2000).
43. J. E. Reynolds, A. Kaminski, A. R. Carroll, B. E. Clarke, D. J. Rowlands, and R. J. Jackson, Internal initiation of translation of hepatitis C virus RNA: the ribosome entry site is at the authentic initiation codon. *RNA* **2**, 867–878 (1996).
44. J. Sasaki and N. Nakashima, Translation initiation at the CUU codon is mediated by the internal ribosome entry site of an insect picorna-like virus in vitro. *J. Virol.* **73**, 1219–1226 (1999).
45. L. L. Domier, N. K. McCoppin, and C. J. D'Arcy, Sequence requirements for translation initiation of Rhopalosiphum padi virus ORF2. *Virology* **268**, 264–271 (2000).
46. J. E. Wilson, M. J. Powell, S. E. Hoover, and P. Sarnow, Naturally occurring dicistronic cricket paralysis virus RNA is regulated by two internal ribosome entry sites. *Mol. Cell. Biol.* **20**, 4990–4999 (2000).
47. J. E. Wilson, T. V. Pestova, C. U. Hellen, and P. Sarnow, Initiation of protein synthesis from the A site of the ribosome. *Cell* **102**, 511–520 (2000).
48. R. Panniers, E. B. Stewart, W. C. Merrick, and E. C. Henshaw, Mechanism of inhibition of polypeptide chain initiation in heat-shocked Ehrlich cells involves reduction of eukaryotic initiation factor 4F activity. *J. Biol. Chem.* **260**, 9648–9653 (1985).
49. A. M. Bonneau, A. Darveau, and N. Sonenberg, Effect of viral infection on host protein synthesis and mRNA association with the cytoplasmic cytoskeletal structure. *J. Cell Biol.* **100**, 1209–1218 (1985).
50. M. S. Sheikh and A. J. Fornace, Jr., Regulation of translation initiation following stress. *Oncogene* **18**, 6121–6128 (1999).
51. J. Bernstein, I. Shefler, and O. Elroy-Stein, The translational repression mediated by the platelet-derived growth factor 2/c-sis mRNA leader is relieved during megakaryocytic differentiation. *J. Biol. Chem.* **270**, 10559–10565 (1995).
52. C. Nanbru, I. Lafon, S. Audigier, M. C. Gensac, S. Vagner, G. Huez, and A. C. Prats, Alternative translation of the proto-oncogene c-myc by an internal ribosome entry site. *J. Biol. Chem.* **272**, 32061–32066 (1997).
53. M. Stoneley, F. E. Paulin, J. P. Le Quesne, S. A. Chappell, and A. E. Willis, C-Myc 5′ untranslated region contains an internal ribosome entry segment. *Oncogene* **16**, 423–428 (1998).
54. G. Akiri, D. Nahari, Y. Finkelstein, S. Y. Le, O. Elroy-Stein, and B. Z. Levi, Regulation of vascular endothelial growth factor (VEGF) expression is mediated by internal initiation of translation and alternative initiation of transcription. *Oncogene* **17**, 227–236 (1998).

55. I. Huez, L. Creancier, S. Audigier, M. C. Gensac, A. C. Prats, and H. Prats, Two independent internal ribosome entry sites are involved in translation initiation of vascular endothelial growth factor mRNA. *Mol. Cell. Biol.* **18,** 6178–6190 (1998).
56. D. L. Miller, J. A. Dibbens, A. Damert, W. Risau, M. A. Vadas, and G. J. Goodall, The vascular endothelial growth factor mRNA contains an internal ribosome entry site. *FEBS Lett.* **434,** 417–420 (1998).
57. I. Stein, A. Itin, P. Einat, R. Skaliter, Z. Grossman, and E. Keshet, Translation of vascular endothelial growth factor mRNA by internal ribosome entry: implications for translation under hypoxia. *Mol. Cell. Biol.* **18,** 3112–3119 (1998).
58. S. Vagner, B. Galy, and S. Pyronnet, Irresistible IRES: Attracting the translation machinery to internal ribosome entry sites. *EMBO Rep.* **2,** 893–898 (2001).
59. S. Cornelis, Y. Bruynooghe, G. Denecker, S. Van Huffel, S. Tinton, and R. Beyaert, Identification and characterization of a novel cell cycle-regulated internal ribosome entry site. *Mol. Cells* **5,** 597–605 (2000).
60. S. Pyronnet, L. Pradayrol, and N. Sonenberg, A cell cycle-dependent internal ribosome entry site. *Mol. Cells* **5,** 607–616 (2000).
61. S. Vagner, C. Touriol, B. Galy, S. Audigier, M. C. Gensac, F. Amalric, F. Bayard, H. Prats, and A. C. Prats, Translation of CUG- but not AUG-initiated forms of human fibroblast growth factor 2 is activated in transformed and stressed cells. *J. Cell Biol.* **135,** 1391–1402 (1996).
62. M. J. Coldwell, M. L. deSchoolmeester, G. A. Fraser, B. M. Pickering, G. Packham, and A. E. Willis, The p36 isoform of BAG-1 is translated by internal ribosome entry following heat shock. *Oncogene* **20,** 4095–4100 (2001).
63. S. A. Chappell, G. C. Owens, and V. P. Mauro, A 5′ Leader of Rbm3, a Cold Stress-induced mRNA, Mediates Internal Initiation of Translation with Increased Efficiency under Conditions of Mild Hypothermia. *J. Biol. Chem.* **276,** 36917–36922 (2001).
64. J. Fernandez, I. Yaman, R. Mishra, W. C. Merrick, M. D. Snider, W. H. Lamers, and M. Hatzoglou, Internal ribosome entry site-mediated translation of a mammalian mRNA is regulated by amino acid availability. *J. Biol. Chem.* **276,** 12285–12291 (2001).
65. M. Holcik, C. Yeh, R. G. Korneluk, and T. Chow, Translational upregulation of X-linked inhibitor of apoptosis (XIAP) increases resistance to radiation induced cell death. *Oncogene* **19,** 4174–4177 (2000).
66. S. Henis-Korenblit, N. L. Strumpf, D. Goldstaub, and A. Kimchi, A novel form of DAP5 protein accumulates in apoptotic cells as a result of caspase cleavage and internal ribosome entry site-mediated translation. *Mol. Cell. Biol.* **20,** 496–506 (2000).
67. M. Stoneley, S. A. Chappell, C. L. Jopling, M. Dickens, M. MacFarlane, and A. E. Willis, c-Myc protein synthesis is initiated from the internal ribosome entry segment during apoptosis. *Mol. Cell. Biol.* **20,** 1162–1169 (2000).
68. T. Subkhankulova, S. A. Mitchell, and A. E. Willis, Internal ribosome entry segment-mediated initiation of c-Myc protein synthesis following genotoxic stress. *Biochem. J.* **359,** 183–192 (2001).
69. B. Galy, A. Maret, A. C. Prats, and H. Prats, Cell transformation results in the loss of the density-dependent translational regulation of the expression of fibroblast growth factor 2 isoforms. *Cancer Res.* **59,** 165–171 (1999).
70. M. J. West, N. F. Sullivan, and A. E. Willis, Translational upregulation of the c-myc oncogene in Bloom's syndrome cell lines. *Oncogene* **11,** 2515–2524 (1995).
71. M. J. West, M. Stoneley, and A. E. Willis, Translational induction of the c-myc oncogene via activation of the FRAP/TOR signalling pathway. *Oncogene* **17,** 769–780 (1998).
72. S. A. Chappell, J. P. LeQuesne, F. E. Paulin, M. L. deSchoolmeester, M. Stoneley, R. L. Soutar, S. H. Ralston, M. H. Helfrich, and A. E. Willis, A mutation in the c-myc IRES leads to enhanced

internal ribosome entry in multiple myeloma: a novel mechanism of oncogene de-regulation. *Oncogene* **19,** 4437–4440 (2000).
73. L. Creancier, D. Morello, P. Mercier, and A. C. Prats, Fibroblast growth factor 2 internal ribosome entry site (IRES) activity ex vivo and in transgenic mice reveals a stringent tissue-specific regulation. *J. Cell Biol.* **150,** 275–281 (2000).
74. L. Creancier, P. Mercier, A. C. Prats, and D. Morello, c-myc Internal ribosome entry site activity is developmentally controlled and subjected to a strong translational repression in adult transgenic mice. *Mol. Cell. Biol.* **21,** 1833–1840 (2001).
75. B. Galy, L. Creancier, L. Prado-Lourenco, A. C. Prats, and H. Prats, p53 directs conformational change and translation initiation blockade of human fibroblast growth factor 2 mRNA. *Oncogene* **20,** 4613–4620 (2001).
76. B. Galy, L. Creancier, C. Zanibellato, A. C. Prats, and H. Prats, Tumour suppressor p53 inhibits human fibroblast growth factor 2 expression by a post-transcriptional mechanism. *Oncogene* **20,** 1669–1677 (2001).
77. J. Futterer, Z. Kiss-Laszlo, and T. Hohn, Nonlinear ribosome migration on cauliflower mosaic virus 35S RNA. *Cell* **73,** 789–802 (1993).
78. A. Yueh and R. J. Schneider, Selective translation initiation by ribosome jumping in adenovirus-infected and heat-shocked cells. *Genes Dev.* **10,** 1557–1567 (1996).
79. A. Yueh and R. J. Schneider, Translation by ribosome shunting on adenovirus and hsp70 mRNAs facilitated by complementarity to 18S rRNA. *Genes Dev.* **14,** 414–421 (2000).
80. E. Arnaud, C. Touriol, C. Boutonnet, M. C. Gensac, S. Vagner, H. Prats, and A. C. Prats, A new 34-kilodalton isoform of human fibroblast growth factor 2 is cap dependently synthesized by using a non-AUG start codon and behaves as a survival factor. *Mol. Cell. Biol.* **19,** 505–514 (1999).
81. I. Huez, S. Bornes, D. Bresson, L. Creancier, and H. Prats, A CUG start codon initiates a new VEGF isoform by a cap-independent mechanism. *Mol. Endocrinol.* **18,** 6178–6190 (2001).
82. P. Barraille, P. Chinestra, F. Bayard, and J. C. Faye, Alternative initiation of translation accounts for a 67/45 kDa dimorphism of the human estrogen receptor ERalpha. *Biochem. Biophys. Res. Commun.* **257,** 84–88 (1999).
83. C. L. Jopling and A. E. Willis, N-myc translation is initiated via an internal ribosome entry segment that displays enhanced activity in neuronal cells. *Oncogene* **20,** 2664–2670 (2001).
84. R. Weiss, N. Teich, H. Varmus, and J. Coffin, in "RNA Tumor Viruses" (C. S. H. L. Press, ed.), pp. 369–512. Cold Spring Harbor Laboratory Press, Cold Spring Harbor, New York, 1984.
85. F. A. Pillemer, D. A. Kooistra, O. N. Witte, and I. L. Weissman, Monoclonal antibody to the amino-terminal L sequence of murine leukemia virus glycosylated gag polyproteins demonstrates their unusual orientation in the cell membrane. *J. Virol.* **57,** 413–421 (1986).
86. C. J. Sherr, A. Sen, G. J. Todaro, A. Sliski, and M. Essex, Pseudotypes of feline sarcoma virus contain an 85,000-dalton protein with feline oncornavirus-associated cell membrane antigen (FOCMA) activity. *Proc. Natl. Acad. Sci. USA* **75,** 1505–1509 (1978).
87. S. P. Becerra, J. A. Rose, M. Hardy, B. M. Baroudy, and C. W. Anderson, Direct mapping of adeno-associated virus capsid proteins B and C: a possible ACG initiation codon. *Proc. Natl. Acad. Sci. USA* **82,** 7919–7923 (1985).
88. A. C. Prats, *Etude de l'expression génétique et de la constitution des particules virales infectieuses chez le rétrovirus murin MuLV*, Virology. Life Sciences, Université Paul Sabatier, Toulouse, France, (1988).
89. T. P. Makela, K. Saksela, and K. Alitalo, Two N-myc polypeptides with distinct amino termini encoded by the second and third exons of the gene. *Mol. Cell. Biol.* **9,** 1545–1552 (1989).
90. P. Acland, M. Dixon, G. Peters, and C. Dickson, Subcellular fate of the int-2 oncoprotein is determined by choice of initiation codon. *Nature* **343,** 662–665 (1990).

91. H. Dosaka-Akita, R. K. Rosenberg, J. D. Minna, and M. J. Birrer, A complex pattern of translational initiation and phosphorylation in L-myc proteins. *Oncogene* **6,** 371–378 (1991).
92. P. Lock, S. Ralph, E. Stanley, I. Boulet, R. Ramsay, and A. R. Dunn, Two isoforms of murine hck, generated by utilization of alternative translational initiation codons, exhibit different patterns of subcellular localization. *Mol. Cell. Biol.* **11,** 4363–4370 (1991).
93. S. R. Hann, K. Sloan-Brown, and G. D. Spotts, Translational activation of the non-AUG-initiated c-myc 1 protein at high cell densities due to methionine deprivation. *Genes Dev.* **6,** 1229–1240 (1992).
94. W. Bruening and J. Pelletier, A non-AUG translational initiation event generates novel WT1 isoforms. *J. Biol. Chem.* **271,** 8646–8654 (1996).
95. G. Packham, M. Brimmell, and J. L. Cleveland, Mammalian cells express two differently localized Bag-1 isoforms generated by alternative translation initiation. *Biochem. J.* **328,** 807–813 (1997).
96. M. Meiron, R. Anunu, E. J. Scheinman, S. Hashmueli, and B. Z. Levi, New isoforms of VEGF are translated from alternative initiation CUG codons located in its 5'UTR. *Biochem. Biophys. Res. Commun.* **282,** 1053–1060 (2001).
97. A. Corbin, A. C. Prats, J. L. Darlix, and M. Sitbon, A nonstructural gag-encoded glycoprotein precursor is necessary for efficient spreading and pathogenesis of murine leukemia viruses. *J. Virol.* **68,** 3857–3867 (1994).
98. A. Kaminski, G. J. Belsham, and R. J. Jackson, Translation of encephalomyocarditis virus RNA: parameters influencing the selection of the internal initiation site. *EMBO J.* **13,** 1673–1681 (1994).
99. T. Nishimura, Y. Utsunomiya, M. Hoshikawa, H. Ohuchi, and N. Itoh, Structure and expression of a novel human FGF, FGF-19, expressed in the fetal brain. *Biochim. Biophys. Acta.* **1444,** 148–151 (1999).
100. T. Yamashita, M. Yoshioka, and N. Itoh, Identification of a novel fibroblast growth factor, FGF-23, preferentially expressed in the ventrolateral thalamic nucleus of the brain. *Biochem. Biophys. Res. Commun.* **277,** 494–498 (2000).
101. M. A. Nugent and R. V. Iozzo, Fibroblast growth factor-2. *Int. J. Biochem. Cell Biol.* **32,** 115–120 (2000).
102. J. Kandel, E. Bossy-Wetzel, F. Radvanyi, M. Klagsbrun, J. Folkman, and D. Hanahan, Neovascularization is associated with a switch to the export of bFGF in the multistep development of fibrosarcoma. *Cell* **66,** 1095–1104 (1991).
103. A. Bikfalvi, S. Klein, G. Pintucci, and D. B. Rifkin, Biological roles of fibroblast growth factor-2. *Endocr. Rev.* **18,** 26–45 (1997).
104. S. Ortega, M. Ittmann, S. H. Tsang, M. Ehrlich, and C. Basilico, Neuronal defects and delayed wound healing in mice lacking fibroblast growth factor 2. *Proc. Natl. Acad. Sci. USA* **95,** 5672–5677 (1998).
105. F. P. Eckenstein, Fibroblast growth factors in the nervous system. *J. Neurobiol.* **25,** 1467–1480 (1994).
106. C. Grothe and K. Wewetzer, Fibroblast growth factor and its implications for developing and regenerating neurons. *Int. J. Dev. Biol.* **40,** 403–410 (1996).
107. J. D. Houle and J. H. Ye, Survival of chronically-injured neurons can be prolonged by treatment with neurotrophic factors. *Neuroscience* **94,** 929–936 (1999).
108. M. Renko, N. Quarto, T. Morimoto, and D. B. Rifkin, Nuclear and cytoplasmic localization of different basic fibroblast growth factor species. *J. Cell. Physiol.* **144,** 108–114 (1990).
109. B. Bugler, F. Amalric, and H. Prats, Alternative initiation of translation determines cytoplasmic or nuclear localization of basic fibroblast growth factor. *Mol. Cell. Biol.* **11,** 573–577 (1991).
110. B. Couderc, H. Prats, F. Bayard, and F. Amalric, Potential oncogenic effects of basic fibroblast

growth factor requires cooperation between CUG and AUG-initiated forms. *Cell Regul.* **2**, 709–718 (1991).
111. N. Quarto, D. Talarico, R. Florkiewicz, and D. B. Rifkin, Selective expression of high molecular weight basic fibroblast growth factor confers a unique phenotype to NIH 3T3 cells. *Cell Regul.* **2**, 699–708 (1991).
112. P. Mignatti, T. Morimoto, and D. B. Rifkin, Basic fibroblast growth factor (bFGF) released by single isolated cells stimulates their migration in an autocrine manner. *Proc. Natl. Acad. Sci. USA* **88**, 11007–11011 (1991).
113. E. Cohen-Jonathan, C. Toulas, S. Monteil, B. Couderc, A. Maret, J. J. Bard, H. Prats, N. Daly-Schveitzer, and G. Favre, Radioresistance induced by the high molecular forms of the basic fibroblast growth factor is associated with an increased G2 delay and a hyperphosphorylation of p34CDC2 in HeLa cells. *Cancer Res.* **57**, 1364–1370 (1997).
114. A. Yayon, M. Klagsbrun, J. D. Esko, P. Leder, and D. M. Ornitz, Cell surface, heparin-like molecules are required for binding of basic fibroblast growth factor to its high affinity receptor. *Cell* **64**, 841–848 (1991).
115. D. M. Ornitz, FGFs, heparan sulfate and FGFRs: complex interactions essential for development. *Bioessays* **22**, 108–112 (2000).
116. G. Bouche, N. Gas, H. Prats, V. Baldin, J. P. Tauber, J. Teissie, and F. Amalric, Basic fibroblast growth factor enters the nucleolus and stimulates the transcription of ribosomal genes in ABAE cells undergoing G0–G1 transition. *Proc. Natl. Acad. Sci. USA* **84**, 6770–6774 (1987).
117. G. Bouche, V. Baldin, P. Belenguer, H. Prats, and F. Amalric, Activation of rDNA transcription by FGF-2: key role of protein kinase CKII. *Cell. Mol. Biol. Res.* **40**, 547–554 (1994).
118. A. Bikfalvi, S. Klein, G. Pintucci, N. Quarto, P. Mignatti, and D. B. Rifkin, Differential modulation of cell phenotype by different molecular weight forms of basic fibroblast growth factor: possible intracellular signaling by the high molecular weight forms. *J. Cell Biol.* **129**, 233–243 (1995).
119. B. R. Henderson and P. Percipalle, Interactions between HIV Rev and nuclear import and export factors: the Rev nuclear localisation signal mediates specific binding to human importin-beta. *J. Mol. Biol.* **274**, 693–707 (1997).
120. A. C. Prats, S. Vagner, H. Prats, and F. Amalric, cis-acting elements involved in the alternative translation initiation process of human basic fibroblast growth factor mRNA. *Mol. Cell. Biol.* **12**, 4796–4805 (1992).
121. A. Lazaris-Karatzas, K. S. Montine, and N. Sonenberg, Malignant transformation by a eukaryotic initiation factor subunit that binds to mRNA 5' cap. *Nature* **345**, 544–547 (1990).
122. B. L. Walter, J. H. Nguyen, E. Ehrenfeld, and B. L. Semler, Differential utilization of poly(rC) binding protein 2 in translation directed by picornavirus IRES elements. *RNA* **5**, 1570–1585 (1999).
123. O. Sella, G. Gerlitz, S. Y. Le, and O. Elroy-Stein, Differentiation-induced internal translation of c-sis mRNA: analysis of the cis elements and their differentiation-linked binding to the hnRNP C protein. *Mol. Cell. Biol.* **19**, 5429–5440 (1999).
124. Y. K. Kim, B. Hahm, and S. K. Jang, Polypyrimidine tract-binding protein inhibits translation of bip mRNA. *J. Mol. Biol.* **304**, 119–133 (2000).
125. S. A. Mitchell, E. C. Brown, M. J. Coldwell, R. J. Jackson, and A. E. Willis, Protein factor requirements of the Apaf-1 internal ribosome entry segment: roles of polypyrimidine tract binding protein and upstream of N-ras. *Mol. Cell. Biol.* **21**, 3364–3374 (2001).
126. C. Touriol, A. Morillon, M. C. Gensac, H. Prats, and A. C. Prats, Expression of human fibroblast growth factor 2 mRNA is post-transcriptionally controlled by a unique destabilizing element present in the 3'-untranslated region between alternative polyadenylation sites. *J. Biol. Chem.* **274**, 21402–21408 (1999).

127. C. Touriol, M. Roussigne, M. C. Gensac, H. Prats, and A. C. Prats, Alternative translation initiation of human fibroblast growth factor 2 mRNA controlled by its 3'-untranslated region involves a Poly(A) switch and a translational enhancer. *J. Biol. Chem.* **275,** 19361–19367 (2000).
128. S. Ogawa, A. Oku, A. Sawano, S. Yamaguchi, Y. Yazaki, and M. Shibuya, A novel type of vascular endothelial growth factor, VEGF-E (NZ-7 VEGF), preferentially utilizes KDR/Flk-1 receptor and carries a potent mitotic activity without heparin-binding domain. *J. Biol. Chem.* **273,** 31273–31282 (1998).
129. N. Ferrara, VEGF: an update on biological and therapeutic aspects. *Curr. Opin. Biotechnol.* **11,** 617–624 (2000).
130. M. Clauss, Molecular biology of the VEGF and the VEGF receptor family. *Semin. Thromb. Hemost.* **26,** 561–569 (2000).
131. E. Tischer, R. Mitchell, T. Hartman, M. Silva, D. Gospodarowicz, J. C. Fiddes, and J. A. Abraham, The human gene for vascular endothelial growth factor. Multiple protein forms are encoded through alternative exon splicing. *J. Biol. Chem.* **266,** 11947–11954 (1991).
132. T. Couffinhal, M. Kearney, B. Witzenbichler, D. Chen, T. Murohara, D. W. Losordo, J. Symes, and J. M. Isner, Vascular endothelial growth factor/vascular permeability factor (VEGF/VPF) in normal and atherosclerotic human arteries. *Am. J. Pathol.* **150,** 1673–1685 (1997).
133. P. Carmeliet, Y. S. Ng, D. Nuyens, G. Theilmeier, K. Brusselmans, I. Cornelissen, E. Ehler, V. V. Kakkar, I. Stalmans, V. Mattot, J. C. Perriard, M. Dewerchin, W. Flameng, A. Nagy, and F. Lupu, Impaired myocardial angiogenesis and ischemic cardiomyopathy in mice lacking the vascular endothelial growth factor isoforms VEGF164 and VEGF188. *Nat. Med.* **5,** 495–502 (1999).
134. T. Veikkola and K. Alitalo, VEGFs, receptors and angiogenesis. *Semin. Cancer Biol.* **9,** 211–220 (1999).
135. M. Shibuya, Structure and function of VEGF/VEGF-receptor system involved in angiogenesis. *Cell Struct. Funct.* **26,** 25–35 (2001).
136. I. Stein, M. Neeman, D. Shweiki, A. Itin, and E. Keshet, Stabilization of vascular endothelial growth factor mRNA by hypoxia and hypoglycemia and coregulation with other ischemia-induced genes. *Mol. Cell. Biol.* **15,** 5363–5368 (1995).
137. A. P. Levy, N. S. Levy, and M. A. Goldberg, Post-transcriptional regulation of vascular endothelial growth factor by hypoxia. *J. Biol. Chem.* **271,** 2746–2753 (1996).
138. E. Tischer, D. Gospodarowicz, R. Mitchell, M. Silva, J. Schilling, K. Lau, T. Crisp, J. C. Fiddes, and J. A. Abraham, Vascular endothelial growth factor: a new member of the platelet-derived growth factor gene family. *Biochem. Biophys. Res. Commun.* **165,** 1198–1206 (1989).
139. D. L. Bentley and M. Groudine, Novel promoter upstream of the human c-myc gene and regulation of c-myc expression in B-cell lymphomas. *Mol. Cell. Biol.* **6,** 3481–3489 (1986).
140. S. R. Hann, M. Dixit, R. C. Sears, and L. Sealy, The alternatively initiated c-Myc proteins differentially regulate transcription through a noncanonical DNA-binding site. *Genes Dev.* **8,** 2441–2452 (1994).
141. G. D. Spotts, S. V. Patel, Q. Xiao, and S. R. Hann, Identification of downstream-initiated c-Myc proteins which are dominant-negative inhibitors of transactivation by full-length c-Myc proteins. *Mol. Cell. Biol.* **17,** 1459–1468 (1997).
142. C. Gazin, M. Rigolet, J. P. Briand, M. H. Van Regenmortel, and F. Galibert, Immunochemical detection of proteins related to the human c-myc exon 1. *EMBO J.* **5,** 2241–2250 (1986).
143. C. Nanbru, A. C. Prats, L. Droogmans, P. Defrance, G. Huez, and V. Kruys, Translation of the human c-myc P0 tricistronic mRNA involves two independent internal ribosome entry sites. *Oncogene* **20,** 4270–4280 (2001).
144. A. E. Willis, F. E. Paulin, M. J. West, and R. L. Whitney, Investigation of aberrant translational control of c-myc in cell lines derived from patients with multiple myeloma. *Curr. Top. Microbiol. Immunol.* **224,** 269–276 (1997).

145. R. Dono, G. Texido, R. Dussel, H. Ehmke, and R. Zeller, Impaired cerebral cortex development and blood pressure regulation in FGF-2-deficient mice. *EMBO J.* **17,** 4213–4225 (1998).
146. F. M. Vaccarino, M. L. Schwartz, R. Raballo, J. Nilsen, J. Rhee, M. Zhou, T. Doetschman, J. D. Coffin, J. J. Wyland, and Y. T. Hung, Changes in cerebral cortex size are governed by fibroblast growth factor during embryogenesis. *Nat. Neurosci.* **2,** 246–253 (1999).
147. A. C. Davis, M. Wims, G. D. Spotts, S. R. Hann, and A. Bradley, A null c-myc mutation causes lethality before 10.5 days of gestation in homozygotes and reduced fertility in heterozygous female mice. *Genes Dev.* **7,** 671–682 (1993).
148. B. M. Fontoura, E. A. Sorokina, E. David, and R. B. Carroll, p53 is covalently linked to 5.8S rRNA. *Mol. Cell. Biol.* **12,** 5145–5151 (1992).
149. V. Marechal, B. Elenbaas, J. Piette, J. C. Nicolas, and A. J. Levine, The ribosomal L5 protein is associated with mdm-2 and mdm-2-p53 complexes. *Mol. Cell. Biol.* **14,** 7414–7420 (1994).
150. B. M. Fontoura, C. A. Atienza, E. A. Sorokina, T. Morimoto, and R. B. Carroll, Cytoplasmic p53 polypeptide is associated with ribosomes. *Mol. Cell. Biol.* **17,** 3146–3154 (1997).
151. P. Oberosler, P. Hloch, U. Ramsperger, and H. Stahl, p53-catalyzed annealing of complementary single-stranded nucleic acids. *EMBO J.* **12,** 2389–2396 (1993).
152. M. E. Ewen, C. J. Oliver, H. K. Sluss, S. J. Miller, and D. S. Peeper, p53-dependent repression of CDK4 translation in TGF-beta-induced G1 cell-cycle arrest. *Genes Dev.* **9,** 204–217 (1995).
153. J. Mosner, T. Mummenbrauer, C. Bauer, G. Sczakiel, F. Grosse, and W. Deppert, Negative feedback regulation of wild-type p53 biosynthesis. *EMBO J.* **14,** 4442–4449 (1995).
154. T. Ueba, T. Nosaka, J. A. Takahashi, F. Shibata, R. Z. Florkiewicz, B. Vogelstein, Y. Oda, H. Kikuchi, and M. Hatanaka, Transcriptional regulation of basic fibroblast growth factor gene by p53 in human glioblastoma and hepatocellular carcinoma cells. *Proc. Natl. Acad. Sci. USA* **91,** 9009–9013 (1994).
155. E. Shaulian, D. Resnitzky, O. Shifman, G. Blandino, A. Amsterdam, A. Yayon, and M. Oren, Induction of Mdm2 and enhancement of cell survival by bFGF. *Oncogene* **15,** 2717–2725 (1997).

Structure and Function of φ29 Hexameric RNA That Drives the Viral DNA Packaging Motor: Review

PEIXUAN GUO

Department of Pathobiology
and Purdue Cancer Center
Purdue University
West Lafayette, Indiana 47907

I. Introduction	416
II. Approaches and Strategies for the Study of pRNA	419
A. Production of Infectious Virions *in Vitro* with Synthetic and Purified Components: A Sensitive System for the Assay of pRNA	420
B. Constructions of Active Circularly Permuted pRNA (cp-pRNA)	421
C. Determination of the Stoichiometry of pRNA	423
D. Construction of a pRNA Inactive in DNA Packaging but Competent in Procapsid Binding	430
III. Studies of pRNA Structure	430
A. Complementary Modification	430
B. Phylogenetic Analysis	431
C. Photoaffinity Crosslinking with GMPS/Aryl Azide	431
D. Crosslinking by Psoralen	433
E. Crosslinking by Phenphi	435
F. Chemical Modification	435
G. Chemical Modification Interference	436
H. Ribonuclease Probing	438
I. Genetic Analysis by Truncation, Insertion, Deletion, and Mutation	438
J. Selex	439
K. Cryo-Atomic Force Microscopy	440
L. Modeling the Three-Dimensional Structure of pRNA	440
M. Conformational Change of pRNA	443
IV. pRNA–Protein Interactions	445
A. Strong Binding of pRNA to Connector but Undetectable Binding to Capsid Proteins	445
B. Footprinting and Chemical Probing of Procapsid-Bound pRNA	447
C. Connector Domain and Sequence for pRNA Binding	447
D. Can pRNA Interact with gp16?	448
V. Effect of Mono- and Divalent Cations on pRNA Dimer Formation, Procapsid Binding, and Viral Assembly	448
VI. Functions of pRNA	449
A. Separation of the pRNA into Two Domains: Procapsid-Binding Domain and DNA-Packaging Domain	449
B. pRNA Remains Associated with DNA-Filled Capsids During DNA Packaging	450

C. pRNA Leaves DNA-Filled Capsid after the Addition of the Tail
   Protein gp9 450
D. Dimer as the Building Block in pRNA Formation 450
E. Sequence and Size Requirement of the Loops in Dimer Formation... 452
F. Minimal RNA Size Requirement for Dimer Formation............. 453
G. Minimal Sequence Requirement for Procapsid Binding
   and DNA Packaging 453
H. Mechanism and Role of pRNA in phi29 DNA Translocation......... 453
VII. Significance and Application of the Study of pRNA 457
A. A Model for the Study of RNA Dimers and Trimers 457
B. A Model for the Study of Viral DNA Packaging and for the Design
   of New Antiviral Strategies................................... 458
C. Similar Mechanism between Viral DNA Translocation and Other
   Nucleic Acid Sliding/Riding Processes......................... 459
D. Hints for Macromolecular Translocation across Cell Membranes
   or through Barrier Walls...................................... 461
VIII. Concluding Remarks.. 461
References... 462

One notable feature of linear dsDNA viruses is that, during replication, their lengthy genome is squeezed with remarkable velocity into a preformed procapsid and packed into near crystalline density. A molecular motor using ATP as energy accomplishes this energetically unfavorable motion tack. In bacterial virus phi29, an RNA (pRNA) molecule is a vital component of this motor. This 120-base RNA has many novel and distinctive features. It contains strong secondary structure, is tightly folded, and unusually stable. Upon interaction with ion and proteins, it has a knack to adapt numerous conformations to perform versatile function. It can be easily manipulated to form stable homologous monomers, dimers, trimers and hexamers. As a result, many unknown properties of RNA have been and will be unfolded by the study of this extraordinary molecule. This article reviews the structure and function of this pRNA and focuses on novel methods and unique approaches that lead to the illumination of its structure and function. © 2002, Elsevier Science (USA).

I. Introduction

A virus is composed of both protein and nucleic acid, either DNA or RNA. Their genomic DNA or RNA is enclosed in a protein shell or capsid. During replication, the synthesis of viral proteins and genomic DNA or RNA is accomplished by two different sets of machinery. After the synthesis of structural proteins and the viral genome, these components must interact with each other to form a complete virion, a process referred to as virus assembly (1–4). One striking feature in the assembly of many linear double-stranded DNA (dsDNA)

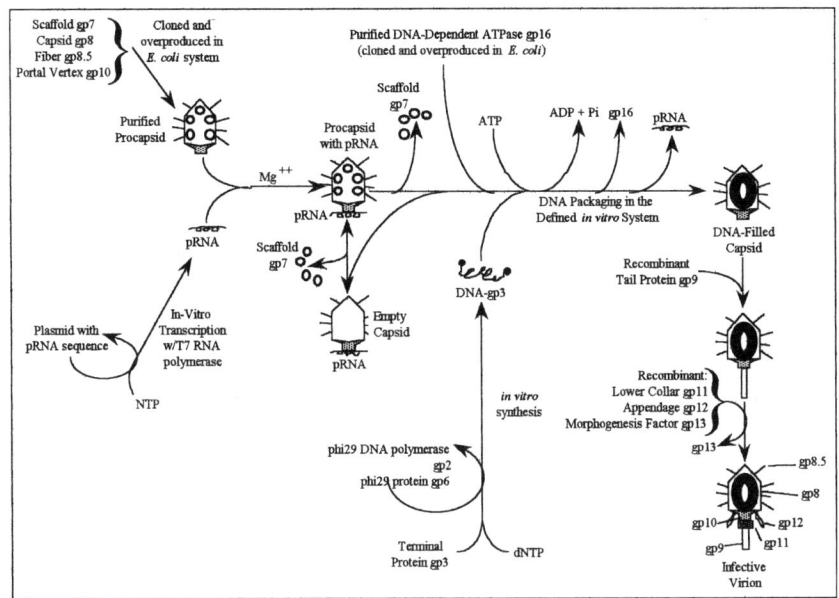

FIG. 1. Steps in the production of infectious virion phi29 in vitro using purified and synthetic components. Reproduced from Ref. 30 with permission.

viruses, including adenovirus (5–9), herpesvirus (10–12), poxvirus (13–15), and bacteriophages T1 (16), T3 (17), T4 (18), T5 (19), T7 (20, 21), P1 (22), P2 (23), P4 (24), P22 (25–27), Mu (28), phi21 (29), phi29 (30–32, 47–48), SP01 (33, 34), SPP1 (35, 36), λ (37–40), and their relatives, is that the viral genome is inserted into a preformed shell called a procapsid during maturation (Fig. 1) (1–3, 41–43). Procapsids of dsDNA phages contain a connector (or portal vertex) that serves as part of the DNA translocation machinery (1, 20, 32, 35, 41). Interestingly, a connector with 12 protein subunits and with a function similar to that of dsDNA phage connectors has been found recently in herpes simplex virus type 1 (44). The translocation of DNA into the procapsid is energetically unfavorable due to the very compact DNA arrangement within the capsid (42, 45, 46). Therefore ATP hydrolysis is required for packaging (4, 43, 47–49). Quantification of ATP consumption in DNA packaging with purified components revealed that one ATP is needed for the translocation of two base pairs of DNA into the procapsid (49, 50).

An exciting aspect of phi29 DNA packaging is the discovery that a small viral RNA, called pRNA, has a novel and essential role in viral DNA packaging (51) (Fig. 2). This pRNA is not a structural component of the mature virion (51); rather, it is a transient factor that serves as a vital component of the viral motor that drives genomic DNA into the viral procapsid (52). Since the discovery

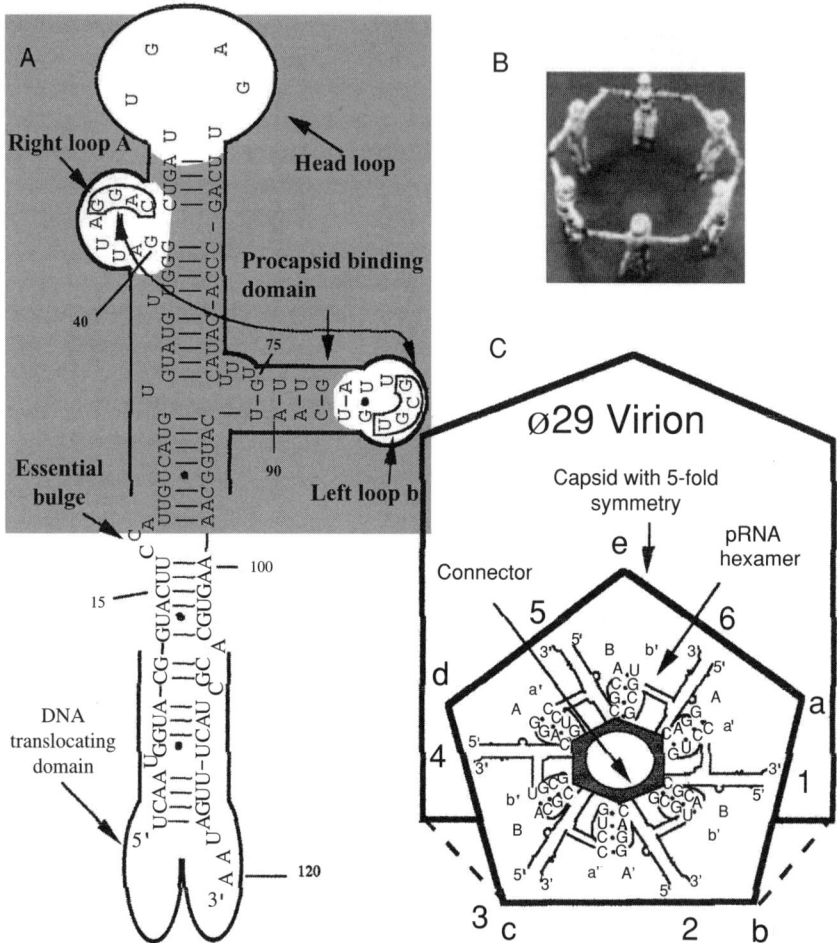

FIG. 2. Secondary structure, domain, and location of pRNA on the phi29 viral particle. (A) Secondary structure of pRNA A–b′. The binding domain (shaded area) and the DNA translocation domain are marked with bold lines. The four bases in the right and left loops, which are responsible for inter-RNA interactions, are boxed. (B) Power Rangers depict the formation of pRNA hexamer by hand-in-hand interaction. (C) Diagrams depicting the formation of a pRNA hexameric ring and its location in the phi29 DNA-packaging machine. Hexamer formation is via right- and left-loop sequence interactions of pRNA A–b′ and B–a′. The shaded hexagon stands for the connector and the surrounding pentagon stands for the fivefold-symmetric capsid vertex. The six protrusions stand for six pRNAs with the central region bound to the connector and the 5′/3′-paired region extending outward. Adapted from Ref. 97 with permission.

of RNA catalytic activities in 1982 (53–55) there has been a revolution in our understanding of RNA function, in that it is now known that RNA can act like an enzyme, thus the term "ribozyme." Except for the newly emerged telomerase, which is the ribonucleoprotein enzyme responsible for the replication of chromosome ends (56), most ribozymes, including RNase P (57), self-splicing introns (53), the hepatitis delta ribozyme (58–60), as well as the hammerhead (61–64) and hairpin ribozymes (65, 66), are all involved in catalytic reactions involving RNA substrates. In contrast, the phi29 pRNA provides a model for a ribozyme that acts on DNA.

The phi29 pRNA was first discovered in experiments showing that the packaging of phi29 DNA was sensitive to treatment with RNase A or T1 (51). DNA packaging was not inhibited by RNase that was pretreated with RNase inhibitor, indicating that the inhibitory effect on packaging was due specifically to RNase. Electrophoresis of purified procapsids revealed the presence of a 120-base polynucleotide that was resistant to DNase (51). This pRNA can be isolated from purified procapsids by incubation with EDTA, and the purified pRNA can bind to RNA-free procapsids in the presence of Mg^{2+}. pRNA-free procapsids are inactive in DNA packaging. However, the addition of pRNA to procapsids reconstitutes DNA-packaging activity, and the subsequent DNA-filled capsid can be converted into infectious virus.

RNA–DNA hybridization analysis was used as a first step in determining the origin of the pRNA (51). The pRNA purified from procapsids was labeled with ^{32}P and used as a probe in RNA–DNA hybridization with restriction fragments of phi29 DNA. The pRNA hybridized specifically to restriction fragments from the left end of the phi29 DNA, which was the end packaged first during DNA encapsidation. RNA sequence analysis (69) confirmed that the pRNA was transcribed from bases 320–147 of the phi29 genome, driven by the early promotor PE1 (A1) (67, 68), with transcription being from right to left. The pRNA was transcribed primarily as a 174-base molecule in phi29-infected cells. However, nuclease cleavage during procapsid purification eliminated 54 bases from its 3' end (70). The 174- and 120-base RNA molecules have similar activity in DNA packaging in vitro (69, 70).

This review will focus on the work done in the author's laboratory during the last 12 years. Some related work from other laboratories will also be discussed. The chapter will deal with the structure and function of the pRNA, as well as the methods and approaches that were used to investigate pRNA structure and function.

II. Approaches and Strategies for the Study of pRNA

Four major approaches have been undertaken in this laboratory to study the structure and function of pRNA. The first one is based on the development of a highly sensitive *in vitro* phi29 virion assembly system (*30*) for the assay of pRNA activity. With this system, up to 5×10^9 infectious virions per milliliter can be obtained in the presence of pRNA, with not a single infectious virion detected in the absence of the pRNA. Therefore, a system with nine-order-of-magnitude sensitivity can be used for the analysis of pRNA structure and function. The second approach involves the construction of circularly permuted pRNA (cp-pRNA) (*87*), in which any internal base of the pRNA can be reassigned to serve as a new 5' or 3' terminus. The circular permutation system greatly facilitates the construction of mutant pRNA via polymerase chain reaction (PCR) and makes it possible to label any specific internal base by radioisotopes or photoaffinity agents. The third approach to analyzing pRNA is based on new methods for studying the stoichiometry of pRNA in DNA packaging (*93*). A fourth approach uses the construction of mutant pRNAs that are inactive in DNA packaging, but are competent to compete with-wild type pRNA for procapsid binding (*86, 93, 102, 103*).

A. Production of Infectious Virions *in Vitro* with Synthetic and Purified Components: A Sensitive System for the Assay of pRNA

Assembly of infectious phi29 virus requires the following components: a procapsid into which the viral DNA genome will be inserted, a viral DNA genome, DNA-packaging protein gp16, and pRNA. These viral components allow formation of DNA-filled capsids. DNA-filled capsids can be converted into infectious virions with the addition of three additional structural proteins, gp9 (tail), gp11 (collar), and gp12 (anti-receptor), along with a factor gp13, which is required for the assembly of gp11 and gp12 onto the DNA-filled capsid. ATP serves as an energy source for DNA translocation and magnesium is required for the appropriate folding of pRNA.

The procapsid of the *Bacillus subtilis* bacteriophage phi29 consists of the major capsid protein gp8 (235 copies), head fiber protein gp8.5 (47 copies), scaffolding protein gp7 (26 copies), and connector protein gp10 (12 copies) (*41, 72–75*). The genome of phi29 is a linear dsDNA of 19,285 base pairs (bp). A virus-encoded terminal protein, gp3, is covalently linked to each 5' end of the genome through the formation of a phosphodiester bond between the —OH group of serine residue 232 of gp3 and the 5' dAMP of genomic DNA (*75–78*). The terminal protein gp3 functions both as a "primer" in the initiation of DNA replication (*76, 79–81*), and as an enhancer for DNA packaging (*47, 82*). Full-length phi29 genomic DNA has been synthesized *in vitro* (*30, 79–81*). The gp16 protein is part of an ATPase complex (*49, 83*).

The genes coding for the structural proteins gp7, gp8, and gp10 of phi29 procapsid have been cloned and overexpressed in *Escherichia coli*, allowing the

TABLE I
IN VITRO ASSEMBLY OF INFECTIOUS phi29 VIRIONS[a]

Experiment	DNA–gp3	Procapsid	pRNA	gp16	gp9	gp11–13	gp12	pfu/ml
1	+	+	+	+	+	+	+	5×10^8
2	−	+	+	+	+	+	+	0
3	+	−	+	+	+	+	+	0
4	+	+	−	+	+	+	+	0
5	+	+	+	−	+	+	+	0
6	+	+	+	+	−	+	+	0
7	+	+	+	+	+	−	+	0
8	+	+	+	+	+	+	−	0

[a]Reprinted from Ref. 30 with permission.

assembly and purification of procapsids that are competent in *in vitro* DNA packaging (52, 84). Other components, gp16 (48), gp9, gp11, gp12, and gp13 (30, 85) are all purified from cloned and overexpressed genes.

To perform *in vitro* phi29 assembly, purified procapsids are dialyzed against an EDTA-containing buffer at ambient temperature to unblock the binding sites for pRNA. pRNA is subsequently added, and the mixtures are then dialyzed against magnesium-containing buffer to allow pRNA to attach to the procapsid. The pRNA-enriched procapsids are mixed with gp16, DNA–gp3, and a reaction buffer (containing ATP) to complete the DNA-packaging reaction. After DNA packaging, gp9, gp11, gp12, and gp13 are sequentially added to the DNA packaging reactions to convert the DNA-filled capsids into infectious virions (Fig. 1). The yield is then assayed by standard plaque formation.

Up to 5×10^9 plaque-forming units (pfu)/ml of the infectious phi29 virions has been assembled *in vitro* with the exclusive use of 10 purified components (Table I). Omitting the pRNA resulted in no plaque formation. Electron microscopy and genome restriction mapping have confirmed the identity of the infectious phi29 virions synthesized in this system (30).

Due to the absence of any background, the generation of a single infectious virion could be detected. With this system, it has been possible to detect the activity of mutant pRNAs with 10^5-fold reductions in DNA-packaging efficiency (85), which is too low to be detected by other assay methods available (48, 82, 83). The system has been used extensively to study the structure and function of pRNA, including functional domains (86–88) and stoichiometry.

B. Constructions of Active Circularly Permuted pRNA (cp-pRNA)

Utilizing 5′- to 3′-end ligation, Pan *et al.* (89) were able to construct circular tRNA, which was subsequently cleaved with limited alkaline hydrolysis to generate one random break per molecule, and generated RNA molecules that were used for structural analysis. Similarly, Nolan *et al.* (90) synthesized tRNAs with native 5' and 3' ends linked by a synthetic loop and new termini in the interior of the native sequence, which they used to map tRNA-binding sites on RNase P. The feasibility of constructing circularly permuted RNAs rests on the close proximity of the native RNA 5' and 3' ends. The phi29 pRNA 5' and 3' ends are in close proximity (86) and hence a series of pRNA molecules with circular permutations have been constructed (Fig. 3).

To construct cp-pRNA, two tandem pRNA-coding sequences separated by a 3-base or 17-base loop sequence are cloned into a plasmid (Fig. 3). PCR primer pairs, such as P6/P5, complementary to various locations within pRNA coding sequences, can be designed to synthesize PCR fragments for transcription of cp-pRNAs. It has been shown that neither a small nor a large linker loop interferes with the biological activity of the pRNA molecule (87). In addition,

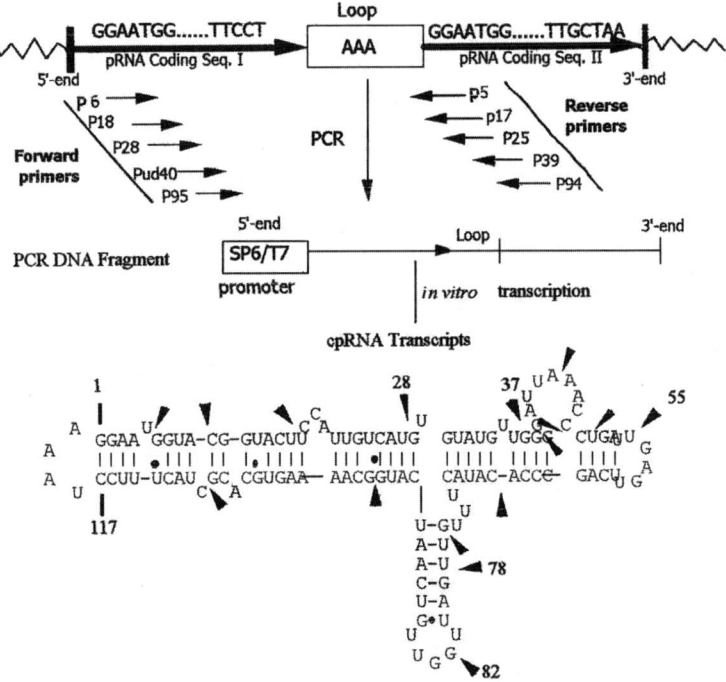

FIG. 3. Construction of tandem DNA templates for the synthesis of circularly permuted pRNA (cp-pRNA). Two pRNA-coding sequences are cloned into a plasmid with head-to-tail linkage via three nucleotides, AAA. Each pairs of PCR primers, such as P6/P5, containing SP6 or T7 promoter, are used to generate a DNA fragment for RNA transcription. Arrows indicate positions of new 5'/3' termini that have been used. Adapted from [87, 91] with permission.

most of the bases, especially the nonessential bases, can be used as new termini for constructing active cp-pRNA (91).

With the synthesis of biologically active cp-pRNA, the analysis of specific interactions of unique internal bases with components of the phi29 DNA-packaging complex is now possible. By labeling the new 5′ or 3′ terminus with photoaffinity cross-linking agents, specific interactions of the modified bases with proteins, DNA, or RNA can be elucidated (90, 119, 105).

C. Determination of the Stoichiometry of pRNA

To elucidate the role of the pRNA in DNA packaging, it is crucial to know how many copies of the pRNA are involved in each DNA-packaging event. Three approaches have been developed to determine the stoichiometry of the pRNA; these have led to the conclusion that six pRNAs are present in each DNA-translocating motor. The quantification of pRNA has been investigated by several approaches (70, 92–94). This section will focus on innovative methods developed in this laboratory.

1. DETERMINATION OF THE STOICHIOMETRY BY USE OF A BINOMIAL DISTRIBUTION

The first approach to pRNA stoichiometry determination involves the use of a binomial distribution (93, 95). pRNAs with mutations in the 5′/3′-paired region retain their procapsid binding capacity, but lose their DNA-packaging function (Sections II.D and VI.C). When mutant pRNA and wild-type pRNA are mixed at various ratios in *in vitro* assembly assays (i.e., 10% mutant and 90% wild type, etc.), the probability of procapsids possessing a certain amount of mutant (M) and a certain amount of wild type can be determined by the expansion of a binomial:

$$(p+q)^Z = \binom{Z}{0} p^Z + \binom{Z}{1} p^{Z-1} q + \binom{Z}{2} p^{Z-2} q^2 + \cdots \binom{Z}{Z-1} pq^{Z-1}$$

$$+ \binom{Z}{Z} q^Z = \sum_{M=0}^{Z} \binom{Z}{M} p^{Z-M} q^M$$

where

$$\binom{Z}{M} = \left(\frac{Z!}{M!(Z-M)!} \right)$$

and Z represents the total number of pRNA per procapsid and p and q represent the percentages of mutant and wild-type pRNA, respectively, which are used in the reaction mixture. For example, if the total number of pRNA per procapsid required for DNA packaging (Z) is 3, then the probability of all combinations of mutant (M) and wild type (N) pRNAs on a given procapsid can be determined by the expansion of the binomial $(p+q)^3 = p^3 + 3p^2q + 3pq^2 + q^3 = 100\%$. That is, in the procapsid population, there is the probability of a procapsid possessing either three copies of mutant pRNA, two copies of mutant and one copy of wild type, one copy of mutant and two copies of wild type, or three copies of wild type, which are written as p^3, $3p^2q$, $3pq^2$, or q^3, respectively. Suppose that there are 70% mutant and 30% wild-type pRNA in a reaction mixture; then the percentage of procapsids that possess one copy of mutant and two copies of wild type is $3pq^2$, or $3(0.7)(0.3)^2 = 19\%$. If only one mutant pRNA per procapsid is sufficient to render the procapsid unable to package DNA, then only those procapsids with 3 wild-type pRNAs bound (0 mutant pRNA, q^3) would package DNA. The yield of virions from the empirical data is plotted and compared to a series of predicted curves to find a best fit (Fig. 4). It is found that the experimental inhibition curve most closely resembles, in both slope and magnitude, the predicted curves where $Z = 5$ or 6 (93).

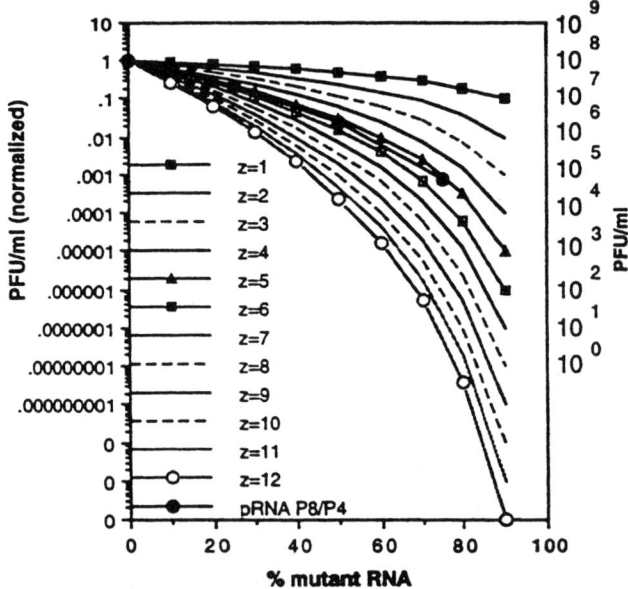

FIG. 4. Stoichiometry determination by binomial distribution. Here Z represents the hypothetical total number of pRNAs per procapsid, varying from 1 to 12, supposing that the minimal number of bound mutant pRNAs required to block DNA packaging is 1. Percentage of theoretical and empirical inactive mutants in each case is plotted against the yield of infectious virions in *in vitro* assembly assays. The empirical curve (pRNA P8/P4) falls between the theoretical curves of $Z = 5$ and 6. Reprinted from Ref. 93 with permission.

2. Determination of the Stoichiometry by Comparison of Slopes of Concentration Dependence (30, 93)

The dose–response curves of *in vitro* phi29 assembly versus concentration of various assembly components have been used to approximate the stoichiometry of pRNA. Phi29 assembly components with known stoichiometry served as standard controls (Fig. 5). In these dose–response curves, the greater the stoichiometry of the component, the more dramatic is the influence of the dilution factor on the reaction. A slope of 1 indicates that one copy of the component is involved in the assembly of one virion, as is the case for genomic DNA in phi 29 assembly. A slope larger than 1 indicates multiple-copy involvement. By this method, the stoichiometries of DNA–gp3, gp11, gp12, and gp16 have been compared (Fig. 5). The result is consistent with other approaches, in that the stoichiometry of gp11 and gp12 is greater than the stoichiometry of gp16 and pRNA. It is known that the stoichiometry of gp11 and gp12 in phi29 assembly is 12, whereas the stoichiometry of pRNA is between 5 and 6.

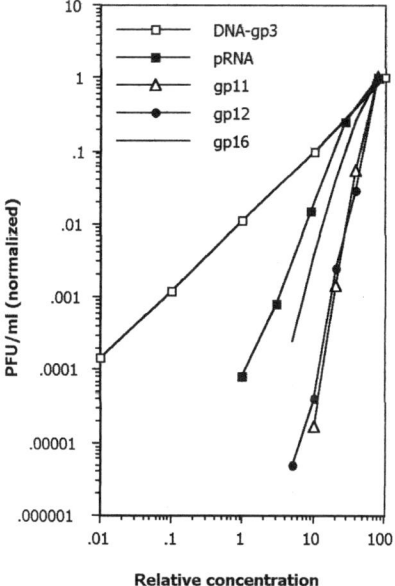

FIG. 5. Dose–response curves on a log/log plot showing concentration dependence. The slope of the curves for DNA–gp3 is 1 (45°), suggesting that one DNA is needed for assembly of one virion. The assembly units for both gp11 and gp12 are 12, with a slope higher than 1. The curves for pRNA and gp16 fall between the curves for DNA–gp3 and gp11/gp12, suggesting that the stoichiometry for pRNA and gp16 is between 1 and 12. As can be seen, the length of each unit for the x axis must be so drawn to equal the length of each unit of the y axis in slope determination. Reprinted from Ref. 93 with permission.

3. Determination of the Stoichiometry by Finding the Common Multiple of 2 and 3

The stoichiometry of pRNA has also been investigated by mixing two inactive mutant pRNAs with loops complementary intermolecularly and then assaying the activity of such mixtures in DNA packaging (Tables II, III, IV) (94, 96). The predicted secondary structure of the pRNA (Fig. 2) reveals two loops, called left- and right-hand loops (97). Sequences of these two loops (bases 45–48 of the right-hand loop and bases 85–82 of the left-hand loop) are complementary and were originally proposed to form a pseudoknot (98). However, extensive studies have revealed that these two sequences interact intermolecularly, allowing the formation of pRNA oligomers. Several pRNAs with mutated left- and right-hand loop sequences were constructed. To simplify the description of these mutants, uppercase and lowercase letters are used to represent the right- and left-hand loop sequences of the pRNA, respectively (Fig. 2). The same letter in upper case and lower case symbolizes a pair of complementary loops. For example, in pRNA A–a′ (Fig. 2), the right loop A (5′–GGAC$_{48}$) and the left loop a′ (3′-CCUG$_{82}$) are complementary, whereas in pRNA A–b′, the four bases of the right loop A are not complementary to the sequence of the left loop b′ (3′-UCCG$_{82}$). Mutant pRNAs with complementary loop sequences (such as pRNA A–a′) are active in phi29 DNA packaging, whereas mutants with noncomplementary loops (such as pRNA A–b′) are inactive (Tables II–IV).

TABLE II
Two Interlocking pRNAs[a]

pRNAs	Predicted hexamer	pRNAs	Predicted hexamer
I–i′ (wild-type pRNA)	8×10^7	I–j′ (unpaired loop)	0
J–j′ (compensatory modification)	7×10^7	(I–j′) + (J–p′) (miss one link)	0
J–i′ (unpaired loop)	0	(I–j′) + (J–i′) (compensatory pair)	4×10^7

[a]Adapted from Ref. 94 with permission.

TABLE III
Three Interlocking pRNAs[a]

pRNAs	Predicted hexamer	pRNAs	Predicted hexamer
J–p' (unpaired loop)	[hexagon diagram, value 0]	(P–k') + (K–j') (miss one link)	[hexagon diagram, value 0]
P–k' (unpaired loop)	[hexagon diagram, value 0]	(K–j') + (J–p') (miss one link)	[hexagon diagram, value 0]
K–j' (unpaired loop)	[hexagon diagram, value 0]	(J–p') + (P–k') + (K–j') (compensatory trimer)	[hexagon diagram, value 1×10^7]

[a]Adapted from Ref. 94 with permission.

a. A Set of Two Interlocking pRNAs (94). A key finding in pRNA research was that the mixing of two mutant pRNAs with *trans*-complementary loops rescued DNA-packaging activity (Table II). For example, inactive pRNA I–j' can rescue inactive pRNA J–i' for DNA packaging. This result can be explained by the *trans*-complementarity of pRNA loops, that is, the right-hand loop I of pRNA I–j' can pair with the left-hand loop i' of pRNA J–i'. Since mixing two inactive pRNAs with interlocking loops resulted in production of infectious virions, the

TABLE IV
Six Interlocking pRNAs[a]

pRNAs	Predicted hexamer
(A–b')+(B–c')+(C–d')+(D–e')+(E–f')+(F–a')	[hexagon diagram, value 2×10^7]

[a]Adapted from Ref. 94 with permission.

stoichiometry of the pRNA is predicted to be a multiple of two. Together with the results from binomial distribution and serial dilution analyses, it has been confirmed that the stoichiometry of pRNA is six.

b. A Set of Three Interlocking pRNAs (94). Another set of mutants is composed of three pRNAs, J–p′, P–k′, and K–j′. This set is expected geometrically to be able to form a 3-, 6-, 9-, or 12-mer ring consisting of each of the three mutants. Each individual pRNA are inactive when used alone (Table II). When any two of the three mutants are mixed, no activity is found. However, when all three pRNAs are mixed in a 1:1:1 ratio, DNA-packaging activity is rescued. The lack of activity in mixtures of only two mutant pRNAs and the restored activity in mixtures of three mutant pRNAs is expected, since this set is designed in such a way that all three RNAs are needed to produce a closed ring. It suggests that the stoichiometry of pRNAs is a multiple of 3 in addition to being a multiple of 2 (Table II), that is 6 (or 12, but this number has been excluded by the approach using a binomial distribution (Section II.C.1).

c. A Set of Six Interlocking pRNAs (94). DNA-packaging activity is also achieved by mixing six mutant pRNAs, each of which is inactive when used alone (Table IV). Thus, an interlocking hexameric ring can be predicted to form by the base pairing of the interlocking loops.

4. Is There a Transition from Hexamer to Pentamer Inferred from Cryo-EM Imaging?

Two laboratories using cryo-electronic microscopy (EM) have also reported the stoichiometry of the pRNA on procapsids. Both compared the images of pRNA-free procapsids to images of pRNA-bound procapsids. After extracting the image of the pRNA-free procapsid from the procapsid with pRNA bound, Carrascosa and coworkers revealed a pRNA hexamer bound to procapsids (Fig. 6A) (99). In contrast, using the same techniques and approaches, Rossmann and coworkers found a pRNA pentamer on phi29 procapsids (Fig. 6B) (100). One explanation by Rossmann and coworkers is that pRNA hexamers are formed initially, but after binding to procapsids, one of the pRNAs is dissociated from the procapsids, leaving five pRNAs still bound. This discrepancy is not resolved; however, it is worth noting that due to the low electron density of RNA molecules and the resolution limit of EM along with the interference of the fivefold symmetrical environment of the procapsid shell around the pRNA complex, EM image reconstruction for such a molecule is difficult. Other biochemical and biophysical approaches are needed to clarify the discrepancy. It would be very interesting to see how the fivefold-symmetric pRNA pentamer could interact with a sixfold-symmetric connector. The other reason for Rossmann and coworkers

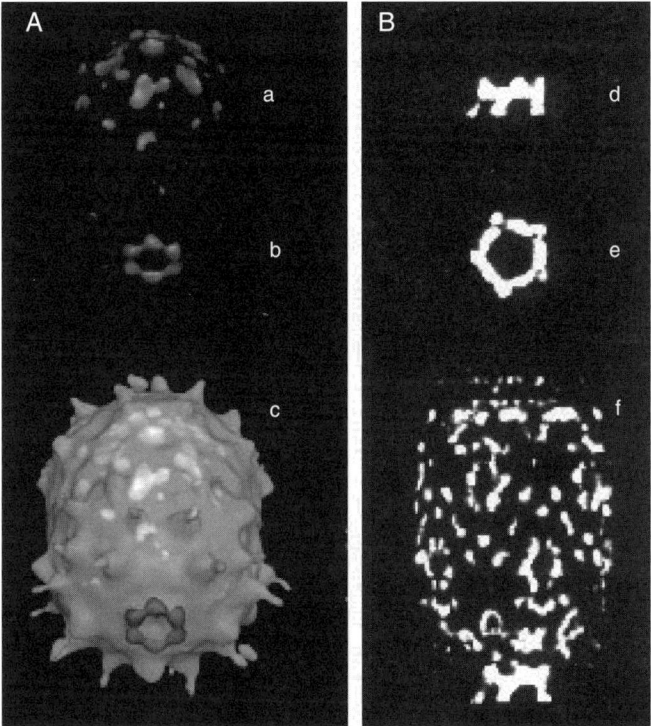

FIG. 6. Cryo-EM images showing that the pRNA forms (A) hexamer (101) and (B) pentamer (102), from two different laboratories using similar approaches. Reprinted by permission from (A) Ref. 101, copyright 2000, Academic Press; and (B) Ref. 100, copyright 2000, Macmillan Magazines.

to believe a pentamer is that they found the pRNA hexamer, constructed by Dr. Mayor and coworkers (96), could not fit into the structural environment of the connector embedded in the procapsid (100). If pRNA is indeed a pentamer, it would have a favorable energy for binding to the capsid protein gp8 with a fivefold symmetry rather than binding to the 12- or 6-fold-symmetric connector. However, extensive investigation by filter binding assay, sedimentation analysis, and genetic studies concludes that pRNA binds to the connector (Section IV.A). If a shift in binding target does occur, it might represent another novel mechanism in pRNA function. A possibility might be that six pRNAs bind to procapsid in such a way that two of them lie on top of each other to gear the DNA translocation motor, resulting in giving the appearance of a pRNA pentamer.

D. Construction of a pRNA Inactive in DNA Packaging but Competent in Procapsid Binding

Extensive investigation has confirmed that pRNA contains two functional domains (Section VI.A). The procapsid-binding domain is located in the middle part of the molecule, and the DNA-translocating domain is located at the 5'/3'-paired ends. Deletion or mutation of the DNA-translocating domain generates a mutant pRNA inactive in DNA packaging while having a procapsid-binding efficiency equivalent to wild-type pRNA (*93, 101*). This special feature of mutant pRNA has greatly facilitated the study of pRNA structure and function. As described in the previous and following sections, this mutant has been used for competitive inhibition in procapsid binding (*101*), stoichiometry determination using a binomial distribution (*93*), design of antiviral approaches (*101, 102*), model simulation that leads to the finding of the mechanism of sequential action (*104*), and design of chemical modification interference (*104*).

III. Studies of pRNA Structure

The astonishing diversity in RNA function is attributed to the flexibility in RNA structure. The question of how some RNA molecules perform their versatile and novel functions is largely unanswered. To elucidate this question, it is crucial to understand the principles and rules that regulate RNA structure building. Due to its complexity and versatility, the mechanisms of RNA folding remain to be elucidated. The very limited number of reported RNA structures solved by NMR or X-ray crystallography (*105–112*) compels the use of alternative approaches to obtain information on RNA structures.

A. Complementary Modification

Computer analysis (*113*) of pRNA secondary structure shows that the 5' and 3' ends are paired. Complementary modification was used to confirm the pRNA secondary structure in this and other areas of the molecule (*91, 98, 114*). An extensive series of helix disruptions by base substitutions almost always resulted in the loss of DNA-packaging activity. Additional compensatory mutations that restored the predicted base pairings rescued the activity of pRNA. Such complementary modification has led to the conclusion that bases 1–3 are paired with bases 117–115; bases 7–9 are paired with bases 112–110; bases 14–16 are paired with bases 103–101; and bases 76–78 are paired with bases 90–88 (Fig. 2). This second site suppression confirmed the existence of a helical structure and that this helical structure is essential for pRNA function.

Complementary modification has also been applied to the study of loop/loop interaction in dimer formation (*94, 96, 97*)(Section II.C.3). The observed activity of a mixture of two inactive mutants (Table II) confirmed that the right loop interacted with the left loop of a different RNA, and was involved in dimer formation.

B. Phylogenetic Analysis

The DNA packaging RNA of phages SF5, phi29, PZA, M2, NF, GA1 (*116*) and B103 (*115*) shows very low sequence identity but have similar secondary structures (*97, 116*) (Fig. 7). The requirement for pRNA in phi29 assembly appears to be very specific, in that pRNAs from other phages cannot replace the phi29 pRNA in *in vitro* packaging assays (*116*), and that a single base mutation can render the pRNA completely inactive (*114*). All seven pRNAs of these phages contain complementary right- and left-hand loops. The number of paired bases ranges from five for SF5 to two for GA1 (*97*). Phylogenetic analysis led to the conclusion that there must be at least one G/C pair in the pairing of the right- and left-hand loops, and that two G/C pairs are sufficient to mediate pRNA interactions (*94, 96, 97*).

C. Photoaffinity Crosslinking with GMPS/Aryl Azide

Aryl azides can be incorporated into RNA molecules to obtain three-dimensional structural information (*89*). These reagents contain groups that are chemically inert in the absence of light, but can be converted to a reactive nitrene by UV irradiation at long wavelengths (*117, 118*). Aryl azides have been specifically attached to the 5′ end of the pRNA or to cp-pRNAs. To accomplish this 5′-end labeling, 5′-thiophosphate pRNA or cp-pRNA is synthesized by *in vitro* transcription in the presence of excess guanine-monophosphorothioate (GMPS) over GTP (*118*). GMPS is an efficient primer of RNA synthesis by T7 RNA polymerase, but cannot be used for chain elongation. The 5′-thio-pRNA and the 5′-thio-cp-pRNAs are then treated with azidophenacyl bromide to produce the 5′-azido-pRNA and 5′-azido-cp-pRNAs, respectively, by the nucleophilic displacement of bromine (*118*). The azido group is converted to a reactive nitrene by long-wavelength UV irradiation, which then inserts into nearby bonds resulting in covalent crosslinks (*117*). Previous work showed that it is possible to generate active cp-pRNAs by assigning certain internal sites of the pRNA as new 5′ and 3′ termini (Section II.B) (*87, 91*). Therefore, with the use of cp-pRNA, specific internal bases of the pRNA were uniquely labeled with photoaffinity crosslinking agents to analyze inter/intramolecular interactions. When necessary, the 5′ end of the RNA can also be labeled with ^{32}P. Crosslinked RNAs can be separated from uncrosslinked RNAs by denaturing gel electrophoresis and crosslink sites determined by primer extension (*105, 119, 123*). Bases identified as crosslink sites by primer extension indicate bases that are in close proximity to the photoagent-labeled base. The use of cp-pRNAs allows the identification of intramolecular contacts throughout the pRNA molecule. Crosslinking data have been used as distance constraints in molecular modeling studies (*199*)(Section III.L).

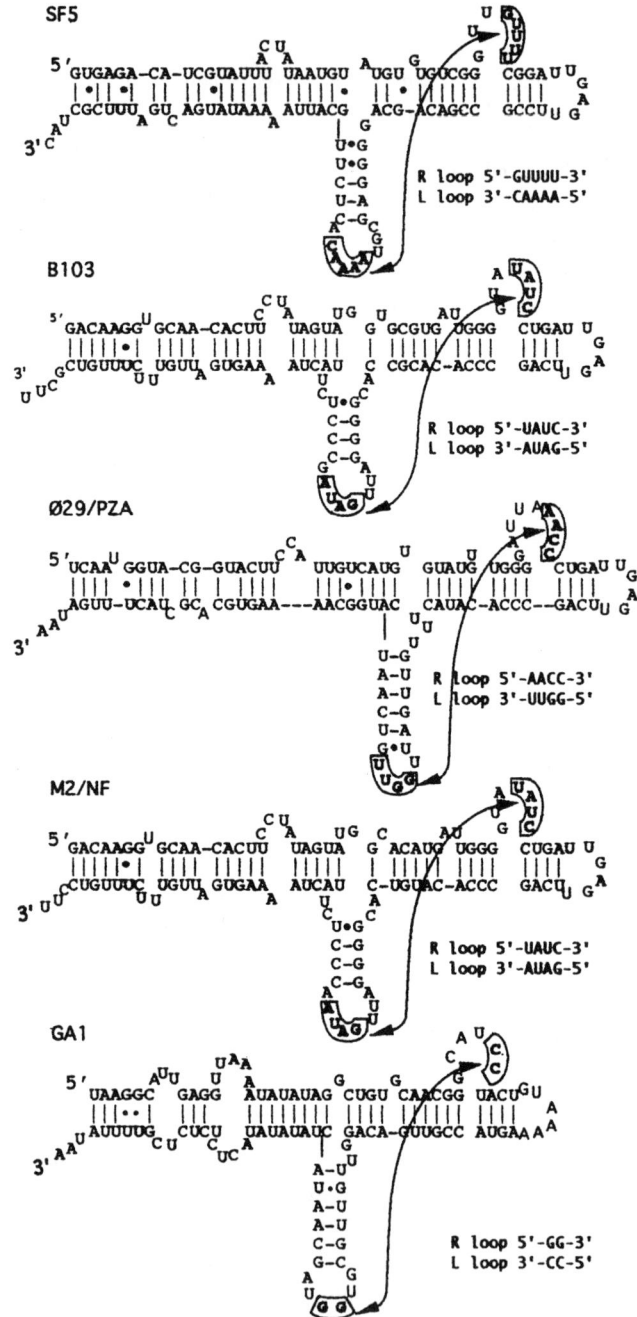

FIG. 7. Phlyogenetic analysis and predicted secondary structures of pRNAs in phages. The right and left loops for inter-RNA interaction are in boxes. Reprinted from Ref. 97 with permission.

1. INTRAMOLECULAR CROSSLINKING OF MONOMERS (106)

Three nucleotides have been selected as new 5' termini for labeling with aryl azides (Table V). One of the new 5' termini, G^{108}, is located within the 5'/3'-end helix necessary for DNA packaging, whereas two of the other sites, G^{75} and G^{78}, are located within the interior sequences involved in procapsid binding. The photosensitive group was coupled to the 5'-thiophosphate of each cp-pRNA. Each photoagent-coupled cp-pRNA is fully functional in DNA packaging. After irradiation, the crosslinked pRNAs are isolated from denaturing gels. Crosslinked pRNA migrates more slowly than uncrosslinked pRNA. Bands representing intramolecular and intermolecular crosslinks have been identified and purified. Crosslinked sites are mapped by primer extension. Reverse transcription stops occur one base prior to the crosslinked bases. For example, a stop at base 32 would mean that base 31 is crosslinked. G^{108} is crosslinked to C^{10} and G^{11}; G^{75} is crosslinked to A^{26}, U^{27}, G^{28}, U^{29}, and G^{30}; and G^{78} is crosslinked to U^{31}. The azidophenacyl group is only 9 Å, but experimental data have demonstrated that the cross-linking group can reach distances of 12 Å (N. Pace, personal communications). The data suggest that G^{108} is proximate to C^{10} and G^{11}; G^{75} is proximate to bases 26–30; and G^{78} is proximate to U^{31}. These distances have been used as constraints in computer modeling of pRNA structure (199) (Section III.L; see Fig. 10).

2. INTERMOLECULAR CROSSLINKING OF DIMERS (120)

The methods used for crosslinking of dimers are basically the same as those used for monomers crosslinking work. The cp-pRNA B–a' was labeled with aryl azides on G^{82} and incubated with an equimolar amount of unlabeled pRNA A–b'. The left and right loop sequences of these two pRNAs are complementary, and therefore they are able to form dimers. After UV crosslinking, dimers were isolated from the gel and the crosslinking site was identified by reverse transcriptase primer extension using a pRNA A–b'-specific primer. G^{82} was found crosslinked to G^{39}, G^{40}, A^{41}, C^{49}, G^{62}, C^{63}, and C^{64} (119), indicating that all eight bases are in close proximity as dimers, as seen in computer models (199) (Fig. 10).

D. Crosslinking by Psoralen

Psoralen has been used as a photoactive probe for nucleic acid structure because it intercalates into RNA or DNA helices and, upon irradiation with 320- to 400-nm light, freezes the uridine of RNA or the thymidine of DNA, which are in close proximity (helix or pseudoknot), by covalent attachment (120–122). The sites of crosslinks can be determined by primer extension (123) and/or mung bean nuclease treatment (124). The psoralen derivative 4'-aminomethyl-4,5', 8-trimethyl psoralen (AMT) has been selected due to its solubility (123). One advantage of psoralen crosslinking is that it crosslinks RNA or DNA, but not

TABLE V
Covalently Linked pRNA Dimers Active in DNA Packaging[a]

pRNAs	pfu/ml	Ref.
Wild type	2.0×10^7	138
Fused dimer	2.2×10^5	138
Tandem dimer	Ii'/Ii', 5.1×10^5; Ab'/Ab', 0; Ab'/Ba', 2.0×10^5	138
Dimer crosslinked with GMPS/aryl azide	Ba'/Ab', 2.2×10^6; Ia'/Ab', 0	119
No pRNA	0	

[a]Adapted from Ref. 138 with permission.

protein. This is different from the azido group described above, which crosslinks nonspecifically to both protein and nucleic acids. Psoralen, however, can induce intramolecular crosslinks within the pRNA even in the presence of other proteins, such as procapsid or gp16. Thus, pRNA conformations in different environments can be detected. Psoralen crosslinks can also be reversed by 254-nm irradiation. With the use of two-dimensional gel (*120, 122, 125*) and 5'-end-radiolabeled cp-pRNAs, pRNA conformational change in the presence of different packaging components can be investigated. The results reveal that pRNA has at least two conformations, one which is able to bind procapsid and the other which is not (Section III.M). It has been found that in the absence of Mg^{2+}, the region of C^{67} to A^{70} is proximate to U31 to U36, since these two areas can be crosslinked by psoralen (*123*).

E. Crosslinking by Phenphi

Unlike psoralen, phenphi [(*cis*-Rh(phen)(phi)Cl_2^+ (phen = 1,10-phenanthroline and phi = 9,10-phenanthrenequinone diimine)] induces covalent bonds between guanosine bases upon UV activation. Phenphi has also been shown to crosslink pRNA and has revealed the proximity of G^{75} with G^{28}, and G^{30} of pRNA (*126*).

F. Chemical Modification

Various chemicals alter specific reactive groups of RNA and thus provide information regarding base pairing, base stacking, and the tertiary interactions of specific bases of RNA. The modifying agents include dimethyl sulfate (DMS), which methylates A at N1, G at N7, and C at N3 (*128, 129*); kethoxal, which modifies G at N1 and N2 (*129*); and 1-cyclohexyl-3-(2-morpholinoethyl)-carbodiimide metho-*p*-toluene sulfonate (CMCT), which attacks U at N3 and G at N1 (*127–129*). Locations of modified bases can be identified by primer extension with reverse transcriptase (*127, 130*). Reverse transcriptase does not stop at methylated N7 of G. Thus, aniline-induced chain scission is required in this case prior to extension (*127, 129*). It is common to find that some nonspecific stops occur in primer extension of unmodified RNA; thus, it is critical to use unmodified RNA as negative controls in primer extension. When pRNA/protein mixtures are being analyzed, reactions are phenol-extracted, and the RNA is precipitated prior to the extension reaction. Chemical modification of a base is a good indication that the base is unpaired. Lack of modification will most likely be due to base pairing, especially in helical regions. Lack of modification may also be the result of tertiary interactions or noncanonical base–base, base–sugar, or base–phosphate interactions (*129*). Such reactions often occur in phylogenetically predicted stem–loop or bulge regions of RNA. Chemical modification data can provide information on base accessibility, which can be used to assess predicted secondary structures and evaluate molecular models of the tertiary structure.

Phi29 pRNAs have been modified with DMS, CMCT, and kethoxal (*131, 132*). Reaction conditions were obtained empirically to produce, on the average, one modification per molecule by titrating the concentration of the modifying reagents needed.

Chemical modification showed that the sequence $C^{18}C^{19}A^{20}$ is accessible to chemicals in monomers and dimers as well as procapsid-bound pRNA (*131, 132*). The results indicate that CCA, although not involved in procapsid binding (*91, 133*), is present on the surface of the pRNA as a bulge, perhaps to interact with other DNA-packaging components (*132*). This conclusion is supported by studies of pRNAs with mutations in the CCA bulge (Section III.I).

Bases $U_{85}U_{84}G_{83}G_{82}$ and $A_{45}A_{46}C_{47}C_{48}$ in monomer form are accessible to chemicals, but in dimer form they are not accessible to modification. These bases compose the right-hand and left-hand loops. Results confirm that these bases are involved in inter-pRNA interaction (*94, 96*). Bases G_{57}, A_{56}, and G_{55} are also protected from chemical modification in dimer form, but can be modified in monomer form. Comparison of the modification patterns of monomer and dimer revealed that the major loops—the right-hand, left-hand, and head loops—are all involved in pRNA/pRNA contact to form the dimer, since all of them are strongly modified in monomers but protected in dimers.

All five single-base bulges in wild-type pRNA are modified fairly strongly. Each of the five single-base bulges is dispensable for pRNA activity (*91, 98*). In fact, only base U^5 has mild negative effect on pRNA activity when deleted, and the detrimental effect is marginal (a 10-fold reduction) (*114*).

G. Chemical Modification Interference

Chemical modification interference has been performed to determine which pRNA bases are involved in dimer formation (Fig. 8) Normally, RNA B–a' would be able to interact with A–b' to form dimers. The monomer RNA B–a' was treated with either DMS or CMCT and then mixed with unmodified monomer A–b' in order to test their competency in dimer formation. If the modified base is involved in dimer formation, pRNA B–a' carrying this modified base would not be able to form a dimer with pRNA A–b', and thus this pRNA will be present in the fast-migrating band representing monomers. After being isolated from the gel, both monomers and dimers were subjected to primer extension to identify the modified bases. The concentration of the chemicals was pretitrated to ensure that on the average only one base per RNA was modified (*105*). The results of chemical modification interference analysis revealed that bases U^{54}, G^{55}, U^{59}, C^{65}, A^{66}, A^{68}, U^{69}, A^{70}, C^{71}, C^{84}, C^{85}, C^{88}, A^{89}, A^{90}, and C^{92} were involved in dimerization, whereas bases 72–81 were not (*105*).

A Chemical Modification of Monomer (pRNA 7/11) with Mg^{++}

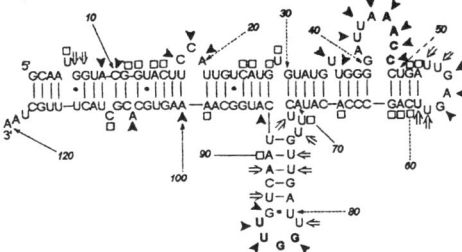

B Chemical Modification of Dimer (pRNA 5'/3' B-a') with Mg^{++}

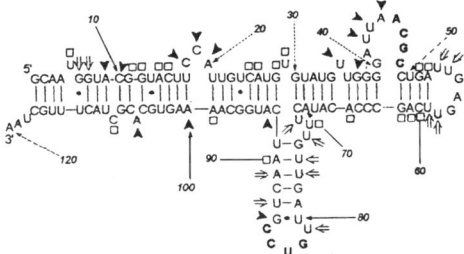

C Chemical Modification Interference of pRNA 5'/3' B-a'

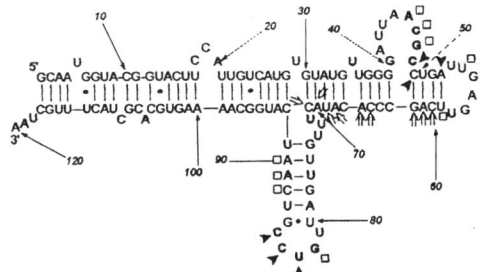

D Formation of Dimer

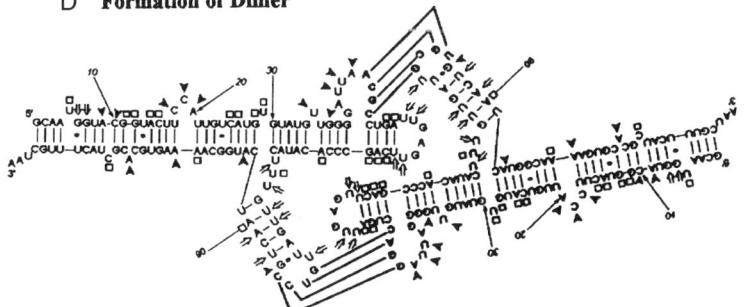

FIG. 8. Summary of (A, B) chemical modification and (C) chemical modification interference. Black arrowheads, double-lined arrows, and empty squares indicate strong, moderate, and weak modification, respectively. (A, B) The chemical modification of monomer and dimer, respectively. (D) RNA/RNA interaction in dimer formation based on the data from A–C. Adapted from Refs. *131* and *105* with permission.

H. Ribonuclease Probing

Some ribonucleases are sensitive to RNA secondary structure. For example, RNases T1 (specific for GpN linkages), U2 (specific for ApN linkages), and S1 prefer to cleave single-stranded RNA. Nuclease V1 is specific for double-stranded RNA. End-labeled pRNA and cpRNA in various solutions containing Mg^{2+}, procapsid, or gp16, individually or in combination, have been probed by T1, U2, or V1 (*92, 116, 123*).

T1 and V1 are used to distinguish the loops and helices of four RNAs with similar functions (*116*). Footprinting has also been conducted using ribonuclease to detect the sequence that binds procapsid (*92*). T1 has been used to study the change of pRNA conformations (*123*). Since the activity of RNase T1 is Mg^{2+}-independent, this enzyme can be used to investigate the conformational change of pRNA induced by ATP in the presence or absence of Mg^{2+} (*123*).

Mg^{2+}-induced pRNA conformational change has been verified by T1 ribonuclease probing (*123*). The pattern of partial digestion of pRNA by T1 provided strong evidence for the presence of two conformations, depending on the presence or absence of Mg^{2+}. Without Mg^{2+}, strong cleavages by T1 were seen in G_{28}, G_{30}, and G_{34}. With Mg^{2+}, these three bases became more resistant to T1 attack, indicating a conformation change or refolding of pRNA caused by Mg^{2+}.

I. Genetic Analysis by Truncation, Insertion, Deletion, and Mutation

The availability of the highly sensitive pRNA assay procedure using the *in vitro* phi29 assembly system (Section II.A) greatly facilitated the genetics analysis. Taking advantage of the circularly permuted pRNA system (Section II.B) and the technique of two-step PCR, one can easily construct mutant pRNA with truncation, deletion, insertion, and mutation targeting at any defined position. Plasmids with pRNA-coding sequence were used as templates to generate PCR DNA fragments with primer pairs containing either the T7 or SP6 promoter and mutations to pRNA. The PCR fragments were used to transcribe mutant or circularly permuted pRNAs *in vitro* with either T7 or SP6 RNA polymerase. By the use of this procedure in combination with the aforementioned *in vitro* assembly assay system, tens of mutant pRNAs can be obtained and tested in 1 or 2 weeks. Complementary modification (*134*) is one of the mutation studies that has been used to identify helix regions and intermolecular loop/loop interaction in dimer formation (*91, 96–98, 114*) (Section III.A).

Nucleotides $U^{72}U^{73}U^{74}$ are predicted to constitute a bulge located at a three-helix junction (Fig 2). Mutation studies have shown that the deletion of these three nucleotides abolishes the activity of the mutant pRNA F5, which has native 5'/3' ends (*98*). However, a circularly permuted pRNA 75/71 that had deletion of these three nucleotides but then had new 5' and 3' ends located at bases 75 and

71, respectively, was fully active in *in vitro* phi29 assembly (*91*). This suggested that the UUU bulge in this area provided flexibility to the pRNA. In pRNA F5 with normal 5' and 3' ends, deletion of the UUU bulge eliminated the flexibility in folding of the three-way junction, thereby creating an inactive mutant. In cpRNA 75/71, this flexibility was compensated for by the discontinuity of the phosphodiester bond in this area. The result suggested that the UUU presents as a bulge at the three-way junction to provide flexibility in folding and serve as a hinge for the twisting of the lower stem–loop.

Deletion (*98*) and insertion (*91*) have been used to study the function of the $C^{18}C^{19}A^{20}$ bulge. A pRNA with three bases 3'-GGU-5' inserted between A^{99} and A^{100} to pair with the bases $C^{18}C^{19}A^{20}$ in the bulge generates the pRNA 7/GGU (*91*). This pRNA was fully competent to form dimers and bind procapsids, however, its activity in DNA packaging and virion assembly was completely lost (*91*). A pRNA with a deletion of all three bases of the CCA bulge (*133*) exhibited the same biological activity as pRNA 7/GGU concerning procapsid binding, DNA packaging, and virion assembly, indicating that the CCA bulge, though not involved in procapsid binding, likely interacts with other DNA-packaging components (*132*).

J. Selex

In vitro evolution is a powerful tool for studying consensus elements of RNA structure and function. Starting with a library containing pRNA sequences with random mutations within a defined region, *in vitro* evolution techniques allow the selection of pRNA variants that are able to bind a specific ligand. Such selection for interacting species is based on different primary structures that would adopt the same structural feature as wild-type RNA. SELEX (systematic evolution of ligands by exponential enrichment) allows screening for covariation of several nucleotides and can be used to reveal noncanonical interactions that are difficult to prove by classical genetic and biochemical approaches (*135, 136*). SELEX has been used for the selection of pRNA sequences that bind procapsids (*137*). Using a DNA template pool with random mutations covering residues 30–91, it was found that after five rounds of selection, most clones selected were wild type. Covariation is found in the predicted helix region, supporting the prediction of helical structures in these regions (*137*).

A second pool of pRNA sequences was also constructed to test the loop/loop interaction in pRNA dimer formation (*137*). Using randomized mutation from bases 45–62, which cover the right-hand loop, and bases 81–84, which cover the left-hand loop, it was found that after five rounds of selection, the pRNA pool can compete with wild-type pRNA for procapsid binding. It was concluded that the wild-type pRNA sequence was in all probability the most suitable sequence for procapsid binding. It is important to point out that special care must be taken when considering dimer formation with regard to the data in this experiment.

A finding of unpaired right-hand and left-hand loops from clones selected from the pool that can also bind procapsid is by no means an indication that pairing is not absolutely necessary for procapsid binding. For example, if mutant pRNA A–b' and B–a' are present in the pool, both of them could pass the screening and thus be selected by binding assay since A–b' will interact with B–a' to form a dimer and will bind procapsids, and be selected. However, the isolation of A–b' from the pool does not mean that A–b' by itself could bind procapsids. It is probable that A–b' was selected based on its ability to bind procapsids in combination with a *trans*-complementary pRNA, such as B–a'.

K. Cryo-Atomic Force Microscopy

Atomic force microscopy (AFM) has been used by several investigators to detect images of RNA in a denatured conformation. As the first attempt to test whether this technique can be used to detect the three-dimensional structure of RNA in native conformation, cryo-AFM has been performed on phi29 pRNA (*105, 131, 138*).

1. Cryo-AFM Images of the Monomer (*132, 139*)

Cryo-AFM imaging reveals that the pRNA monomer folded into a "check mark"-shaped structure. The illumination and brightness in Fig. 9 indicates the thickness or height of the image, but does not reflect the atom density observed end-on. The brighter or whiter the illumination, the thicker is the surface; the darker the illumination, the thinner is the surface. The illumination and contrast of the image clearly indicate that the area around the head loop (the elbow of the "check mark") is the thickest/tallest (Fig. 9), as seen in the computer model (Fig. 10, see color insert).

2. Cryo-AFM Images of Dimers (*105, 131, 138*)

Cryo-AFM has been used to directly visualize purified pRNA dimers. The native dimers have an elongated shape (Fig. 9) (*131*). It appears that head-to-head contact is involved in dimer formation, resulting in a complex almost twice as long as the monomer. The conclusion of head-to-head contact agrees with chemical modification results showing that nucleotides of the head loop in dimers were protected from chemical attack. Cryo-AFM images of the fused dimer, a pRNA construct consisting of two tandemly linked pRNAs (Table V), exhibit a similar structure to the non-covalently linked pRNA dimer (*105*). The dimensions of the covalently linked fused dimer are comparable to those of the native dimer (*131, 138*).

L. Modeling the Three-Dimensional Structure of pRNA

The goal of modeling pRNA is to organize collections of structural data from crosslinking, chemical or ribonuclease probing, cryo-AFM, and other genetic

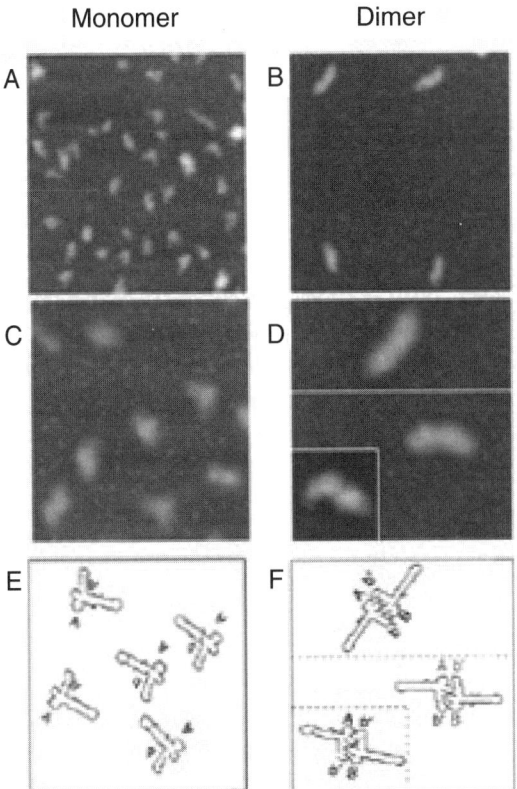

FIG. 9. Cryo-AFM showing monomers (left column) and dimers (right column) of pRNA. The magnification in A and B is ×5,000,000. C and D, enlarged images of monomers and dimers, respectively. (E, F) Illustrations of monomers and dimers respectively. Reproduced from Ref. *131* with permission.

data into a three-dimensional form. Since a large number of structural constraints are available, computer programs can successfully construct three-dimensional (3D) structures (*112, 139, 140, 142*). The first attempt to construct the 3D structure of the pRNA monomer and hexamer was based only on the information available for pRNA secondary structure (*96, 141*) and the finding of the intermolecular interaction of two loops with four bases. Considering the construction of a RNA 3D structure without 3D structure data, these models are plausible. Based on the subsequent extensive investigation of the pRNA 3D structure, new models of the 3D structure of pRNA monomer and hexamer have been constructed (*199*). A computer model of pRNA dimer has also been constructed based on the constraints obtained from chemical modification, chemical modification interference, crosslinking, and cryo-AFM data (Fig. 10; see color insert) (*199*).

New models of pRNA monomer, dimer, and hexamer (Fig. 10) were produced on Silicon Graphics Octane and Indigo^2 computers running IRIX 6.2 or 6.5, using the programs NAHELIX, MANIP, PRENUC, NUCLIN, and NUCMULT (*142, 112*). The modeling was performed based on the following assumptions: (1) All helices are modeled as a regular A-form double helix. (2) Internal loops and mismatched bases are constructed by maintaining the integrity of the double helix while optimizing base pairing and stacking inside the loop, as suggested by most structural data from X-ray and NMR analysis. (3) A general rule for the modeling of the RNA hairpin loop has been proposed (*143*), which involves maximal stacking on the 3' side of the stem and enough nucleotides stacked on the 5' side to allow loop closure, as found in the anticodon loop of tRNA. (4) Bulges less than four bases are modeled either sticking out from stems to avoid helical distortion, or within the helical domain, causing the helical axis to bend. Parameters for stacking energy are considered to decide whether they are protruding or within the helical stems (*144*). (5) Helix untwisting or twisting, helix–helix interactions, or other higher order structures have been built into the model at constant geometrical distances, but allowing certain torsion angle variation. A program regarding RNA flexibility has been applied to the construction of the UUU bulge at the three-way junction. These three-base bulges have been found to provide flexibility for the appropriate folding of the three-way junction. Conventional computer algorithms involving the minimization of empirical energy functions have been considered. Twelve angstroms has been considered as a maximum-distance constraint when bases are crosslinked by GMPS/aryl azide. Modified distance geometry and molecular mechanics algorithms have been considered for generating structures consistent with data from crosslinking, chemical modification, and chemical modification interference. A constraint satisfaction algorithm is used to refine the disturbed area in order to ensure the regular arrangement of all atoms.

Crystal structure reveals that the connector contains three sections, a narrower end, a central section, and a wider end with a diameter of 6.6 nm, 9.4 nm and 13.8 nm, respectively (Fig. 10D) (*100, 101*). The hexamer model by Hoeprich and Guo (*100*) contains a central channel with a diameter of 7.6 nm, that perhaps can sheath onto the narrowed end of the connector and anchored by the connector central section, (Fig. 10E & F).

As noted earlier (Section VI,A), pRNA contains two functional domains (Fig. 10), one is for connector binding and the other is for DNA translocation. The connector binding domain is located at the middle, bases 23–97, and the DNA translocation domain is located in the 5'/3' paired ends. It has been predicted that the connector protein (gp10) contains a conserved RNA recognition motif (RRM), residues 148–214, located at the narrow end of the

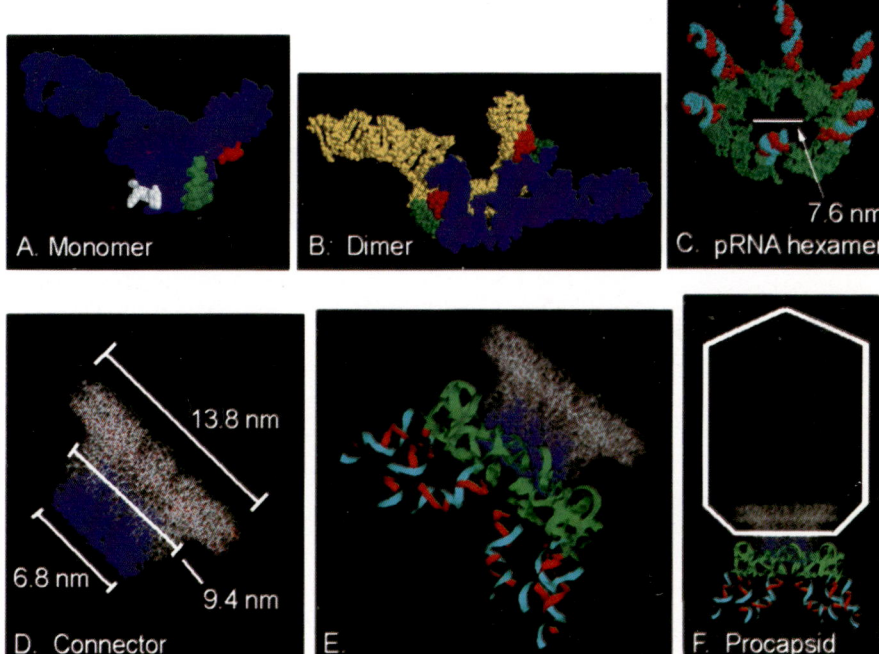

FIG. 11.10. Computer models showing the 3-D structure of pRNA monomer, dimer, hexamer, and pRNA/connector complex. (A) The model of monomer in spacefill format showing the $U^{72}U^{73}U^{74}$ bulge (in white), the right (in red), and left- (in green) hand loop. (B) The model of dimer in spacefill format with one unit in blue and the other unit in yellow. The right and left-hand loops are highlighted in red and green, respectively. (C) The model of hexamer in spacefill format showing the procapsid binding domain in green, and the DNA translocating domain in red (the 5′-end) and cyan (the 3′-end). The DNA translocating domain of the 5′/3′-paired region points up. (D) The crystal structure of the connector (Simpson, Tao, et al., 2000) in wire-frame format. The RNA Recognition Motif (RRM) (Grimes & Anderson, 1990; Guo, 2002) is in blue. (E) Docking of the pRNA hexamer to the RNA binding domain (RRM) of the connector. The connector-binding domain is in green and the DNA translocating domain is in red (the 3′-end) and cyan (the 5′-end). (F) Illustration of phi29 viral particle with the DNA-packaging motor composed of pRNA hexamer and connector (From Hoeprich & Guo, 2001, with permission from *J. Biol Chem.*).

connector that extends from the procapsid (152, 160). The hexamer model by Hoeprich and Guo (199) complies with the aforementioned data showing that the bases 23-97 (colored green in Fig. 10C, E & F), which is the connector binding domain, interact with the predicted RRM of connector (Fig. 10E and F in blue), while the 5'/3' paired ends (Fig. 10E in red and cyan), which is the DNA translocation domain, protrudes from the connector.

M. Conformational Change of pRNA

The ability of RNA to perform diverse biological functions despite being made up of only four different building blocks (A,C, G, and U) can be attributed to the flexibility in RNA folding. In two instances of pRNA action, the tasks performed could not be accomplished without considering a conformational change. The first instance is the transition from pRNA dimers to hexamers. The other is the execution of the physical motion required to translocate phi29 genomic DNA into empty procapsids. It has also been demonstrated that pRNA monomers possess two conformations.

Conformational changes in pRNA have been demonstrated by assessing the structure of pRNA in the presence or absence of Mg^{2+} and other DNA-packaging components. The conformation of pRNA in the presence and absence of Mg^{2+} has been investigated with psoralen crosslinking (123), nuclease probing (123), and chemical modification (131). The conformation of the pRNA bound to procapsids has been analyzed by RNase footprinting (92) and chemical probing (141).

1. TRANSITION FROM MONOMER TO DIMER: CONFORMATIONAL CHANGE INDUCED BY Mg^{2+}

Free pRNA in solution can adopt two conformations. Mg^{2+} induces a pRNA conformational change, as revealed by pRNA migration rate changes in EDTA and Mg^{2+} gels, by T1 ribonuclease probing, and by psoralen crosslinking (123). In the presence of 1 mM Mg^{2+} and absence of EDTA, pRNA migrates faster in gel, indicating a conformational change. Indeed, an Mg^{2+} concentration of 1 mM was found to be sufficient to induce this conformational change, and this concentration is the same as that required for pRNA/procapsid binding and DNA packaging. Moreover, this concentration was also close to the Mg^{2+} concentration in normal host cells. With Mg^{2+}, pRNA seems to adopt a more compact conformation, which may facilitate its attachment to procapsid.

Ribonuclease T1 digestion confirms the presence of two conformations. In the absence of Mg^{2+}, strong cleavages by T1 are seen in pRNA G_{28}, G_{30}, and G_{34}. With Mg^{2+}, these three bases become more resistant to T1 attack.

A conformational change in pRNA induced by Mg^{2+} has been further investigated by psoralen crosslinking (Section III.D) (123). In the absence of Mg^{2+}, pRNA is crosslinked and shows a slower migration rate in gel. When 1 mM

Mg^{2+} is present, the crosslinked band is undetectable. It is known that Mg^{2+} does not affect the crosslinking efficiency of RNA with psoralen (120). The induced conformational change is reversible by additions of EDTA. Nucleotide U^{69} has been confirmed to crosslink to U^{31}, U^{33}, and U^{35}. At concentrations of 5 mM, Mn^{2+}, Ca^{2+}, and Co^{2+}, can also inhibit the crosslinking by psoralen. This is in agreement with later experiments showing that Mn^{2+} and Ca^{2+} could replace Mg^{2+} in pRNA dimer formation and procapsid binding (138).

Chemical modification of the pRNA in the absence of Mg^{2+} has also been performed. Both the right-hand loop and the head loop show a relative lack of modification in the absence of Mg^{2+} compared to when Mg^{2+} is present. A similar reduction in modification occurs in the stem of the head-stem loop. The presence of drastically altered chemical modification patterns within this procapsid-binding domain suggests that Mg^{2+} plays a crucial role in pRNA folding to generate the appropriate right-hand and head loop structures, and explains why Mg^{2+} is needed for dimer formation, procapsid binding, DNA packaging, and phi29 assembly.

Other evidence points to a conformational shift from monomer to dimer. For example, in pRNA monomers, base A^{56} in the head loop is accessible to chemicals, whereas in dimers, A^{56} in the head loop is not modified by DMS (131). This is an indication of a conformational change in the head loop area during the formation of dimers—the building blocks of the hexamer motor complex (138).

A conformational change in the pRNA has also been observed upon procapsid binding. It was found that the cleavage of four guanines, $G^{37}G^{38}G^{39}G^{40}$, by ribonucleases was enhanced in procapsid-bound pRNA compared to cleavage of free pRNA. Such a result implies that pRNA undergoes a conformational change upon binding to procapsid (92).

2. TRANSITION FROM DIMER TO HEXAMER

It is concluded that dimers are the building blocks in hexamer assembly with a pathway of dimer → tetramer → hexamer (138). Closed dimers, two molecules linked together by the holding of two pairs of hands, are active in procapsid binding and DNA packaging, whereas open dimers formed by the holding of only one pair of hands, are unstable in solution (138). (Section VI.D). Both tandem and fused pRNA dimers (Table V) with complementary loops designed to form even-numbered rings are active in DNA packaging, whereas those without complementary loops are inactive (138). These results imply that the true pRNA intermediates in hexamer assembly is the closed dimer with the holding of two pairs of hands (i.e., right- and left-hand loops of each pRNA paired with those of the other), and that the two interacting loops of one dimer play a key role in recruiting an incoming dimer (94, 97). In dimers, each pRNA holds hands with only one additional pRNA. However, in hexamers, each pRNA holds hands with two additional pRNAs. It seems paradoxical, then, that a dimer with no unpaired loops would be the precursor to hexamers where each pRNA interacts with two others. Thus, how is the conclusion that the dimer is the building block for pRNA hexamer to be interpreted? Recent results reveal that the pRNA has a strong tendency to form a circular ring by hand-in-hand contact, regardless of whether the pRNA is in dimer, trimer, or hexamer form (results to be published), suggesting that the structure of pRNA is bendable. Therefore, a conformational shift is expected during the transition from dimer to hexamer. It is reasonable to believe that dimer formation is a prerequisite for the generation of an appropriate 3D interface for procapsid binding. In such a scenario, one of the hands of the dimer that is already paired with its complementary loop sequence on the other pRNA would release after binding to the procapsid. The dimer with a released hand is similar to the open (linear) dimer, which has been demonstrated to be unstable in solution, yet still active in procapsid binding and DNA packaging (*97*). The open hand may serve as a welcoming hand to recruit an incoming dimer to the procapsid (Fig. 12). Indeed, pRNA conformational changes before and after binding to procapsid have been documented by nuclease probing (*92*) and chemical modification (*141*) (Section III.F).

3. CONFORMATIONAL SHIFT IN EXECUTING THE MOTION TASK

Many macromolecules undergo a conformational change in executing tasks involving motion, rotation, or migration. In the phi29 system, procapsids contain a sixfold-symmetric connector embedded in a protein shell with a fivefold rotational symmetry. The relative motion of two rings could allow the motion of DNA translocation to occur (*104, 146*). The finding that a hexameric pRNA complex is bound to the connector and that six pRNAs work sequentially is potentially relevant to this model (Section VI.H and Fig. 13). It has been shown that pRNA contains two domains: one for connector binding and the other for DNA translocation (Section VI.C). The pRNA binds to the connector, leaving the 5'/3' domain free to interact with other DNA-packaging components. It is proposed that the pRNA is part of an ATPase and possesses at least two conformations, a relaxed form and a contracted form. Alternating between contraction and relaxation, each member of the hexameric RNA complex, driven by ATP hydrolysis, helps rotate the DNA translocation machine (*104*).

IV. pRNA–Protein Interactions

A. Strong Binding of pRNA to Connector but Undetectable Binding to Capsid Proteins

One important aspect in investigating the role of an enzyme is to find out where it binds. Cosedimentation of the pRNA with purified connector (portal vertex) in sucrose gradients, and retention of radioactive-labeled pRNA with

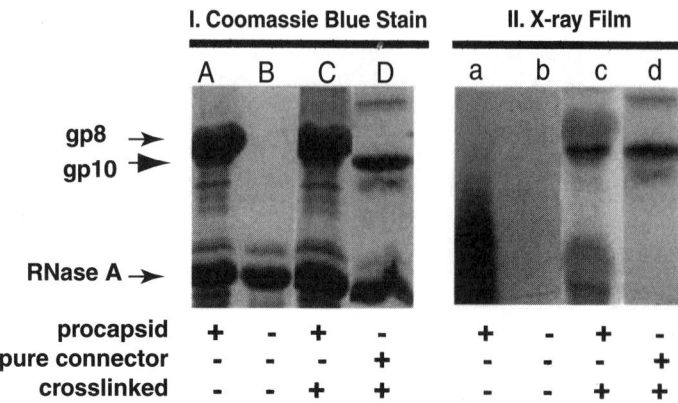

FIG. 11. Specific binding of radiolabeled pRNA to connector demonstrated by crosslinking. I and II are Coomassie blue stain and X-ray film, respectively of the same SDS polyacrylamide gel. Lanes a–d correspond to A–D, respectively. In addition to the components noted on the bottom of the figures, all lanes contain [^{32}P]pRNA. Reaction mixtures were treated with RNase A before loading to the gel. The smears in lane a suggests that the RNase A could not digest pRNA completely due to the presence of a high concentration of procapsid in the reaction. Adapted from Ref. 148 with permission.

this connector protein in nitrocellulose binding assays, indicate that the pRNA is attached to the connector and is part of the translocating machinery (69). Moreover, the isometric particle, a defective procapsid lacking the portal vertex protein, does not contain pRNA (69, 147).

It would be interesting to know whether pRNA binds to the connector gp10 only, or whether it binds to the capsid protein gp8 to bridge the connector and capsid. UV crosslinking has been performed to answer these questions (148). Briefly, pRNA was made in the presence of alpha-^{32}P[UTP] and 5′-iodo-UTP. After binding, the RNA/procapsid complex was irradiated with short-wavelength UV light. When [^{32}P]pRNA was crosslinked with procapsid, the band in the sodium dodecyl sulfate (SDS) gel corresponding to gp10 was specifically labeled and the gp8 was not, indicating that pRNA bound selectively to gp10 (Fig. 11) (148). 5S rRNA did not compete with radioactive-labeled pRNA for gp10 binding, but unlabeled pRNA did, indicating the binding is specific. Carrascosa and coworkers, who found that 5S RNA was able to produce some DNA packaging in the in vitro assay (149, 150), published reports later that the specific pRNA was able to discriminate and select phi29 DNA for packaging from a pool containing both bacterial and phage DNAs (151).

The nonappearance of pRNA/gp8 binding in cross-linking experiment does not completely exclude the possibility of transient interaction of pRNA with capsid protein. The data can only conclude that the affinity of pRNA to connector is much stronger than to capsid, and connector is the foothold for pRNA.

B. Footprinting and Chemical Probing of Procapsid-Bound pRNA

Footprinting of the pRNA/procapsid complex with nuclease A, T1 and V1 reveals that bases 22–84 are protected from enzyme digestion (92). This is an indication that the region from bases 22 to 84 contacts with connector. Chemical probing of procapsid-bound pRNA has also been performed to investigate the pRNA/procapsid interaction (141). The chemicals DMS, CMCT, and kethoxal used in this experiment are much smaller than ribonucleases; thus, a finer mapping of procapsid-protected bases can be achieved than by RNase probing.

The most striking feature of the modification of procapsid-bound pRNA was the relative lack of modification of the RNA compared to wild-type pRNA alone. The entire left stem–loop was completely unreactive. Additionally, only base A^{56} of the head loop was modified. In the right loop, which was modified strongly and completely in wild-type pRNA, only bases 42–44 were significantly modified and base 45 was only moderately modified.

The helical regions throughout the molecule were completely uncreative, with the exception of base U^{36}, which was strongly modified in both bound and free pRNAs. The bases C^{18}, C^{19}, A^{20}, and U^{35} were all strongly modified in both procapsid-bound and free pRNAs, suggesting that this three-base bulge is extended from the procapsid. The bases U^5 and U^{73} were moderately modified in free pRNA and strongly modified in procapsid-bound pRNA. Also, bases U^{17} and U^{21} were moderately modified in procapsid-bound pRNA, but not in free pRNA, suggesting that the binding of pRNA to procapsids changed the conformation of pRNA.

C. Connector Domain and Sequence for pRNA Binding

By sequence-homologous comparison, it was predicted that the connector protein gp10 contains three conserved RNA recognition motifs (RRMs). The first motif was predicted to be residues 7–57 (152), and the second one, residues 22–95. However, by comparison with the crystal structure of the purified connector (32, 100, 101, 153, 154), these two motifs fall into the upper half of the wider connector end, which is embedded in the procapsid. Since this region is not exposed, it is impossible for pRNA to bind if the crystal structure of the purified connector possesses an identical conformation to the one embedded in the procapsid. The third motif, residues 148–214, is located at the narrow end of the connector, which protrudes from the procapsid. It is possible that this might represent the true RRM for pRNA binding, based on the assumption that the conformation of the purified connector is identical to the one embedded in the procapsid.

Site-directed mutation of several charged and aromatic residues of connector protein gp10 render the procapsid unable to package DNA (155). The RNA-binding sequence of gp10 is predicted to be the amino acid residues 21–94. Specific antibodies against polypeptides covering different stretches of the gp10 sequence have been generated to investigate the RNA-binding region of the connector (156). In combination with immunoelectron microscopy, the RNA-binding domain has been mapped to be at the end of the narrow side of the connector (99). Again, the crystallographic structure of a purified connector reveals that residues 21–94 are located at the wider end, which is embedded in the procapsid. It would not be possible for pRNA to bind residues 21–94. Several domains of the connector have been mapped by antibody studies (156, 157). The domains and sequence mapped by the antibody (156) do not agree with X-ray crystallography structural data (158). But the immunolabelling detection of the pRNA site, other than sequence, at the narrow end of the connector is fully consistent with the crystallographic data. The global connector structure obtained from EM studies (157, 159–161) was very similar to the structure solved by crystallography (100).

D. Can pRNA Interact with gp16?

A filter binding assay was performed to study protein binding of pRNA (69). Glass filters could bind proteins nonspecifically, but did not bind RNA. When radioactive-labeled pRNA binds to protein, pRNA attaches to the filter due to the retention of the protein on the filter. Both connector and gp16 have been tested for pRNA binding. gp16 does not bind to purified pRNA, but the connector does. Later, it was found that gp16 could protect the procapsid-bound pRNA from nuclease digestion (152) by the addition of a gp16-containing cell extract to the reaction, but not or little protection was observed by the addition of purified gp16. The difference in these two gp16s is that the cell extract contains a large amount of unspecific proteins. The RNA recognition motif of gp16 has also been predicted, though not proven, to be residues 47–119. Whether gp16, after binding to the procapsid, interacts with pRNA remains an interesting unsolved problem. It is worth noting that due to the hydrophobicity, gp16 has a strong tendency to self-aggregation. This special feature of gp16 must be dealt with carefully in order to avoid nonspecific binding due to the strong self-aggregation.

V. Effect of Mono- and Divalent Cations on pRNA Dimer Formation, Procapsid Binding, and Viral Assembly (123, 138)

Mg^{2+} is required for dimer formation. For circularly permuted RNAs with new termini at base 71 and 75, the Mg^{2+} concentration required for 50% dimer formation is 4 mM; for pRNAs with wild-type 5'/3' ends it is 0.4 mM (138). However, the mechanism behind this difference is unknown.

Effects of other cations on dimer formation, procapsid binding, and viral assembly have also been investigated. At least a 1 M concentration of monovalent ions is needed for pRNA dimerization, whereas as low as 5 mM of divalent ions is sufficient. Spermidine, a positively charged compound, can also stimulate dimer formation at a concentration of 5 mM, indicating that dimerization is a result of a cation effect. $CoCl_2$ and $NiCl_2$ did not promote dimer formation, whereas $FeCl_2$, $ZnCl_2$, and $CdCl_2$ caused the precipitation of pRNA. These data suggest that pRNAs form dimers in the presence of positively charged cations, including mono- or divalent cations, as well as spermidine. Dimer formation is an intrinsic feature of pRNAs, and cations play a facilitating role.

Contrary to the lack of ion specificity for dimer formation, the binding of dimers to procapsid showed a specific requirement for divalent cations. Neither monovalent cations nor spermidine could stimulate the binding of dimers to procapsids. Only Mg^{2+} and Ca^{2+} could, whereas Mn^{2+} showed a reduced efficiency. These data indicate that dimer formation is not sufficient for procapsid binding, DNA packaging, and viral assembly. Divalent cations, Mg^{2+}, Ca^{2+}, and Mn^{2+}, play other roles in addition to promoting dimer formation.

VI. Functions of pRNA

A. Separation of the pRNA into Two Domains: Procapsid-Binding Domain and DNA-Packaging Domain

Extensive investigation has provided strong evidence that the pRNA molecule contains two functional domains (Fig. 2). This conclusion comes from the results of different approaches, including (a) base deletion and mutation (86, 91, 98, 114, 162), (b) ribonuclease probing (92, 123), (c) oligo targeting (88, 103), (d) competition assays to inhibit phage assembly (93, 102, 103), (e) crosslinking to portal protein by UV (148), and (f) psoralen crosslinking and primer extension (123). Mutations of more than two bases at the 5' or 3' end of the pRNA to noncomplementarity within the 5'/3'-end terminal helix resulted in complete loss of DNA-packaging activity. However, these mutants were able to bind procapsids with an affinity similar to wild-type pRNA (86, 93, 103, 148). Chemical and ribonuclease probing revealed that the binding pattern of these mutant pRNAs could not be distinguished from that of wild-type pRNA (131, 141). These mutants were able to compete with wild-type pRNA for procapsid binding and also strongly inhibited phage assembly *in vitro*. Antisense oligos targeting the procapsid-binding domain blocked the binding of pRNA to procapsid, and therefore completely inhibited viral assembly *in vitro* (88). Antisense oligos targeting the 5'/3' DNA-translocating domain (Fig. 2) completely inhibited viral assembly, but did not block procapsid binding (88). A truncated pRNA, made up of bases 28–91, can still be UV-crosslinked to the phi29 connector protein gp10 specifically (147). A 75-base RNA segment, made up of bases 23–97, was able to form dimers, interlock into hexamers, compete with full-length pRNA for procapsid binding, and therefore inhibit phi29 assembly *in vitro* (138). All of the above data support a conclusion that pRNA contains two functional domains, one for procapsid binding and the other for an as-yet-undefined role in DNA translocation. The procapsid-binding domain is located in the central part of the molecule (92, 138, 148), bases 23–97, and the DNA-translocation domain is located in the 5'/3'-paired ends (86).

B. pRNA Remains Associated with DNA-Filled Capsids During DNA Packaging (104)

One question that remains to be answered is whether the pRNA is needed only for the initiation of DNA packaging, or whether it participates in the DNA translocation process. [^{3}H]pRNA was used to package unlabeled genomic DNA–gp3 *in vitro*. Following sucrose sedimentation to separate DNA-filled capsid from empty procapsids, it was found that [^{3}H]pRNA was present in the radioactive peak corresponding to the location of DNA-filled capsids. These results suggest that the pRNA is associated with the DNA-packaging machinery during the DNA translocation.

C. pRNA Leaves DNA-Filled Capsid after the Addition of the Tail Protein gp9 (104)

Although the evidence suggests that pRNA participates in the DNA-packaging process of phage phi29, it is not known when the pRNA leaves the procapsid. Sucrose gradient sedimentation has been used to detect assembly intermediates. After the addition of purified gp9 following DNA packaging, the pRNA is no longer associated with the DNA-filled capsids.

D. Dimer as the Building Block in pRNA Formation

It has been shown that the hexamer is built from the dimer (94, 96, 138). The pathway of building a hexamer is dimer → tetramer → hexamer (Fig. 12) (138). This conclusion comes from the following evidence.

1. Homogeneous molecules with two complementary loops, such as pRNA I–i', can form dimers (94, 96, 119, 138). Two heterogeneous molecules such as A–b' plus B–a' can also form stable pRNA dimers in solution. Both dimers have been purified from native gel and/or sucrose gradient. The purified dimers are active in procapsid binding and phi29 DNA packaging (119, 138).

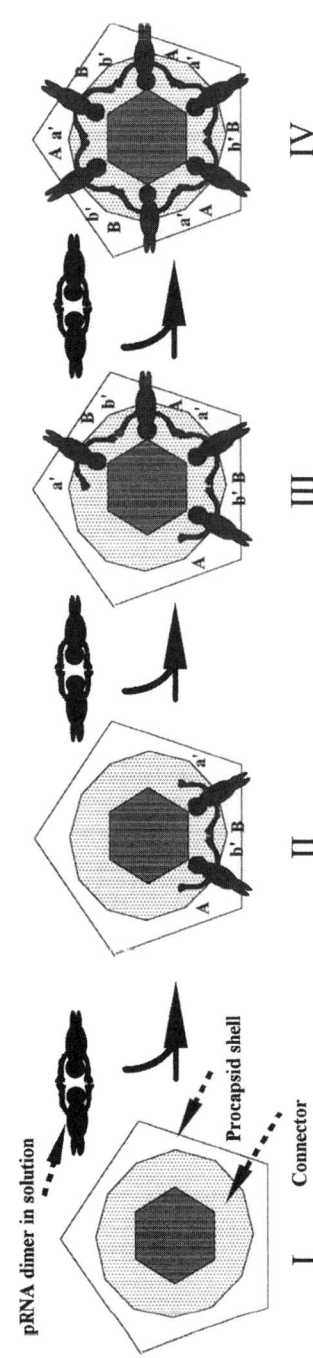

FIG. 12. The pathway of pRNA hexamer formation. The pentagon represents the procapsid shell with fivefold symmetry. The dodecagon represents the connector with a sixfold central hole for DNA entry (32, 100, 101, 161, 200). The black elliptical particles represent the pRNA. Reproduced from Ref. 199 with permission.

2. Homologous pRNAs with noncomplementary right and left loops, such as A′–b′, appear predominantly as monomers. Such monomeric pRNAs cannot bind procapsid nor package DNA (*94, 96, 138*). Monomeric pRNAs are unable to compete with dimers formed by homologous pRNA containing complementary loops, for example, I–i′, for procapsid binding, whereas dimers containing two heterogeneous pRNAs are potent competitors. Monomeric pRNA is also unable to inhibit dimeric pRNA for DNA packaging (*94, 138*).

3. Cryo-AFM also directly confirms the existence of dimers (*105, 131, 138*). The detection of dimeric pRNAs in solution suggests that it might be a stable intermediate during assembly of pRNA hexamer.

4. The fact that the minimum size of pRNA for procapsid binding is the same as the minimum size for dimer formation (*97, 148*) also supports the conclusion that dimers are the binding units for hexamer assembly.

5. Covalently linked dimeric pRNAs are able to package DNA *in vitro*.

The strongest evident to support the conclusion that dimers are the building blocks of the RNA hexamer comes from the finding that covalently linked dimers are active in phi29 DNA packaging. Three kinds of dimers, including cross-linked dimers (*119*), fused dimers, and tandem dimers (*138*), have been constructed and all of them are active in phi29 DNA packaging (Table V).

Crosslinked dimers are generated by linking pRNA A–b′ to a pRNA with a photosensitive agent attached at G_{82} of pRNA B–a′ and then linking it to A–b′ by UV activation and purified by gel (*119*) (Table II). Fused pRNA dimers are generated by merging the right loop from one pRNA with the left loop sequences of the other pRNA. The merging site is located at the crosslinking sites (Table V). Tandem pRNA dimers are generated by joining two pRNAs together via head-to-tail linkage (Table V).

If dimers are the building blocks, there are two possible pathways to assembling a hexamer. The first is via the independent addition of individual dimers ($2 \times 3 = 6$). The second is via the cooperative addition of dimers through hand-in-hand interaction to recruit the incoming dimers ($2 + 2 + 2 = 6$). pRNA-binding analysis reveals that the Hill coefficient (*164*) is 2.5, strongly suggesting that on the connector there are three binding sites for dimer binding and that the binding process is cooperative. Therefore, the sequence of hexamer formation is ($2 + 2 + 2 = 6$)(*138*).

E. Sequence and Size Requirement of the Loops in Dimer Formation

As described earlier, four bases in each of the two wild-type pRNA loops are involved in RNA/RNA interactions to form a hexagonal complex (Section II.C.3). It has been suggested that two GC pairs are sufficient for loop/loop interaction

(96, 97). The size and sequence requirements for such loop/loop interactions in dimer formation have been investigated thoroughly (97). When all four bases are paired, at least one G/C pair is needed. The maximum number of base pairings between the two loops to give full activity is five, and the minimum loop length is five bases for one loop and three bases for the other. Phylogenetic analysis of seven pRNAs, from phages SF5, B103, phi29, PZA, M2, NF, and GA1, supports these conclusions (97, 116, 138).

Truncation, deletion, mutation, and phylogenetic analysis have been performed to determine the sequence requirement for loop/loop interaction in dimer formation. Without considering the tertiary interaction, in some cases only two G/C pairs between the interacting loops could provide certain pRNAs with activity (96). When all four bases are paired, at least one G/C pair is required (97). The maximum number of base pairings between the two loops to allow pRNA to retain wild-type activity is five; the minimum number is five for one loop and three for the other. These findings are supported by phylogenetic analysis of seven pRNAs from different phages (97).

F. Minimal RNA Size Requirement for Dimer Formation

A 75-base RNA segment, bases 23–97, is able to form dimer, interlock into the hexamer, compete with full-length pRNA for procapsid binding, and therefore inhibit phi29 assembly *in vitro*. This suggests that segments 23–97 are a self-folded independent domain involved in procapsid binding and RNA/RNA interaction in dimer and hexamer formation, whereas the paired ends (bases 1–22 and 98–120) are involved in DNA translocation, but dispensable for RNA/RNA interaction (97). Therefore, this 75-base RNA could be a model for structural studies of RNA dimerization.

G. Minimal Sequence Requirement for Procapsid Binding and DNA Packaging

It has also been reported (148) that a 55-base RNA (bases 37–91) is able to bind procapsid specifically, but the efficiency is much lower. Bases 6–113 (105 nucleotides with the additional deletion of two nonessential bases C_{109} and A_{106}) are the minimum sequence required for full procapsid-binding activity as in wild type. Bases 1–117 are the minimum sequence needed for full DNA-packaging activity.

H. Mechanism and Role of pRNA in phi29 DNA Translocation

1. SEQUENTIAL ACTION OF SIX pRNAs

In 1997, a comprehensive paper was published describing the role of phi29 pRNA in DNA packaging (104). This was the first paper to describe the specific

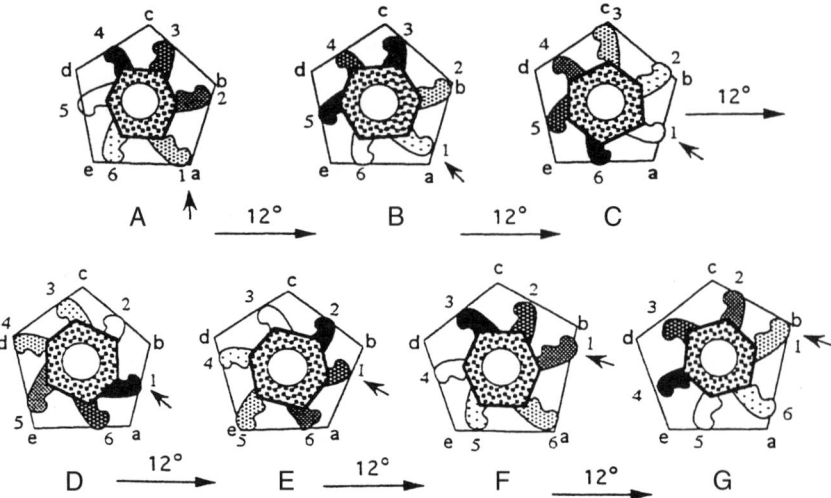

FIG. 13. A model showing the sequential action of pRNA in phi29 DNA packaging motor. The hexagon symbolizes the connector. The surrounding pentagon symbolizes the viral capsid. Six protrusions symbolize six pRNAs. The variable shapes and patterns portray the pRNA in serial energetic states. pRNA 4 in A is in a contracted conformation, pRNA 1 in A is in extended conformation. Arrows point to the different states of pRNA 1 (A–G) shows six steps of rotation. Each step rotates 12°, since a five- to sixfold symmetry mismatch generates 30 equivalent orientations (360°/30 = 12°). The connector turns 72° after six steps. For example, pRNA 1 moves from vertex a (A) to vertex b (G) and rotates 72°. Each step requires one ATP to initiate one pRNA conformation change. Six ATPs are needed to move from one vertex to another; 30 ATPs are needed for one 360° rotation. Reprinted from Ref. *104* with permission.

role of pRNA in phi29 DNA packaging and to provide strong evidence for concluding that six pRNAs work sequentially in driving the DNA translocation motor. This paper also provided an elegant model for elucidating the mechanism of rotation. A quantification of energy (ATP) usage was also provided (*104*). It suggested that phi29 DNA packaging is accomplished by a mechanism similar to driving a bolt with a hex nut, and involves six DNA-packaging pRNAs (Fig. 13). Recently, phi29 DNA packaging has become a popular subject. Publications in this area have affirmed that the aforementioned inference concerning the consecutive action of six pRNAs in driving the DNA rotation machine is convincing (*100, 104, 165, 166*).

All icosahedral viruses contain a fivefold-symmetric capsid face. Six copies of pRNA bind to a connector (portal vertex), which is embedded in such a fivefold-symmetric area (*104, 146*). Rotation of the hexameric pRNA within a fivefold-symmetric environment could constitute a mechanical motor in which the relative motion of the two rings produces a driving force to translocate the

DNA into the procapsid. Analogous to the sequential action of six cylinders in a engine, sequential action of six pRNAs is predicted to achieve the turning of this DNA-packaging motor. If the six cylinders fired synchronously, the engine would not run continuously (Fig. 13). The pRNA may collaborate with gp16 to perform this rotation job. Sequential action means that six pRNAs appear to act in a step-by-step process, with each pRNA exerting its individual effect alternatively (Fig. 13).

The pRNAs contains two functional domains: one for connector binding and the other for DNA translocation (Section VI.C). pRNA binds to the connector, leaving the 5'/3' essential domain free for interaction with other components. It has been reported that the $C^{18}C^{19}A^{20}$ bulge is located on the surface of a pRNA-bound procapsid (*131, 141*) and is essential for DNA translocation but dispensable for procapsid binding (*91, 97, 104, 133, 141*). The $C^{18}C^{19}A^{20}$ bulge might be directly involved in interacting with ATP, gp16, or DNA–gp3. It is predicted that pRNA is part of an ATPase and possesses at least two conformations (Section III.M), a relaxed and a contracted one. Alternating between contraction and relaxation, powered by ATP hydrolysis, each member of the hexameric RNA complex helps rotate the DNA packaging motor.

The requirement of loop/loop interaction in hexamer formation (*94, 96*) supports the pRNA sequential action model. The pRNAs may need to communicate with each other to ensure that the progress of the motion is consecutive. Inter-pRNA interactions via loops might serve as a link to pass a signal to adjacent pRNAs. Thus, base pairing between the right- and left-hand loops might be necessary to transfer a conformational change from one pRNA to an adjacent one.

2. EVIDENCE TO SUPPORT SEQUENTIAL ACTION

The conclusion of sequential consecutive action of pRNA comes from the following evidence (*104*).

a. DNA Packaging Is Blocked by Only One Inactive pRNA in the Chain. When any one of six pRNAs was replaced with an inactive pRNA, DNA packaging was completely blocked (*93, 102, 103*). This is an indication that six pRNAs work in a pattern similar to a serial, not parallel, circuit. It strongly supports the speculation that individual pRNAs take turns mediating successive steps of DNA packaging.

b. pRNA Is Very Sensitive to Mutation. A single base change can render the pRNA completely inactive in DNA packaging (*98, 114*).

c. Complementation Cannot Rescue Inactive pRNA with a Single Base Mutation. Extensive investigation shows that not a single plaque can be produced in any complementation test with any two inactive pRNAs with mutations at two different locations (104). Moreover, the yield of a complementation test with two partially active pRNAs has never been higher than the yield produced with one of the more active pRNAs alone. The result suggests that the action of pRNA is cooperative and synergetic.

d. pRNA Mutation Cannot be Compensated by Wild-Type pRNA. Compensation tests showed that the wild-type phenotype pRNA could not compensate for the defect of any one of these mutants (104). The failure of wild-type pRNA to compensate for the defect in mutant pRNAs is clearly documented by the results of inhibition assays in which the activity of wild-type pRNA is strongly inhibited by mutant pRNA (93, 102, 104, 138). The failure of wild-type pRNA to compensate for mutant pRNAs and the success of mutant pRNAs in inhibiting wild-type pRNA are also demonstrated by an almost perfect match upon comparing the empirical with the theoretical yield predicted based on the sequential model (Fig. 14) (104).

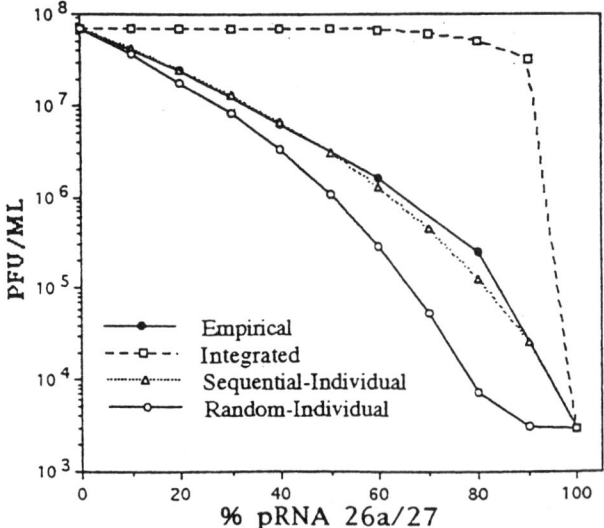

FIG. 14. Comparison of the empirical curve with computer simulation curves representing models of sequential action of six pRNAs. The semilog plot of the empirical data is compared with the plots predicted by the integrated, the sequential, and the random models of pRNA action, simulated by a binomial distribution. Curves represent the yield of virion production versus percentage of mutant pRNA in the assembly mixture. Wild-type pRNA was mixed with pRNA 26A/27 in certain ratio, and assayed for phi29 assembly.

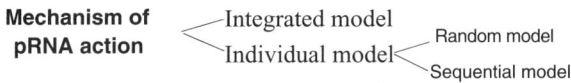

FIG. 15. Models for simulating the mechanism of pRNA action.

e. Mathematical Simulation and Modeling. The strongest support for sequential action comes from the extensive quantification and mathematical simulation as described below.

Three models, the integrated, the sequential individual, and the random individual models, have been used to simulate the mechanism of pRNA action (Fig. 15). Compensation and complementation studies did not support the integrated model, but favored individual models (*104*). Of the two individual models, competitive inhibition excluded the random individual model. Therefore, only the sequential model is favored.

To investigate the role of pRNA in DNA packaging, it is crucial to determine whether the DNA-packaging machine is an entirety and should be studied as an intact assemblage (an integrated model) or whether it can be dissected into the functions of each individual pRNA molecule. If six pRNA molecules work individually, then do the six pRNAs work consecutively (sequential individual model) or does each pRNA contribute one step of each reaction cycle consisting of six steps but in random order (random individual model)? These three models have been proposed to distinguish these possibilities. The principles behind the procedure for discriminatin, among models and the mathematical formulas for quantification have been described in detail (*104*).

The models were simulated using binomial distributions (Section II.C.1) and compared with empirical curves. Our results agreed with the sequential individual model, since the empirical curve overlapped with the curve predicted with this model (Fig. 14).

VII. Significance and Application of the Study of pRNA

A. A Model for the Study of RNA Dimers and Trimers

Stable dimers and trimers of the pRNA in the absence of protein can be isolated (*138*). The dimerization/trimerization assay provides a simple and stable system for studying the structure of pRNA complexes and investigating the mechanism of RNA/RNA interactions. RNA dimerization or oligomerization has been found to play a vital role in certain living systems. Dimerization of retroviral genomic RNA is critical for the retroviral life cycle, including translation, reverse transcription, RNA encapsidation, and virion assembly (*167, 168*). An RNA–RNA intermolecular interaction is required for the formation of a specific ribonucleoprotein particle, *bicoid* mRNA 3'-UTR-STAUFEN, which determines the formation of the anterior pattern of the *Drosophila* embryo (*169*). RNA–RNA interaction is one essential step in the cleavage reaction of RNase P on tRNA (*54, 170, 171*). Replications of the plasmid ColE1 are regulated by plasmid-specified small RNAs (RNA I and RNA II), which form complexes by complementary RNA stem–loop interactions (*172*). The system which we have developed is a very simple and stable system that could be used as a model for the study of RNA/RNA intermolecular interactions.

B. A Model for the Study of Viral DNA Packaging and for the Design of New Antiviral Strategies

Extensive investigation of DNA packaging of the dsDNA viruses documents certain common features in this step of the viral life cycle. The commonalties include the use of a pair of noncapsid enzymes to translocate the viral DNA into a procapsid coupled with the hydrolysis of ATP (*37, 43, 49, 173–178*). Similarities in DNA packaging between dsDNA bacteriophages and herpes viruses, adenovirus, pavovirus, and poxviruses (see Section I) justify the use of phi29 as model system for the design of new antiviral strategies. Using phi29 pRNA as a target, several methods have been modeled (*102, 103*) for the inhibition of viral assembly.

1. INHIBITION OF phi29 ASSEMBLY BY ANTISENSE DNA TARGETING pRNA *IN VITRO*

Antisense oligos include antisense RNA and antisense DNA. They are small, regulatory, single-stranded polynucleotides that bind to complementary regions on a specific target in order to control their biological function. Antisense DNA could bind to the pRNA and cause a change in the pRNA electrophoretic mobility (*88*). Antisense DNA oligos were shown to hybridize to phi29 pRNA by gel shift assays (*88*).

Phi 29 pRNA contains two functional domains (Section VI.C). One domain, located in the central region, is required for procapsid binding. The other domain, consisting of the paired 5' and 3' ends, is needed for DNA translocation into procapsids, but is not required for procapsid binding. Oligo P6, targeting the left-hand loop of the procapsid-binding domain, could block the binding of pRNA to procapsids, resulting in the inhibition of phi29 assembly *in vitro*. Oligos P11 and P15, targeting either the 5' or the 3' end, respectively, of the pRNA did not prevent pRNA from binding to procapsid, but strongly inhibited DNA packaging (*86, 88*).

2. COMPLETE INHIBITION OF phi29 ASSEMBLY WITH MUTANT pRNA *IN VIVO*

Phi29 is used as a model to explore new avenues in antiviral research beyond the *in vitro* inhibition with antisense oligos targeting pRNA. As already noted, pRNA contains two functional domains (Section VI.C). Sequence alterations that disrupt the base pairing at the 5'/3' ends of the DNA-translocating domain result in mutant pRNAs that are completely inactive in phage assembly *in vitro*, but retain wild-type procapsid-binding affinity (*86, 93, 102, 148*). These mutant pRNAs are able to compete with wild-type pRNAs for procapsid binding and efficiently inhibit viral assembly *in vitro* and *in vivo* (*102*).

For *in vivo* studies, a plasmid expressing a pRNA with a four-base mutation at the 3' end was constructed and transformed into host cells. Cells harboring this plasmid were completely resistant to plaque formation by wild-type phi29. The novelty here is the result of *complete* inhibition. It indicates that factors involved in viral assembly can be targeted for efficient and specific antiviral treatment. The high efficiency was due to two pivotal features. First, the pRNA contains two domains, one for procapsid binding and the other for DNA packaging. Mutation of the DNA-packaging domain resulted in a pRNA with no DNA-packaging activity, but intact procapsid-binding competence. Second, six pRNAs were involved in the packaging of one genome. This higher order dependence of pRNA concomitantly resulted in a higher order of inhibitory effect due to the fact that blocking only one of the six positions occupied by the mutant pRNA could result in the complete block of virion assembly (*93*). The principle of using molecules containing two functional domains and requiring multiple-copy involvement as inhibitors could be applied to gene therapy, intracellular immunization, antiviral drug design, or construction of transgenic plants resistant to viral infection, using certain viral structural proteins, enzymes, and other RNAs involved in the viral life cycle.

C. Similar Mechanism between Viral DNA Translocation and Other Nucleic Acid Sliding/Riding Processes

There is a group of nucleic acid-binding proteins that plays a similar role in DNA or RNA riding or sliding related to DNA replication, translocation, and recombination and RNA transcription (Fig. 16) (*94, 179, 180*). They can be categorized into two subsets, proteins that bind nucleic acids and proteins that act on nucleic acids. The common feature of these two subsets is that they interact with RNA or DNA in a polymer conformation with a ring-shaped morphology. Most members of this group are hexamers.

The subset that acts on nucleic acids includes helicase (*181, 183*), the *E. coli* transcription termination protein Rho (*183, 184*), the yeast DNA polymerase processivity factor (*185*), bovine papillomavirus E1 replication initiator (*186*), and the *E. coli* DNA polymerase III holoenzyme (*187, 188*), which exist as hexamers (*181, 189–192*). Although the mechanisms of this kind of DNA–protein interaction remain to be elucidated, the finding that six copies of pRNA are

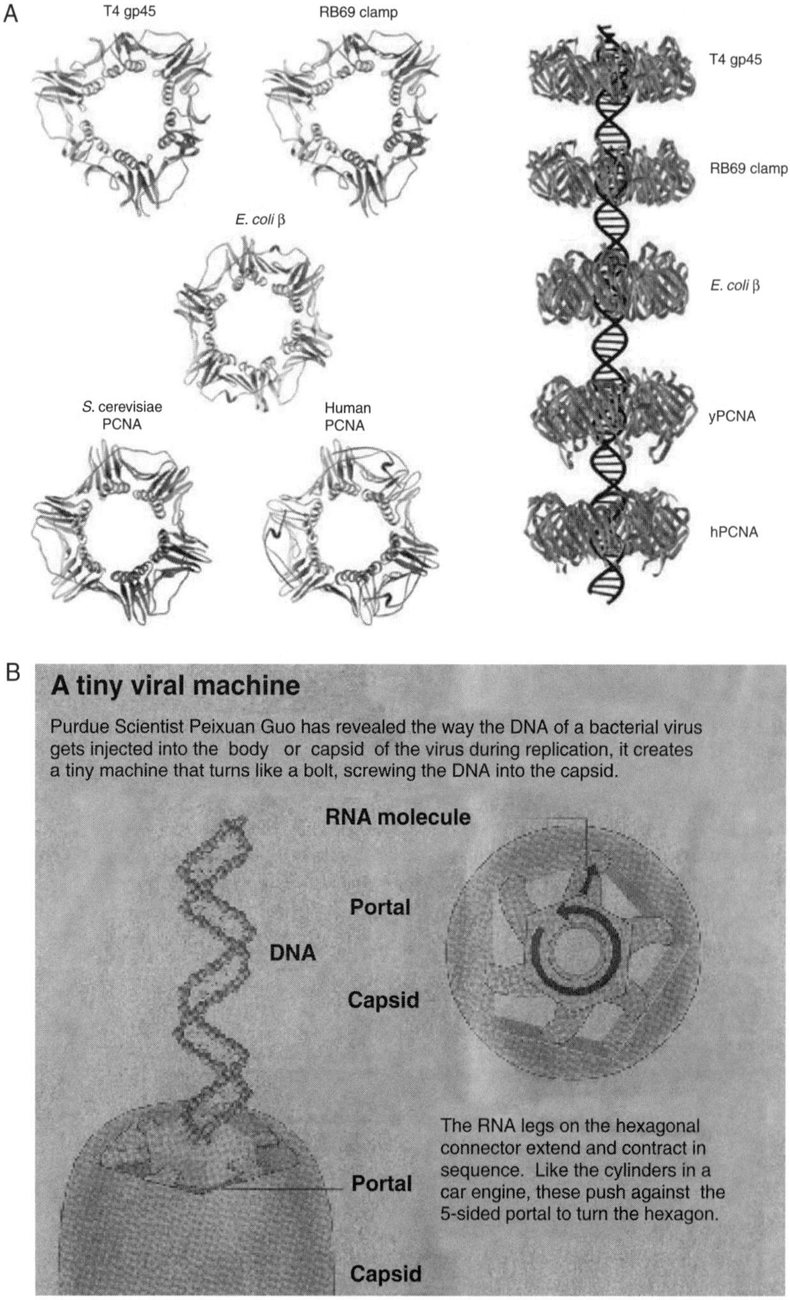

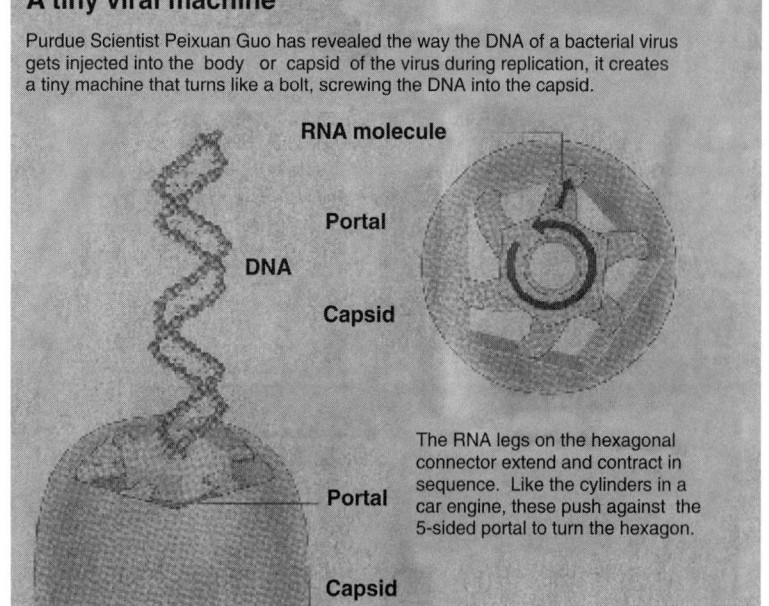

FIG. 16. Comparison of phi29 DNA-packaging machine (B) with the DNA replication apparatus (A). Reprinted (A) from Ref. *180* with permission of Elsevier Science and (B) from Ref. *201* with permission of the newspaper *Indianapolis Star*.

attached to a sixfold-symmetric connector during DNA translocation indicates they might have something in common. The process of both viral DNA packaging and DNA replication (or RNA transcription) involve the relative motion of two components, one of which is a nucleic acid. It would be intriguing to show how the pRNA may play a role similar to that of protein enzymes such as DNAS helicases or the termination factor Rho.

D. Hints for Macromolecular Translocation across Cell Membranes or through Barrier Walls

Migration or transportation of macromolecules, such as proteins or nucleic acids, through a barrier or cell membrane is a common process in living systems. After transcription in the nucleus, mRNA and tRNA must pass through the nuclear membrane to reach the translation machinery in the cytoplasm. Similarly, nuclear proteins must pass from the cytoplasm, where they are synthesized, to the nucleus, where they function. After infection or transfection, most viral or plasmid DNAs must pass through the nuclear membrane to serve as templates for gene expression (reviewed in Ref. *193*). The Rev protein of HIV helps in the translocation of viral mRNA from the nucleus to the cytoplasm through a nuclear pore (*194, 195*). Varieties of proteins and other elements migrate into and out of the cell and nucleus to perform their respective functions. Molecular migration or translocation also is manifest in the tracking and rail-riding of enhancers or transcription factors along DNA (*191*), the translocation of the transcription termination protein Rho along RNA (*192*), and the migration of helicases along single-stranded DNA during DNA replication (*181, 196–198*) (see C).

One of the most complex and intricate translocation processes is viral DNA encapsidation. Our study of DNA encapsidation could provide hints for macromolecular translocation through cell membranes.

VIII. Concluding Remarks

The phi29 DNA-packaging system exhibits an extraordinary number of unique features. The *in vitro* DNA-packaging efficiency is the highest in comparison with all *in vitro* systems available. All the components required for the assembly of the infectious virion and for interaction with pRNA have been cloned in and purified from *E. coli*. Infectious virions can assemble *in vitro* with the exclusive use of all purified components and synthetic pRNA with zero background. This is a highly sensitive system, with a nine-order-of-magnitude for the assay of pRNA function. The DNA-packaging motor is the strongest nanometer-size motor with the highest stalling force. The crystal structure of one of the important motor components, the connector where the pRNA binds, has been solved. This motor, of size less than $1/10^{18}$ in.^{3}, can be made with other, artificially designed purified chimeric components (unpublished data). Therefore, this is an unparalleled system and model for studying viral DNA packaging, making function analyses of molecular motors, designing nanomotors and other machines, characterizing biochemical energy transformation into motion, deciphering the process of DNA or RNA riding or sliding enzymes such as helicases and bioclamps, understanding the mechanism of macromolecules crossing cell membranes and other barriers, and developing antiviral therapies to block DNA packaging, a unique process that is not used by eukaryotic cells. The illustration of the role of pRNA in this fascinating biological process would certainly provide new insights in modern biology.

This pRNA is a novel and intriguing molecule. It is unusually stable, with a very compact and tightly folded structure. It interacts with ions and proteins to adapt numerous conformations depending on the environment and solution conditions. It can be manipulated to form homologous monomers, dimers, trimers, and hexamers; all of the oligomers have a strong tendency to form a circle by hand-in-hand interaction intra- or intermolecularly. Therefore, many as-yet-unknown properties of RNA can be studied by the investigation of this unusual molecule.

ACKNOWLEDGMENTS

I thank Drs. Dwight Anderson, Mark Trottier and Chaoping Chen for critical comments and review of the manuscript, and Jane Kovach, Dan Shu, and Steve Hoeprich for aid in its preparation. I thank Dr. Jose Carrascosa for making available information on his research data. I appreciate the permission and support of Drs. Znifeng Shao (cryo-AFM images), Michael Rossmann (connector structure), T. M. Baker (EM image), Jose Carrascosa (EM image), M. O' Donnell (slinding clamps), Eric Schoch (newspaper drawing) for the use of their figures. The work in the author's laboratory is supported by NIH grants GM59944, GM60529, GM48159, and GM46490 as well as NSF grant MCB9723923.

REFERENCES

1. C. Bazinet and J. King, The DNA translocation vertex of dsDNA bacteriophages. *Annu. Rev. Microbiol.* **39**, 109–129 (1985).
2. S. Casjens and R. Hendrix, Control mechanisms in dsDNA bacteriophage assembly, in "The Bacteriophages," (R. Calendar, ed.), pp. 15–92. Plenum Press, New York, 1988.
3. P. Guo, Introduction: Principles, perspectives, and potential applications in viral assembly. *Sem. Virol.* **5**, 1–3 (1994).
4. W. C. Earnshaw and S. R. Casjens, DNA packaging by the double-stranded DNA bacteriophages, *Cell* **21**, 319–331 (1980).
5. K. E. Gustin and M. J. Imperiale, Encapsidation of viral DNA requires the adenovirus L1 52/55-kilodalton protein, *J. Virol.* **72**, 7860–7870 (1998).

6. S. I. Schmid and P. Hearing, Bipartite structure and functional independence of adenovirus type 5 packaging elements. *J. Virol.* **71,** 3375–3384 (1997).
7. J. C. D'Halluin, M. Milleville, P. A. Boulanger, and G. R. Martin, Temperature-sensitive mutants of adenovirus type 2 blocked in virion assembly: accumulation of light intermediate particles. *J. Virol.* **26,** 344–356 (1978).
8. C. L. Cepko and P. A. Sharp, Assembly of adenovirus major capsid protein is mediated by a nonvirion protein, *Cell* **31,** 407–415 (1982).
9. B. Edvardsson, E. Everitt, E. Joernvall, L. Prage, and L. Philipson, Intermediates in adenovirus assembly. *J. Virol.* **19,** 533–547 (1976).
10. F. J. Rixon and D. McNab, Packaging-competent capsids of a herpes simplex virus temperature-sensitive mutant have properties similar to those of in vitro-assembled procapsids, *J. Virol.* **73,** 5714–5721 (1999).
10a. M. L. Perdue, J. C. Cohen, M. C. Kemp, C. C. Randall, and D. J. O'Callaghan (1975). Characterization of three species of nucleocapsids of equine herpesvirus type-1 (EHV-1) *Virology* **64,** 187–204.
11. J. Y. Lee, A. Irmiere, and W. Gibson, Primate cytomegalovirus assembly: evidence that DNA packaging occurs subsequent to B capsid assembly, *Virology* **167,** 87–96 (1988).
12. A. Dasgupta and D. W. Wilson, ATP Depletion Blocks Herpes Simplex Virus DNA Packaging and Capsid Maturation *J. Virol.* **73**(3), 2006–2015 (1999).
13. B. Moss, Replication of Poxviruses, in *"Virology"* (B. N. Fields *et al.,* eds.), pp. 685–703. Raven Press, New York, 1985.
14. A. M. DeLange, M. Reddy, D. Scraba, C. Upton, and G. McFadden, Replication and resolution of cloned poxvirus telomeres *in vivo* generates linear minichromosomes with intact viral hairpin termini *J. Virol.* **59,** 249–259 (1986).
15. B. L. Parsons and D. J. Pickup, Transcription of orthopoxvirus telomeres at late times during infection *Virology* **175,** 69–80 (1990).
16. H. Drexler, Initiation by bacteriophage T1 of DNA packaging at a site between the P and Q genes of bacteriophage 1, *J. Virol.* **49,** 754–759 (1984).
17. H. Shibata, H. Fujisawa, and T. Minagawa, Early events in a defined *in vitro* system for packaging of bacteriophage T3 DNA, *Virology* **159,** 250–258 (1987).
18. V. B. Rao and L. W. Black, Evidence that a phage T4 DNA packaging enzyme is a processed form of the major capsid gene product, *Cell* **42,** 967–977 (1985).
19. R. D. Everett, DNA replication of bacteriophage T5. 3. Studies on the structure of concatemeric T5 DNA, *J. Gen. Virol.* **52,** 25–38 (1981).
20. M. E. Cerritelli and F. W. Studier, Purification and characterization of T7 head-tail connectors expressed from the cloned gene, *J. Mol. Biol.* **285,** 299–307 (1996).
21. M. Sun, D. Louie, and P. Serwer, Single-event analysis of the packaging of bacteriophage T7 DNA concatemers in vitro, *Biophys. J.* **77**(3), 1627–1637 (1999).
22. K. Skorupski, B. Sauer, and N. Sternberg, Faithful cleavage of the P1 packaging site (pac) requires two phage proteins, PacA and PacB, and two Escherichia coli proteins, IHF and HU, *J. Mol. Biol.* **243,** 268–282 (1994).
23. G. Pruss and R. Calendar, Maturation of bacteriophage P2 DNA, *Virology* **86,** 454–467 (1978).
24. S. Rishovd, A. Holzenburg, B. V. Johansen, and B. H. Lindqvist, Bacteriophage P2 and P4-morphogenesis: structure and function of the connector. *Virology* **245,** 11–17 (1998).
25. B. Greene and J. King, In vitro unfolding/refolding of wild type phage P22 scaffolding protein reveals capsid-binding domain, *J. Biol. Chem.* **274,** 16135–16140 (1999).
26. S. D. Moore and P. E. Prevelige, Jr., Structural transformations accompanying the assembly of bacteriophage P22 portal protein rings in vitro, *J. Biol. Chem.* **276,** 6779–6788 (2001).
27. K. Eppler, E. Wyckoff, J. Goates, R. Parr, and S. Casjens, Nucleotide sequence of the bacteriophage P22 genes required for DNA packaging, *Virology* **183,** 519–538 (1991).

28. C. M. Burns, H. L. B. Chan, and M. S. DuBow, In vitro maturation and encapsidation of the DNA of transposable μ-like phage D108. *Proc. Natl. Acad. Sci. USA* **87**, 6092–6096 (1990).
29. M. P. Smith and M. Feiss, Sequence analysis of the phage 21 genes for prohead assembly and head completion, *Gene* **126**, 1–7 (1993).
30. C. S. Lee and P. Guo, In vitro assembly of infectious virions of ds-DNA phage φ29 from cloned gene products and synthetic nucleic acids, *J. Virol.* **69**, 5018–5023 (1995).
31. C. Gutierrez, R. Freire, M. Salas, and J. M. Hermoso, Assembly of phage phi29 genome with viral protein p6 into a compact complex, *EMBO J.* **13**(1) 269–276 (2001).
32. J. M. Valpuesta, J. J. Fernandez, J. M. Carazo, and J. L. Carrascosa, The three-dimensional structure of a DNA translocating machine at 10 A resolution, *Struct. Fold. Des.* **7**, 289–296 (1999).
33. M. H. Levner and N. R. Cozzarelli, Replication of viral DNA in SPO1-infected Bacillus subtilis, I. Replicative intermediates, *Virology* **48**, 402–416 (1972).
34. L. P. Gage and E. P. Geiduschek, RNA synthesis during bacteriophage SPO1 development: six classes of SPO1 RNA, *J. Mol. Biol.* **57**, 279–297 (1971).
35. P. Dubé, P. Tavares, R. Lurz, and M. van Heel, The portal protein of bacteriophage SPP1: a DNA pump with 13-fold symmetry. *EMBO J.* **12**, 1303–1309 (1993).
36. A. Gual, A. G. Camacho, and J. C. Alonso, Functional analysis of the terminase large subunit, G2P, of Bacillus subtilis bacteriophage SPP1. *J. Biol. Chem.* **275**, 35311–35319 (2000).
37. A. Becker and M. Gold, Prediction of an ATP reactive center in the small subunit, gpNul of phage lambda terminase enzyme, *J. Mol. Biol.* **199**, 219–222 (1988).
38. T. Dokland and H. Murialdo, Structural transitions during maturation of bacteriophage lambda capsids. *J Mol. Biol* **233**, 682–694 (1993).
39. L. Woods and C. E. Catalano, Kinetic characterization of the GTPase activity of phage lambda terminase: evidence for communication between the two "NTPase" catalytic sites of the enzyme. *Biochemistry* **38**(44), 14624–14630 (1999).
40. J. Q. Hang, C. E. Catalano, and M. Feiss, The Functional Asymmetry of cosN, the Nicking Site for Bacteriophage lambda DNA Packaging, Is Dependent on the Terminase Binding Site, cosB, *Biochemistry.* **40**, 13370–13377 (2001).
41. D. L. Anderson and B. Reilly, in Morphogenesis of bacteriophage φ29, "*Bacillus subtilis* and Other Gram-Positive Bacteria: Biochemistry, Physiology, and Molecular Genetics" (A. L. Sonenshein, J. A. Hoch, and R. Losick, eds.), pp. 859–867. American Society for Microbiology, Washington, D.C. (1993).
42. L. W. Black, DNA Packaging in dsDNA bacteriophages. *Ann. Rev. Microbiol.* **43**, 267–292 (1989).
43. P. Guo and M. Trottier, Biological and biochemical properties of the small viral RNA (pRNA) essential for the packaging of the double-stranded DNA of phage φ29. *Semin. Virol.* **5**, 27–37 (1994).
44. W. W. Newcomb, R. M. Juhas, D. R. Thomsen, F. L. Homa, A. D. Burch, S. K. Weller, and J. C. Brown, The UL6 Gene Product Forms the Portal for Entry of DNA into the Herpes Simplex Virus Capsid, *J. Virol.* **75**, 10923–10932 (2001).
45. T. Hohn, M. Wurtz, and B. Hohn, Capsid transformation during packaging of bacteriophage λ DNA, *Philos. Trans. R. Soc. Lond. B* **276**, 51–61 (1976).
46. T. Hohn, Packaging of genomes in bacteriophages: a comparison of ssRNA bacteriophages and dsDNA bacteriophages. *Philos. Trans. R. Soc. Lond. B* **276**, 143–150 (1976).
47. M. A. Bjornsti, B. E. Reilly, and D. L. Anderson, Morphogenesis of bacteriophage φ29 of *Bacillus subtilis:* oriented and quantized in vitro packaging of DNA protein gp3. *J. Virol.* **45**, 383–396 (1983).
48. P. Guo, S. Grimes, and D. Anderson, A defined system for in vitro packaging of DNA-gp3 of the *Bacillus subtilis* bacteriophage φ29, *Proc. Natl. Acad. Sci. USA* **83**, 3505–3509 (1986).

49. P. Guo, C. Peterson, and D. Anderson, Prohead and DNA-gp3-dependent ATPase activity of the DNA packaging protein gp16 of bacteriophage ϕ29, *J. Mol. Biol.* **197**, 229–236 (1987).
50. M. Morita, M. Tasaka, and H. Fujisawa, DNA packaging ATPase of bacteriophage T3, *Virology* **193**, 748–752 (1993).
51. P. Guo, S. Erickson, and D. Anderson, A small viral RNA is required for *in vitro* packaging of bacteriophage ϕ29 DNA, *Science* **236**, 690–694 (1987).
52. P. Guo, B. Rajogopal, D. Anderson, S. Erickson, and C.-S. Lee, sRNA of bacteriophage ϕ29 of *B. subtilis* mediates DNA packaging of ϕ29 proheads assembled in *E. coli*, *Virology* **185**, 395–400 (1991).
53. K. Kruger, P. J. Grabowski, A. J. Zaug, J. Sands, D. E. Gottschling, and T. R. Cech, Self-splicing RNA: autoexcision and autocyclization of the ribosomal RNA intervening sequence of Tetrahymena, *Cell* **31**, 147–157 (1982).
54. C. Guerrier-Takada, K. Gardiner, T. Marsh, N. Pace, and S. Altman, The RNA moiety of ribonuclease P is the catalytic subunit of the enzyme, *Cell* **35**, 849–857 (1983).
55. T. R. Cech, The chemisty of self-splicing RNA and RNA enzymes. *Science* **236**, 1532–1539 (1987).
56. S. Aigner, J. Lingner, K. J. Goodrich, C. A. Grosshans, A. Shevchenko, M. Mann, and T. R. Cech, Euplotes telomerase contains an La motif protein produced by apparent translational frameshifting, *EMBO J.* **19**, 6230–6239 (2000).
57. D. J. Lane, B. Pace, G. L. Olsen, D. A. Stahl, M. L. Sogin, and N. R. Pace, Rapid-determination of 16s ribosomal-RNA sequences for phylogenetic analysis. *Proc. Natl. Acad. Sci. USA* **82**, 6955–6959 (1985).
58. D. W. Lazinski and J. M. Taylor, Intracellular Cleavage and Lifation of Hepatitis Delta Virus Genomic RNA: Regulation of Ribozyme Activity by cis-Acting Sequences and Host Factors, *J. Virol.* **69**, 1190–1120 (1995).
59. T. B. Macnaughton, Y. J. Wang, and M. M. Lai, Replication of hepatitis delta virus RNA: effect of mutations of the autocatalytic cleavage sites, *J. Virol.* **67**, 2228–2234 (1993).
60. M. M. Lai, The molecular biology of hepatitis delta virus. *Annu. Rev. Biochem.* **64**, 259–286 (1995).
61. N. A. Sarver, E. M. Cantin, P. S. Chang, J. A. Zaia, P. A. Ladne, D. A. Stephens, and J. J. Rossi, Ribozymes as potential anti-HIV-1 therapeutic agents, *Science* **247**, 1222–1225 (1990).
62. A. Woisard, J. L. Fourrey, and A. Favre, Multiple folded conformations of a hammerhead ribozyme domain under cleavage conditions, *J. Mol. Biol.* **239**, 366–370 (1994).
63. J. B. Murray, C. J. Adams, J. R. P. Arnold, and P. G. Stockley, The roles of the conserved pyrimidine bases in hammerhead ribozyme catalysis: evidence for a magnesium ion-binding site. *Biochem. J.* **311**, 487–494 (1995).
64. A. L. Feig, W. G. Scottand, and O. C. Uhlenbeck, Inhibition of the Hammerhead Ribozyme Cleavage Reaction by Site-Specific Binding of Tb(III), *Science* **279**, 81–84 (1998).
65. B. M. Chowrira, A. Berzal-Herranz, and J. M. Burke, Novel guanosine requirement for catalysis by the hairpin ribozyme, *Nature* **354**, 320–322 (1991).
66. S. A. Strobel and S. P. Ryder, Biological catalysis, The hairpin's turn. *Nature* **410**, 761–763 (2001).
67. F. Kawamura and J. Ito, Transcription of the genome of bacteriophage ϕ29: Isolation and mapping of the major early mRNA synthesized *in vivo* and *in vitro*, *J. Virol.* **23**, 562–577 (1977).
68. J. M. Sogo, M. R. Inciarte, J. Corral, E. Vinuela, and M. Salas, RNA polymerase binding sites and transcription map of the DNA of a *Bacillus subtilis* phage ϕ29, *J. Mol. Biol.* **127**, 411–436 (1979).
69. P. Guo, S. Bailey, J. W. Bodley, and D. Anderson, Characterization of the small RNA of the bacteriophage ϕ29 DNA packaging machine, *Nucleic Acids Res.* **15**, 7081–7090 (1987).

70. J. Wichitwechkarn, S. Bailey, J. W. Bodley, and D. Anderson, Prohead RNA of bacteriophage ϕ29: size, stoichiometry and biological activity, *Nucleic Acids Res.* **17**(9), 3459–3468 (1989).
71. A. McGregor, M. V. Rao, G. Duckworth, P. G. Stockley, and B. A. Connolly, Preparation of oligoribonucleotides containing 4-thiouridine using Fpmp chemistry. Photo-crosslinking to RNA binding proteins using 350 nm irradiation, *Nucleic Acids Res.* **24**(16), 3173–3180 (1996).
72. M. E. Tosi, B. E. Reilly, and D. L. Anderson, Morphogenesis of bacteriophage ϕ29 of Bacillus subtilis: cleavage and assembly of the neck appendage protein, *J. Virol.* **16**, 1282–1295 (1975).
73. E. W. Hagen, B. E. Reilly, M. E. Tosiand, and D. L. Anderson, Analysis of gene function of bacteriophage ϕ29 of Bacillus subtilis: identification of cistrons essential for viral assembly, *J. Virol.* **19**(2), 501–517 (1976).
74. E. Vinuela, A. Camacho, F. Jimenez, J. L. Carrascosa, G. Ramirez, and M. Salas, Structure and assembly of phage phi29, *Philos. Trans. R. Soc. Lond. B Biol. Sci.* **276**, 29–35 (1976).
75. A. Camacho and M. Salas, Assembly of Bacillus Subtillis phage phi29, mutants in the cistrons coding for the structural proteins, *Eur. J. Biochem.* **73**, 39–55 (1977).
76. M. Salas, Protein-Priming of DNA Replication, *Annu. Rev. Biochem.* **60**, 39–71 (1991).
77. D. L. Anderson, H. H. Hickman, and B. E. Reilly, Structure of *Bacillus subtilis* bacteriophage ϕ29 and the length of ϕ29 deoxyribonucleic acid, *J. Bacteriol.* **91**, 2081–2089 (1966).
78. J. Ortín, E. Viñuela, M. Salas, and C. Vásquez, DNA-protein complex in circular DNA from phage ϕ29, *Nat. New Biol.* **234**, 275–277 (1971).
79. J. Méndez, L. Blanco, J. A. Esteban, A. Bernad, and M. Salas, Initiation of ϕ29 DNA replications occurs at the second 3+ nucleotide of the linear template: A sliding-back mechanism for protein-primed DNA replication, *Proc. Natl. Acad. Sci. USA* **89**, 9579–9583 (1992).
80. L. Blanco, A. Bernad, J. M. Lázaro, G. Martin, C. Garmendia, and M. Salas, Highly efficient DNA synthesis by the phage ϕ29 DNA polymerase symmetrical mode of DNA replication, *J. Biol. Chem.* **264**, 8935–8940 (1989).
81. L. Blanco, J. M. Lazaro, M. de Vega, A. Bonnin, and M. Salas, Terminal protein-primed DNA amplification, *Proc. Natl. Acad. Sci. USA* **91**, 12198–12202 (1994).
82. S. Grimes and D. Anderson, In Vitro Packaging of Bacteriophage ϕ29 DNA Restriction Fragments and the Role of the Terminal Protein gp3, *J. Mol. Biol.* **209**, 91–100 (1989).
83. M. A. Bjornsti, B. E. Reilly, and D. L. Anderson, In vitro assembly of the Bacillus subtilis bacteriophage ϕ29, *Proc. Natl. Acad. Sci. USA* **78**, 5861–5865 (1981).
84. P. Guo, S. Erickson, W. Xu, N. Olson, T. S. Baker, and D. Anderson, Regulation of the phage ϕ29 prohead shape and size by the portal vertex, *Virology* **183**, 366–373 (1991).
85. C. S. Lee and P. Guo, A highly sensitive system for the *in vitro* assembly of bacteriophage ϕ29 of *Bacillus subtilis*. *Virology* **202**, 1039–1042 (1994).
86. C. L. Zhang, C.-S. Lee, and P. Guo, The proximate 5' and 3' ends of the 120-base viral RNA (pRNA) are crucial for the packaging of bacteriophage ϕ29 DNA, *Virology* **201**, 77–85 (1994).
87. C. L. Zhang, M. Trottier, and P. X. Guo, Circularly permuted viral pRNA active and specific in the packaging of bacteriophage ϕ29 DNA, *Virology* **207**, 442–451 (1995).
88. C. L. Zhang, K. Garver, and P. Guo, Inhibition of phage ϕ29 assembly by antisense oligonucleotides targeting viral pRNA essential for DNA packaging, *Virology* **211**, 568–576 (1995).
89. T. Pan, R. R. Gutell, and O. C. Uhlenbeck, Folding of Circularly Permuted Transfer RNAs, *Science* **254**, 1361–1364 (1991).
90. J. M. Nolan, D. H. Burke, and N. R. Pace, Circularly Permuted tRNAs as Specific Photoaffinity Probes of Ribonuclease P RNA Structure, *Science* **261**, 762–765 (1993).
91. C. L. Zhang, T. Tellinghuisen, and P. Guo, Use of circular permutation to assess six bulges and four loops of DNA-Packaging pRNA of bacteriophage ϕ29, *RNA* **3**, 315–322 (1997).
92. R. J. D. Reid, J. W. Bodley, and D. Anderson, Characterization of the prohead-pRNA interaction of bacteriophage ϕ29, *J. Biol. Chem.* **269**, 5157–5162 (1994).

93. M. Trottier and P. Guo, Approaches to determine stoichiometry of viral assembly components *J. Virol.* **71,** 487–494 (1997).
94. P. Guo, C. Zhang, C. Chen, M. Trottier, and K. Garver, Inter-RNA interaction of phage phi29 pRNA to form a hexameric complex for viral DNA transportation, *Mol. Cells* **2,** 149–155 (1998).
95. C. Chen, M. Trottier, and P. Guo, New approaches to stoichiometry determination and mechanism investigation on RNA involved in intermediate reactions, *Nucleic Acids Symp. Ser.* **36,** 190–193 (1997).
96. F. Zhang, S. Lemieux, X. Wu, S. St.-Arnaud, C. T. McMurray, F. Major, and D. Anderson, Function of hexameric RNA in packaging of bacteriophage phi29 DNA in vitro, *Mol. Cells* **2,** 141–147 (1998).
97. C. Chen, C. Zhang, and P. Guo, Sequence requirement for hand-in-hand interaction in formation of pRNA dimers and hexamers to gear phi29 DNA translocation motor, *RNA* **5,** 805–818 (1999).
98. R. J. D. Reid, F. Zhang, S. Benson, and D. Anderson, Probing the structure of bacteriophage ϕ29 prohead RNA with specific mutations, *J. Biol. Chem.* **269,** 18656–18661 (1994).
99. B. Ibarra, J. R. Caston, O. Llorca, M. Valle, J. M. Valpuesta, and J. L. Carrascosa, Topology of the components of the DNA packaging machinery in the phage phi29 prohead, *J. Mol. Biol.* **298,** 807–815 (2000).
100. A. A. Simpson, Y. Tao, P. G. Leiman, M. O. Badasso, Y. He, P. J. Jardine, N. H. Olson, M. C. Morais, S. Grimes, D. L. Anderson, T. S. Baker, and M. G. Rossmann, Structure of the bacteriophage phi29 DNA packaging motor, *Nature* **408,** 745–750 (2000).
101. A. A. Simpson, P. G. Leiman, Y. Tao, Y. He, M. O. Badasso, P. J. Jardine, D. L. Anderson, and M. G. Rossmann (2001). Structure determination of the head-tail connector of bacteriophage phi29. *Acta Cryst* **D57,** 1260–1269.
102. M. Trottier, C. L. Zhang, and P. Guo, Complete inhibition of virion assembly *in vivo* with mutant pRNA essential for phage ϕ29 DNA packaging, *J. Virol.* **70,** 55–61 (1996).
103. M. Trottier, K. Garver, C. Zhang, and P. Guo, DNA-packaging pRNA as target for complete inhibition of viral assembly in vitro and in vivo, *Nucleic Acids Symp. Ser.* **36,** 187–189 (1997).
104. C. Chen and P. Guo, Sequential action of six virus-encoded DNA-packaging RNAs during phage phi29 genomic DNA translocation, *J. Virol.* **71,** 3864–3871 (1997).
105. Y. Mat-Arip, K. Garver, C. Chen, S. Sheng, Z. Shao, and P. Guo, Three-dimensional Interaction of Phi29 pRNA Dimer Probed by Chemical Modification Interference, Cryo-AFM, and Crosslinking. *J. Biol. Chem.* **276,** 32575–32584 (2001).
106. P. Ayback, A. Sandstrom, S. Yamakage, C. Sund, C. Glemarec, and J. Chattopadtryaya, Solution structure of lariat RNA by 500MHz NMR spectroscopy and molecular dynamics studies in water, *J. Biochem. Biophys. Meth.* **27,** 229–259 (1993).
107. R. D. Peterson, D. P. Bartel, J. W. Szostak, and J. Feigon, 1H NMR studies of the high-affinity Rev binding site of the Rec responsive element of HIV-1 mRNA base pairing in the core binding element, *Biochemistry* **33,** 5357–5366 (1994).
108. J. H. Cate, A. R. Gooding, E. Podell, K. Zhou, B. L. Golden, C. E. Kundrot, T. R. Cech, and J. A. Doudna, Crystal structure of a group I ribozyme domain: primciples of RNA packaging, *Science* **273,** 1678–1685 (1996).
109. H. W. Pley, K. M. Flaherty, and D. B. McKay, Three-dimensional structure of a hammerhead ribozym, *Nature* **372,** 68–74 (1994).
110. J. L. Sussman, S. R. Holbrook, W. R. Warrant, G. M. Church, and S. H. Kim, Crystal structure of yeast phenylalanine transfer RNA. I. Crystallograhpic refinement, *J. Mol. Biol.* **123,** 607–630 (1978).
111. B. Hingerty, R. S. Brown, and A. Jack, Further refinement of the structure of yeast tRNA^{phe}, *J. Mol. Biol.* **124,** 523 (1987).

112. E. Westhof, P. Dumas, and D. Moras, Crystallographic refinement of yerst aspartic acid transfer RNA, *J. Mol. Biol.* **184**, 119 (1985).
113. M. Zuker, On finding all suboptimal foldings of an RNA molecule, *Science* **244**, 48–52 (1989).
114. C. L. Zhang, T. Tellinghuisen, and P. Guo, Conformation of the helical struture of the 5′/3′ termini of the essential DNA packaging pRNA of phage ϕ29, *RNA* **1**, 1041–1050 (1995).
115. T. Pecenkova, V. Benes, J. Paces, R. Vlcekand, and V. Paces, Bacteriophage B103: complete DNA sequence of its genome and relationship to other Bacillus phages, *Gene* **199**, 157–163 (1997).
116. S. Bailey, J. Wichitwechkarn, D. Johnson, B. Reilly, D. Anderson, and J. W. Bodley, Phylogenetic analysis and secondary structure of the *Bacillus subtilis* bacteriophage RNA required for DNA packaging, *J. Biol. Chem.* **265**, 22365–22370 (1990).
117. S. H. Hixson and S. S. Hixson, p-Azidophenacyl bromide, a versatile photolabile bifunctional reagent. Reaction with glyceraldehyde-3-phosphate dehydrogenase. *Biochemistry* **14**, 4251–4254 (1975).
118. A. B. Burgin and N. R. Pace, Mapping the active site of ribonuclease P RNA using a substrate containing a photoaffinity agent. *EMBO J.* **9**, 4111–4118 (1990).
119. K. Garver and P. Guo, Mapping the inter-RNA interaction of phage phi29 by site-specific photoaffinity crosslinking, *J. Biol. Chem.* **275**(4), 2817–2824 (2000).
120. D. A. Wassarman, Psoralen crosslinking of small RNAs *in vitro*, *Mol. Biol. Rep.* **17**, 143–151 (1993).
121. K. Tyc and J. A. Steitz, A new interaction between the mouse 5′ external transcribed spacer of pre-rRNA and U3 snRNA detected by psoralen crosslinking, *Nucleic Acids Res.* **20**, 5375–5382 (1992).
122. G. D. Cimino, H. B. Gamper, S. T. Isaacs, and J. E. Hearst, Psoralens as photoactive probes of nucleic acid structure and function: organic chemistry, photochemistry, and biochemistry, *Annu. Rev. Biochem.* **54**, 1151–1193 (1985).
123. C. Chen and P. Guo, Magnesium-induced conformational change of packaging RNA for procapsid recognition and binding during phage phi29 DNA encapsidation, *J. Virol.* **71**, 495–500 (1997).
124. C. F. Hui and C. R. Cantor, Mapping the location of psoralen crosslinks on RNA by mung bean nuclease sensitivity of RNA-DNA hybrids, *Proc. Natl. Acad. Sci USA* **82**, 1381–1385 (1985).
125. D. A. Wassarman and J. A. Steitz, Interactions of small nuclear RNA's with precursor messenger RNA during in vitro splicing, *Science* **257**, 1918–1925 (1992).
126. T. Mohammad, C. Chen, P. Guo, and H. Morrison, Photoinduced cross-linking of RNA by cis-Rh(phen)2Cl2+ and cis-Rh(phen)(phi)Cl2+: a new family of light activatable nucleic acid cross-linking agents, *Bioorg. Med. Chem. Lett.* **9**, 1703–1708 (1999).
127. U. Moazed, S. Stern, and H. F. Noller, Rapid chemical probing of conformation in 16S ribosomal RNA and 30S ribosomal subunits using primer extension, *J. Mol. Biol.* **187**, 399–416 (1986).
128. L. P. Yap and K. Musier-Forsyth, Transfer RNA aminoacylation: Identification of a critical ribose 2′-hydroxyl-base interaction, *RNA* **1**, 418–424 (1995).
129. C. Ehresmann, F. Baudin, M. Mougel, P. Romby, J.-P. Ebel, and B. Ehresmann, Probing the structure of RNAs in solution, *Nucleic Acids Res.* **15**, 9109–9128 (1987).
130. S. Stern, D. Moazed, and H. F. Noller, Structural analysis of RNA using chemical and enzymatic probing monitored by primer extension, *Meth. Enzymol.* **164**, 481–489 (1988).
131. M. Trottier, Y. Mat-Arip, C. Zhang, C. Chen, S. Sheng, Z. Shao, and P. Guo, Probing the structure of monomers and dimers of the bacterial virus phi29 hexamer RNA complex by chemical modification, *RNA* **6**, 1257–1266 (2000).

132. C. Zhang, M. Trottier, C. Chen, and P. Guo, Chemical modification patterns of active and inactive as well as procapsid-bound and unbound DNA-packaging RNA of bacterial virus Phi29, *Virology* **281**, 281–293 (2001).
133. R. J. D. Reid, J. W. Bodley, and D. Anderson, Identification of bacteriophage ϕ29 prohead RNA (pRNA) domains necessary for *in vitro* DNA-gp3 packaging, *J. Biol. Chem.* **269**, 9084–9089 (1994).
134. D. G. Knorre and V. V. Vlassov, Complementary-addressed (sequence-specific) modification of nucleic acids, *Prog. Nucleic Acid Res. Mol. Biol.* **32**, 291–320 (1985).
135. A. D. Ellington and J. W. Szostak, *In vitro* selection of RNA molecules that bind specific ligands, *Nature* **346**, 818–822 (1990).
136. G. Tuerk and L. Gold, Systematic evolution of ligands by exponential enrichment: RNA ligands to bacteriophage T4 DNA ploymerase, *Science* **249**, 505–510 (1990).
137. F. Zhang and D. Anderson, *In vitro* selection of Bacteriophage phi29 prohead RNA aptamers for prohead binding, *J. Biol. Chem.* **273**, 2947–2953 (1998).
138. C. Chen, S. Sheng, Z. Shao, and P. Guo, A Dimer as a Building Block in Assembling RNA. A hexamer that gears bacterial virus phi29 DNA-translocating machinery, *J. Biol. Chem.* **275**(23), 17510–17516 (2000).
139. M. S. Babcock, E. P. D. Pednault, and W. K. Olson, Nucleic Acid Structure Analysis. Mathematics for local cartesian and helical structure parameters that are truly comparable between structures, *J. Mol. Biol.* **237**, 125–156 (1994).
140. D. Gautheret and R. Cedergren, Modeling the three-dimensional structure of RNA, *FASEB J.* **7**, 97–105 (1993).
141. C. Zhang, M. Trottier, C. Chen, and P. Guo, Chemical Modification Patterns of Active and inactive as Well as Procapsid-Bound and Unbound DNA-Packaging RNA of Bacterial Virus Phi29. *Virology* **281**, 281–293 (2001).
142. C. Massire and E. Westhof, MANIP: an interactive tool for modelling RNA, *J. Mol. Graph. Model.* **16**, 197 (1998).
143. C. A. G. Haasnoot, C. W. Hilbers, G. A. van der Marel, J. H. van Broom, U. C. Singh, N. Pattabiraman, and P. A. Kollman, On loop folding in nucleic acid hairpin-type structures, *J. Biomol. Struct. Dyn.* **3**, 843–857 (1986).
144. D. H. Turner, N. Sugimoto, and S. M. Freier, RNA structure prediction, *Annu. Rev. Biophys. Chem.* **17**, 167–192 (1988).
145. F. Michel, A. D. Ellington, S. Couture, and J. W. Szostak, Phylogenetic and genetic evidence for base-triples in the catalytic domain of group I introns, *Nature* **34**, 578–580 (1990).
146. R. W. Hendrix, Symmetry mismatch and DNA packaging in large bacteriophages, *Proc. Natl. Acad. Sci. USA* **75**, 4779–4783 (1978).
147. Y. Tao, N. H. Olson, W. Xu, D. L. Anderson, M. G. Rossmann, and T. S. Baker, Assembly of a Tailed Bacterial Virus and Its Genome Release Studied in Three Dimensions, *Cell* **95**, 431–437 (1998).
148. K. Garver and P. Guo, Boundary of pRNA functional domains and minimum pRNA sequence requirement for specific connector binding and DNA packaging of phage phi29, *RNA* **3**, 1068–1079 (1997).
149. L. E. Donate and J. L. Carrascosa, Characterization of a versatile *in vitro* DNA packaging system based on hybrid λ/ϕ29 proheads, *Virology* **182**, 534–544 (1991).
150. L. E. Donate, H. Murialdo, and J. L. Carrascosa, Production of λ-phi29 Chimeras, *Virology* **179**, 936–940 (1990).
151. J. M. Valpuesta, L. E. Donate, C. Mier, L. Herranz, and J. L. Carrascosa, RNA-mediated specificity of DNA packaging into hybrid λ/ϕ29 proheads, *EMBO J.* **12**, 4453–4459 (1993).

152. S. Grimes and D. Anderson, RNA Dependence of the Bateriophage phi29 DNA Packaging ATPase, *Mol. Biol.* **215**, 559–566 (1990).
153. M. O. Badasso, P. G. Leiman, Y. Tao, Y. He, D. H. Ohlendorf, M. G. Rossmann, and D. Anderson, Purification, crystallization and initial X-ray analysis of the head-tail connector of bacteriophage phi29, *Acta Crystallogr. D Biol. Crystallogr.* **56**(Pt 9), 1187–1190 (2000).
154. A. Guasch, A. Parraga, J. Pous, J. M. Valpuesta, J. L. Carrascosa, and M. Coll, Purification, crystallization and preliminary X-ray diffraction stydies of the bacteriophage phi 29 connector particle, *FEBS* **430**, 283–287 (1998).
155. L. E. Donate, J. M. Valpuesta, C. Mier, F. Rojo, and J. L. Carrascosa, Characterization of an RNA-binding domain in the bacteriophage phi29 connector, *J. Biol. Chem.* **268**, 20198–20204 (1993).
156. M. Valle, L. Kremer, A. Martinez, F. Roncal, J. M. Valpuesta, J. P. Albar, and J. L. Carrascosa, Domain architecture of the bacteriophage phi29 connector protein, *J. Mol. Biol.* **288**, 899–909 (1999).
157. J. L. Carrascosa and J. M. Valpuesta, Bacteriophage connectors: Structural features of a DNA translocating motor, *Rec. Res. Dev. Virol.* **1**, 449–465 (1999).
158. M. C. Morais, Y. Tao, N. H. Olsen, S. Grimes, P. J. Jardine, D. Anderson, T. S. Baker, and M. G. Rossmann, Cryoelectron-Microscopy Image Reconstruction of Symmetry Mismatches in Bacteriophage phi29, *J. Struct. Biol.* **135**, 38–46 (2001).
159. J. M. Valpuesta and J. Carrascosa, Structure of viral connectors and their funciton in bacteriophage assembly and DNA packaging, *Q. Rev. Biophys.* **27**, 107–155 (1994).
160. J. M. Carazo, L. E. Donate, L. Herranz, J. P. Secilla, and J. L. Carrascosa, Three-dimensional reconstruction of the connector of bacteriophage ϕ29 at 1.8 nm resolution, *J. Mol. Biol.* **192**, 853–867 (1986).
161. J. Jimenez, A. Santisteban, J. M. Carazo, and J. L. Carrascosa, Computer graphic display method for visualizing three-dimensional biological structures, *Science* **232**, 1113–1115 (1986).
162. J. Wichitwechkarn, D. Johnson, and D. Anderson, Mutant prohead RNAs in the in vitro packaging of bacteriophage phi 29 DNA-gp3, *J. Mol. Biol.* **223**, 991–998 (1992).
163. M. L. Perdue, J. C. Cohen, M. C. Kemp, C. C. Randall, and D. J. O'Callaghan, Characterization of three species of nucleocapsids of equine herpesvirus type-1 (EHV-1), *Virology* **64**, 187–204 (1975).
164. J. N. Weiss, The Hill equation revisited: uses and misuses, *FASEB J.* **11**, 835–841 (1997).
165. R. J. Davenport, Crossover Research Yield Scents and Sensitivity—Watching a virus get stuffed, *Science* **291**, 2071–2072 (2001).
166. D. E. Smith, S. J. Tans, S. B. Smith, S. Grimes, D. L. Anderson, and C. Bustamante, The bacteriophage phi29 portal motor can package DNA against a large internal force, *Nature* **413**, 748–752 (2001).
167. E. Skripkin, J. C. Paillart, R. Marquet, B. Ehresmann, and C. Ehresmann, Identification of the primary site of the human immunodeficiency virus type 1 RNA dimerization in vitro, *Proc. Natl. Acad. Sci. USA* **91**, 4945–4949 (1994).
168. J. C. Paillart, E. Skripkin, B. Ehresmann, C. Ehresmann, and R. Marquet, A loop-loop "kissing" complex is the essential part of the dimer linkage of genomic HIV-1 RNA, *Proc. Natl. Acad. Sci. USA* **93**, 5572–5577 (1996).
169. D. Ferrandon, I. Koch, E. Westhof, and C. Nusslein-Volhard, RNA-RNA interaction is required for the formation of specific bicoid mRNA 3′ UTR-STAUFEN ribonucleoprotein particles, *EMBO J.* **16**, 1751–1758 (1997).
170. B. K. Oh and N. R. Pace, Interaction of the 3′-end of tRNA with ribonuclease P RNA, *Nucleic Acids Res.* **22**, 4087–4094 (1994).

171. M. F. Baer, R. M. Reilly, G. M. McCorkle, T. Y. Hai, S. Altman, and U. L. Rajbhandary, The recognition by Rnase P of precursor tRNAs, *J. Biol. Chem.* **263**, 2344–2351 (1988).
172. Y. Eguchi and J. Tomizawa, Complex formed by complementary RNA stem-loops and its stabilization by a protein: function of ColE1 Rom protein, *Cell* **60**, 199–209 (1990).
173. M. Feiss, J. Sippy, and G. Miller, Processive action of terminase during sequential packaging of bacteriophage λ chromosomes, *J. Mol. Biol.* **186**, 759–771 (1985).
174. R. R. Higgins, H. J. Lucko, and A. Becker, Mechanism of cos DNA cleavage by bacteriophage lambda terminase-multiple roles of ATP, *Cell* **54**, 765–775 (1988).
175. J. R. Panuska and D. A. Goldthwait, A DNA dependent ATPase from T4-infected *E. coli*. Purification and properties of a 63,000-dalton enzyme and its conversion to a 22,000-dalton form, *J. Biol. Chem.* **255**, 5208–5214 (1980).
176. V. S. Manne, V. B. Raoand, and L. W. Black, A bacteriophage T4 DNA packaging related DNA-dependent ATPase-endonuclease. *J. Biol. Chem.* **257**, 13223–13232 (1982).
177. K. Hamada, H. Fujisawa, and T. Minagawa, Characterization of ATPase activity of a defined *in vitro* system for packaging of bacteriophage T3 DNA, *Virology* **159**, 244–249 (1987).
178. P. Serwer, The Source of Energy for Bacteriophage DNA Packaging: An Osmotic Pump Explains the Data, *Biopolymers* **27**, 165–169 (1988).
179. V. Ellison and B. Stillman, Opening of the clamp: an intimate view of an ATP-driven biological machine, *Cell* **106**, 655–660 (2001).
180. M. M. Hingorani and M. O'Donnell, Sliding Clamps: a (tail)ored fit. *Curr. Biol.* **10**, 25–29 (2000).
181. S. C. West, DNA helicases: New breeds of translocating motors and molecular pumps, *Cell* **86**, 177–180 (1996).
182. T. Niedenzu, D. Roleke, G. Bains, E. Scherzinger, and W. Saenger, Crystal structure of the hexameric replicative helicase RepA of plasmid RSF1010, *J. Mol. Biol.* **306**, 479–487 (2001).
183. E. P. Gogol, S. E. Seifried, and P. H. von Hippel, Structure and assembly of the Escherichia coli transcription termination factor rho and its interaction with RNA. I. Cryoelectron microscopic studies, *J. Mol. Biol.* **221**, 1127–1138 (1991).
184. B. R. Burgess and J. P. Richardson, RNA passes through the hole of the protein hexamer in the complex with the Escherichia coli Rho factor, *J. Biol. Chem.* **276**, 4182–4189 (2001).
185. J. Bowers, P. T. Tran, A. Joshi, R. M. Liskay, and E. Alani, MSH-MLH complexes formed at a DNA mismatch are disrupted by the PCNA sliding clamp, *J. Mol. Biol.* **306**, 957–968 (2001).
186. J. Sedman and A. Stenlund, The papillomavirus E1 protein forms a DNA-dependent hexameric complex with ATPase and DNA helicase activities, *J. Virol.* **72**, 6893–6897 (1998).
187. F. P. Leu and M. O'Donnell, Interplay of a clamp loader subunits in opening the (beta) sliding clamp of E. coli DNA polymerase III holoenzyme, *J. Biol. Chem.* **276**, 47185–47194 (2001).
188. M. S. Song, H. G. Dallmann, and C. S. McHenry, Carboxyl-terminal Domain III of the delta ' Subunit of the DNA Polymerase III Holoenzyme Binds delta, *J. Biol. Chem.* **276**, 40668–40679 (2001).
189. E. P. Geiduschek, Riding the (mono)rails: the structure of catenated DNA-tracking proteins, *Chem. Biol.* **2**, 123–125 (1997).
190. M. C. Young, D. E. Schultz, D. Ring, and P. H. von Hippel, Kinetic Parameters of the Translocation of Bacteriophage T4 Gene 41 Protein Helicase on Single-stranded DNA, *J. Mol. Biol.* **235**, 1447–1458 (1994).
191. D. R. Herendeen, G. A. Kassavetis, and E. P. Geiduschek, A transcriptional enhancer whose function imposes a requirement that proteins track along DNA, *Science* **256**, 1298–1303 (1992).
192. J. Geiselmann, Y. Wang, S. E. Seifried, and P. H. von Hippel, A physical model for the translocation and helicase activities of Escherichia coli transcription termination protein rho, *Proc. Natl. Acad. Sci. USA* **90**, 7754–7758 (1993).
193. L. I. Davis, The nuclear pore complex, *Annu. Rev. Biochem.* **64**, 865–896 (1995).

194. R. M. Krug, The regulation of export of mRNA from nucleus to cytoplasm, *Curr. Opin. Cell Biol.* **5,** 944–949 (1993).
195. K. Pfeifer, B. E. Weiler, D. Ugarkovic, M. Bachmann, H. C. Schroder, and W. E. Muller, Evidence for a direct interaction of Rev protein with nuclear envelope mRNA-translocation system, *Eur. J. Biochem.* **199,** 53–64 (1991).
196. M. Young, S. Kuhl, and P. von Hippel, Kinetic theory of ATP-driven translocases on one-dimensional polymer lattices, *J. Mol. Biol.* **235,** 1436–1446 (1994).
197. E. H. Egelman, Homomorphous hexameric helicases: tales from the ring cycle, *Structure* **4,** 759–762 (1996).
198. M. C. San Martin, C. Gruss, and J. M. Carazo, Six molecules of SV40 large T antigen assemble in a propeller-shaped particle around a channel, *J. Mol. Biol.* **268,** 15–20 (1997).
199. S. Hoeprich and P. Guo, Computer modeling of three-dimensional structuring of DNA-packaging RNA (pRNA) Monomer, Dimer, and Hexamer of Phi29 DNA packaging motor. *J. Biol. Chem.* 7;**277**(23), 20794–20803 (2002).
200. A. Guasch, J. Pous, A. Parraga, J. M. Valpuesta, J. L. Carrascosa, and M. Coll, Crystallographic analysis reveals the 12-fold symmetry of the bacteriophage phi29 connector particle, *J. Mol. Biol.* **281,** 219–225 (1998).
201. E. B. Schoch. Discovery may help fight viral illnesses. Indianapolis Star, Sept. 8, (1998).

Index

A

ABA, wound response, 190
Adenovirus, shunt, 17–19
AdoMet, $nrdG$ protein, 119
Albumin, mRNA, stability regulation, 142–143
Allene oxide cyclase, 175–176
Allene oxide synthase, 173–175
Allosteric mechanism, ribonucleotide reductases, 108–114
$\alpha_2\beta_2$ complex, $nrdG$ protein, 115–116
Amino acids
 CAD, TAF interaction, 281
 RNA contacts, 229–230
Anaerobic ribonucleotide reductases
 gene organization and regulation, 120
 multicomponent system definition, 99
 $nrdD$ protein
  allosteric regulation, 108–114
  glycyl radical enzyme definition, 99–101
  site-directed mutagenesis studies, 101–103
  substrate reduction, 106–108
  3D structure, 103–106
 $nrdG$ protein
  activation reaction, 116–119
  $\alpha_2\beta_2$ complex, 115–116
  iron–sulfur protein definition, 114–115
  iron–sulfur protein prototype, 119
 overview, 96–98
 RNA–DNA link, 120–124
Anaphase-promoting complex, proteolysis in mitosis, 52
Animal viruses, shunt
 adenovirus, 17–19
 internal initiation combinations, 21
 papillomaviruses, 20–21
 Sendai virus, 19–20
Annealing, nucleic acid chaperone proteins, 235–237, 245
Antiviral strategies, models, 458–459
AOC, see Allene oxide cyclase

AOS, see Allene oxide synthase
APC, see Anaphase-promoting complex
ARE-binding proteins, unstable mRNAs, 134–135
Assays
 CAD in pol II, 289–291
 nucleic acid chaperone proteins, 237–239
 pRNA, phi29, 420–421
Ataxia telangiectasia and Rad-3-related, 77
Ataxia telangiectasia-mutated, 77, 79–80
ATM, see Ataxia telangiectasia-mutated
ATP, allosteric regulation of RNRs, 108–114
ATR, see Ataxia telangiectasia and Rad-3-related
AUG codon
 cap-dependent scanning, 4–5
 internal translation initiation, 6
 leaky scanning, 5
 MuLV genomic mRNA, 380–382
 Sendai virus shunt, 19–20
AU-rich elements, unstable mRNAs, 134–135
AUU codon, shunting ribosome initiation, 15

B

Bacilliform Caulimoviridae, 22–23
Basic fibroblast growth factor, see Fibroblast growth factor 2
Binomial distribution, pRNA stoichiometry, 423–424
BRCT family, DNA replication, 74–77
Budding yeast
 Cdc6/Cdc18 protein, 55–56
 cyclin-dependent kinases, 43–44
 proteolysis at G1–S transition, 51–52
 SBF and MBF, 48–49

C

CAD, see Constitutive activation domain
Calcium ion, pRNA effect, 449

CaMV, see Cauliflower mosaic virus
Cap-dependent scanning, eukaryote translation initiation, 4–5
Capsid proteins
    pRNA association, 450
    pRNA binding, 445–447
    pRNA departure, 450
CAT, see Chloramphenicol acetyl transferase
Cauliflower mosaic virus
    RNA translation–packaging interplay, 9
    sORF-dependent shunt
        adenovirus comparison, 17–19
        other initiation events, 17
        overview, 12
        shunting as reinitiation, 16
        take-off and landing sites, 14–15
Caulimoviridae, sORF-dependent shunt
    other initiation events, 17
    overview, 12–14
    shunting as reinitiation, 16
    shunting ribosome initiation, 15–16
    take-off and landing sites, 14–15
Cdc, see Cell division cycle proteins
Cdk, see Cyclin-dependent kinases
Cdt1p, chromatin binding, 56–57
Cell cycle
    control by checkpoints
        coiled-coil proteins, 77
        effector kinases, 78
        overview, 73–74
        phosphoinositol-3-kinase-like complexes, 77
        protein network, 78–80
        sensors, 74–77
    eukaryotic, see Eukaryotic cell cycle
Cell division cycle proteins
    Cdc6 and Cdc18, chromosomal DNA replication, 55–56
    Cdc45, G1–S phase transition, 61–63
Cell membranes, macromolecular translocation, 461
Cell stress, RNA destruction, 136–138
Cell transformation
    FGF-2 isoforms, 390–391
    IRES activity, 398–399
Cellular messenger RNA, IRESs, 374–376
c-fos, unstable mRNAs, 135
Chemical modification, pRNA, 435–436
Chemical modification interference, pRNA, 436

Chemical probing, procapsid-bound pRNA, 447
Chk1, cell cycle control, 78
Chk2, cell cycle control, 78
Chloramphenicol acetyl transferase
    FGF-2 mRNA expression, 392–393
    FGF-2 translational silencing, 401
Chromatin
    binding by Cdt1p, geminin, and Mcm10p, 56–57
    eukaryotic, usage, 2
    replication-initiation factors organization
        activation of origins, 71
        elongation reaction, 72–73
        leading and lagging strands, 72
        overview, 69–70
        prereplicative complex assembly, 70–71
Circularly permuted pRNA, 421–423
Cis-acting elements, FGF-2 translation, 387
C-myc, mRNA
    overview, 396–397
    P0 leader, 397–398
    P2, IRES identification, 397
COB, RNA folding, 231–232
Coiled-coil proteins, cell cycle, 77
Complementary DNA, self-primed synthesis, 241–242
Conformational change, pRNA
    dimer–hexamer transition, 444–445
    monomer–dimer transition, 443–444
    motion task, 445
    overview, 443
Connector
    domain for pRNA binding, 447–448
    pRNA binding, 445–447
Constitutive activation domain, CREB
    definition, 274–279
    transcription
        CAD:TAF interaction, 291–292
        overview, 284–287
        polymerase complex assay, 289–291
        transcription factor interactions, 279–284
cp-pRNA, see Circularly permuted pRNA
CREB, see Cyclic AMP response element-binding protein
Crosslinking
    photoaffinity, pRNA, 431–433
    pRNA, phenphi, 435
    pRNA, psoralen, 433–435
Cryo-AFM, see Cryo-atomic force microscopy

INDEX

Cryo-atomic force microscopy, pRNA
  dimer images, 440
  monomer images, 440
Cryo-electron microscopy, pRNA, 428–429
Cryo-EM, see Cryo-electron microscopy
Crystal structure
  nrdD protein, 103–106
  pRNA, 442
CTD, see C-terminal domain phosphatase
C-terminal domain phosphatase
  dephosphorylation and transcript
    termination, 354
  discovery, 335–337
  early elongation complexes, 349–350
  elongation complex RNAP II, 355–357
  mRNA processing regulation, 353–354
  recombinant, cloning and expression,
    342–346
  RNAP II mobilization, 347–348
  TFIIF and TFIIB regulation, 341–342
  transcript elongation link, 350–351
  transcript elongation turnover, 351–352
  transcription cycle, 346–347
  yeast and mammalian, 337–341
CUG codon
  FGF-2 mRNA, 383–385
  GCSA synthesis, 378–379
  L-VEGF identification, 394–396
  MuLV genomic mRNA, 380–382
  survival factor-like FGF-2 isoform, 386–387
Cyclic AMP response element-binding
  protein
  activation domain
    CAD and KID definitions, 274–279
    CAD–transcription factor interactions,
      279–284
    constitutive and kinase-inducible
      activities, 273–274
    experimental model, 272–273
  basic function, 270
  CAD and P-KID, transcription
    CAD:TAF interaction, 291–292
    KID phosphorylation, 293–296
    overview, 284–287
    polymerase complex assay, 289–291
    phosphorylated, transcription model,
      296–298
Cyclin-dependent kinases, eukaryotic cell
  cycle, 43–45
CYT-18, RNA folding, 232–233

D

Dbf4-dependent kinase, DNA replication, 45
Dbf4p–Cdc7p kinase, DNA replication
  origins, 45
DDK, see Dbf4-dependent kinase
DEAD box proteins
  characteristics, 308
  eIF4A, see eIF4A
  research overview, 328–329
DEAD box sequences, eIF4A, 323–324
Deadenylation
  mRNA decay, 131–133
  mRNA degradation coordinate regulation,
    138–139
Deletion, pRNA genetic analysis, 438–439
Deoxyribonucleoside triphosphates, allosteric
  regulation of RNRs, 108–114
Dephosphorylation, CTD, transcript
  termination, 354
Direct defense, plant–insect interactions,
  190–192
Dissociation model, adenovirus shunt, 19
Dithiothreitol, nrdD protein, 100
Divalent cations, pRNA effect, 448–449
DNA
  anaerobic ribonucleotide reductases,
    120–124
  self-primed synthesis, 241–242
  viral composition, 416–417
DNA-binding proteins, replication protein A,
  63–65
DNA-filled capsids, pRNA, 450
DNA oligonucleotides
  annealing assay, 235–237
  strand exchange assay, 237–239
DNA packaging
  inactive pRNA, 430
  phi29, pRNA role, 417–419
  pRNA domain, 449–450
  pRNA minimal sequence, 453
  viral, model, 458–459
DNA polymerase α–primase
  characteristics and function, 65–68
  replication initiation, 72
DNA replication
  BRCT family, 74–77
  Dbf4p–Cdc7p kinase, 45
  eukaryotic, see Eukaryotic DNA replication
  initiation, leading and lagging strands, 72

DNA targeting, phi29 pRNA, 458
DNA topoisomerase I, 68-69
DNA translocation
   nucleic acid sliding/riding processes, 459-461
   phi29 pRNA role, 453-457
Dose-response curves, pRNA stoichiometry, 424-425
DTT, see Dithiothreitol

E

E2F, pRB cooperation, 46-48
Effector kinases, cell cycle control, 78
eIF4A
   biology, 312-313
   DEAD box sequences role, 323-324
   helicase activity, 324-325
   80S initiation pathway, 326-328
   mRNA binding, 314-315
   original purification, 308
   other protein effects, 319-321
   research data inconsistencies, 321-323
   research overview, 328-329
   substrate specificity, 316
   unwinding assay, 316-321
Elicitors, jasmonate levels, 166-167
Elongation complexes
   early, CTD role, 349-350
   RNAP II, CTD phosphatase, 355-357
Elongation reaction
   eukaryotic DNA replication, 72-73
   transcript, CTD phosphate turnover, 351-352
   transcript, CTD phosphorylation link, 350-351
Endonuclease cleavage, mRNA
   coordinate regulation, 138-139
   other systems, 139-142
   site regulation, 138
Endonucleases, mRNA cleavage, cell stress, 136-138
Estrogen
   inducible vitellogenin mRNA, 145-154
   vitellogenin mRNA stabilization, 155-156
Ethylene, wound response, 189
Eukaryotic cell cycle

cyclin-dependent kinases, 43-45
Dbf4p-Cdc7p kinase, 45
E2F-pRM cooperation, 46-48
overview, 42-43
protein phosphatase 2A, 45-46
proteolysis at G0-G1 transition, 49-51
proteolysis at G1-S transition, 51-52
proteolysis in mitosis, 52
SBF and MBF, 48-49
Eukaryotic DNA replication
   elongation reaction, 72-73
   initiation
      Cdc6/Cdc18 protein, 55-56
      Cdt1p, geminin, and Mcm10p, 56-57
      DNA polymerase α-primase, 65-68
      DNA topoisomerase I, 68-69
      G1-S phase transition, 61-63
      Mcm2p-7p complex, 57-61
      origin recognition complex, 54-55
      replication protein A, 63-65
   origins, 53-54
Eukaryotic translation initiation
   alternative translations, 377-378
   cap-dependent scanning, 4-5
   internal initiation, 6
   IRESs in cellular mRNAs, 374-376
   IRESs in viral mRNAs, 372-374
   leaky scanning, 5
   reinitiation, 6
   ribosome shunt, 376-377
   scanning model, 371-372
   shunting, 6
   viral vs. cellular IRESs, 376

F

FGF-2, see Fibroblast growth factor 2
Fibroblast growth factor 2
   IRES-binding trans-acting factors, 391
   isoform localization, 385
   isoform modes of action, 385-386
   isoforms in transformed and stressed cells, 390-391
   mRNA CUG start codons, 383-385
   mRNA expression control, 392-393
   mRNA IRES identification, 387-390
   survival factor-like isoform, 386-387
   translational silencing by p53, 400-403
   translation regulation, 387

INDEX

Fission yeast, cyclin-dependent kinases, 43–44
Flower development, jasmonates and octadecanoids, 198–200
FMRP, ribonucleoparticle association, 257
Footprinting, procapsid-bound pRNA, 447

G

G0–G1 transition, proteolysis, 49–51
G1–S transition
  Cdc45 protein, 61–63
  proteolysis, 51–52
Gag–Pol translation, retroelement, 10–12
GCSA, see Gross cell surface antigen
Geminin, chromatin binding, 56–57
Gene economy, reverse-transcribing elements, 2–4
Gene organization, anaerobic ribonucleotide reductases, 120
General transcription factors
  CREB CAD interactions, 279–284
  CREB CAD and P-KID roles, 284–287, 289–290
Genes
  jasmonate-induced regulation
    downregulation, 181
    responsive promoters, 184–185
    upregulation, 181–184
  vitellogenin, 143–144
Genetic analysis, pRNA, 438–439
Germination, plant, jasmonates and octadecanoids, 196–197
Glycyl radical
  nrdD protein, 99–101
  protein α, 116–119
GMPS/aryl azide, see Guanine-monophosphorothioate/aryl azide
Gross cell surface antigen
  CUG initiation codon, 378–379
  MuLV dissemination and pathogenicity, 380
GTFs, see General transcription factors
GTP, allosteric regulation of RNRs, 108–114
Guanine-monophosphorothioate/aryl azide, 431–433

H

Hammerhead ribozyme cleavage, 239–241
HBV, see Hepatitis B virus
HCV NS3, see Hepatitus C virus NS3 proteins
Helicase, eIF4A, 324–325
Helix, destabilizing activity of chaperones, 242–245
Hepadnaviruses
  Gag–Pol translation, 11
  gene economy, 2
Hepatitis B virus, 7–9
Hepatitis C virus NS3 proteins, 311–312
HsCdc25p phosphatases, 44–45
Hydroxylation, JA metabolism, 178

I

Icosahedral Caulimoviridae, polycistronic translation
  TAV control, 23–24
  TAV–host interactions, 24–26
Indirect defense, plant–insect interactions, 192–193
Induced-systemic resistance, plant–pathogen interactions, 194
Infectious virions, pRNA assay, 420–421
Initiation complex, activation of origins, 71
Initiation pathway, 80S, eIF4A, 326–328
Insertion, pRNA genetic analysis, 438–439
Intermolecular crosslinking, pRNA dimers, 433
Internal ribosome entry site
  alternative translation initiations, 377–378
  cellular mRNAs, 374–376
  dependent translation, generalization, 382–383
  discovery, 370–371
  FGF-2 mRNA, 387–390
  Mo-MuLV, PTB identification, 382
  MuLV genomic mRNA, 380–382
  P2 c-myc mRNA, 397
  RNA translation–packaging interplay, 10
  tissue specificity in vivo, 399–400
  transformed cells, 398–399
  VEGF mRNA, 393–394
  viral vs. cellular IRESs, 376
  viral mRNAs, 372–374

Internal ribosome entry site *trans*-acting
  factors, FGF-2, 391
Intramolecular crosslinking, pRNA monomers,
  433
IRES, *see* Internal ribosome entry site
Iron–sulfur proteins
  center, protein β, 116–119
  nrdG protein definition, 114–115
  nrdG protein as prototype, 119
ISR, *see* Induced-systemic resistance
ITAFs, *see* Internal ribosome entry site
  trans-acting factors

J

JA, *see* Jasmonic acid
Jasmonate-responsive promoters, 184–185
Jasmonates
  external stimuli effects, 166–167
  gene expression
    downregulation, 181
    jasmonate-responsive promoters, 184–185
    upregulation, 181–184
  LOX pathway
    AOC, 175–176
    AOS, 173–175
    biosynthesis regulation, 176–177
    JA biosynthesis, transgenic manipulation, 180
    JA metabolism, 177–178
    mutants, 178–180
    OPR and β-oxidation, 176
    overview, 171–173
  mycorrhiza, 193–194
  natural occurrence, 168–171
  plant development
    flowers, 198–200
    germination and seedlings, 196–197
    senescence, 202–205
    tendril coiling, 202
    tuberization, 200–202
  plant–insect interactions
    direct defense, 190–192
    indirect defense, 192–193
  plant–pathogen interactions, 194–196
  plant–plant interactions, 193
  wound response pathway
    ABA, 190
    ethylene, 189
    intracellular signal transduction, 188
    local generation, 187–188
    overview, 185
    SA, 190
    systemic response, 189
Jasmonic acid
  biosynthesis, transgenic manipulation, 180
  biosynthesis regulation, 176–177
  identification, 166
  metabolism, 177–178
  natural occurrence, 168–171
  signal tranduction pathways, 167–168
Jasmonic acid- and elicitor-responsive
  elements, 184
Jasmonic acid-induced protein
  gene downregulation, 181
  gene upregulation, 181–184
JEREs, *see* Jasmonic acid- and
  elicitor-responsive elements
JIP, *see* Jasmonic acid-induced protein

K

KH domains, vigilin interaction model,
  151–153
KID, *see* Kinase-inducible domain
Kinase-inducible domain, CREB
  definition, 274–279
  KID phosphorylation, 293–296
  polymerase complex assay, 290–291
  transcription overview, 284–287
Kinetics, RNA misfolding, 228

L

Leaky scanning
  eukaryote translation initiation, 5
  polycistronic RNA translation, 22–23
Lentiviruses, IRES-dependent translation,
  382–383
Lipoxygenase
  jasmonate and octadecanoids, 171–173
  jasmonates
    AOC, 175–176
    AOS, 173–175
    biosynthesis regulation, 176–177
    JA biosynthesis, transgenic manipulation,
    180

JA metabolism, 177–178
  mutants, 178–180
  OPR and β-oxidation, 176
LOX, see Lipoxygenase

M

Macromolecular translocation, cell membranes and barrier walls, 461
Magnesium-agarose gel system, TBP recruitment, 289–290
Magnesium ion
  pRNA conformational change, 443–444
  pRNA effect, 448–449
Mammals, CTD phosphatase, 337–341
Manganese, pRNA effect, 449
Mathematical simulation, phi29 pRNA, 457–458
MBF, see MluI cell cycle box-binding factor
Mcm2p, see Minichromosome maintenance protein 2
Mcm7p, see Minichromosome maintenance protein 7
Mcm10p, see Minichromosome maintenance protein 10
Messenger RNA
  binding, eIF4A role, 314–315
  cellular, shunting candidates, 21
  c-myc
    overview, 396–397
    P0, IRES identification, 397–398
    P2, IRES identification, 397
    coordinate regulation, 138–139
    CTD phosphatase role, 353–354
    deadenylation-dependent decay, 131–133
    endonucleolytically cleaved
      cell stress, 136–138
      other systems, 139–142
      site regulation, 138
    error-prone, removal, 135–136
    FGF-2
      CUG start codons, 383–385
      expression control, 392–393
      IRES identification, 387–390
      translational silencing by p53, 400–403
    life and half-life, 130–131
    MuLV, IRES, 380–382
    polycistronic, translation
      basic strategies, 21–22

Icosahedral Caulimoviridae, 23–26
leaky scanning, 22–23
TAV-activated reinitiation model, 27–28
survival factor-like FGF-2 isoform, 386–387
unstable, decay regulation, 134–135
5'-UTR, cis-acting elements, FGF-2, 387
VEGF, distinct IRES identification, 393–394
vitellogenin
  3'-UTR, vigilin binding, 148–151
  vigilin functions, 153–154
  vigilin identity, 147
  vigilin interaction models, 151–153
  vigilin localization and regulation, 147–148
  vigilin overview, 145–147
vitellogenin, stability
  basic features, 144–145
  precursor processing, 144
  transcription, 143
  vigilin role, 154–156
vitellogenin and albumin, stability regulation, 142–143
Metabolic transformations, JA metabolism, 178
Metal ions, RNA structures, 228–229
Methylation, JA metabolism, 178
Minichromosome maintenance protein 2, 57–61
Minichromosome maintenance protein 7, 57–61
Minichromosome maintenance protein 10, 56–57
Mitosis, regulated proteolysis, 52
MluI cell cycle box-binding factor, yeast, 48–49
Models
  antiviral strategy design, 458–459
  CREB activation domain, 272–273
  phi29 pRNA, 457
  phosphorylated CREB transcription, 296–298
  pRNA dimers and trimers, 457–458
  pRNA 3D structure, 440–443
  TAV-activated reinitiation, 27–28
  vigilin–single-stranded nucleic acid interaction, 151–153
  viral DNA packaging, 458–459
  vitellogenin mRNA stabilization, 156
Moloney murine leukemia virus
  cell surface antigen synthesis, 378–379
  PTB identification, 382

Mo-MuLV, see Moloney murine leukemia virus
Monovalent cations, pRNA effect, 448–449
Motion task, pRNA conformational change, 445
Multicistronic messenger, c-*myc*, see c-*myc*
MuLV, see Murine leukemia virus
Murine leukemia virus
  dissemination and pathogenicity, 380
  genomic mRNA, 380–382
Mutagenesis, see Site-directed mutagenesis
Mutations
  jasmonate biosynthesis, 178–180
  PEPCK, 272
  pRNA, 438–439, 455–456, 458–459
Mycorrhiza, jasmonates, 193–194

N

NABPs, see Nucleic acid-binding proteins
NBS1, see Nijmegen breakage syndrome 1
Nijmegen breakage syndrome 1, 79
Nondissociating model, adenovirus shunt, 19
nrdD protein, RNR component
  allosteric regulation, 108–114
  glycyl radical enzyme definition, 99–101
  site-directed mutagenesis, 101–103
  substrate reduction, 106–108
  3D structure, 103–106
nrdG protein, RNR component
  activation reaction, 116–119
  $\alpha_2\beta_2$ complex, 115–116
  iron–sulfur protein definition, 114–115
  iron–sulfur protein prototype, 119
Nucleic acid-binding proteins
  RNA contacts, 229–230
  vigilin identity, 147
Nucleic acid chaperone proteins
  annealing assay, 235–237
  annealing *vs.* unwinding, 245
  biochemical property experiments, 234–235
  hammerhead ribozyme cleavage, 239–241
  helix-destabilizing activity, 242–245
  self-primed cDNA synthesis, 241–242
  strand exchange assay, 237–239
Nucleic acids
  eIF4A unwinding assay, 316–321
  vigilin interaction model, 151–153

Nucleic acid sliding/riding processes, viral DNA, 459–461

O

Octadecanoids
  LOX pathway, 171–173
  natural occurrence, 168–171
  plant development
    flowers, 198–200
    germination and seedlings, 196–197
    senescence, 202–205
    tendril coiling, 202
    tuberization, 200–202
ODNs, see DNA oligonucleotides
Oncoretroviruses, IRES-dependent translation, 382–383
OPDA, see *cis*(+)-12-*Oxo*-phytodienoic acid
Open reading frames
  eukaryotic translation initiation, 6
  Gag–Pol translation, 11
  papillomavirus shunt, 20–21
  Sendai virus shunt, 19–20
OPR, see *cis*(+)-12-*Oxo*-phytodienoic acid reductase
ORC, see Origin recognition complex
ORF, see Open reading frames
Origin recognition complex
  DNA replication initiation, 54–55
  prereplicative complex assembly, 70–71
Osmotic stress, jasmonate levels, 166–167
β-Oxidation, LOX pathway, 176
cis(+)-12-*Oxo*-phytodienoic acid
  flower development, 198–199
  natural occurrence, 168–171
cis(+)-12-*Oxo*-phytodienoic acid reductase, LOX pathway, 176

P

P0, c-*myc* IRES, 397–398
P2, c-*myc* mRNA, IRES, 397
p53
  cell cycle control, 79
  FGF-2 translational silencing, 400–403
p97, eIF4A effects, 313
Packaging
  DNA, see DNA packaging

INDEX

DNA-packaging domain, pRNA, 449–450
viral RNA, translation interplay, 7–10
Paip1, *see* Poly(A)-binding protein interacting protein 1
Papillomaviruses, shunt, 20–21
Pararetroviruses
  retroelement Gag–Pol translation, 10–12
  RNA translation–packaging interplay, 7–10
Pathogenicity, MuLV, GCSA role, 380
Pdcd4, eIF4A effects, 313
PEPCK, *see* Phosphoenolpyruvate carboxykinase
Phenphi, pRNA crosslinking, 435
phi29
  assembly, mutant pRNA, 458–459
  cp-pRNA, 421–422
  DNA packaging, 417–419, 452
  DNA translocation, pRNA role, 453–457
  pRNA antisense DNA targeting, 458
  pRNA assay, 420–421
  pRNA dimer images, 440
  pRNA monomer images, 440
  pRNA stoichiometry, 425
Phosphoenolpyruvate carboxykinase, 272
Phosphoinositol-3-kinase-like complexes, 77
Phosphorylation
  dephosphorylation, CTD, 354
  KID, pol II, 293–296
  RNA polymerase II, 334–335
Photoaffinity crosslinking, pRNA, 431–433
Phylogenetic analysis, pRNA, 431
PINs
  plant–insect interactions, 190–192
  systemic wound response, 189
Plant development, jasmonates and octadecanoids
  flowers, 198–200
  germination and seedlings, 196–197
  senescence, 202–205
  tendril coiling, 202
  tuberization, 200–202
Plant–insect interactions
  direct defense, 190–192
  indirect defense, 192–193
Plant–pathogen interactions, jasmonates, 194–196
Plant–plant interactions, jasmonates, 193
PMR-1 nuclease system, vitellogenin mRNA, 155–156
pol II, *see* RNA polymerase II

Poly(A)-binding protein interacting protein 1, 313
Poly(A) sites, FGF-2 mRNA expression control, 392–393
Polycistronic messenger RNA, translation
  basic strategies, 21–22
  leaky scanning, 22–23
  TAV-activated reinitiation model, 27–28
  TAV control, 23–24
  TAV–host interactions, 24–26
PP2A, *see* Protein phosphatase 2A
pRB, E2F cooperation, 46–48
Prereplicative complex
  activation of origins, 71
  assembly, 70–71
pRNA
  assay with phi29, 420–421
  chemical modification, 435–436
  chemical modification interference, 436
  complementary modification, 430–431
  conformational change overview, 443
  crosslinking by phenphi, 435
  crosslinking by psoralen, 433–435
  cryo-AFM, dimer images, 440
  cryo-AFM, monomer images, 440
  dimer as building block, 450–452
  dimer formation, RNA size, 453
  dimer–hexamer transition, 444–445
  dimer and trimer model, 457–458
  DNA-filled capsid, 450
  DNA packaging, minimal sequence, 453
  domain separation, 449–450
  genetic analysis, 438–439
  inactive in DNA packaging, 430
  loops in dimer formation, 452–453
  macromolecular translocation, 461
  mono- and divalent cation effects, 448–449
  monomer–dimer transition, 443–444
  motion task, 445
  mutant, phi29 assembly, 458–459
  nucleic acid sliding/riding processes, 459–461
  phi29 assembly inhibition, 458
  phi29 DNA packaging, 417–419
  phi29 DNA translocation, 453–457
  photoaffinity crosslinking, 431–433
  phylogenetic analysis, 431
  procapsid binding, minimal sequence, 453
  research study overview, 419–420
  ribonuclease probing, 438

RNA–DNA hybridization analysis, 419
SELEX, 439–440
stoichiometry
  binomial distribution, 423–424
  common multiple technique, 426
  concentration dependence slopes, 425
  cryo-EM imaging, 428–429
  six interlocking set, 428
  three interlocking set, 428
  two interlocking set, 427–428
3D structure modeling, 440–443
viral DNA translocation, 459–461
pRNA–protein interactions
  connector and capsid proteins, 445–447
  connector domain and sequence, 447–448
  gp16, 448
  procapsid-bound pRNA, 447
Procapsids, pRNA binding
  active pRNA, 430
  binding domain, 449–450
  footprinting and chemical probing, 447–448
  minimal sequence, 453
  mono- and divalent cation effects, 448–449
Promoter recognition factor TFIID, 280, 282–283
Protein α, glycyl radical, 116–119
Protein β, iron–sulfur center, 116–119
Protein kinase A
  activation, 270–271
  catalytic subunits, 271
  CREB activation domain, 273–274
Protein phosphatase 2A, 45–46
Proteins
  ARE-binding proteins, 134–135
  capsid proteins, 445–447, 450
  cell division cycle proteins, 55–56, 61–63
  coiled-coil proteins, cell cycle, 77
  CREB, see Cyclic AMP response element-binding protein
  DEAD box proteins, 308, 328–329
  DNA-binding proteins, replication protein A, 63–65
  hepatitis C virus NS3 proteins, 311–312
  iron–sulfur proteins, 114–119
  jasmonic acid-induced protein, 181–184
  Mcm2p, 57–61
  Mcm7p, 57–61
  Mcm10p, 56–57
  nrdD protein, 99–114

nrdG protein, 114–119
nucleic acid-binding proteins, 147, 229–230
nucleic acid chaperone proteins, 234–245
poly(A)-binding protein interacting protein 1, 313
pRNA interactions, 445–448
pyrimidine tract-binding protein, 382
replication protein A, 63–65
RNA-binding proteins, 229–233
tail proteins, 448, 450
TATA-binding protein, 280, 289–290
transactivator/viroplasmin protein, 23–28
3′-UTR-binding protein, 145–154
Proteolysis
  G0–G1 transition, 49–51
  G1–S transition, 51–52
  regulated, mitosis, 52
Psoralen, pRNA crosslinking, 433–435
PTB, see Pyrimidine tract-binding protein
Pyrimidine tract-binding protein, 382

R

Reinitiation
  eukaryotic translation, 6
  polycistronic translation
    TAV control, 23–24
    TAV–host interactions, 24–26
    shunting definition, 16
    TAV-activated model, 27–28
Replication factor C, elongation reaction, 72
Replication factors, G1–S phase transition, 61–63
Replication-initiation factors, chromatin
  activation of origins, 71
  elongation reaction, 72–73
  leading and lagging strands, 72
  overview, 69–70
  prereplicative complex assembly, 70–71
Replication protein A, 63–65
Retroviruses
  retroelement Gag–Pol translation, 10–12
  RNA translation–packaging interplay, 7–10
Reverse-transcribing elements, gene economy, 2–4
RFC, see Replication factor C
Ribonuclease probing, pRNA, 438
Ribonucleoparticles
  abundance, 257

INDEX

complexes, 246
FMRP association, 257
functions, 259–261
Ribonucleotide reductases, see Anaerobic ribonucleotide reductases
Ribosomes, shunting, 6, 15–16
Rice tungro bacilliform virus
  polycistronic RNA translation, 22–23
  RNA translation–packaging interplay, 9
  shunting ribosomes, 15–16
RNA
  anaerobic ribonucleotide reductases, 120–124
  cellular, classification, 225
  composition in virus, 416–417
  functional role, 224
  hybridization, chaperones, 259–260
  messenger, see Messenger RNA
  metal ion role, 228–229
  misfolding, 228
  occupancy of RNA chaperones, 246
  secondary and tertiary structures, 226
  size for pRNA dimer formation, 453
  viral translation–packaging interplay, 7–10
RNA binding motifs, RNA chaperones, 233–234
RNA-binding proteins
  amino acid–RNA contacts, 229–230
  cooperativity, 230
  other platforms, 231
  reciprocal induced fit, 230
  RNA folding problem, 231–233
RNA chaperones
  biotechnological applications, 262
  classes, 225–226
  functionality, 246–250
  molecular level, 258–259
  RNA binding motifs, 233–234
  RNA helicase interactions, 261–262
  RNA hybridization, 259–260
  RNA occupancy, 246
  structural and functional diversity, 250, 257–258
RNA–DNA hybridization analysis, pRNA, 419
RNA folding
  large RNAs, 225
  RNA-binding proteins role, 231–233
RNA helicase, RNA chaperone interactions, 261–262
RNAP II, see RNA polymerase II

RNA polymerase II
  basal transcription initiation, 284–285
  CAD role
    assay, 289–291
    CAD:TAF interaction, 291–292
    KID phosphorylation, 293–296
  CTD phosphatase
    dephosphorylation and transcript termination, 354
    discovery, 335–337
    early elongation complexes, 349–350
    elongation complex RNAP II, 355–357
    mobilization role, 347–348
    mRNA processing regulation, 353–354
    purification and characterization, 337–341
    recombinant, cloning and expression, 342–346
    TFIIF and TFIIB regulation, 341–342
    transcript elongation link, 350–351
    transcription cycle, 346–347
    turnover, 351–352
  phosphorylation dynamics, 334–335
RNPs, see Ribonucleoparticles
RNRs, see Anaerobic ribonucleotide reductases
RPA, see Replication protein A
RTBV, see Rice tungro bacilliform virus

S

SA, wound response, 190
SBF, see Swi cell cycle box-binding factor
Scanning model, translation initiation, 371–372
Seedling development, jasmonates and octadecanoids, 196–197
SELEX, pRNA, 439–440
Sendai virus, shunt, 19–20
Senescence, jasmonates and octadecanoids, 202–205
Short open reading frame-dependent shunt
  other initiation events, 17
  overview, 12–14
  shunting as reinitiation, 16
  shunting ribosome initiation, 15–16
  take-off and landing sites, 14–15
Short open reading frames, viral RNA, 9–10
Shunting
  adenovirus, 17–19
  animal virus internal initiation combinations, 21

Shunting (*cont.*)
  cellular mRNAs, candidates, 21
  eukaryote translation initiation, 6
  papillomaviruses, 20–21
  ribosome, translation initiation, 376–377
  Sendai virus, 19–20
  sORF-dependent, Caulimoviridae, 12–17
Signal transduction
  jasmonic acid, 167–168
  systemin via jasmonate, 188
Site-directed mutagenesis, *nrdD* protein, 101–103
sORF, *see* Short open reading frames
SpRad3p, cell cycle control, 77
Strand exchange assay, nucleic acid chaperone proteins, 237–239
Stressed cells, FGF-2 isoforms, 390–391
Survival factors, related FGF-2 isoform, 386–387
Swi cell cycle box-binding factor, yeast, 48–49
Systemic wound response, jasmonates, 189
Systemin, jasmonates, 187–188

T

TAFs, *see* TATA-binding protein-associated factors
Tail proteins
  gp9, pRNA DNA-filled capsid departure, 450
  gp16, pRNA interaction, 448
TATA-binding protein
  CREB CAD interactions, 280
  Mg^{2+}-agarose gel system, 289–290
TATA-binding protein-associated factors
  CAD:TAF in transcription activation, 291–292
  CREB CAD interactions, 280–281
TAV, *see* Transactivator/viroplasmin protein
TBP, *see* TATA-binding protein
Tendril coiling, jasmonates and octadecanoids, 202
TFIIB, *see* Transcription factor IIB
TFIIF, *see* Transcription factor IIF
Thermodynamics, RNA misfolding, 228
Three-dimensional structure
  nrdD protein, 103–106
  pRNA, 440–443
Tissue specificity, IRES *in vivo*, 399–400

Trans-acting factor, PTB identification, 382
Transactivator/viroplasmin protein
  activated reinitiation model, 27–28
  TAV control, 23–24
  TAV–host interactions, 24–26
Transcription
  CREB CAD and P-KID roles, 284–287, 289–290
  CTD dephosphorylation, 354
  CTD phosphatase role, 346–347
  phosphorylated CREB, model, 296–298
  vitellogenin mRNA, 143
Transcription factor E2F, pRM cooperation, 46–48
Transcription factor IIB, CTD phosphatase regulation, 341–342
Transcription factor IIF, CTD phosphatase regulation, 341–342
Transcription factors, yeast SBF and MBF, 48–49
Transformations
  cell, FGF-2 isoforms, 390–391
  cell, IRES activity, 398–399
  JA metabolism, 178
Transgenic manipulation, JA biosynthesis, 180
Translation
  FGF-2, 387
  Gag–Pol, retroelement, 10–12
  internal initiation, shunting combinations, 21
  IRES-dependent translation, 382–383, 398–399
  polycistronic mRNA
    basic strategies, 21–22
    leaky scanning, 22–23
    TAV control, 23–24
    TAV–host interactions, 24–26
  viral RNA, packaging interplay, 7–10
Translational silencing, FGF-2, p53, 400–403
Translation enhancer, FGF-2 mRNA expression control, 392–393
Translation initiation
  eIF4A, *see* eIF4A
  eukaryotic, *see* Eukaryotic translation initiation
  FGF-2 mRNA, 383–385
Translocation
  adenovirus shunt model, 19
  cell membranes and barrier walls, 461
  DNA, phi29, pRNA role, 453–457

INDEX

viral DNA, nucleic acid sliding/riding processes, 459–461
Truncation, pRNA genetic analysis, 438–439
TTP, allosteric regulation of RNRs, 108–114
Tuberization, jasmonates and octadecanoids, 200–202
Tumor suppressor, FGF-2 translational silencing, 400–403
Tyrosyl-tRNA synthetase, RNA folding, 232–233

U

Unwinding
  eIF4A-dependent unwinding assay, 316–321
  nucleic acid chaperone proteins, 245
3′-UTR
  translation enhancer, FGF-2 mRNA, 392–393
  vitellogenin mRNA, 148–151, 156
5′-UTR, mRNA, cis-acting elements, FGF-2, 387
3′-UTR-binding protein, vitellogenin mRNA, 145–154

V

Vascular endothelial cell growth factor mRNA, distinct IRES identification, 393–394
  new isoform identification, 394–396
VEGF, see Vascular endothelial cell growth factor
Vigilin
  binding to vitellogenin mRNA 3′-UTR, 148–151
  interaction with single-stranded nucleic acids, 151–153
  localization and regulation, 147–148
  nucleic acid-binding protein identity, 147
  overview, 145–147
  proposed functions, 153–154

vitellogenin mRNA stabilization role, 154–156
Viral assembly, mono- and divalent cation effects, 448–449
Viral RNA
  composition in virus, 416–417
  IRESs, 372–374
  translation–packaging interplay, 7–10
Viruses
  DNA packaging, model, 458–459
  DNA and RNA composition, 416–417
Vitellogenin, mRNA
  stability regulation
    basic features, 144–145
    early work, 142–143
    precursor processing, 144
    transcription, 143
  vigilin
    interaction model, 151–153
    localization and regulation, 147–148
    nucleic acid-binding protein identity, 147
    overview, 145–147
    proposed functions, 153–154
    stabilization role, 154–156
    3′-UTR binding, 148–151

W

Wound response pathway, jasmonates
  ABA, 190
  ethylene, 189
  intracellular signal transduction, 188
  local generation, 187–188
  overview, 185
  SA, 190
  systemic response, 189

Y

Yeast
  Cdc6/Cdc18 protein, 55–56
  CTD phosphatase, 337–341
  cyclin-dependent kinases, 43–44
  proteolysis at G1–S transition, 51–52
  SBF and MBF, 48–49

ISBN 0-12-540072-1